NASA
APOLLO SPACECRAFT
COMMAND AND SERVICE MODULE
NEWS REFERENCE

©2011 Periscope Film LLC
All Rights Reserved
ISBN #978-1-937684-99-0
www.PeriscopeFilm.com

APOLLO SPACECRAFT
NEWS REFERENCE

It took 400 years of trial and failure, from da Vinci to the Wrights, to bring about the first flying machine, and each increment of progress thereafter became progressively more difficult.

But nature allowed one advantage: air. The air provides lift for the airplane, oxygen for engine combustion, heating, and cooling, and the pressurized atmosphere needed to sustain life at high altitude. Take away the air and the problems of building the man-carrying flying machine mount several orders of magnitude. The craft that ventures beyond the atmosphere demands new methods of controlling flight, new types of propulsion and guidance, a new way of descending to a landing, and large supplies of air substitutes.

Now add another requirement: distance. All of the design and construction problems are re-compounded. The myriad tasks of long-distance flight call for a larger crew, hence a greater supply of expendables. The functions of navigation, guidance, and control become far more complex. Advanced systems of communications are needed. A superior structure is required. The environment of deep space imposes new considerations of protection for the crew and the all-important array of electronic systems. The much higher speed of entry dictates an entirely new approach to descent and landing. Everything adds up to weight and mass, increasing the need for propulsive energy. There is one constantly recurring, insistent theme: everything must be more reliable than any previous aerospace equipment, because the vehicle becomes in effect a world in miniature, operating with minimal assistance from earth.

Such is the scope of Apollo.

Appropriately, the spacecraft was named for one of the busiest and the most versatile of the Greek gods. Apollo was the god of light and the twin brother of Artemis, the goddess of the Moon. He was the god of music and the father of Orpheus. At his temple in Delphi, he was the god of prophecy. Finally, he was also known as the god of poetry, of healing, and of pastoral pursuits.

The Apollo Spacecraft News Reference was prepared by the Space Division of North American Rockwell Corp., Downey, Calif., in cooperation with NASA's Manned Spacecraft Center.

The book is arranged in five distinct parts. The first (identified by white tabs) includes general information about the program, the elements of the spacecraft and launch vehicles, and the missions. The second part (blue tabs) is a detailed description of the Apollo modules. The third (tan tabs) contains descriptions of the equipment and operation of major subsystems. The fourth (green tabs) concerns vital operations and support, and the fifth (gray tabs) contains a series of references.

Information on most of the subsystems follows this format: first, a general description of the system, its equipment, and its function; second, an equipment list containing all major data about key equipment; and third, a detailed description of subsystem operation. The general description should provide all the information normally needed about each subsystem; the detailed description is necessarily quite technical and is included in response to requests for this level of detail.

Descriptions and data are taken from the latest available information. As modifications are made to equipment in response to continuing tests, the book will be amended to reflect these changes.

Photographs or illustrations in this volume are available for publication. Prints may be ordered according to the code designation (P-1, P-2, etc.) appearing at the lower left corner of each illustration. Send requests to:

                    Public Relations Department
                    Space Division
                    North American Rockwell Corp.
                    12214 Lakewood Blvd.
                    Downey, Calif. 90241

CONTENTS

PART 1—GENERAL
    Apollo Program   1
    Apollo Spacecraft   3
    Launch Vehicles   9
    Missions   15
    The Moon   25
    Manned Space Program   27
    Apollo Flight Tests   33

PART 2—MODULES
    Command Module   39
    Service Module   53
    Lunar Module   61
    Spacecraft-LM Adapter   65

PART 3—SUBSYSTEMS
    Crew   69
    Displays and Controls—   83
    Docking   87
    Earth Landing   93
    Electrical Power   99
    Environmental Control   117
    Launch Escape   137
    Reaction Control   147
    Service Propulsion   159
    Telecommunications   173
    Guidance and Control   189
    Space Suit   223

PART 4—SUPPORT
    Checkout and Final Test   229
    Safety   235
    Reliability and Training   239
    Apollo Manufacturing   245

PART 5—REFERENCE
    CSM Subcontractors   253
    CSM Contract   257
    Biographical Summaries   265
    Apollo Chronology   273
    Apollo Briefs   277
    Glossary   285
    Abbreviations   299
    Index   303

Saturn V lifts Apollo spacecraft off pad in unmanned flight test

APOLLO PROGRAM

Apollo is the United States program to land men on the moon for scientific exploration and return them safely to earth. It has been described as the greatest scientific, engineering, and exploratory challenge in the history of mankind.

The challenge essentially was to create an artificial world: a world large and complex enough to supply all the needs of three men for two weeks. The world had to contain all of the life-sustaining elements of earth—food, air, shelter—as well as many special complex extras (navigation, propulsion, communications). Perhaps the greatest challenge was that of reliability; everything had to work and keep working no matter what the circumstances. Unlike the previous manned space programs in which crewmen could return to earth almost within minutes if an emergency arose, it could be as much as three days before the Apollo crew can get back to earth from the moon.

A parallel problem was to develop a launch vehicle large enough to put this world into space and to send it on its way to the moon 239,000 miles away. Many different plans were examined before the technique of lunar orbit rendezvous was selected.

NASA announced the Apollo program and its objectives in July of 1960. As President Kennedy pointed out to Congress on May 25, 1961, the overall objective is for this nation "to take a clearly leading role in space achievement which in many ways may hold the key to our future on earth." Of the lunar landing mission in particular, he said: "No single space project in this period will be more exciting, or more impressive to mankind, or more important for the long-range exploration of space; and none will be so difficult or so expensive to accomplish."

On Nov. 28, 1961, after a series of studies on the feasibility of the project, NASA awarded the basic Apollo spacecraft contract to the Space Division of North American Rockwell Corporation (at that time North American Aviation, Inc.). Development of a large carrier rocket—the Saturn program—had begun in late 1958 and in early 1962 was changed and expanded to meet the new goal of a landing on the moon.

Saturn V/Apollo shortly before launch

Briefly, the objective of the program is to send a three-man spacecraft to the moon and into orbit around it, land two of the three men on the moon while the third remains in orbit, provide up to 35 hours on the moon, return the two moon explorers to the orbiting spacecraft, and return all three safely to earth. The entire trip, from launch to earth landing, is expected to last between 8 and 10 days; the Apollo spacecraft has been designed for 14-day operation to give a wide margin of safety.

The program is the most extensive ever undertaken by any nation. During the peak in 1966, more than 20,000 companies and 350,000 persons throughout the country participated directly in it. North American Rockwell Corp.'s Space Division is principal contractor for the spacecraft's command and service modules, the launch escape system, and spacecraft-lunar module adapter, and the Saturn V second stage (the S-II). The rocket engines for all stages are produced by North American Rockwell's Rocketdyne Division. The lunar module (LM) contractor is Grumman Aircraft Engineering Corp. Spacecraft associate contractors include the Massachusetts Institute of Technology and AC Electronics Division of General Motors Corp. for the guidance and navigation subsystem, International Latex Co. for space suits, and United Aircraft Corp. for lunar surface life support equipment. Major North American Rockwell Space Division subcontractors for Apollo (contracts of more than $500,000) are listed in Part 5.

The Saturn program involves three separate launch vehicles. Two of them are used with Apollo spacecraft: the Saturn IB, a two-stage vehicle with a first stage thrust of 1,600,000 pounds, which is used for earth-orbital missions of the Apollo program; and the Saturn V, a three-stage vehicle with a maximum off-the-pad thrust of 7,500,000 pounds, which will be used for some earth-orbital missions and for the lunar mission. The Saturn I launch vehicle was used to develop large rocket engine technology.

The Apollo program is under the management of the Office of Manned Space Flight, Headquarters NASA. The Apollo spacecraft program is directed by NASA's Manned Spacecraft Center in Houston, Tex. The Saturn program is under the management of NASA's Marshall Space Flight Center in Huntsville, Ala. Pre-flight checkout and testing and launch activities are directed by NASA's Kennedy Space Center at Cape Kennedy, Fla.

Artist's conception of Apollo spacecraft in orbit around the moon before lunar landing

APOLLO SPACECRAFT

The Apollo spacecraft is the entire structure atop the launch vehicle. It is 82-feet tall and has five distinct parts: the command module, the service module, the lunar module, the launch escape system, and the spacecraft-lunar module adapter.

The three modules make up the basic spacecraft; the launch escape system and adapter are special-purpose units which are jettisoned early in the mission after they have fulfilled their function. The launch escape system is essentially a small rocket which will thrust the command module—with the astronauts inside—to safety in case of a malfunction in the launch vehicle on the pad or during the early part of boost. The spacecraft-lunar module adapter serves as a smooth aerodynamic enclosure for the lunar module during boost and as the connecting link between the spacecraft and the launch vehicle.

The spacecraft program has been divided into two parts, referred to as Block I (early earth-orbital test) and Block II (lunar mission version). The Block I program has been completed, and all future Apollo spacecraft flights will be with the Block II lunar mission type.

Spacecraft just before mating with Saturn V

	CM	SM	LM
Shape	Cone	Cylinder	Bug-like cab on legs
Height	10 ft, 7 in.	22 ft, 7 in. excluding fairing	22 ft, 11 in. (legs extended)
Diameter	12 ft, 10 in.	12 ft, 10 in.	29 ft, 9 in. (legs extended)
Habitable volume	210 cu ft		160 cu ft (approx.)
Launch weight	13,000 lb (approx.)	53,000 lb (approx.)	32,500 lb (approx.)
Primary material	Aluminum alloy Stainless steel Titanium	Aluminum alloy Stainless steel Titanium	Aluminum alloy

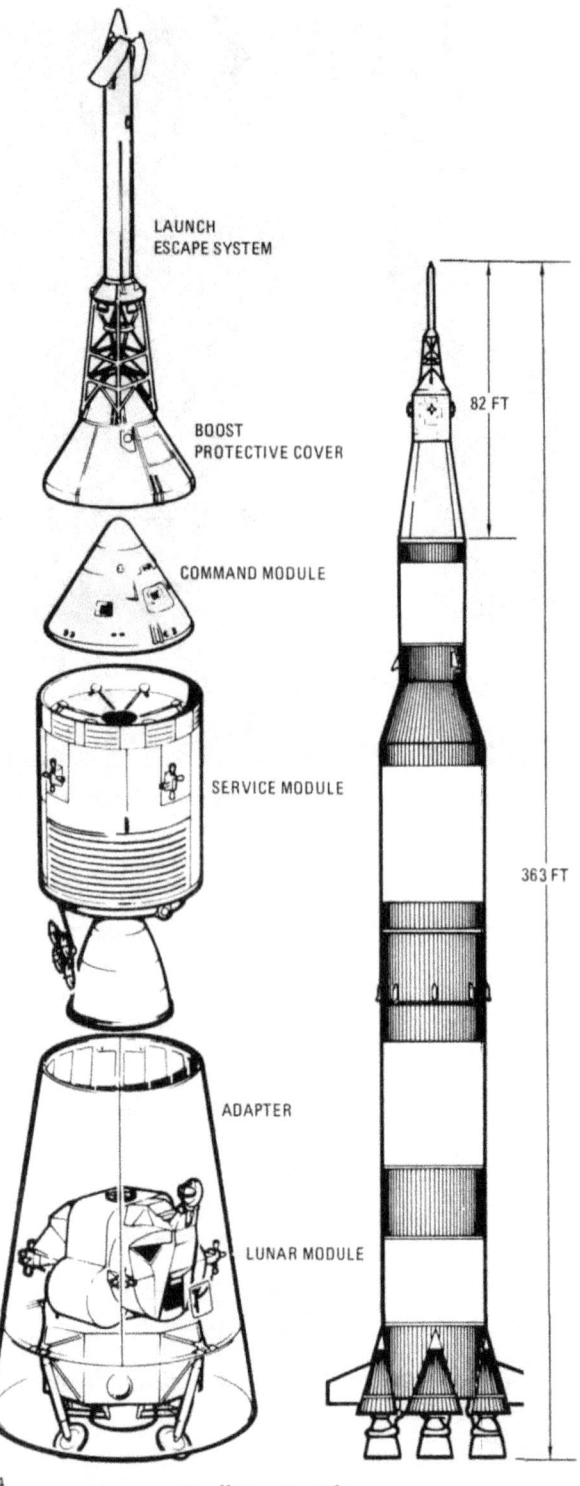

Apollo spacecraft

The basic difference in the two versions was in the addition, in Block II, of some equipment and systems designed specifically for the lunar mission. NASA's purpose in dividing the program was to get basic structure and systems tested in space as quickly as possible, while providing the time and the flexibility to incorporate changes. Thus, in addition to lunar equipment, Block II contains a great number of refinements and improvements of equipment and systems, many the result of continuing research and many evolving from the Block I unmanned flight and ground tests.

The spacecraft and systems described in this book are Block II.

For brevity, abbreviations for a few basic items of the Apollo program will be used throughout the book. For the spacecraft, these are CM for command module, SM for the service module, LM for the lunar module, CSM for the command and service modules together, and SLA for the spacecraft-lunar module adapter.

Abbreviations and acronyms are a key part of the engineering jargon; thousands are used commonly in the Apollo program. Many of the major ones are listed at the end of the glossary. Otherwise, they appear in this book only on a few diagrams or schematics where it was impossible because of limited space to spell them out. If the text does not make it clear what item of equipment is being referred to, a check with the glossary should provide the answer.

COMMAND MODULE

This is the control center for the spacecraft; it provides living and working quarters for the three-man crew for the entire flight, except for the period when two men will be in the LM for the descent to the moon and return. The command module is the only part of the spacecraft that returns to earth from space.

The CM consists of two shells: an inner crew compartment (pressure vessel) and an outer heat shield. The outer shell is stainless steel honeycomb between stainless steel sheets, covered on the outside with ablative material (heat-dissipating material which chars and falls away during earth entry).

The inner shell is aluminum honeycomb between aluminum alloy sheets. A layer of insulation separates the two shells. This construction makes the

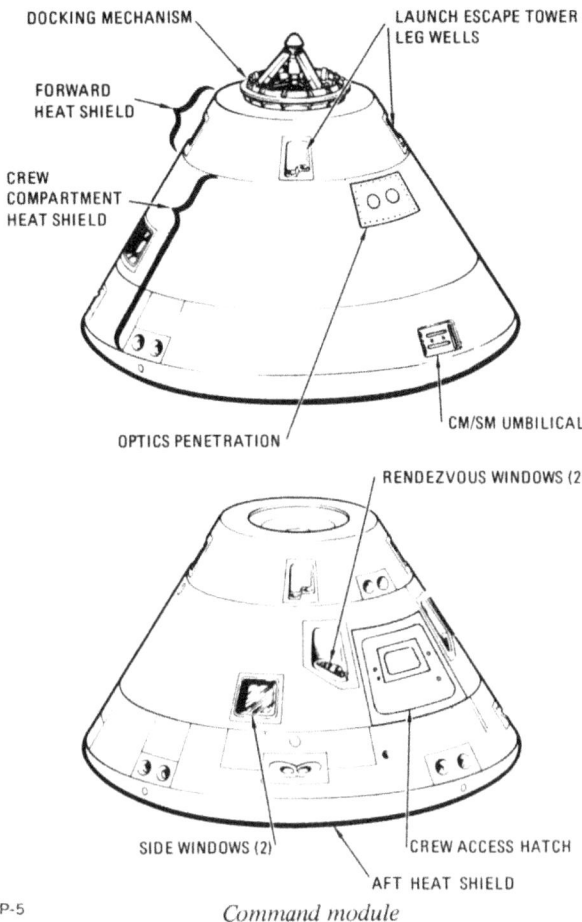

Command module

and the moon orbit rendezvous. These and other subsystems, such as the reaction control, guidance and navigation, earth landing, and parts of the environmental control and electrical power, occupy almost every inch of available space in the module.

Although crewmen can move about from one station to another, much of their time will be spent on their couches. The couches can be adjusted so the crew can stand or move around. Space by the center couch permits two men to stand at one time. The couches are made of steel framing and tubing and covered with a heavy, fireproof fiberglass cloth. They rest on eight crushable honeycomb shock struts which absorb the impact of landing. Control devices are attached to the armrests.

SERVICE MODULE

The service module's function, as its name implies, is to support the command module and its crew. It houses the electrical power subsystem, reaction control engines, part of the environmental control subsystem, and the service propulsion subsystem including the main propulsion engine for insertion into orbit around the moon, for return from the moon, and for course corrections.

CM light as possible yet rugged enough to stand the strain of acceleration during launch, the shock and heat of earth entry, the force of splashdown, and the possible impact of meteorites.

Inside, it is a compact but efficiently arranged combination cockpit, office, laboratory, radio station, kitchen, bedroom, bathroom, and den. Its walls are lined with instrument panels and consoles, and its cupboards (bays) contain a wide variety of equipment. In flight, the cabin is air conditioned to a comfortable 70 to 75 degrees. The atmosphere is 100-percent oxygen, and the pressure is about 5 pounds per square inch (a little better than one-third of sea-level pressure of 14.7 pounds per square inch).

The command module's controls enable the crew to guide it during flight. Test equipment permits checkout of malfunctions in spacecraft subsystems. Television, telemetry and tracking equipment, and two-way radio provide communication with earth and among the astronauts during moon exploration

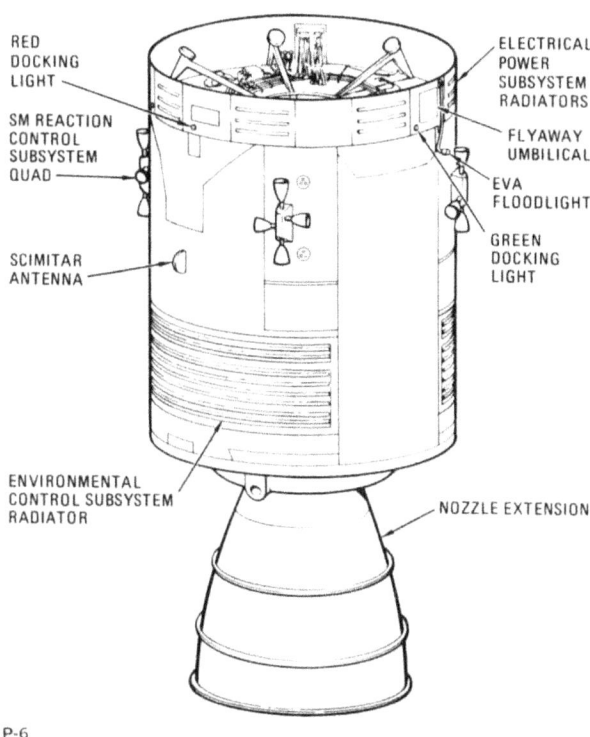

Service module

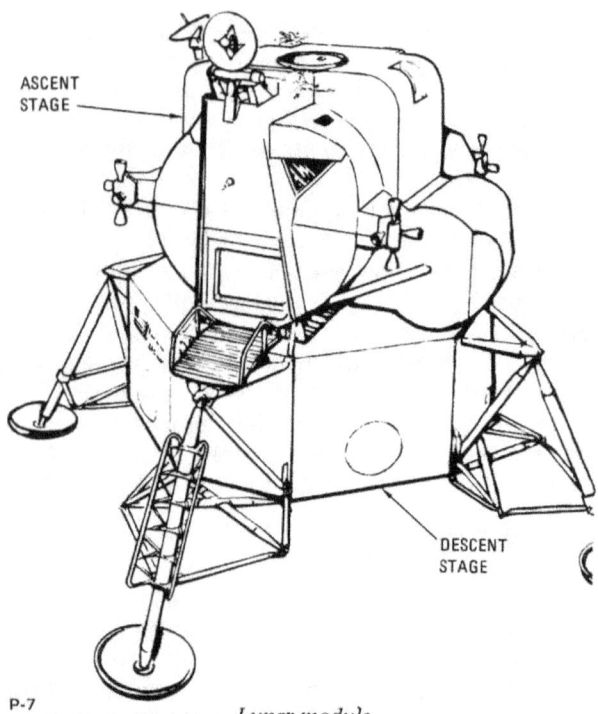

Lunar module

The SM is constructed of aluminum alloy. Its outer skin is aluminum honeycomb between aluminum sheets. Propellants (a combination of hydrazine and unsymmetrical dimethylhydrazine as fuel and nitrogen tetroxide as oxidizer) and various subsystems are housed in six wedge-shaped segments surrounding the main engine.

The service module is attached to the command module until just before earth entry, when the SM is jettisoned.

LUNAR MODULE

The LM will carry two men from the orbiting CSM down to the surface of the moon, provide a base of operations on the moon, and return the two men to a rendezvous with the CSM in orbit. Its odd appearance results in part from the fact that there is no necessity for aerodynamic symmetry; the LM is enclosed during launch by the SLA and operates only in the space vacuum or the hard vacuum of the moon.

The LM structure is divided into two components: the ascent stage (on top) and the descent stage (at the bottom). The descent stage has a descent engine and propellant tanks, landing gear assembly, a section to house scientific equipment for use on the moon, and extra oxygen, water, and helium tanks.

The ascent stage houses the crew compartment (which is pressurized for a shirtsleeve environment like the command module), the ascent engine and its propellant tanks, and all LM controls. It has essentially the same kind of subsystems found in the command and service modules, including propulsion, environmental control, communications, and guidance and control.

Portable scientific equipment carried in the LM includes an atmosphere analyzer, instruments to measure the moon's gravity, magnetic field, and radiation, rock and soil analysis equipment, a seismograph, a soil temperature sensor, and cameras (including television).

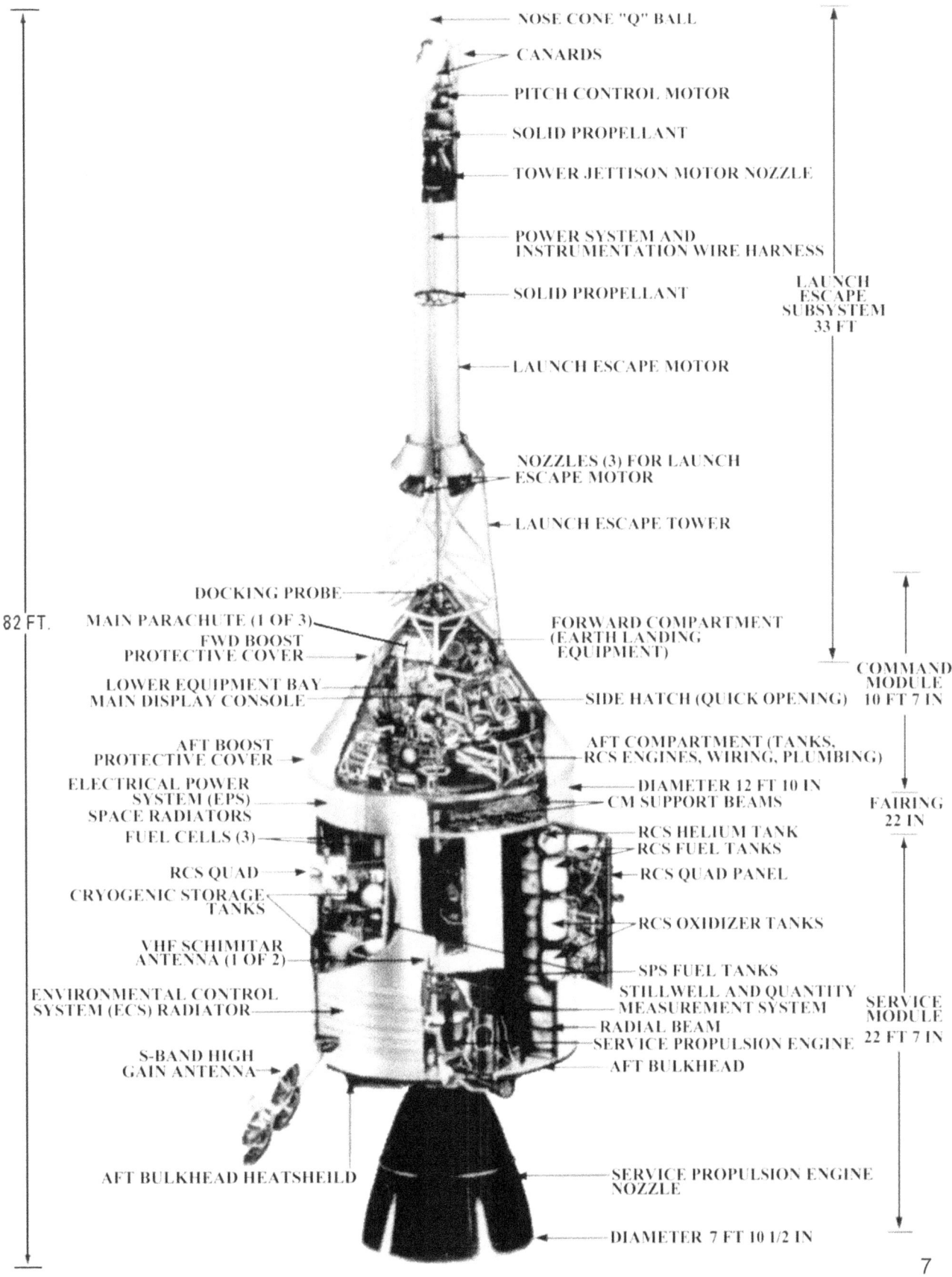

LAUNCH VEHICLES

Development of launch vehicles for the lunar landing mission represents a tremendous stride forward in rocket propulsion: they are bigger, more powerful, and vastly more complex than previous U.S. launch vehicles.

Development of the Saturn family began in late 1958 under the Department of Defense's Advanced Research Projects Agency. The work was conducted by the Army Ballistic Missile Agency at Redstone Arsenal in Huntsville, Ala., which in 1960 was transferred in part to NASA to become the nucleus of the Marshall Space Flight Center.

Studies under Dr. Wernher von Braun aimed at developing a booster with a total thrust of 1.5 million pounds had been conducted in 1957.

There are three launch vehicles in the Saturn family: the Saturn I, which had a perfect record of ten successful flights; the Saturn IB, and the Saturn V. The name Saturn was adopted in 1959 and at that time applied only to the 1.5 million-pound thrust vehicle which became the Saturn I.

This stepping stone approach led to the development of Saturn V in three phases: Saturn I, which used primarily modified existing equipment; Saturn IB, which uses a modified first stage of the Saturn I and a new second stage and instrument unit; and the Saturn V, which uses new first and second stages and the third stage and instrument unit of the Saturn IB.

Development of the engines needed for the Saturn vehicles was begun separately, but much of it was in parallel with the vehicle program. Work started on the F-1 engine, the nation's largest, in 1958 and on the hydrogen-powered J-2 engine in 1960. The J-2, which burns a cryogenic (ultra low temperature) propellant composed of liquid hydrogen and liquid oxygen, was the key to development of the powerful upper stages of the Saturn IB and Saturn V.

The Saturn I program is complete. From 1961 through 1965 it was launched 10 times successfully, putting Apollo boilerplate (test) vehicles and Pegasus meteoroid technology satellites into orbit.

Saturn V launch vehicle (with Apollo)

The first launch of a Saturn IB came in early 1966 and also was the first space test of an Apollo spacecraft. Succeeding tests also have been successful.

The first flight test of the Saturn V was November 9, 1967, when it boosted an Apollo spacecraft into space. A NASA report described the performance of the North American Rockwell-produced Apollo spacecraft as satisfactory in all respects.

The Saturn IB and the Saturn V are the basic heavy launch vehicles of the United States civilian space program. Saturn V will be used for Apollo test flights and for lunar missions. The Saturn IB will be used with smaller payloads.

In the early studies on the Saturn, many configurations were considered. The Saturn I originally was designated C-1; the Saturn V was C-5. Although the other configurations were dropped, the designations "I" and "V" remained.

Specific figures are given on weight, height, and amount of propellant for each Saturn vehicle. However, all are approximate. Like Apollo, changes and improvements affect the figures, particularly those on weight. In addition, each launch vehicle produced and each mission is somewhat different.

SATURN IB

The Saturn IB has two stages and an instrument unit (IU).

The first stage (S-IB) built by the Chrysler Corp., is the same size and shape as the first stage of the Saturn I (S-1), but was redesigned to cut its weight and increase its power.

The second stage (an S-IVB), produced by McDonnell Douglas Co.'s Missile and Space Systems Division, is a large, all-cryogenic booster. The cryogenic propellant—liquid hydrogen at 423° below zero F and liquid oxygen at 297° below zero—provides more energy per pound of weight than the chemical fuels used previously, but posed many problems in insulation, handling, and engine systems.

The instrument unit (IU), a cylindrical-shaped segment mounted atop the second stage, contains equipment for sequencing, guidance and control, tracking, communications, and monitoring. It was designed by NASA and is being produced by IBM's Federal Systems Division.

The engines for the Saturn IB are designated the H-1, used in the first stage, and the J-2, used in the second stage. Both are produced by North American Rockwell's Rocketdyne Division. The H-1 engine was used for the first stage of the Saturn I, but was uprated, first 200,000 pounds, then to 205,000 for the Saturn IB. The stage has eight H-1 engines for a total thrust of 1,600,000 pounds. The J-2 engine used on the second stage is more than 11 feet long and weighs about 3400 pounds. It has a thrust of up to 225,000 pounds at high altitude.

Basic facts about the two-stage Saturn IB:

Height	138 ft (vehicle only)
	224 ft (with spacecraft)
Weight	1,300,000 lb (with propellant)
	153,000 lb. (dry)
Payload	40,000 lb in low earth orbit

FIRST STAGE

Height	80.3 ft
Diameter	21.4 ft
Gross weight	1,000,000 lb
Propellant weight	910,000 lb
Propellant	RP-1 and liquid oxygen
Engines	8 H-1's
Thrust	1,600,000 lb (sea level)

SECOND STAGE

Height	58.4 ft
Diameter	21.7 ft
Gross weight	253,000 lb
Propellant weight	230,000 lb
Propellant	Liquid hydrogen and liquid oxygen

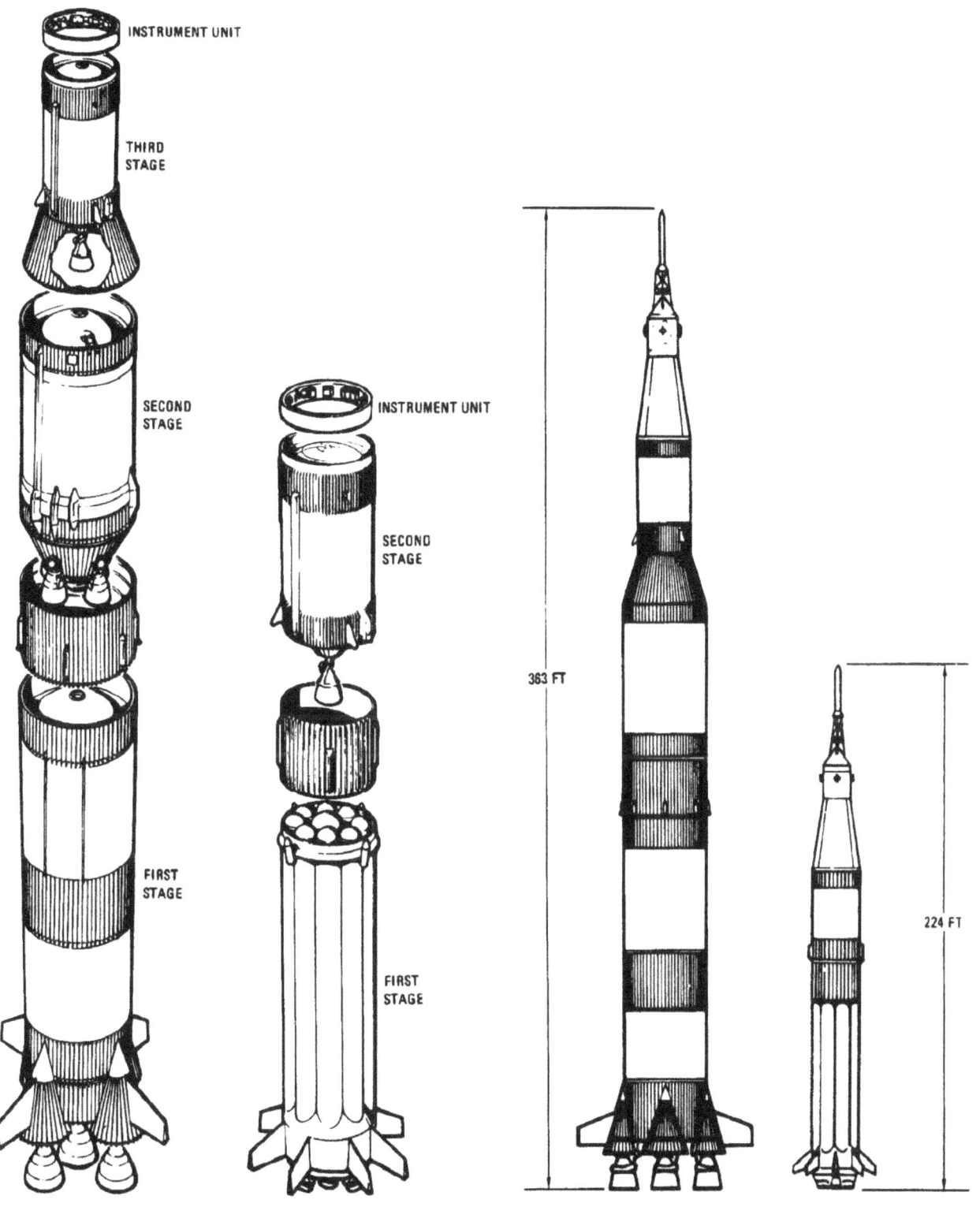

Saturn V (left) and Saturn IB

Engine	1 J-2
Thrust	225,000 lb (vacuum)

INSTRUMENT UNIT

Height	3 ft
Diameter	21.7 ft
Weight	4500 lb (approximate)

SATURN V

The Saturn V is the nation's largest and most powerful launch vehicle. It has three stages and an instrument unit.

The first stage produced by the Boeing Co.'s Launch Systems Branch, is 138 feet high (higher than any entire pre-Saturn launch vehicle) and weighs close to five million pounds when fueled. The function of this stage is to lift the enormous weight (more than 6.2 million pounds) of the Apollo/Saturn V space vehicle off the pad and carry it to an altitude of about 38 miles and a speed of about 6000 miles per hour.

The second stage, built by North American Rockwell's Space Division, is the largest and most powerful hydrogen-fueled stage ever produced. It is 81.5 ft. tall and weighs more than 1 million pounds fueled. It takes over from the first stage and boosts its payload of the third stage and Apollo spacecraft into space (an altitude of about 118 miles) and to a speed of more than 14,000 miles per hour.

The third stage is essentially the same as the second stage of the Saturn IB. On the Saturn V it serves in a double capacity. After the second stage burns out and is jettisoned, the third stage's engine burns briefly, just long enough to increase its velocity to about 17,400 miles per hour and put it and the Apollo into earth orbit. It stays connected to the spacecraft from one to three orbits, then its engine is reignited at the proper moment to power itself and the spacecraft toward the moon.

The instrument unit for Saturn V is essentially the same as on the Saturn IB. Like the third stage, however, it is modified and improved to help it carry out the different missions of the Saturn V.

There are two types of main engines used on the Saturn V, both built by North American Rockwell's Rocketdyne Division. The first stage uses the F-1, the most powerful engine ever produced. It is 19 feet long, weighs about 18,500 pounds and produces 1,500,000 pounds of thrust. The first stage has five F-1 engines for a total thrust of 7,500,000 pounds. Both the second and third stages use the J-2 engine, the largest hydrogen-fueled engine ever built. The J-2 engine produces up to 225,000 pounds of thrust; the second stage uses five J-2 engines producing a maximum of 1,125,000 pounds of thrust. The third stage uses one J-2 engine.

Basic facts about the Saturn V:

Height	282 ft (vehicle only) 363 ft (with spacecraft)
Weight	6,200,000 lb (with propellant) 430,000 lb (dry)
Payload	270,000 lb in low earth orbit 100,000 lb translunar injection

FIRST STAGE

Height	138 ft
Diameter	33 ft
Gross weight	4,792,000 lb
Propellant useable weight	4,492,000 lb
Propellant	RP-1 and liquid oxygen
Engines	5 F-1's
Thrust	7,500,000 lb (1,500,000 lb each engine)
Burning time	150 sec

SECOND STAGE

Height	81.5 ft
Diameter	33 ft

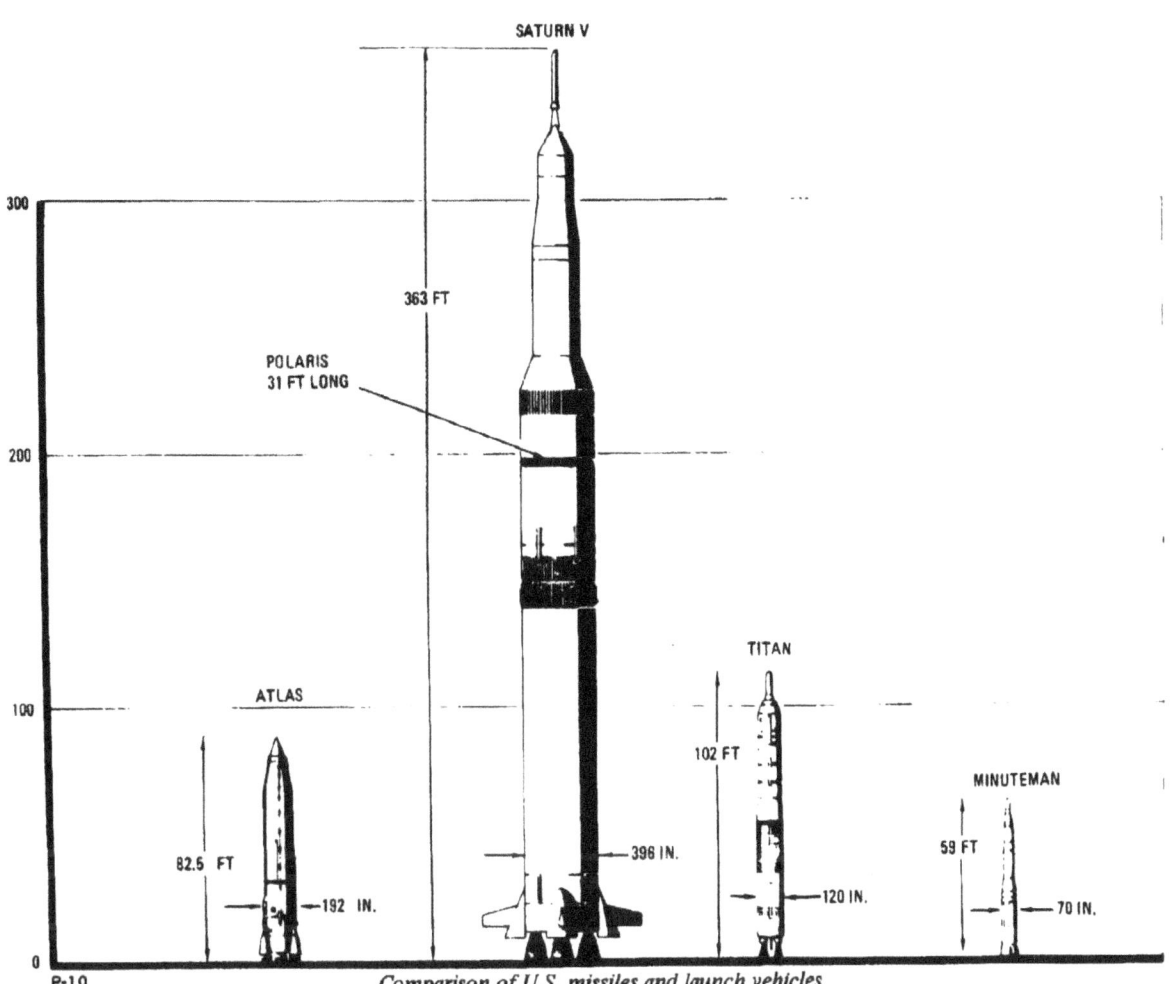

Comparison of U.S. missiles and launch vehicles

Gross weight	1,037,000 lb at liftoff	Gross weight	262,000 lb
Propellant useable weight	942,000 lb	Propellant useable weight	228,000 lb (excluding reserves)
Propellant	Liquid hydrogen and liquid oxygen	Propellant	Liquid hydrogen and liquid oxygen
Engines	5 J-2's	Engine	1 J-2
Thrust	1,125,000 lb (225,000 lb each engine)	Thrust	225,000 lb maximum (at altitude)
Burning time	359 sec	Burning time	480 sec (2 burns)

THIRD STAGE

IU

Height	58.5 ft	Height	3 ft
Diameter	21.7 ft (lower interstage expands to 33 ft)	Diameter	21.7 ft
		Weight	4500 lb (approximate)

Gigantic mobile launcher inches along on way to launch pad carrying Saturn V/Apollo

MISSIONS

Apollo missions fall into two categories: earth-orbital and lunar. Earth-orbital and lunar-orbital missions are part of the flight-testing program to test the spacecraft, the launch vehicles, launch and communications equipment and procedures, and crew operations. The lunar mission calls for the landing of two American astronauts on the moon, exploration of the moon, and return to earth.

EARTH-ORBITAL MISSIONS

NASA's schedule of Apollo development missions is a flexible one that progresses logically toward accomplishing a lunar landing mission. The flight test program consists of unmanned flights, manned flights in earth orbit, a lunar orbital flight, and, finally, the lunar landing flight. Alternative flight plans are prepared for use in the event contingencies arise.

Previous flight tests of the Apollo command, service, and lunar modules--all unmanned--have been successful. CSM tests were aimed primarily at the operation of its subsystems and of man-rating the subsystems (certifying for manned space flight). Particularly important were tests of the heat shield and command module structure, which survived such rigorous conditions as the 5,000-degree heat during atmospheric entry from a lunar mission. The spacecraft's compatibility with the Saturn launch vehicles has been demonstrated.

Primary objective of the manned missions is to determine the proficiency of the crew in the

Apollo and Saturn are checked out before launch

Three astronauts enter the command module

complex tasks required during the lunar missions and to test the operation of the manned space flight network, the communications link that will be used during the lunar mission. Crew tasks to be evaluated are those required for navigation, transposition and docking, rendezvous and docking, major propulsion maneuvers, entry, and recovery.

The first manned mission is designed to test the adequacy and overall performance of all CSM subsystems, including its life support and environmental control subsystems, over a substantial period. In the first mission, only the command and service modules will be launched with a Saturn IB.

The second manned flight, designated Apollo 8, calls for low earth-orbital checkout of the spacecraft

Mobile launcher carries space vehicle to launching pad

and upper portion of the Saturn V launch vehicle. The flight will be in some senses an expansion of the Apollo 7 mission. No lunar module will be flown.

The third manned earth-orbital mission will involve the CSM and LM and is designed to demonstrate the combined operation for the first time of the complete spacecraft.

LUNAR MISSION

A lunar mission will mark the first time that astronauts will not be within minutes of earth. The landing mission will be a milestone in man's history--the first time man will set foot on another celestial body.

For planning purposes, the lunar mission is divided into phases. To gain an understanding of how the mission will be accomplished, each of these phases, with the exception of pre-launch and post-landing, is described.

EARTH ASCENT

At liftoff, the Saturn V's first stage, developing over 7-1/2 million pounds of thrust from its five F-1 engines, lifts the 6.4 million-pound space vehicle off the pad and boosts it on its way. The first stage burns for about 2-1/2 minutes and reaches a velocity of about 5400 miles per hour and an altitude of about 40 miles. After the F-1 engines cut off, retrorockets on the first stage fire to achieve separation from the second stage. Four seconds later, the second stage's five J-2 engines ignite to boost the third stage and spacecraft to an altitude of approximately 114 statute miles. During its approximately 6 minutes of firing, the second

P-15 *First stage lifts huge space vehicle off pad*

P-16 *Second stage ignites after first stage falls away*

P-17 *Launch escape assembly is jettisoned*

P-18 *Third stage ignites as second stage falls away*

stage increases velocity to about 15,000 miles per hour. When its engines cut off, the second stage is jettisoned.

The third stage's single J-2 engine ignites at separation and burns for about 2 minutes to increase speed to about 16,500 miles an hour and put it and the Apollo spacecraft into a near-circular earth orbit at about 115 statute miles.

During ascent, the crew monitors the launch vehicle displays to be prepared for an abort, if necessary; relays information about boost and the spacecraft to the ground; and monitors critical subsystem displays.

EARTH PARKING ORBIT

The spacecraft is inserted into an earth parking orbit to permit checkout of subsystems before it is committed to lunar flight, and to allow for more than one opportunity for translunar injection (instead of a single one in a direct launch).

The mission allows the spacecraft, with the third stage attached, to orbit the earth up to three times (for 4-1/2 hours) before injection into translunar flight. Because injection is desirable as soon as possible after checkout, the translunar injection maneuver probably will be performed during the second orbit. To inject the spacecraft into translunar flight, the crew reignites the third-stage engine.

TRANSLUNAR INJECTION

The translunar injection parameters are computed with the guidance system in the Saturn V's third-stage instrumentation unit. Thus, the third stage is commanded to fire at the proper moment and for the precise length of time necessary to put the spacecraft into a trajectory toward the moon.

P-19
Third stage fires again for translunar injection

This trajectory is nominally one that provides a "free return" to earth; that is, if for any reason the spacecraft is not inserted into an orbit around the moon, the spacecraft will return to earth.

The third-stage engine burns for about 5-1/2 minutes and cuts off at an altitude of about 190 miles and at a velocity of about 24,300 statute miles an hour.

During the engine thrusting, the crewmen remain in their couches and monitor the main display console.

INITIAL TRANSLUNAR COAST

The manned space flight network tracks the spacecraft for about 10 minutes after third-stage engine cutoff to determine whether to proceed with transposition and docking. During the same period, the third stage maneuvers the spacecraft to the attitude programmed for the transposition, docking, and LM-withdrawal maneuvers.

The crewmen position their couches to see out of the docking windows. The commander begins the transposition and docking maneuver by firing the service module reaction control engines. A signal is sent almost simultaneously to deploy and jettison the SLA panels, separate the CSM from the SLA, and deploy the CSM's high-gain antenna. The lunar module remains attached to the adapter.

The commander stops the CSM 50 to 75 feet away from the third stage, turns 180 degrees with a pitch maneuver so the docking windows are facing the LM, rolls the CSM for proper alignment with the LM, closes with the LM, and docks.

After the CSM and LM have docked, the pressure between the CM and the LM is equalized and the CM forward hatch is removed. A check is made to determine that all docking latches are engaged, the CSM-LM electrical umbilicals are connected, and the CM forward hatch is reinstalled. The LM's four connections to the SLA are severed by small explosive charges, and the spacecraft is separated from the SLA and third stage by spring thrusters.

TRANSLUNAR COAST

Now the long journey to the moon begins. It has been three to six hours since liftoff from Kennedy Space Center, depending on the number of earth

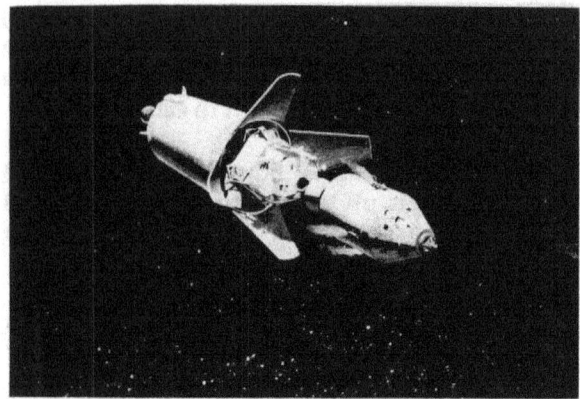

P-20

SLA panels jettison and the CSM pulls away

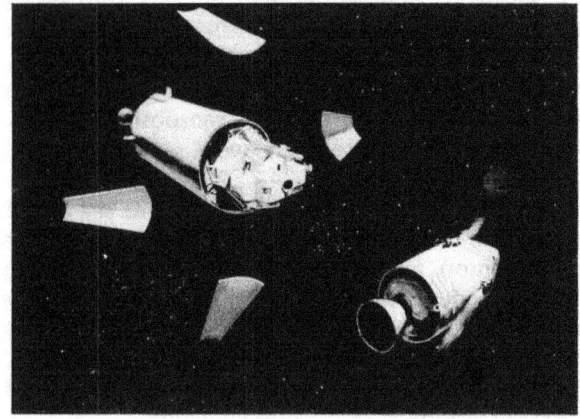

P-21

Commander turns CSM around for docking maneuver

P-22

After docking, spring thrusters separate two craft

parking orbits, and the crew settles down for the 2-1/2- to 3-1/2-day flight.

In this phase of the flight, the spacecraft is coasting. At the time of injection into the translunar trajectory, the spacecraft is traveling almost 24,300 miles per hour with respect to the earth. It begins slowing almost immediately because of the pull of earth's gravity. The speed drops until the spacecraft enters the moon's sphere of influence where it again increases due to the moon's gravitational pull.

Shortly after the coast period begins, the spacecraft is oriented for navigation sightings of stars and earth landmarks. The spacecraft is then put into a slow roll (about 2 revolutions an hour) to provide uniform solar heating. This thermal control rolling is stopped for inertial measurement unit alignment and for course corrections.

If tracking from the ground indicates a course correction is needed during the translunar coast, the correction is made with the service propulsion engine when a large change is indicated or with the SM reaction control engines when the change required is small.

The crew has a number of subsystem duties. Electrical power and environmental control subsystem status checks are conducted. The service propulsion and SM reaction control subsystems are checked. Hydrogen and oxygen purges of the fuel cells are conducted, the lithium hydroxide canisters exchanged and communication with the ground is maintained.

The three astronauts eat in shifts but sleep at the same time. The ground monitors the spacecraft performance continuously and can awaken the crew. Biomedical data is sent to the ground continuously.

P-23

Engine retro-fires to put spacecraft in lunar orbit

LUNAR ORBIT INSERTION

Insertion of the spacecraft into lunar orbit is essentially a braking maneuver in which the spacecraft is transferred from the ellipse of the lunar approach to an orbit around the moon.

The insertion maneuver involves the longest firing of the service propulsion engine and results in a reduction in the craft's velocity with respect to the moon from about 5600 to 3600 miles per hour. The insertion may be a two-stage firing of the service propulsion engine, the first to put the CSM in an elliptical orbit of approximately 70-by-195 statute miles and the second to put the CSM in a circular orbit of about 70 miles. The precise timing of the firing and the exact length of the burn or burns are determined by the Mission Control Center in Houston and are programmed into the CM computer, which automatically fires the engine.

During the firing the spacecraft is out of communication with the ground since it will be passing behind the moon. Communications, which require line-of-sight to earth, are lost for about 45 minutes on each 2-hour lunar orbit.

During the maneuver, the crew monitors the display of the velocity change required, the digital event timer, the flight director attitude indicators, and subsystem status displays.

LUNAR ORBIT COAST

The docked CSM and LM orbit the moon until the LM is checked out for descent to the lunar surface. During this time coarse and fine alignments are made of the CSM inertial measurement unit, as is a series of sightings of landmarks on the lunar surface. These operations, each involving changes in spacecraft attitude, are compared with tracking data from the manned space flight network to determine the spacecraft's precise location in orbit with respect to the landing site on the moon.

The CM-LM tunnel and the LM are pressurized, and the CM hatch, the probe, and drogue are removed. The LM hatch is opened to clear the way into the LM. First to transfer to the LM is the LM pilot, who activates the LM's environmental control, electrical power and communications subsystems.

After the commander has transferred to the LM, he and the LM pilot perform a lengthy series of checks

P-24
Two crewmen transfer to LM to prepare for separation

of the LM subsystems. While they do this, the CM pilot is performing another series of alignments and landmark sightings. The CM controls the attitude of the spacecraft during the lunar orbits and during the coarse alignments of the LM inertial measurement unit. (The fine aligning of the LM inertial measurement unit is done after the CSM and LM have separated.)

The probe and drogue are installed, the 12 docking latches are unlatched, the LM hatch is closed, the CM hatch is installed, the LM landing gear is deployed, and guidance computations are made in the final minutes before separation. Then the CM pilot activates the probe extend/release switch which undocks the LM from the CSM. The LM reaction control system moves the LM away from the CSM a short distance and is oriented so the CM pilot can inspect the LM landing gear. The LM's reaction control system then fires again to separate further the LM and CSM.

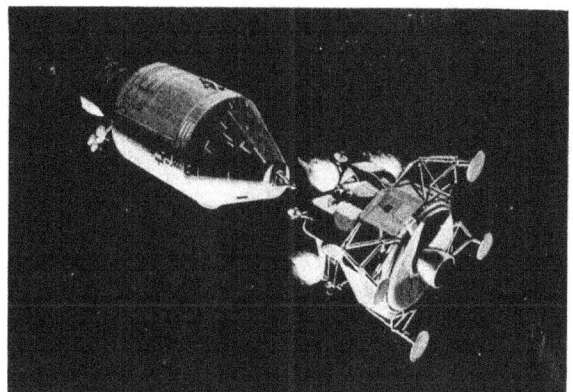

P-25
LM with two men separates from CSM

CSM LUNAR ORBIT

The CSM remains in the 70-mile lunar orbit for about a day and a half, until the LM returns from its moon landing. The CM pilot has many duties during this time and is particularly busy during two periods: LM descent to the moon and LM ascent to rendezvous and docking. His principal jobs during these periods are to monitor the performance of the LM (requiring changes in CSM attitude to keep it in sight), communicate with the LM and with earth, and activate or operate equipment to aid in both the landing and the rendezvous and docking procedures. The CM pilot will have a period to sleep while the LM is on the lunar surface.

After separation, the CSM and LM pass behind the moon, and communications with earth are cut off. During this period, the LM telemeters its data to the CSM, where it is stored and relayed to earth after the CSM emerges from behind the moon.

P-26 *LM descends and CSM stays in orbit*

P-27 *LM's descent engine fires as craft comes in for landing*

LM DESCENT

The descent to the moon takes an hour of complex maneuvering that taxes the skills of the astronauts and the capabilities of the lunar module.

Briefly, the descent will follow this course. The LM fires its descent engine to put it into an elliptical orbit that reaches from about 70 miles to within 50,000 feet of the moon's surface. Near the 50,000 foot altitude at a preselected surface range from the landing site, the engine is fired again in a braking maneuver to reduce the module's speed.

The LM's two-man crew is busy with position and velocity checks, subsystem checks, landing radar test, attitude maneuvering, and preparation of the LM computer or the braking maneuver.

The final approach begins at approximately 9000 feet altitude with a maneuver to bring the landing site into the view of the LM crew. The firing is controlled automatically until the craft reaches an altitude of about 500 feet. When the commander takes over, the LM is pitched at an angle which permits the crewmen to assess the landing site. At about 65 feet altitude, the LM is re-oriented and descends vertically to the surface at about 3 feet per second. The commander shuts off the descent engine as soon as the landing gear touches the moon.

LUNAR STAY

It will be about 4-1/2 hours after landing before the first American steps foot on the moon. Upon landing, the commander and LM pilot first spend

P-28 *One astronaut checks LM, other stays inside at first*

P-29
Two astronauts set up scientific equipment on moon

LM ASCENT

Ascent of the LM from the moon and rendezvous and docking with the orbiting CSM takes three hours.

P-30
Astronaut fires ascent stage engine to leave moon

about two hours checking out the LM ascent stage. Any remaining propellant for the descent engine is vented and the inertial measurement unit is aligned and placed in standby operation. It takes 2 hours for the extravehicular mobility units to be checked out and prepared for use.

Finally, the LM cabin is depressurized, and one of the LM crewmen emerges from the lunar module, descends its ladder, and walks on the moon. He remains alone on the lunar surface for about 20 minutes while he gathers a sample of surface material and transfers it to the crewman in the ascent stage, who has been recording this activity on still and motion picture film.

Following a period of equipment transfer between the astronaut in LM cabin, and the astronaut on the lunar surface, the second crewman descends to the surface and the two inspect the LM to determine the effects of the lunar touchdown on the vehicle. The rest of the time in the first surface excursion is consumed by erecting the S-band surface antenna, collecting a preliminary set of geological samples, and making a TV scan of the landing site.

A second lunar exploration period lasts about three hours, with both astronauts on the surface. They have many tasks to perform, including sample collections, photography, exploration of the lunar surface up to about a quarter-mile from the LM, and erection of a station that will continue to send scientific data to earth after the astronauts leave.

Between the two exploratory periods, the astronauts will have a sleep period. Then, after about 24 to 26 hours on the moon, the astronauts prepare for their return to the CSM.

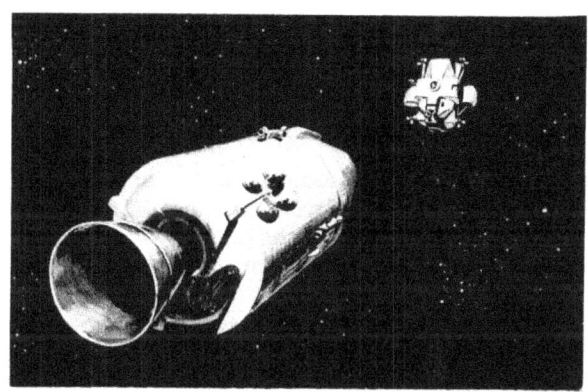
P-31
LM's ascent stage in rendezvous with orbiting CSM

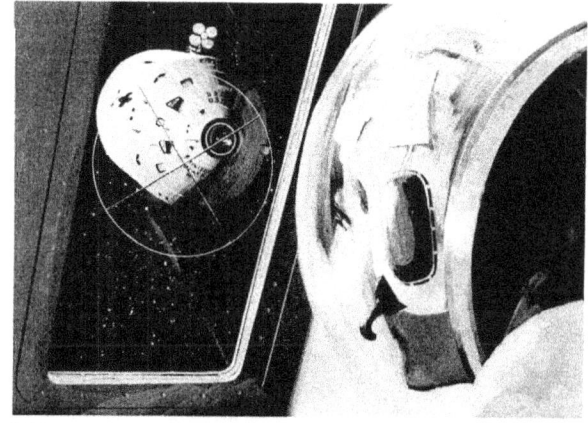
P-32
Commander in LM guides craft to docking with CSM

21

When the ascent engine ignites, the ascent stage of the LM separates from the descent stage, using the latter as a launching platform. The engine boosts the ascent stage off the moon into an elliptical orbit of an estimated 60,000 feet by about 11.4 by 34.5 miles.

The next maneuver places the LM in a circular orbit which has a constant altitude distance from that of the CSM. When the LM's orbit is in the proper phase with the CSM orbit, the LM reaction control engines are fired to raise the LM orbit to that of the CSM, about 70 miles. During these maneuvers the CM pilot tracks the LM.

The LM takes about 30 minutes to intercept the CSM, during which course corrections are made with the reaction control engines. The firings are controlled by the LM computer on the basis of data supplied by the ground. The LM closes on the CSM through a series of short reaction control engine firings. The commander takes over control of the LM and maneuvers it with short bursts of the reaction control engines to a docking with the CSM.

LUNAR ORBIT COAST

After docking, the spacecraft coasts in lunar orbit while the crew transfers equipment and samples into the CSM, returns to the CSM, jettisons the LM ascent stage, and prepares for transearth injection.

The LM crew opens the LM hatch after the CSM and LM pressures have been equalized and the CM pilot removes the CM tunnel hatch. The drogue and probe are removed and stowed in the LM. Lunar samples, film, and equipment to be returned to earth are transferred from the LM to the CM; equipment in the CM that is no longer needed is put into the LM and the LM hatch is closed, the CM hatch is replaced, and the seal checked.

The LM is jettisoned by firing small charges around the CM docking ring. The entire docking mechanism separates from the CM and remains with the LM ascent stage. The SM reaction control engines are fired in a short burst to assure separation and to put the CSM into the lead in the orbit.

TRANSEARTH INJECTION

The service propulsion engine injects the CSM into a trajectory for return to earth. The engine fires for about 2-1/2 minutes to increase the spacecraft's velocity relative to the moon from about 3600 to nearly 5500 miles per hour. This maneuver takes place behind the moon, out of communications with earth. Communication is regained about 20 minutes after the engine has cut off.

P-34

After jettisoning LM, the CSM heads for earth

P-33 *Astronauts return to CM, prepare to jettison LM*

TRANSEARTH COAST

The trip from lunar orbit back to the earth's atmosphere could be the longest phase of the mission, lasting anywhere from 80 to 110 hours. The spacecraft's velocity on the coast back gradually decreases because of the moon's gravitational pull and then increases again when the spacecraft comes into the earth's sphere of influence. When the spacecraft enters the atmosphere, its velocity has increased to about 25,000 miles per hour.

Crew duties during the homeward coast are similar to those of the outbound journey. The spacecraft is again in a slow roll for thermal control. Crewmen make any necessary course corrections, maneuver

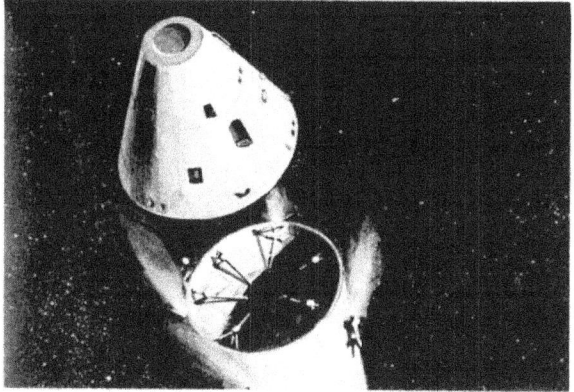

P-35 *Shortly before entry, the CM separates from SM*

P-37 *Aft heat shield chars to absorb 5,000-degree heat*

the spacecraft for inertial measurement unit alignments and regularly check subsystems.

About three and a half hours before entry, the CSM is rotated and held in an attitude that puts the forward heat shield of the CM in shadow. This cooling of the shield lasts for about an hour and a half, after which the attitude must be changed for the final course correction.

Shortly before entry into the earth's atmosphere, the service module is jettisoned. The CM and SM are separated by small explosive devices in the SM. The SM reaction control engines fire simultaneously to increase separation and assure that the two modules will not collide.

ENTRY

The desired entry conditions include the arrival of the CSM at a particular point above earth at a particular time and with a proper flight path angle, neither too steep nor too shallow.

Entry is considered to begin at an altitude of about 400,000 feet, when the CM begins to meet the resistance of the atmosphere. At this point the CM is traveling about 24,500 miles an hour, and the heat generated on its plunge through the atmosphere may reach 5000 degrees Fahrenheit on the blunt aft heat shield.

But despite the heat generated on the outside of the CM, its cabin will remain at 80 degrees. The maximum gravitational forces felt by the astronauts will be a little over 5 G's.

LANDING

The landing is controlled automatically by the earth landing subsystem, although the crew has

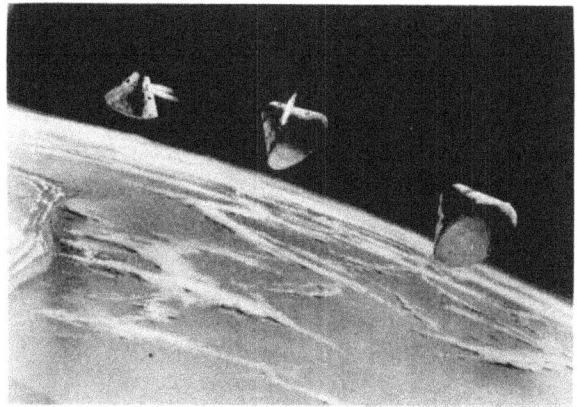

P-36 *CM is oriented blunt end forward for entry*

P-38 *Drogue chutes open to provide initial slowing*

backup controls. At about 24,000 feet, a barometric switch closes to start the subsystem in operation.

The forward heat shield is jettisoned to permit deployment of parachutes, and the drogue parachutes are immediately released. They are deployed reefed (half closed) and open after a few seconds. The drogues orient the CM for main parachute deployment and reduce the CM's speed from an estimated 325 to 125 miles an hour.

At an altitude of about 10,700 feet the drogues are disconnected and the pilot parachutes are deployed. They pull out the main parachutes. The main parachutes are double reefed, which means they open in two stages. They further slow the CM, and final descent and splashdown is made at about 22 miles an hour.

P-40

Recovery helicopters move in as CM floats in water

As soon as the main parachutes are disreefed, the crewmen start burning the remaining reaction control propellant, activate the VHF recovery beacon, adjust their couches for landing, and purge the propellant lines. Final descent on the main parachutes takes about 5 minutes.

In addition to the recovery beacon deployed by the crew, two VHF antennas are deployed automatically shortly after the main parachutes are deployed. These provide voice communication with the recovery forces. The recovery beacon transmits a continuous signal.

The main parachutes are released by the crew at splashdown and postlanding ventilation is turned on.

P-39

CM drifts gently to splashdown on main chutes

THE MOON

The landing of Apollo astronauts, and their return to earth with lunar soil samples, will help solve some of the mysteries of the moon. What is known about the moon, from centuries of astronomical observation and from the recent space mission, is this:

Terrain—Mountainous and crater-pitted, the former rising thousands of feet and the latter ranging from a few inches to 180 miles in diameter. The craters are thought to be formed by the impact of meteorites. The surface is covered with a layer of fine-grained material resembling silt or sand, as well as small rocks.

Environment—No air, no wind, and no moisture. The temperature ranges from 250 degrees in the two-week lunar day to 280 degrees below zero in the two-week lunar night. Gravity is one-sixth that of earth. Micrometeoroids pelt the moon (there is no atmosphere to burn them up). Radiation might present a problem during periods of unusual solar activity.

Dark Side—The dark or hidden side of the moon no longer is a mystery. It was first photographed by a Russian craft and since then has been photographed many times, particularly by NASA's Lunar Orbiter spacecraft.

Origin—There is still no agreement among scientists on the origin of the moon. The three theories: (1) the moon once was part of earth and split off into its own orbit, (2) it evolved as a separate body at the same time as earth, and (3) it formed elsewhere in space and wandered until it was captured by earth's gravitational field.

Possible landing sites for Apollo's lunar module have been under study by NASA's Apollo Site Selection Board for about two years. Thirty sites originally were considered, and these were later narrowed down to eight.

Selection of the final five sites was based on high-resolution photographs returned by Lunar Orbiter, plus close-up photos and surface data provided by Surveyor.

All of the original sites were on the visible side of the moon within 45 degrees east and west of the center of the moon and 5 degrees north and south of its equator.

The final five choices were based on these factors:

- Smoothness (relatively few craters and boulders)

- Approach (no large hills, high cliffs, or deep craters that could cause incorrect altitude signals to the landing radar)

- Propellant (selected sites allow the least expenditure of propellant)

Physical Facts	
Diameter	2,160 miles (about 1/4 that of earth)
Circumference	6,790 miles (about 1/4 that of earth)
Distance from earth	238,857 miles (mean; 221,463 minimum to 252,710 maximum)
Surface temperature	250 (sun at zenith) -280 (night)
Surface gravity	1/6 that of earth
Mass	1/100th that of earth
Volume	1/50th that of earth
Lunar day and night	14 earth days each
Mean velocity in orbit	2,287 miles per hour
Escape velocity	1.48 miles per second
Month (period of rotation around earth)	27 days, 7 hours, 43 minutes

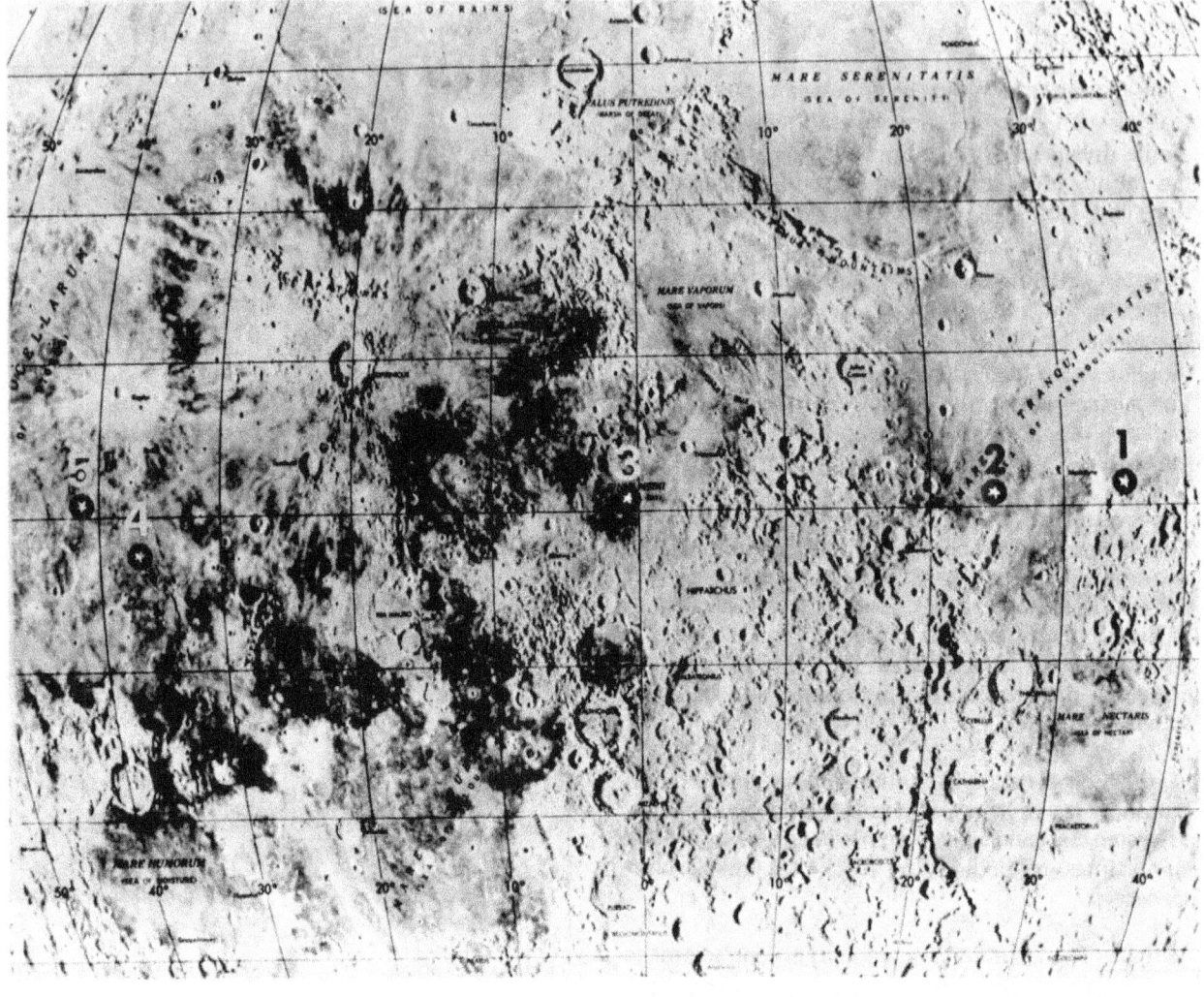

The five Apollo moon landing sites

- Recycling (selected sites allow for necessary recycling time if the Apollo/Saturn countdown is delayed)

- Free return (sites are within reach of the spacecraft in a free-return trajectory)

- Slope (there is little slope—less than 2 degrees—in the landing area and approach path)

Three of the five sites will be chosen for a specific lunar landing mission so that a three-day period each month will be available for the launch.

The Apollo lunar landing sites:

No.	Coordinates	Location
1	34° E, 2 40′N	Sea of Tranquility
2	23° 37′E, 0° 45′N	Sea of Tranquility
3	1° 20′W, 0° 25′N	Central Bay
4	36° 25′W, 3° 30′S	Ocean of Storms
5	41° 40′W, 1° 40′N	Ocean of Storms

MANNED SPACE PROGRAM

The United States manned space program has been conducted in three major phases—Mercury, Gemini, and Apollo. Each manned flight has led to increased knowledge of the systems and techniques needed to operate successfully in space, and each phase represents a significant advancement over the previous one.

The first man in space was the Russian Yuri Gagarin, who made one orbit of the earth in his Vostok 1 spacecraft on April 12, 1961. The first American spaceman was Alan B. Shepard, Jr., who rode his Mercury spacecraft into space atop a Redstone booster on May 5, 1961. The first American to orbit the earth was John H. Glenn, Jr., who made three orbits in a Mercury spacecraft on Feb. 20, 1962.

To date (August 1, 1968), 19 Americans have been in space, and seven of these have made two space flights. There were six manned flights during the Mercury program and 10 manned flights in the Gemini program.

Eleven Russians have been in space in their nine-launch program. The Russian manned program also has involved three spacecraft, with six flights aboard the one-man Vostok, two flights with the Voskhod (one a three-man and the other a two-man craft), and a single flight with the Soyuz spacecraft.

There have been no fatalities in space, but accidents have marred the advanced programs of both the United States and Russia. Three American astronauts—Virgil I. Grissom, Edward H. White, II, and Roger B. Chaffee—died in a fire aboard an Apollo spacecraft on the pad at Kennedy Space Center. Grissom, a veteran of two space flights, was pilot of the second Mercury spacecraft and commander of the first Gemini to go into space. White was aboard the second manned Gemini spacecraft in orbit, and made the historic 21-minute "walk in space." The accident occurred on Jan. 27, 1967, as the three men were rehearsing countdown procedures for what was to have been the first Apollo manned launch.

The Russian tragedy occurred during a space mission, but not in space. Cosmonaut Vladimir Komarov died in the Soyuz 1 spacecraft on April 24, 1967, when it crashed during landing. The Soyuz flight, which lasted about 25 hours, had been characterized as successful by the USSR. It had entered the atmosphere and was at an altitude of about 4.3 miles

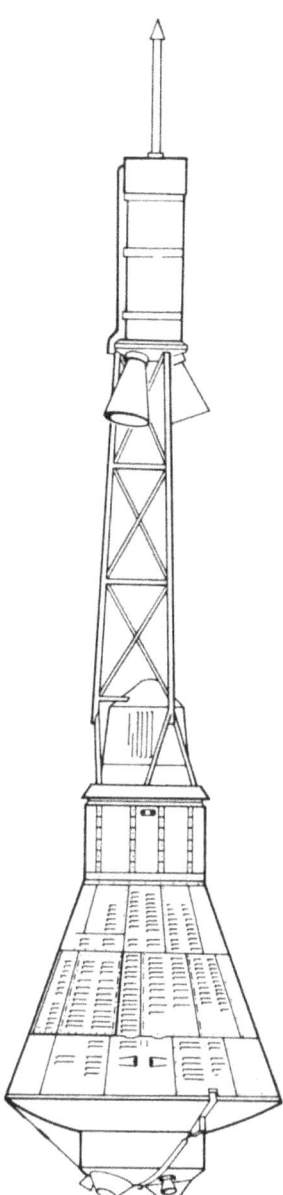

Mercury spacecraft

when its parachutes became fouled and it plunged to earth. Komarov was the first Russian to go into space twice: he was one of the three cosmonauts aboard the Voskhod 1.

MERCURY

Project Mercury was America's first step into space. The one-man Mercury capsules were designed to

MERCURY FLIGHTS

Date	Vehicle	Astronaut	Revolutions	Hours	
May 5, 1961	Mercury-Redstone 3	Alan B. Shepard, Jr.	*	00:15:22	First American in space; Freedom 7
July 21, 1961	Mercury-Redstone 4	Virgil I. Grissom	*	00:15:37	Capsule sank; Liberty Bell 7
Feb. 20, 1962	Mercury-Atlas 6	John H. Glenn, Jr.	3	04:55:23	First American in orbit; Friendship 7
May 24, 1962	Mercury-Atlas 7	M. Scott Carpenter	3	04:56:05	Landed 250 miles from target; Aurora 7
Oct. 3, 1962	Mercury-Atlas 8	Walter M. Schirra, Jr.	6	09:13:11	Landed 5 miles from target; Sigma 7
May 15-16, 1963	Mercury-Atlas 9	L. Gordon Cooper, Jr.	22	34:19:49	First long flight; Faith 7

*Sub-orbital

answer the basic questions about man in space; how he was affected by weightlessness, how he withstood the gravitational forces of boost and entry, how well he could perform. A milestone in applied science and engineering, the Mercury flights proved that man not only could survive, he could greatly increase the knowledge of space.

GEMINI

Gemini was the next step in NASA's program. The goal of these two-man flights was to find out how man could maneuver himself and his craft, and to increase our knowledge about such things as celestial mechanics and space navigation. Gemini has a record of 10 successful manned flights and set many records, including the longest duration (almost 14 days), the first rendezvous by two maneuverable spacecraft, and the first docking.

RUSSIAN MANNED PROGRAM

The Soviet Union opened the space age when it put the first man, Yuri Gagarin, into space in April of 1961. They followed four months later with the 25-hour flight of Gherman Titov.

The Russians waited a year after the Gagarin flight before their next, but that was the first group flight;

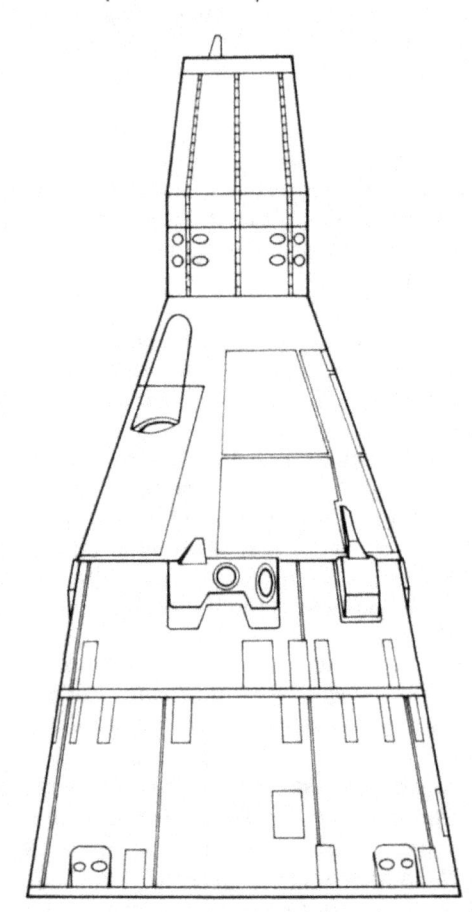

Gemini spacecraft

GEMINI FLIGHTS

Date	Vehicle	Astronauts	Revolutions	Hours	
Mar. 23, 1965	Gemini III	Virgil Grissom, John Young	3	4.9	First manned orbital maneuvers
June 3-7, 1965	Gemini IV	James McDivitt, Edward White	62	97.9	21-minute "space walk" by White
Aug. 21-29, 1965	Gemini V	Gordon Cooper, Charles Conrad	120	190.9	First extended manned flight
Dec. 4-18, 1965	Gemini VII	Frank Borman, James Lovell	206	330.6	Longest space flight; served as Gemini VI-A target vehicle
Dec. 15-16, 1965	Gemini VI-A	Walter Schirra, Tom Stafford	16	25.9	Rendezvous within 1 foot of Gemini VII
Mar. 16-17, 1966	Gemini VIII	Neil Armstrong, David Scott	6.5	10.7	First docking, to Agena target; short circuit cut flight short
June 3-6, 1966	Gemini IX-A	Tom Stafford, Eugene Cernan	45	72.3	Rendezvous, extra-vehicular activity, precision landing
July 18-21, 1966	Gemini X	John Young, Michael Collins	43	70.8	Rendezvous with 2 targets; retrieved package from Agena in space walk
Sept. 12-15, 1966	Gemini XI	Charles Conrad, Richard Gordon	44	71.3	Rendezvous and docking, 161-minute extravehicular activity
Nov. 11-15, 1966	Gemini XII	James Lovell, Edwin E. Aldrin	59	94.6	3 successful extra-vehicular trips, rendezvous and docking, rendezvous with solar eclipse

two spacecraft on successive days. A 10-month lull followed before the second group flight, this time including a woman, Valentina Tereshkova, as one of the cosmonauts.

These six flights were with the Soviet Union's first manned spacecraft, the Vostok. Their second-generation spacecraft, the Voskhod, made only two flights about six months apart. The first Voskhod mission, 16 months after the last Vostok flight, carried a crew of three and was the first spacecraft with more than one passenger. The second Voskhod flight carried only two men but featured the first man to leave his spacecraft and "walk" in space.

The Soviet Union's third-generation spacecraft, Soyuz, made its only flight in April of 1967, when Komarov was killed.

RUSSIAN MANNED FLIGHTS

Date	Spacecraft	Cosmonaut	Revolutions	Hours	
Apr. 12, 1961	Vostok 1	Yuri Gagarin	1	1.8	First manned flight
Aug. 6, 1961	Vostok 2	Gherman Titov	17	25.3	More than 24 hours in space
Aug. 11, 1962	Vostok 3	Andrian Nikolayev	64	94.4	First group flight
Aug. 12, 1962	Vostok 4	Pavel Popovich	48	71.0	Came within 3.1 miles of Vostok 3 on first orbit
June 14, 1963	Vostok 5	Valery Bykovsky	81	119.1	Second group flight
June 16, 1963	Vostok 6	Valentina Tereshkova	48	70.8	Passed within 3 miles of Vostok 5; only woman in space
Oct. 12, 1964	Voskhod 1	Vladimir Komarov K. Feoktistov B. Yegorov	16	24.3	First 3-man craft
Mar. 18, 1965	Voskhod 2	A. Leonov P. Belyayev	17	26.0	Leonov was first man outside spacecraft in 10-minute "walk"
Apr. 23, 1967	Soyuz 1	Vladimir Komarov	17	25.2	Heaviest manned craft; crashed killing Komarov

SPACECRAFT DIFFERENCES

Many differences in the three manned U.S. spacecraft are readily apparent, such as size and weight. The major differences are in the complexity and refinement of subsystems. Apollo's requirement for hardware "maturity" is significantly higher than for earlier spacecraft programs. Each subsystem has become progressively more complex, with many more demands made upon it and a correspondingly greater capability. Only Apollo has its own guidance and navigation system.

Electrical power is a good example of increased system complexity. Electrical power for Mercury was supplied by six batteries; for Gemini, it was supplied by seven batteries and two fuel cell powerplants; for Apollo, it is supplied by five batteries and three fuel cell powerplants. The three systems do not sound too different physically. In operation, however, the differences are considerable.

The greatest demand on the Mercury system was to supply power to sustain the 4,265-pound spacecraft and its single astronaut for a day and a half (the 34-hour flight of Gordon Cooper). In Gemini, the electrical power system had to provide sufficient power to operate a typical 7,000-pound craft containing two astronauts for as long as two weeks (the 14-day flight of Frank Borman and James Lovell). In Apollo, the system is designed to support a 100,000-pound spacecraft carrying three men for up to two weeks.

BASIC SPACECRAFT DIFFERENCES

	Mercury	Gemini	Apollo
Height	26 ft	19 ft	82 ft
Diameter	6.2 ft	10 ft	12 ft 10 in.
Launch weight	4265 lb at launch 2987 lb in orbit 2422 lb at recovery	8360 lb	109,500 lb at launch 100,600 lb injected
Crew	1	2	3
Major components	Manned capsule (6 ft 10 in.) Adapter (4 ft 3 in.) Launch escape tower (16 ft 11 in.)	Entry (manned) module (11 ft 4 in.) Adapter module (7 ft 6 in.)	Command module (10 ft 7 in.) (top of apex cover) Service module (24 ft 2 in.) (top of fairing) Lunar module (22 ft 11 in.) (legs folded) Launch escape system (33 ft) Adapter (28 ft)
Subsystems			
Abort	Launch escape rocket and tower to carry manned capsule to safety	Ejection seat for each astronaut up to about 70,000 ft; malfunction detection system	Launch escape rocket and tower (similar to Mercury but about twice the size); emergency detection system
Communications	UHF and HF for voice; UHF for telemetry; C-band and S-band tracking radar	UHF primary for voice with HF backup; C-band tracking beacon; rendezvous radar; 300 flight measurements telemetered to ground	VHF-AM primary for near earth; S-band primary for deep space; rendezvous radar; 700 flight measurements telemetered to ground
Docking	None	Index bar (to fit in notch on target vehicle) and latches	Probe and docking ring on CM, drogue on LM
Earth Landing	4 chutes: main, drogue, reserve, pilot	3 chutes: main, drogue, pilot, and ejection seats	8 chutes: 3 main, 2 drogue, 3 pilot

	Mercury	Gemini	Apollo
Electrical Power	6 batteries: 3 main auxiliary, 2 stand-by, and 1 isolated	2 small fuel cells; 2 cryogenic tanks, 4 entry batteries, 3 pyro batteries	3 large fuel cells; 4 cryogenic tanks; 3 entry batteries; 2 pyro batteries
Environmental Control	Suit cooling and oxygen loop, cabin cooling loop (water coolant); cabin pressurized to 5 psi; no space radiators	Suit cooling and oxygen loops; redundant cabin cooling loops (silicon ester coolant); cabin pressurized to 5 psi; space radiators, cold-plates for operating equipment; shirt-sleeve environment	Four major loops: oxygen, suit circuit, water, and coolant (water-glycol); space radiators and cold-plates; cabin pressurized to 5 psi; shirtsleeve operations
Guidance and Control	Attitude control equipment (2 attitude gyros, 3 rate gyros, logic and programming circuits); automatic system using 12 small H_2O_2 thrusters and manual system using 6 small H_2O_2 thrusters; horizon sensors, periscope	Small computer (4,000-word memory), horizon sensor, no redundancy; 16 orbital attitude maneuvering system thrusters of 25 to 100 lb (No redundancy); 16 entry control thrusters (redundant systems of 8 25-lb engines); inertial platform, rendezvous radar	Large computer (39,000-word memory), telescope and sextant; semi-automatic operation; optical, inertial, and computer systems; attitude control through stabilization and control systems; 16 SM reaction control engines (100-lb) redundant systems, 12 CM reaction control engines in redundant systems; separate guidance and control systems in LM
Propulsion	3 posigrade rockets for separation from booster, 3 retrograde rockets for entry from orbit	4 retrograde rockets for entry (2,500 lb each)	Restartable service propulsion engine (20,000 lb); liquid propellant rodset propulsion with unlimited restart and thrust vector control (automatic and manual)

APOLLO FLIGHT TESTS

The Apollo flight test program up to September, 1968, included space tests of four command and service modules, one lunar module, and space and atmospheric tests of 10 boilerplate (test) command and service modules. These tests were conducted under the "all-up" philosophy of testing as many things simultaneously as possible and thus minimizing the number of launches, as well as cost and time.

The program is aimed at designing the spacecraft so that all launches contribute to its development. The command and service modules are being developed separately from the lunar module; this permits both modules to be tested on the smaller Saturn IB launch vehicle. The test program depends on the Saturn V only for missions that require its large payload.

Another test program goal has been maximum development on the ground; space flights have been undertaken only with spacecraft with almost all systems aboard and operating.

An example of this philosophy of combining many tests on one flight was the Apollo 6 mission on April 4, 1968. This mission included the second flight of a Saturn V launch vehicle as well as a number of important spacecraft tests.

Although launch vehicle problems caused selection of an alternate mission and prevented achievement of some major objectives, NASA termed the spacecraft's accomplishments impressive. These included the longest single burn in space of the service propulsion engine (7 minutes, 25 seconds), proper control of the engine during this burn by the guidance and navigation subsystem, proper maintenance of spacecraft attitude by the reaction control subsystem during the long cold soak period, and another successful test of the spacecraft's heat shield. This also was the first space test of the new unified crew hatch and seals and they withstood the mission in good condition.

The first flight of the Saturn V was on Nov. 9, 1967, in the Apollo 4 mission, which also was a major test of the CM's heat shield, service propulsion subsystem, guidance and navigation equipment, and environmental control subsystem. The major objectives of Apollo 4, all fulfilled, were: the first launch of the Saturn V first stage, the first

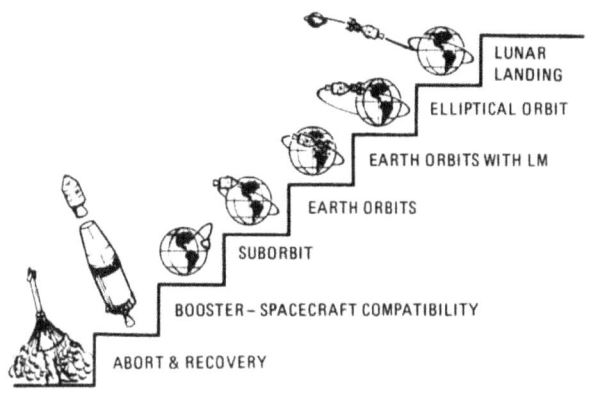

P-44 *Apollo development sequence*

flight of the hydrogen-powered second stage, restart of the third stage in earth orbit, restart of the service propulsion engine in space and its record firing for nearly 5 minutes, a hot and cold soak of the spacecraft far out in space, and entry under the severest conditions yet encountered by a spacecraft (a velocity of 24,913 miles per hour and a heat shield temperature of about 5,000 degrees F).

The Apollo 4 results were impressive. There was no structural damage to the command module and no areas of burn-through on the heat shield. The environmental control subsystem kept the cabin temperature between 60 and 70 degrees even through the fiery entry. Cabin pressure remained between 5.6 and 5.8 psia during the entire mission, indicating negligible leakage rate. Fuel cells and subsystems using cryogenics operated satisfactorily, as did all other operating subsystems.

The first space test of the lunar module came Jan. 22, 1968, on the Apollo 5 mission. The LM was launched by a Saturn IB, with the apex of the vehicle covered by an aerodynamic shroud. The shroud was jettisoned and then the spacecraft-LM adapter panels deployed as on a lunar mission. The lunar module's descent engine was burned three times and performed as expected. At the end of the third burn, a "fire-in-the-hole" abort—in which the LM's ascent and descent stages separate, the ascent engine begins to burn and simultaneously the descent engine stops firing—was performed successfully. A second ascent engine burn was performed later in the mission. Data telemetered to the ground indicated that all other subsystems of the module operated satisfactorily.

The Apollo 4, 5, and 6 missions were part of the earth-orbital phase of the flight test program. (There were no Apollo 1, 2, or 3 missions.) The program is divided into two blocks with interrelated phases: launch abort, sub-orbital, and earth-orbital (Block I) and earth-orbital and lunar (Block II).

For economy, boilerplate spacecraft are used in the program where an actual spacecraft is not required. Boilerplates are research and development vehicles that simulate production modules in size, shape, structure, mass, and center of gravity. Each boilerplate has instruments to record data for engineering evaluation.

The sub-orbital flights tested the heat shield and the operation of subsystems. The earth-orbital portion of the flight test program tests further the operational abilities of subsystems, the Saturn I, the Saturn IB, and Saturn V operation and compatibility, and operations during earth orbit, and also develops qualified teams for checkout, launch, flight operations, mission support, recovery, and flight analysis.

P-45

Command and service modules mounted on Little Joe II booster at White Sands, N.M., for test of launch escape subsystem

APOLLO FLIGHT TESTS

DATE	SITE		SPACECRAFT	RESULT
Apr. 4, 1968	Kennedy Space Center	Apollo 6: Second flight of Saturn V; launch vehicle engine problems caused spacecraft to go into alternate mission; service propulsion engine burned for record length, other subsystems performed well	SC020	Partial success
Jan. 22, 1968	Kennedy Space Center	Apollo 5: First space flight of lunar module; tested ascent and descent engines and ability to abort lunar landing and return to orbit; Saturn IB was launch vehicle	LM-1	Successful
Nov. 9, 1967	Kennedy Space Center	Apollo 4: First Saturn V launch; spacecraft entered atmosphere at almost 25,000 mph; heat shield temperatures reached about 5000°F; first test at lunar return speed	SC017	Successful
Aug. 25, 1966	Kennedy Space Center	Second flight of unmanned Apollo spacecraft to test command module's ability to withstand entry temperatures under high heat load; Saturn IB was launch vehicle	SC011	Successful
Feb. 26, 1966	Kennedy Space Center	First flight of unmanned Apollo spacecraft to test command module's ability to withstand entry temperatures; determine CM's adequacy for manned entry from low orbit test CM and SM reaction control engines and test service propulsion engine firing and restart; this was also first flight of the Saturn IB	SC009	Successful (Service module engine produced slightly less thrust than expected, resulting in slightly lower reentry speed and temperatures.)

DATE	SITE		SPACECRAFT	RESULT
Jan. 20, 1966	White Sands	Final abort test utilizing actual spacecraft to test escape in high tumbling region; this completed the abort test phase, qualifying the astronaut escape system for manned flights; Little Joe II was booster	SC002	Successful
July 30, 1965	Kennedy Space Center	Third Pegasus meteoroid detection satellite; launched by Saturn I; Apollo spacecraft shell and spacecraft-LM adapter housed and protected the Pegasus payload until reaching orbit where SLA panels opened, permitting the satellite to deploy	BP 9A	Successful
June 29, 1965	White Sands	Pad abort: Second test of the launch escape system's ability to work in emergency before launch and while still on the pad; canards, boost protective cover, jettisonable apex cover, and dual reefed drogue chutes were tested	BP 23A	Successful
May 25, 1965	Kennedy Space Center	Second Pegasus meteoroid detection satellite; Saturn I was launch vehicle	BP 26	Successful
May 19, 1965	White Sands	Planned high-altitude launch escape system test to determine performance of launch escape vehicle canard subsystem, and to demonstrate orientation of launch escape vehicle (Little Joe II)	BP 22	Partially successful (boost vehicle guidance malfunctioned causing premature low-altitude abort; Apollo systems functioned perfectly, pulling command module away from debris and lowering it safely to earth)
Feb. 16, 1965	Kennedy Space Center	First Pegasus micrometeoroid detection satellite; Saturn I was launch vehicle	BP 16	Successful

DATE	SITE		SPACECRAFT	RESULT
Dec. 8, 1964	White Sands	High Q abort test launch escape, earth landing systems and canard subsystems; Little Joe II was booster	BP 23	Successful
Sept. 18, 1964	Kennedy Space Center	Determined space vehicle launch exit environment on Saturn I	BP 15	Successful
May 28, 1964	Kennedy Space Center	Proved spacecraft compatibility with Saturn I launch vehicle; went into earth orbit	BP 13	Successful
May 13, 1964	White Sands	Transonic abort test utilizing Little Joe II to simulate a Saturn V launch vehicle; abort was performed at high speed with high loads	BP 12	Successful
Nov. 7, 1963	White Sands	Pad abort; tested the launch escape system's ability to perform an abort before launch while on the pad	BP 6	Successful

Recovery of first unmanned Apollo spacecraft by U.S.S. Boxer Feb. 26, 1966 in South Atlantic following launch by Saturn IB. During suborbital flight, spacecraft traveled 300 miles up.

Saturn IB lifts CSM off pad on early Apollo flight test

COMMAND MODULE

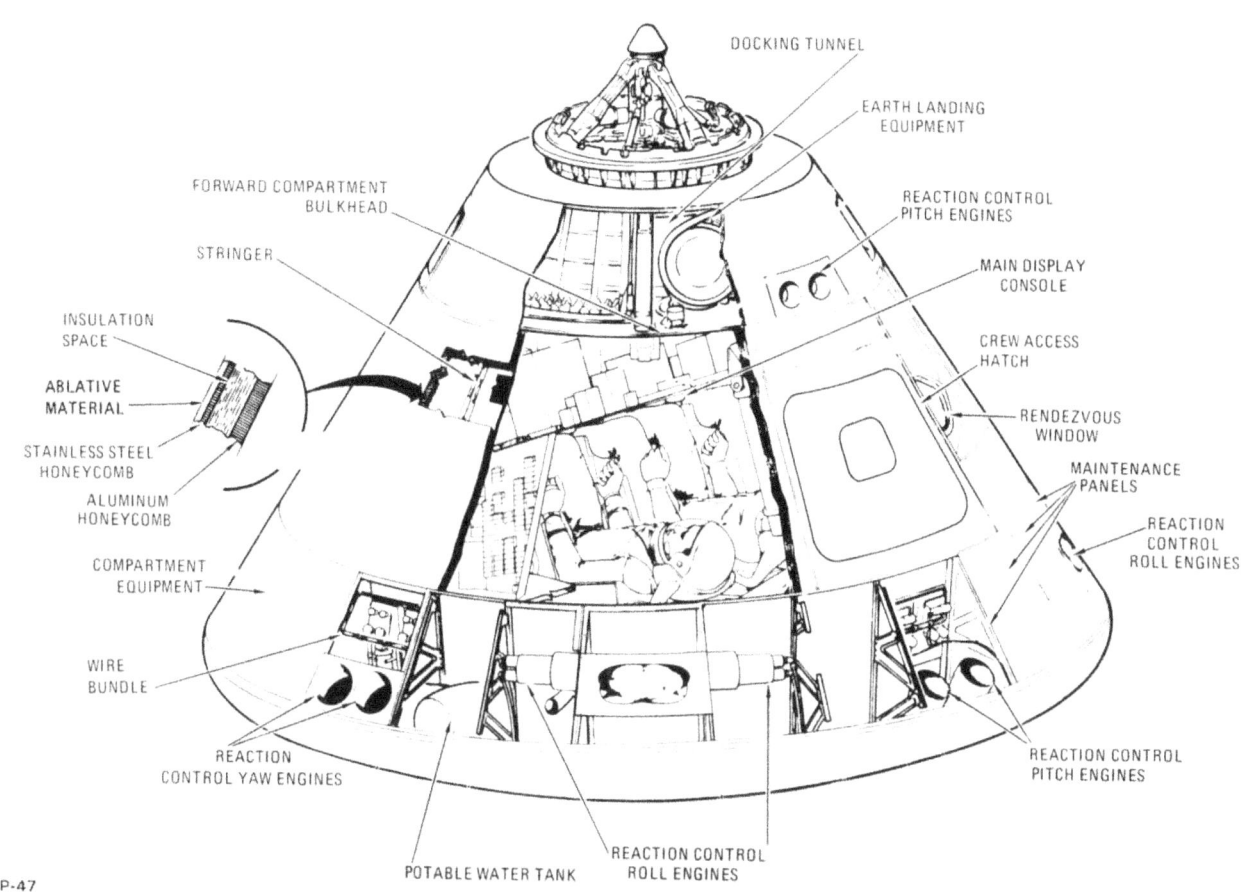

Dimensions

Height	10 ft 7 in.
Diameter	12 ft 10 in.
Weight (including crew)	13,000 lb
Weight (splashdown)	11,700 lb

Propellant

Reaction control subsystem (fuel—monomethylhydrazine; oxidizer—nitrogen tetroxide)	270 lb

Function

The command module is the control center and living quarters for most of the lunar mission; one man will spend the entire mission in it and the other two will leave it only during the lunar landing. It is the only part of the spacecraft recovered at the end of the mission.

Major Subsystems

Communications
Earth landing
Electrical power
Environmental control
Guidance and navigation
Launch escape
Reaction control
Service propulsion
Stabilization and control
Thermal protection (heat shields)

Command module, its checkout complete, is taken from test stand to be prepared for shipment to Florida

The CM is divided into three compartments: forward, crew, and aft. The forward compartment is the relatively small area at the apex of the module, the crew compartment occupies most of the center section of the structure, and the aft compartment is another relatively small area around the periphery of the module near the base.

During boost and entry the CM is oriented so that its aft section is down, like an automobile resting on its rear bumper. In this position the astronauts are on their backs; the couches are installed so that the astronauts face the apex of the module. In the weightlessness of space the orientation of the craft would make little difference except in maneuvers like docking, where the craft is moved forward so that the probe at the CM's apex engages the drogue on the LM. Generally, however, the module will be oriented in space so that its apex is forward.

Crewmen will spend much of their time on their couches, but they can leave them and move around. With the seat portion of the center couch folded, two astronauts can stand at the same time. The astronauts will sleep in two sleeping bags which are mounted beneath the left and right couches. The sleeping bags attach to the CM structure and have restraints so that a crewman can sleep either in or out of his space suit.

Food, water, clothing, waste management, and other equipment are packed into bays which line the walls of the craft. The cabin normally will be pressurized to about 5 pounds per square inch (about a third of sea level pressure) and the temperature will be controlled at about 75°F. The pressurization and controlled atmosphere will enable the three crewmen to spend much of their time out of their suits. They will be in their space suits, however, during critical phases of the mission such as launch, entry, docking, and crew transfer.

The astronaut in the left-hand couch is the spacecraft commander. In addition to the duties of command, he will normally operate the spacecraft's flight controls. The astronaut in the center couch is the CM pilot; his principal task is guidance and navigation, although he also will fly the spacecraft at times. On the lunar mission, he is the astronaut who will remain in the CM while the other two descend to the surface of the moon. The astronaut in the right-hand couch is the LM pilot and his principal task is management of spacecraft subsystems.

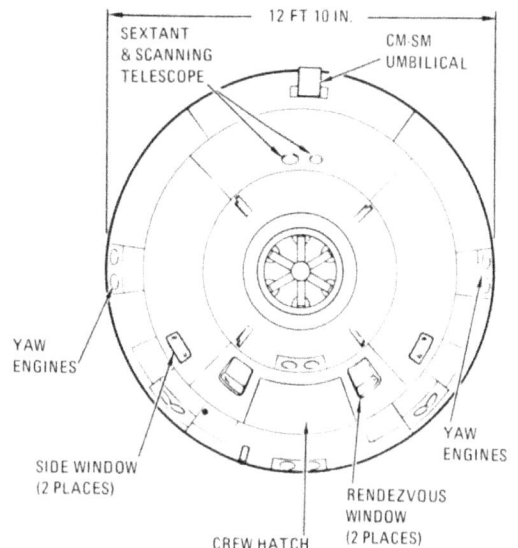

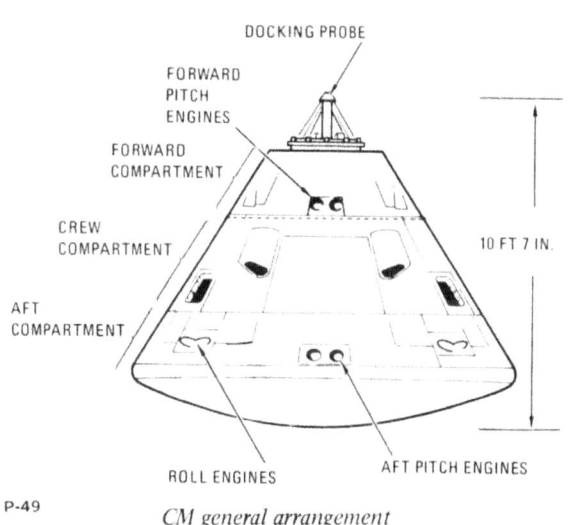

CM general arrangement

Although each has specific duties, any of the astronauts can take over the duties of another. The command module has been designed so that one astronaut can return it safely to earth.

STRUCTURE

The CM consists of two basic structures joined together: the inner structure (pressure shell) and the outer structure (heat shield).

The inner structure is of aluminum sandwich construction which consists of a welded aluminum inner skin, adhesively bonded aluminum honeycomb core and outer face sheet. The thickness of the honeycomb varies from about 1-½ inches at the base to about ¼ inch at the forward access tunnel. This inner structure—basically the crew compartment—is the part of the module that is pressurized and contains an atmosphere.

The outer structure is the heat shield and is made of stainless steel brazed honeycomb brazed between steel alloy face sheets. It varies in thickness from ½ inch to 2-½ inches.

Part of the area between the inner and outer shells is filled with a layer of fibrous insulation as additional heat protection.

THERMAL PROTECTION (HEAT SHIELDS)

The interior of the command module must be protected from the extremes of environment that will be encountered during a mission. These include the heat of boost (up to 1200°F), the cold of space and the heat of the direct rays of the sun (about 280° below zero on the side facing away from the sun and 280° above zero on the other side), and—most critical—the intense temperatures of entry (about 5000°F).

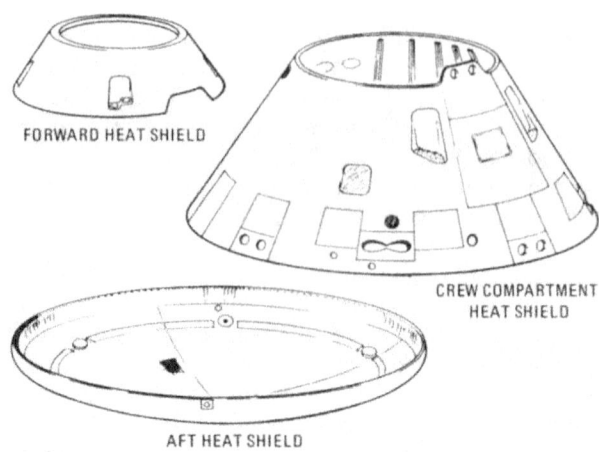

Heat shields

The heat of launch is absorbed principally through the boost protective cover, a fiberglass structure covered with cork which fits over the command module like a glove. The boost protective cover weighs about 700 pounds and varies in thickness from about 3/10 of an inch to about 7/8 of an inch (at the top). The cork is covered with a white reflective coating. The cover is permanently attached to the launch escape tower and is jettisoned with it at approximately 295,000 feet during a normal mission.

The insulation between the inner and outer shells, plus temperature control provided by the environmental control subsystem, protects the crew and sensitive equipment during the CM's long journey in space.

The principal task of the heat shield that forms the outer structure is to protect the crew from the fiery heat of entry—heat so intense that it melts most metals. The ablative material that does this job is a phenolic epoxy resin, a type of reinforced plastic.

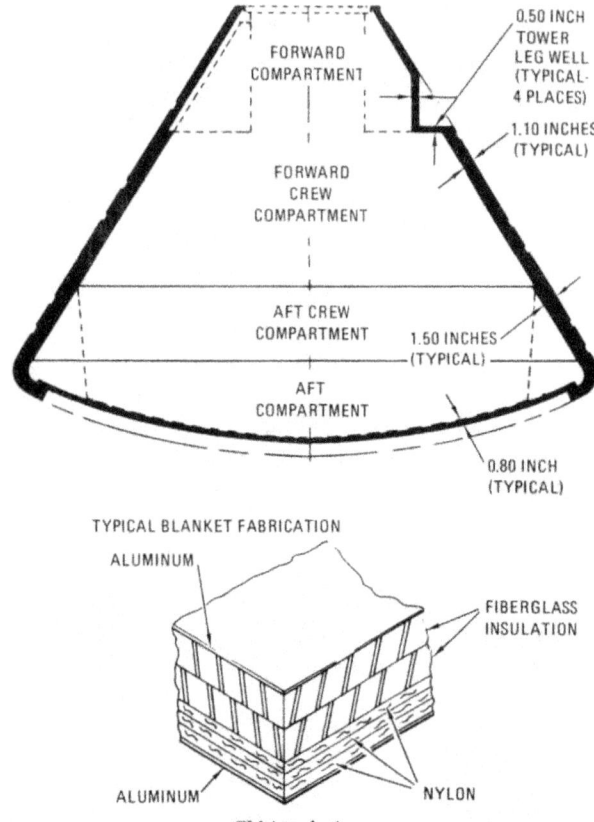

CM insulation

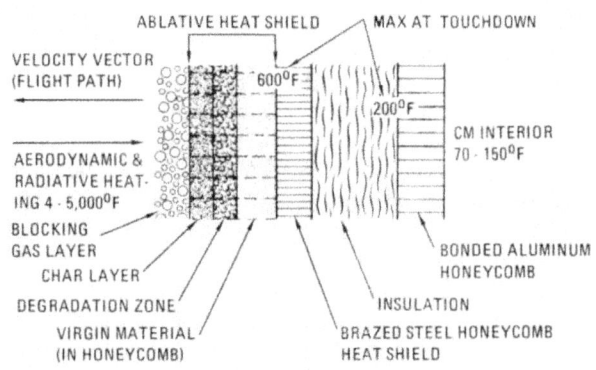

Mechanics of ablative material

Special ablative material is injected into each cell of heat shield by technicians at Avco plant in Lowell, Mass.

This material turns white hot, chars, and then melts away, but it does it in such a way that the heat is rejected by the shield and does not penetrate to the surface of the spacecraft.

The ablative material controls the rate of heat absorption by charring or melting rapidly. This dissipates the heat and keeps it from reaching the inner structure.

The command module enters the atmosphere with its base down; this is covered by the aft heat shield which is the thickest portion.

The heat shield varies in thickness: the aft portion is 2 inches and the crew compartment and forward portions are ½ inch. Total weight of the shield is about 3,000 pounds. The heat shield has several outer coverings: a pore seal, a moisture barrier (a white reflective coating), and a silver Mylar thermal coating that looks like aluminum foil.

The heat shield panels are produced by Aeronca Manufacturing Co., Middletown, Ohio, and the ablative coating was developed and applied by Avco Corp., Lowell, Mass.

IMPACT ATTENUATION

During a water impact the CM deceleration force will vary from 12 to 40 G's, depending on the shape of the waves and the CM's rate of descent. A major portion of the energy (75 to 90 percent) is absorbed by the water and by deformation of the CM structure. The module's impact attenuation system reduces the forces acting on the crew to a tolerable level.

The impact attenuation system is part internal and part external. The external part consists of four crushable ribs (each about 4 inches thick and a foot in length) installed in the aft compartment. The ribs are made of bonded laminations of corrugated aluminum which absorb energy by collapsing upon themselves at impact. The main parachutes suspend the CM at such an angle that the ribs are the first point of the module that hits the water.

The internal portion of the system consists of eight struts which connect the crew couches to the CM structure. These struts (two each for the Y and Z axes and four for the X axis) absorb energy by deforming steel wire rings between an inner and an outer piston. The struts vary in length from 34 to 39 inches and have a diameter of about 2½ inches.

The axes of the spacecraft are three straight lines, each at a right angle to the other two. They are used for reference and to describe the spacecraft's movements. The X axis is the line running from the apex of the command module through its base; the Y axis is the line running laterally, or from side to side through the couches; the Z axis is the line running up and down, or from the head to the feet of the astronauts in their couches. The command module's movement about the X axis is called roll, about the Y axis is called pitch, and about the Z axis is called yaw.

FORWARD COMPARTMENT

The forward compartment is the area around the forward (docking) tunnel. It is separated from the crew compartment by a bulkhead and covered by the forward heat shield. The compartment is divided into four 90-degree segments which contain earth landing equipment (all the parachutes, recovery antennas and beacon light, and sea recovery sling), two reaction control engines, and the forward heat shield release mechanism.

The forward heat shield contains four recessed fittings into which the legs of the launch escape tower are attached. The tower legs are connected to the CM structure by frangible (brittle) nuts which contain small explosive charges. When the launch escape subsystem is jettisoned, these charges are fired, breaking the nuts and separating the tower from the module.

At about 25,000 feet during entry, the forward heat shield is jettisoned to expose the earth landing equipment and permit deployment of the parachutes.

AFT COMPARTMENT

The aft compartment is located around the periphery of the command module at its widest part, just forward of (above) the aft heat shield. The compartment is divided into 24 bays by the 24 frames of the structure. In these bays are 10 reaction control engines; the fuel, oxidizer, and helium tanks for the CM reaction control subsystem; water tanks; the crushable ribs of the impact attenuation system; and a number of instruments. The CM-SM umbilical—the point where wiring and plumbing runs from one module to the other—also is in the aft compartment. The panels of the heat shield around the aft compartment are removable for maintainance of the equipment before flight.

CREW COMPARTMENT

The crew compartment is a sealed cabin with a habitable volume of 210 cubic feet. Pressurization and temperature are maintained by the environmental control subsystem. In it are the controls and

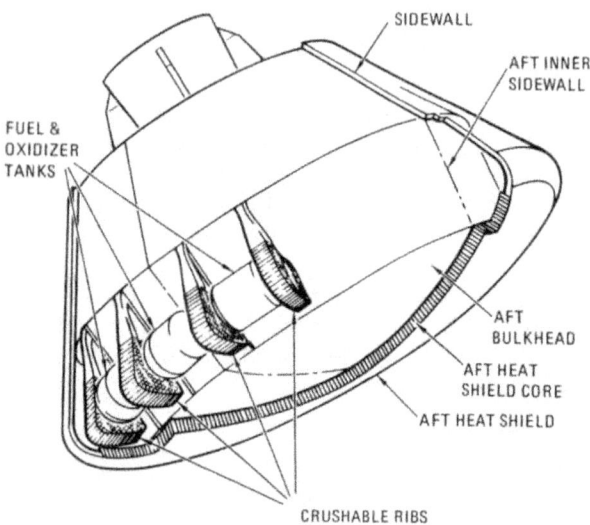

External attenuation system

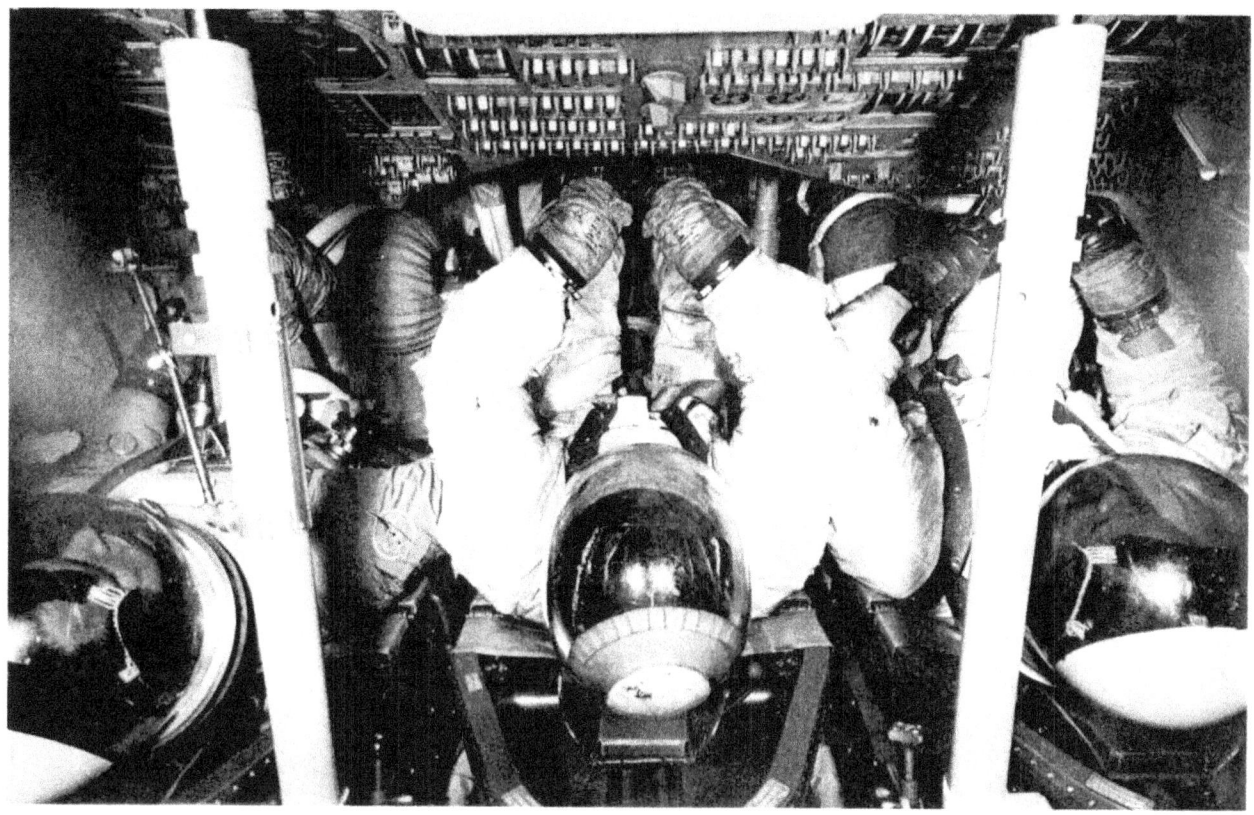

Astronauts check out crew compartment, train in use of controls

displays for operation of the spacecraft, crew couches, and all the other equipment needed by the crew. It contains two hatches, five windows, and a number of bays or cupboards packed with equipment.

HATCHES

The two CM hatches are the side hatch, used for getting in and out of the module, and the forward hatch, used to transfer to and from the lunar module when the two modules are docked.

The side hatch is a single integrated assembly which opens outward and has primary and secondary thermal seals. It is about 29 inches high and 34 inches wide. The hatch normally contains a small (about 9 inches in diameter) window, but has provisions for installation of an airlock. The hatch weighs about 225 pounds; with the airlock it weighs about 245 pounds.

The hatch normally is operated by a handle which the crewman pumps back and forth. The handle drives a ratchet mechanism which opens or closes the 12 latches around the periphery of the hatch. The latches are so designed that pressure exerted against the hatch serves only to increase the locking pressure of the latches. If the latch gear mechanism should fail, it can be disconnected and the latches opened or closed manually.

The hatch also can be opened from the outside by a tool that is part of the crew's tool set and is carried by ground personnel. The tool is the emergency wrench, essentially a modified allenhead L-wrench. It is 6-1/4 inches long and has a 4-1/4-inch drive shaft.

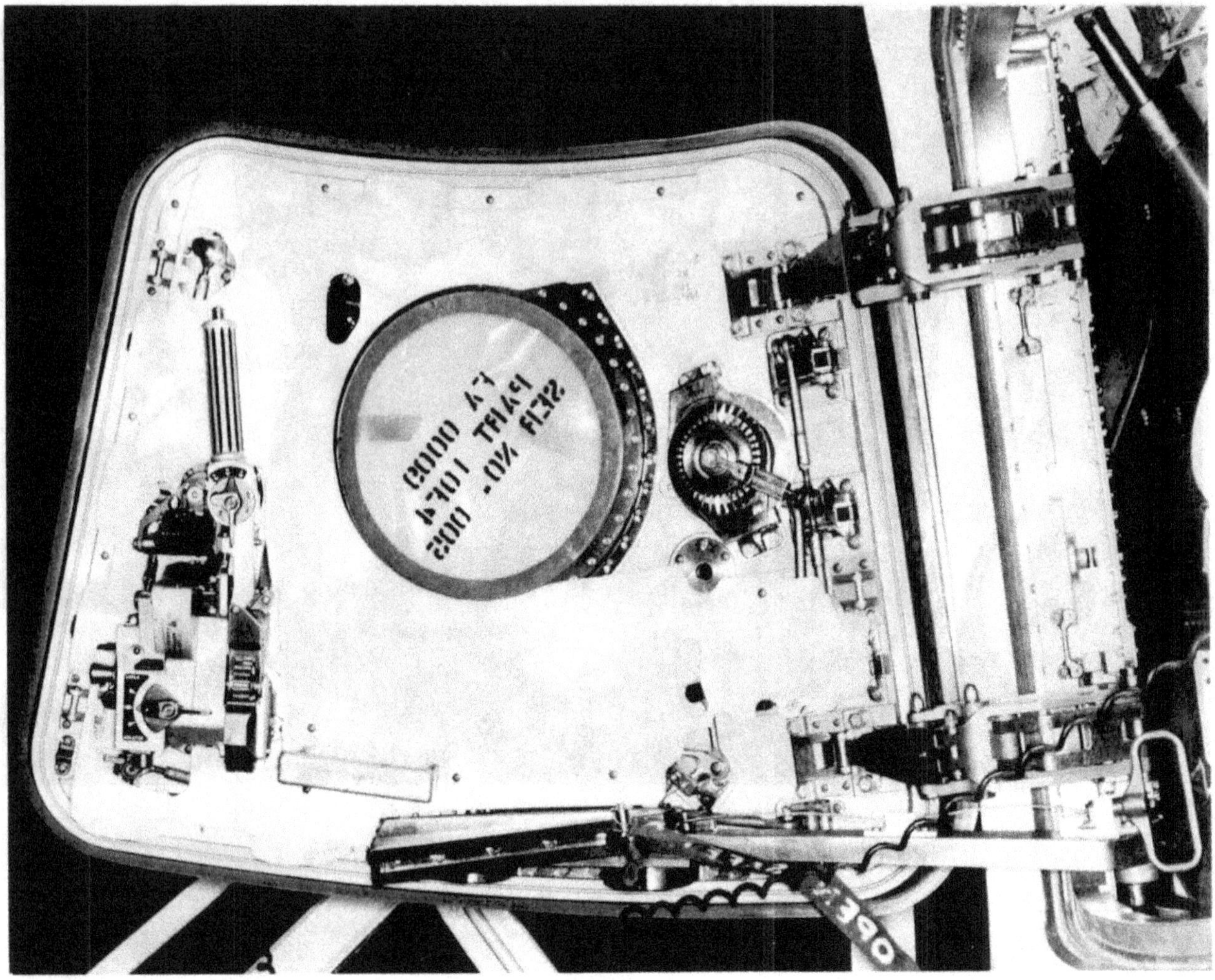

Side (unified) hatch with hard plastic cover that protects components during ground handling; cover is removed before flight

P-56

The hatch handle mechanism also operates the mechanism which opens the access hatch in the boost protective cover. A counterbalance assembly enables the hatch and boost protective cover hatch to be opened easily. This consists of two nitrogen bottles and a piston assembly. Each nitrogen bottle contains about 5-1/2 cubic inches of gas under a pressure of 5,000 pounds per square inch. One of the bottles is punctured on the launch pad, permitting the gas to stroke the piston and force the door open when the latches are released. A pressure of about 2,200 pounds per square inch is needed to open the door.

The ground crew can easily close the hatch by pushing it. In the weightlessness of space, the crew

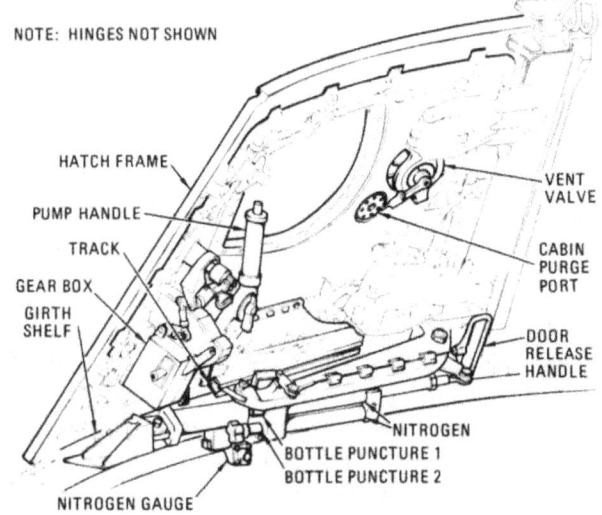

P-57 *Major components of unified hatch*

46

can close the hatch from the inside by pulling on a handle near the lower hinge which swings the hatch inward. Another handle near the opposite edge of the door is provided primarily for the use of astronauts engaged in extravehicular activity.

The piston cylinder and nitrogen bottle can be vented after launch since the counter-balance assembly is not needed in space. If it is vented, the second nitrogen bottle can be used to open the hatch after landing. A knob on this bottle is used to puncture it and release the gas into the system.

In case some deformation or other malfunction prevented the latches from engaging, three jackscrews are provided in the crew's tool set to hold the door closed. The jackscrews engage three catches on the hatch and parallel catches on the inner structure. The jackscrews are then tightened, pulling the hatch closed tight enough to withstand the thermal load of entry.

The side hatch also contains a valve to vent the cabin pressure. The valve handle normally is locked in the down position. The crewman simply releases the lock, pulls the handle up and winds it in a clockwise direction, opening the valve. The valve will vent the cabin pressure in one minute. The valve also can be operated from the outside by using the same tool used to open the door.

P-57a *Forward (tunnel) hatch from inside CM*

The forward (docking) hatch is a combined pressure and ablative hatch mounted at the top of the docking tunnel. It is about 30 inches in diameter and weighs about 80 pounds. The exterior or upper side of the hatch is covered with a half-inch of insulation and a layer of aluminum foil.

The forward hatch has a six-point latching arrangement operated by a pump handle similar to that on the side hatch except that only one stroke is needed to open the latches. The handle is offset so that it can be reached easily by a crewman standing in the tunnel. This hatch also can be opened from the outside. It has a pressure equalization valve so that the pressure in the tunnel and that in the LM can be equalized before the hatch is removed. The valve is similar to the vent valve in the side hatch. There are also provisions for opening the latches manually if the handle gear mechanism should fail.

WINDOWS

The CM has five windows: two side, two rendezvous, and a hatch window. On some missions the

P-58 *Foil covers exterior of forward hatch*

hatch window may be replaced by a scientific airlock. The side windows, about 13 inches square, are positioned at the side of the left and right couches and are used for observation and photography. The triangular rendezvous windows, about 8 by 13 inches, face the left and right couches and permit a view forward (toward the apex of the module). They are used to aid in the rendezvous and docking maneuvers as well as for observation. The hatch window is over the center couch.

The windows each consist of inner and outer panes. The inner windows are made of tempered silica glass with ¼-inch thick double panes, separated by a tenth of an inch. The outer windows are made of amorphous-fused silicon with a single pane seven-tenths of an inch thick. Each pane has an anti-reflecting coating on the external surface and a blue-red reflective coating on the inner surface to filter out most infrared and all ultraviolet rays. The outer window glass has a softening temperature of 2800°F and a melting point of 3110°F. The inner window glass has a softening temperature of 2000°F.

Each window has a shade which can be installed to cut off all outside light. The shades are made of aluminum sheet, have a non-reflective inner surface, and are held in place by wing levers.

EQUIPMENT BAYS

The interior of the command module is lined with equipment bays or cupboards which contain all of the items needed by the crew for up to 14 days, as well as much of the electronics and other equipment needed for operation of the spacecraft. The bays are named according to their position with reference to the couches. The left-hand and right-hand bays are beside the left and right couches, respectively; the left-hand forward and right-hand forward bays are to the side and in front of the left and right couches, respectively; the lower bay is at the foot of the center couch; and the aft bay is on the aft bulkhead beneath the couches.

The lower equipment bay is the largest and contains much critical equipment. It is the navigation station and contains most of the guidance and navigation electronics, as well as the sextant and telescope, the computer, and a computer keyboard. Most of the telecommunications subsystem electronics are in this bay, as well as the five batteries, inverters, and battery charger of the electrical power subsystem. Stowage areas in the bay contain food supplies, scientific instruments, and other astronaut equipment.

The left-hand equipment bay contains key elements of the environmental control subsystem, including the environmental control unit, the oxygen surge tank, and displays and controls. Space is provided on this bay for stowing the forward hatch when the command and lunar modules are docked and the tunnel between the craft is open.

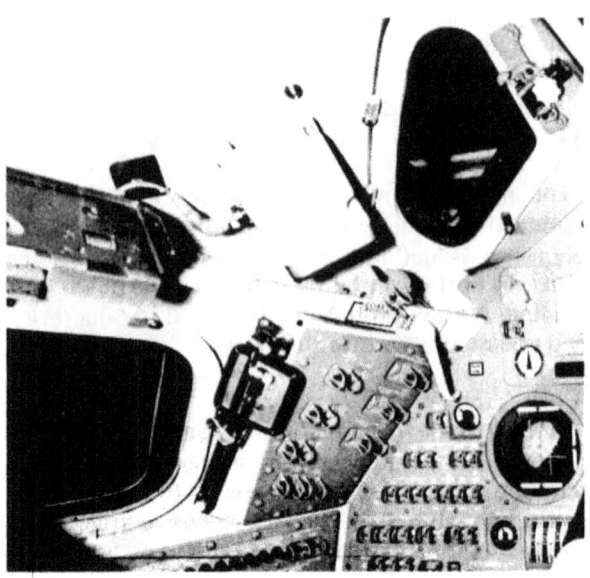

P-59

Side (left) and rendezvous windows

The left-hand forward equipment bay also contains environmental control subsystem equipment, as well as the water delivery unit and clothing storage.

The right-hand equipment bay contains waste management system controls and equipment, electrical power equipment, and a variety of electronics, including sequence controllers and signal conditioners. Food also is stored in a compartment in this bay.

The right-hand forward equipment bay is used principally for stowage and contains such items as survival kits, medical supplies, optical equipment, the LM docking target, and bio-instrumentation harness equipment.

The aft equipment bay is used for storing space suits and helmets, life vests, the fecal canister, portable life support systems (backpacks), and other equipment, and includes space for stowing the probe and drogue assembly.

PROTECTION PANELS

Protection panels are mounted throughout the interior of the command module as additional protection for spacecraft equipment. The panels are made of aluminum in varying thicknesses and are used to cover wiring and equipment and to smooth out irregular surfaces of the cabin. The panels prevent loose equipment or debris from getting into crevices of the spacecraft during preparation for flight. They also help suppress fire by sealing off areas and protect critical parts from damage during work by ground personnel.

MIRRORS

Internal and external mirrors are provided to aid astronauts' visibility. The internal mirrors (4 by 6 inches) include one for each couch and are designed to help the astronaut see to adjust his restraint harness and locate his couch controls while in his pressurized suit. The external mirrors are 2-1/2 by 3-1/2 inches and are located at the top and bottom of the right-hand rendezvous window so that the astronaut in the right-hand couch can verify parachute deployment during entry.

CM-SM CONNECTION

For most of an Apollo mission, the command and service modules are attached; they separate only a short time before the command module enters the atmosphere.

The two modules are connected by three tension ties which extend from the CM's aft heat shield to six compression pads on the top of the SM. The tension ties are essentially stainless steel straps about 2-1/2 inches wide and 4 inches long bolted at one end to the CM and at the other to the SM. The CM rests base down on the six compression pads, which are circular metal "cups," three of them about 4 inches in diameter and three about 6 inches in diameter. The areas in the heat shield which rest on the pads are reinforced with laminated fiberglass.

The two modules also are connected through the CM-SM umbilical, an enclosure protruding from the CM on the side opposite the side hatch. The umbilical is the wiring and tubing through which vital power, water, oxygen, and water-glycol flows from one module to the other. These connections are covered by an aluminum fairing about 18 inches wide and 40 inches long.

At separation, electrical circuits are deadfaced (power cut off) and valves closed at the umbilical,

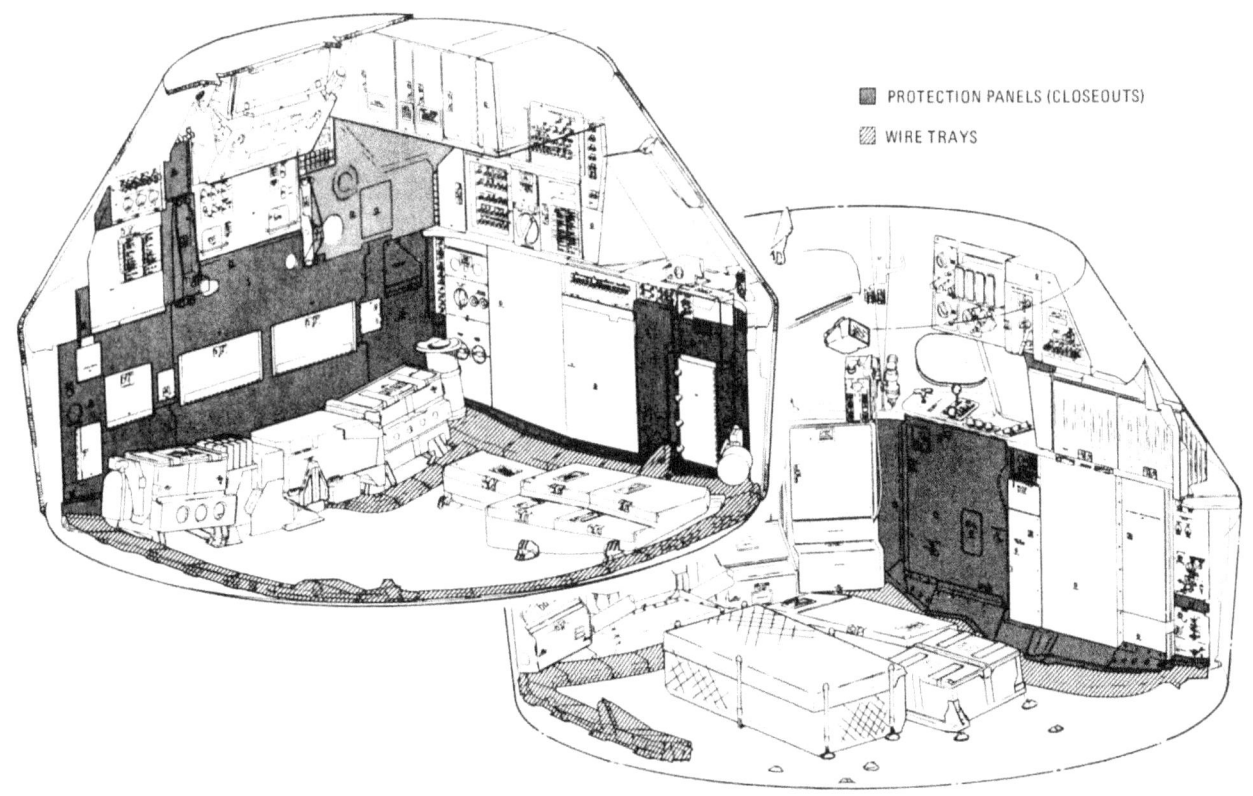

■ PROTECTION PANELS (CLOSEOUTS)
▨ WIRE TRAYS

P-60

Protection panels and wire trays

stainless steel blades, either one of which will cut all the connections. The guillotine is driven by redundant detonating cord charges. The tension ties are severed by linear-shaped charges set off by detonators. The signals that set off the detonators, the detonators themselves, and the charges are all redundant.

The area between the bottom (aft) of the CM and the top of the SM where the two modules are joined is enclosed by a fairing 26 inches high. This fairing is part of the service module and contains space radiators for the electrical power subsystem.

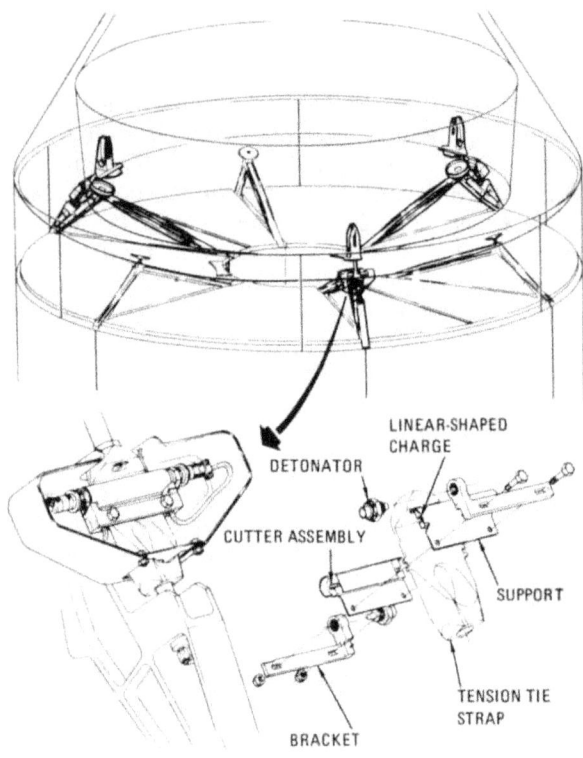

P-61

CM-SM separation system

a guillotine mechanism cuts the connecting wires and tubing, and small charges sever the tension ties. The umbilical firing pulls away from the CM and remains attached to the SM. The guillotine that severs the wires and tubes consists of two

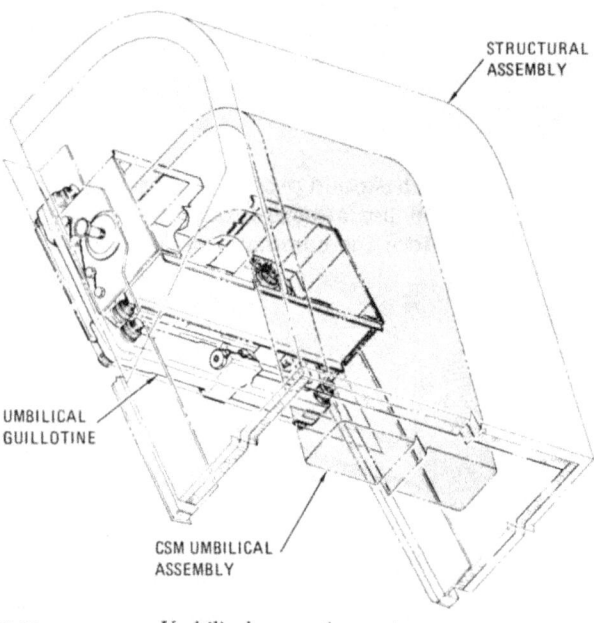

P-62

Umbilical separation system

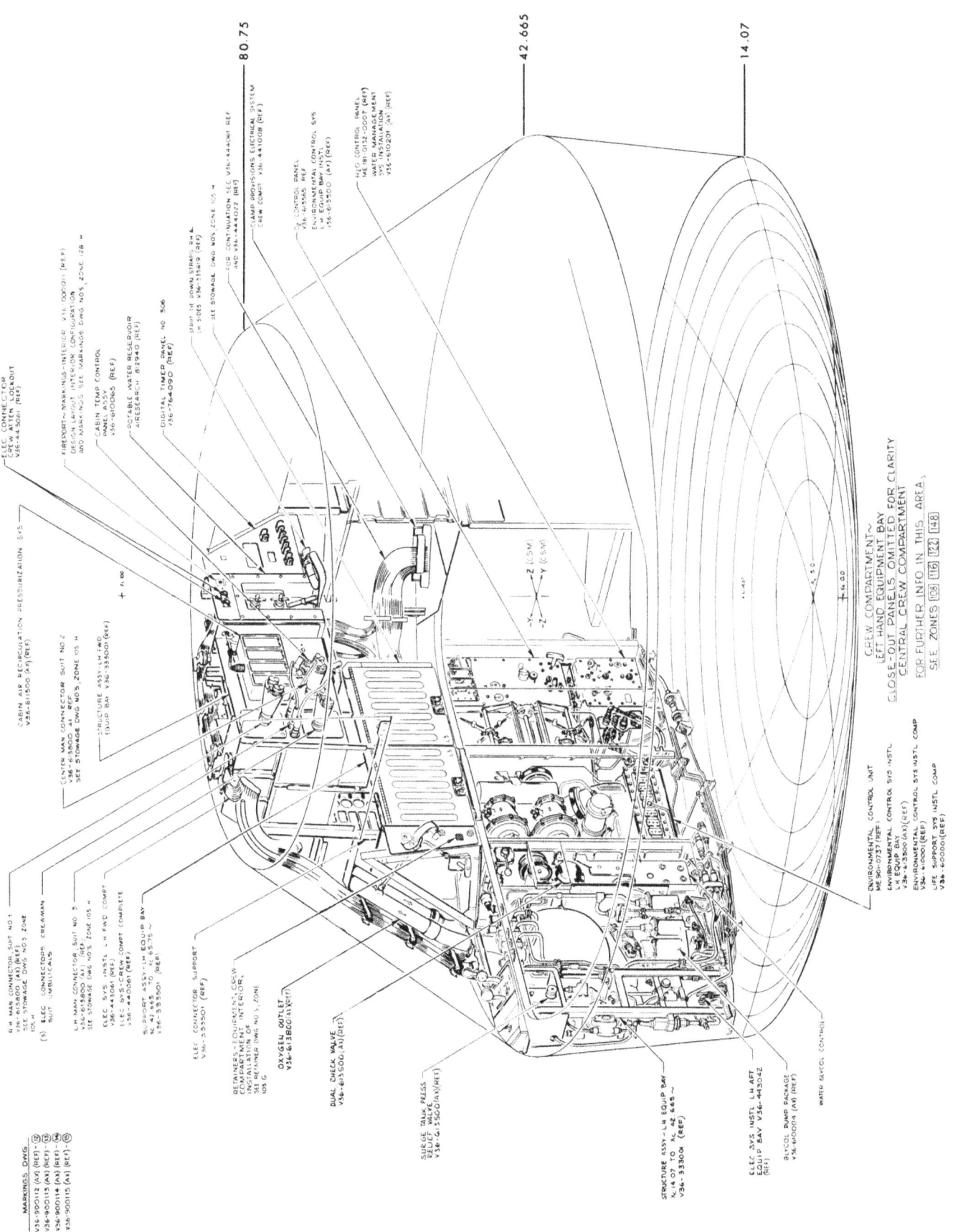

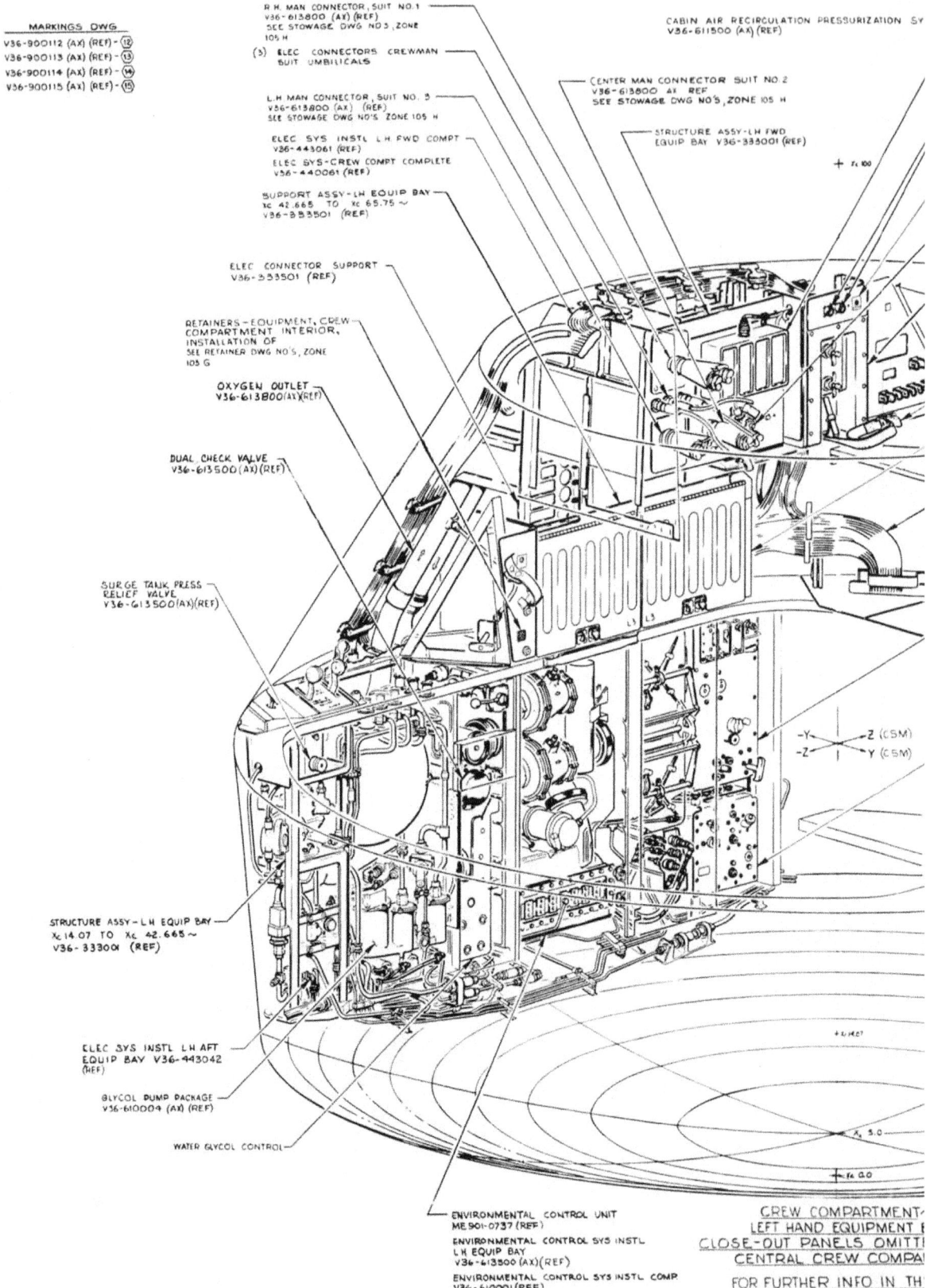

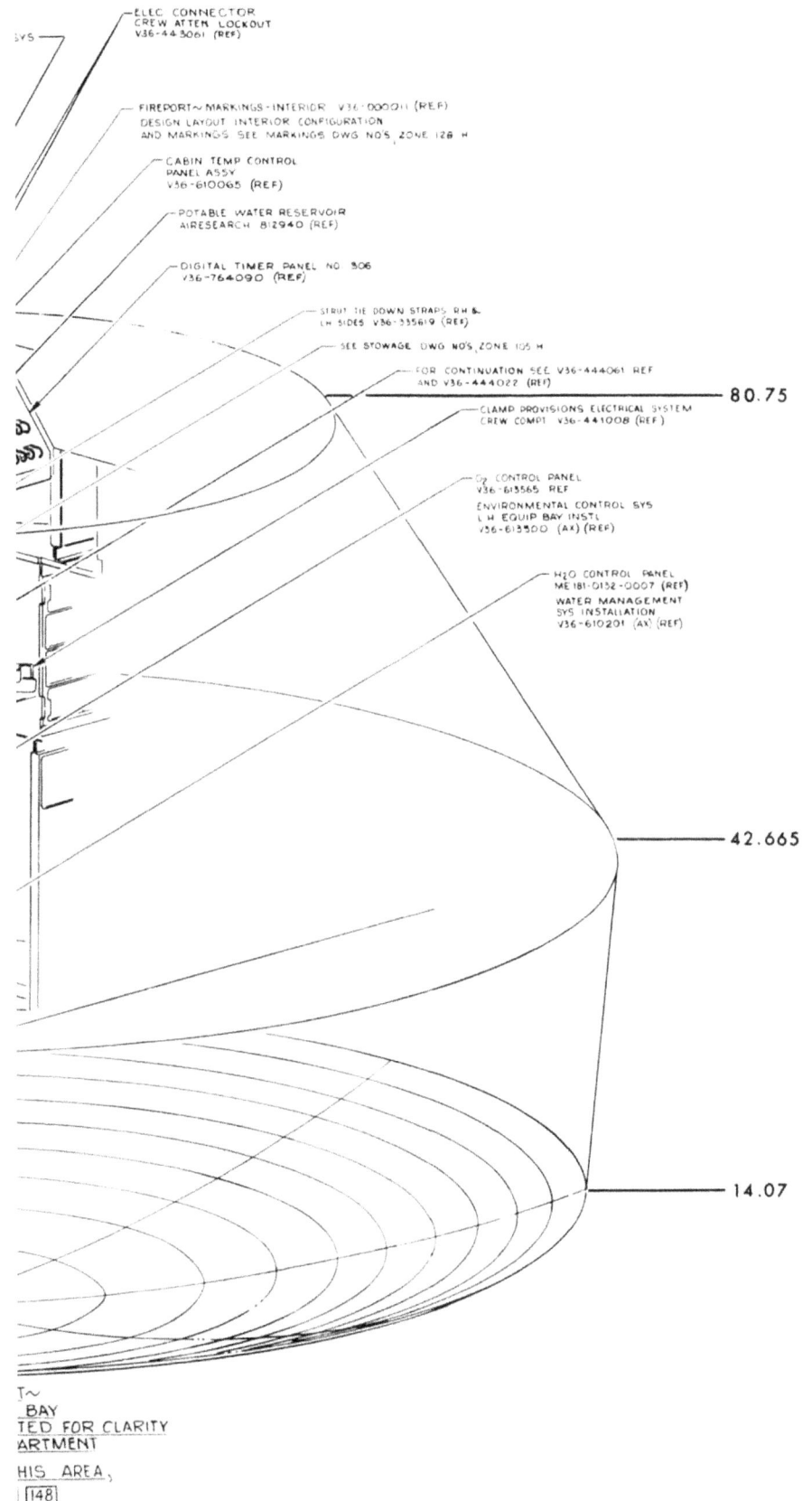

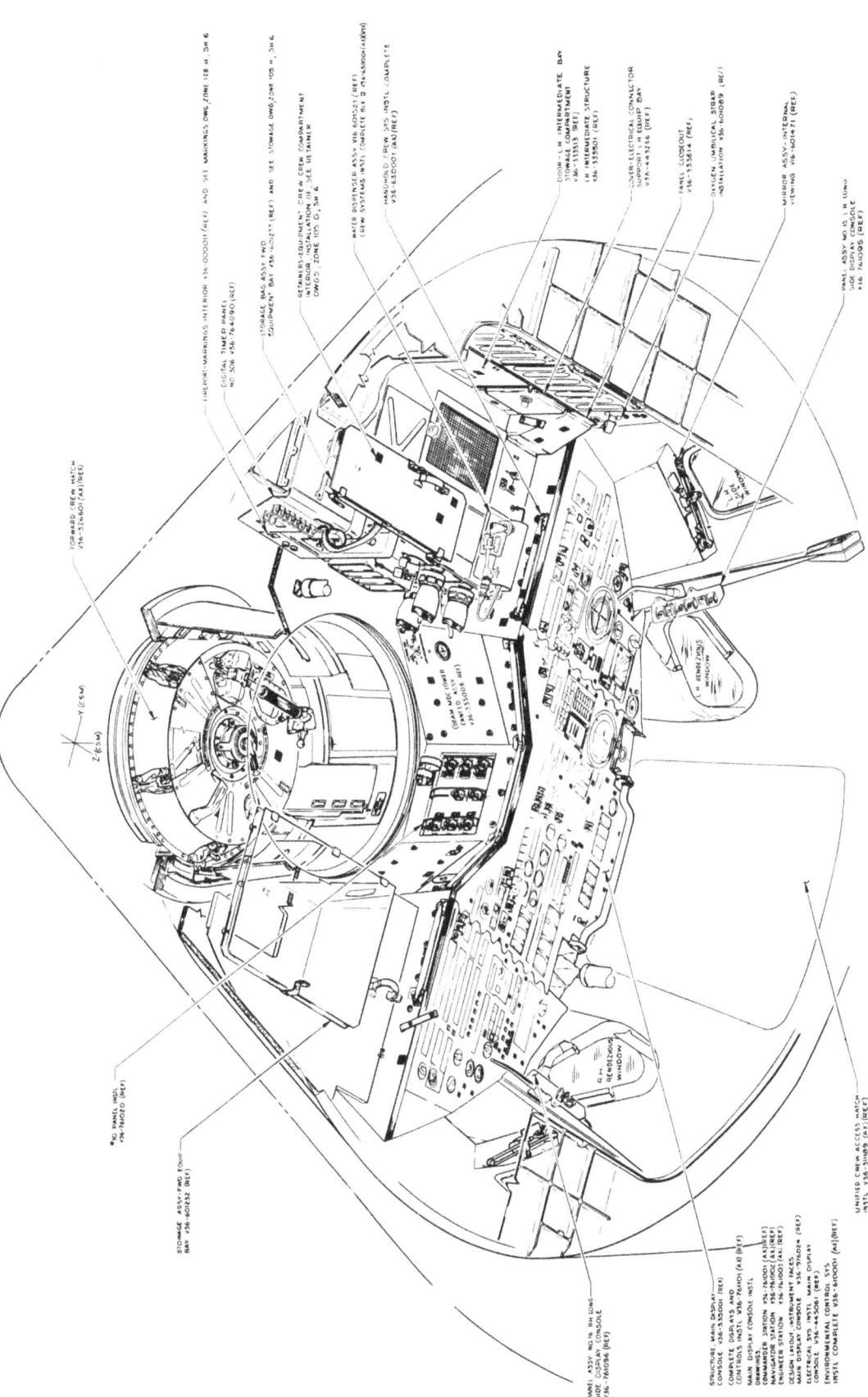

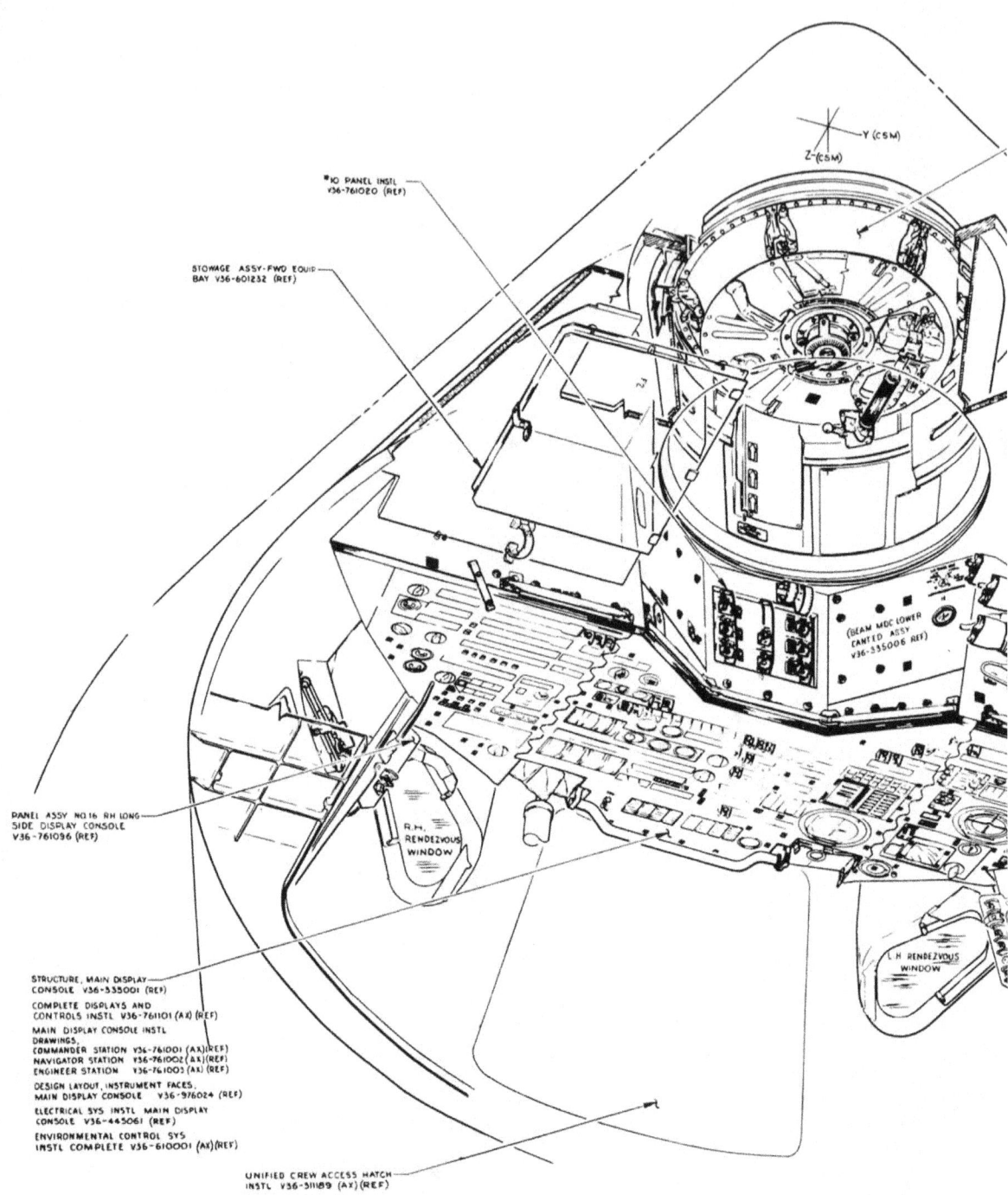

VIEW LOOKING FWD AT MAIN DISPLAY CONSOLE, FORWARD BULKHEAD, FWD HATCH, UPPER SIDE WALL, MAIN CREW HATCH, SIDE WINDOWS, L.H. FWD AND INTERMEDIATE EQUIP BAYS

- FORWARD CREW HATCH
  V36-526601 (AX)(REF)

- FIREPORT-MARKINGS-INTERIOR V36-000011 (REF) AND SEE MARKINGS DWG, ZONE 128 H, SH 6
- DIGITAL TIMER PANEL
  NO 306 V36-764090 (REF)
- STORAGE BAG ASSY FWD
  EQUIPMENT BAY V36-601277 (REF) AND SEE STOWAGE DWG, ZONE 105 H, SH 6
- RETAINERS-EQUIPMENT CREW CREW COMPARTMENT
  INTERIOR, INSTALLATION OF, SEE RETAINER
  DWGS, ZONE 105 G, SH 6
- WATER DISPENSER ASSY V16-601521 (REF)
  CREW SYSTEMS INSTL COMPLETE BLK II V36-630001 (AX)(REF)
- HANDHOLD CREW SYS INSTL COMPLETE
  V36-630001 (AX)(REF)

- DOOR - L.H. INTERMEDIATE BAY
  STOWAGE COMPARTMENT
  V36-333513 (REF)
  L H INTERMEDIATE STRUCTURE
  V36-333501 (REF)
- COVER-ELECTRICAL CONNECTOR
  SUPPORT L H EQUIP BAY
  V36-443236 (REF)
- PANEL CLOSEOUT
  V36-333614 (REF)
- OXYGEN UMBILICAL STRAP
  INSTALLATION V36-601089 (REF)
- MIRROR ASSY-INTERNAL
  VIEWING V16-601471 (REF)
- PANEL ASSY NO 15 L H LONG
  SIDE DISPLAY CONSOLE
  V36-761095 (REF)

FOR FURTHER INFO IN THIS AREA
SEE ZONES 21 36 108 116 124 132 140

Upper Equipment Bay Forward
2 Unified Hatch
3 Floodlight Fixture

Upper Equipment Bay Aft
4 Emergency Oxygen & Repressurization System
5 Stowage Locker

Aft Bulkhead
6 Electrical Cableway
7 Suit Stowage Bags
8 Stowage Locker
9 Stowage Locker

Left Hand Forward Equipment Bay
10 Communication Cable Connectors
11 Oxygen Hose Connections
12 Food Preparation Water Unit
13 Timer Panel

Left Hand Equipment Bay
14 Optical Alignment Sight
15 Cabin Pressure Relief Control Panel (325)
16 Oxygen & Glycol Control Panel (326)
17 Stowage Compartment
18 Oxygen Surge Tank Pressure Relief
19 Oxygen Demand Regulator Valve Panel (380)
20 PLVC Control Panel (376)
21 Environmental Control Filter Access Panel (350)
22 Oxygen Control Panel (351)
23 Seawater Plug Access Panel
24 Water-Glycol Accumulator Fill Control Panel (379)
25 Water-Glycol Accumulator Isolation Control Panel (378)
26 Coolant Control Panel (352)
27 Water Control Panel (352)

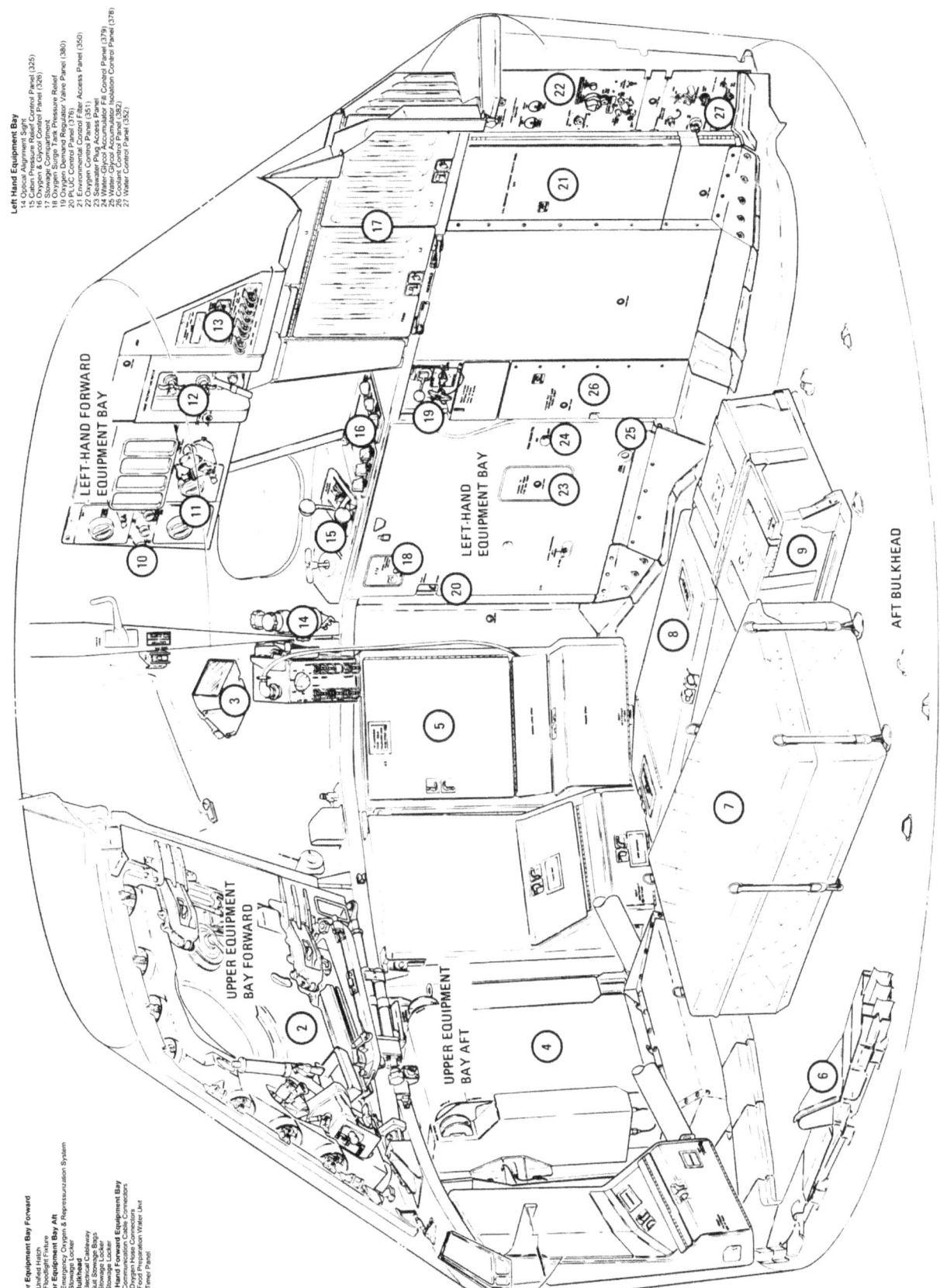

AFT BULKHEAD, UPPER, & LEFT-HAND EQUIPMENT BAYS

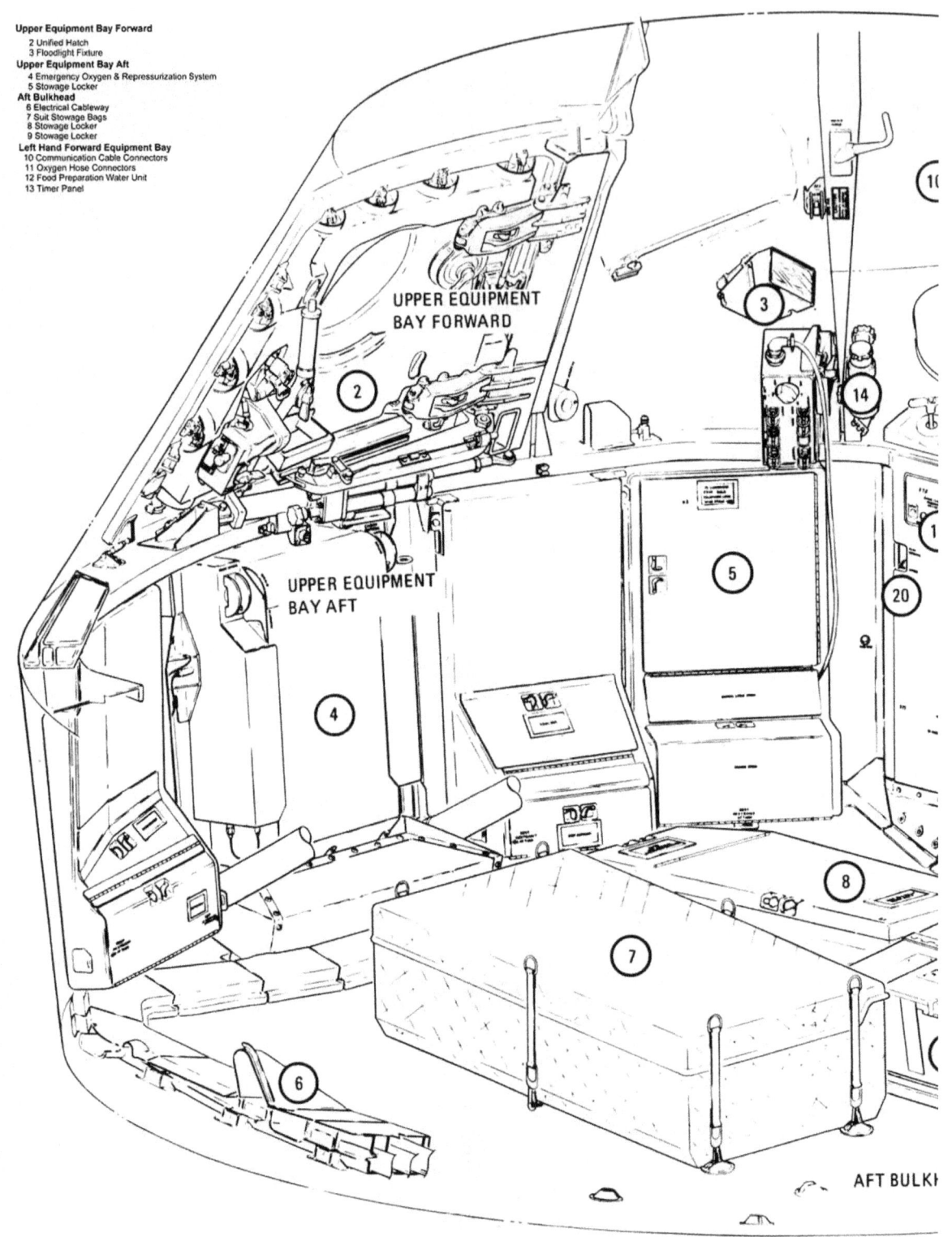

Upper Equipment Bay Forward
 2 Unified Hatch
 3 Floodlight Fixture

Upper Equipment Bay Aft
 4 Emergency Oxygen & Repressurization System
 5 Stowage Locker

Aft Bulkhead
 6 Electrical Cableway
 7 Suit Stowage Bags
 8 Stowage Locker
 9 Stowage Locker

Left Hand Forward Equipment Bay
 10 Communication Cable Connectors
 11 Oxygen Hose Connectors
 12 Food Preparation Water Unit
 13 Timer Panel

AFT BULKHEAD, UPPER, & LEFT-HAND EQ

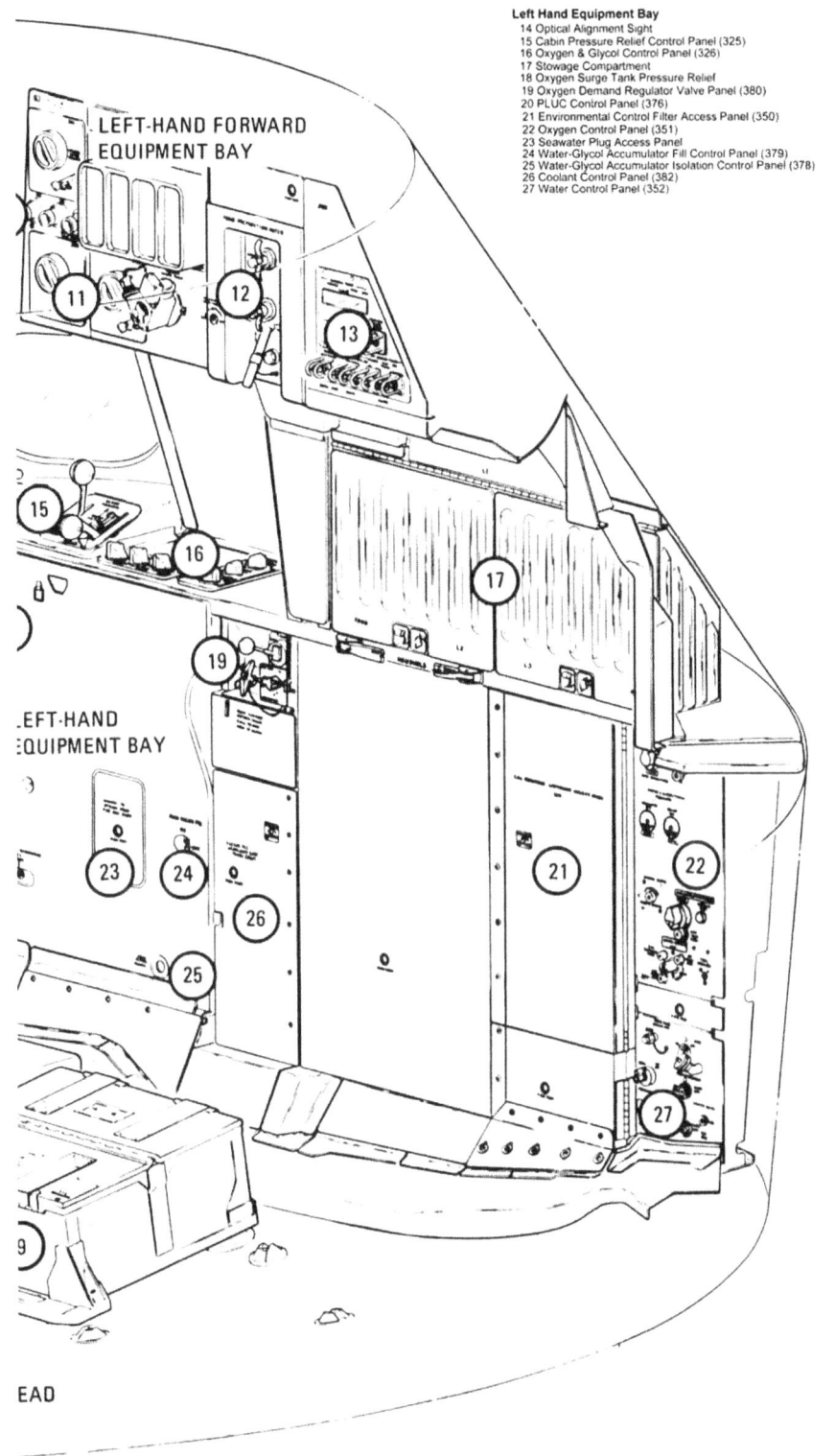

Left Hand Equipment Bay
14 Optical Alignment Sight
15 Cabin Pressure Relief Control Panel (325)
16 Oxygen & Glycol Control Panel (326)
17 Stowage Compartment
18 Oxygen Surge Tank Pressure Relief
19 Oxygen Demand Regulator Valve Panel (380)
20 PLUC Control Panel (376)
21 Environmental Control Filter Access Panel (350)
22 Oxygen Control Panel (351)
23 Seawater Plug Access Panel
24 Water-Glycol Accumulator Fill Control Panel (379)
25 Water-Glycol Accumulator Isolation Control Panel (378)
26 Coolant Control Panel (382)
27 Water Control Panel (352)

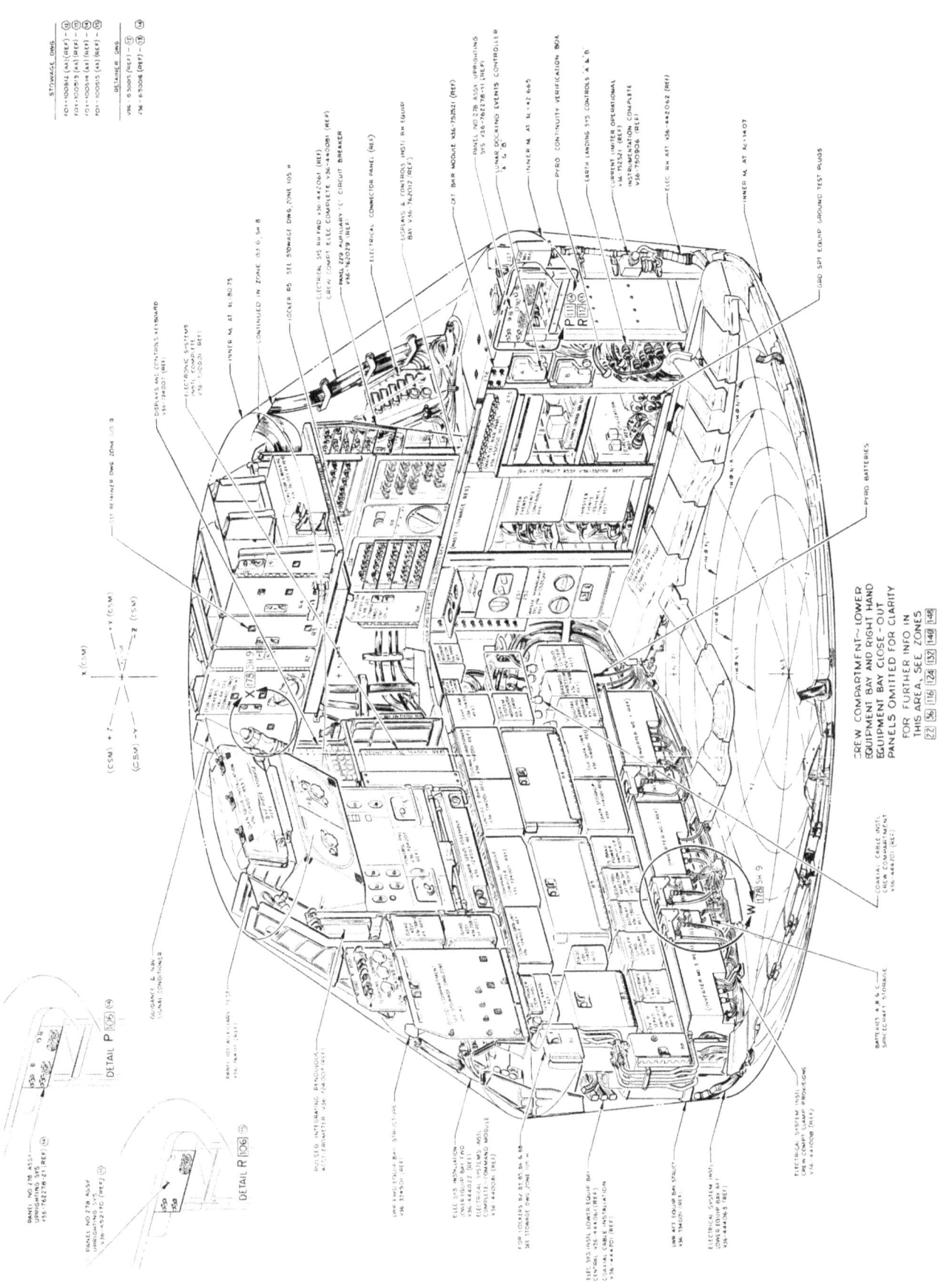

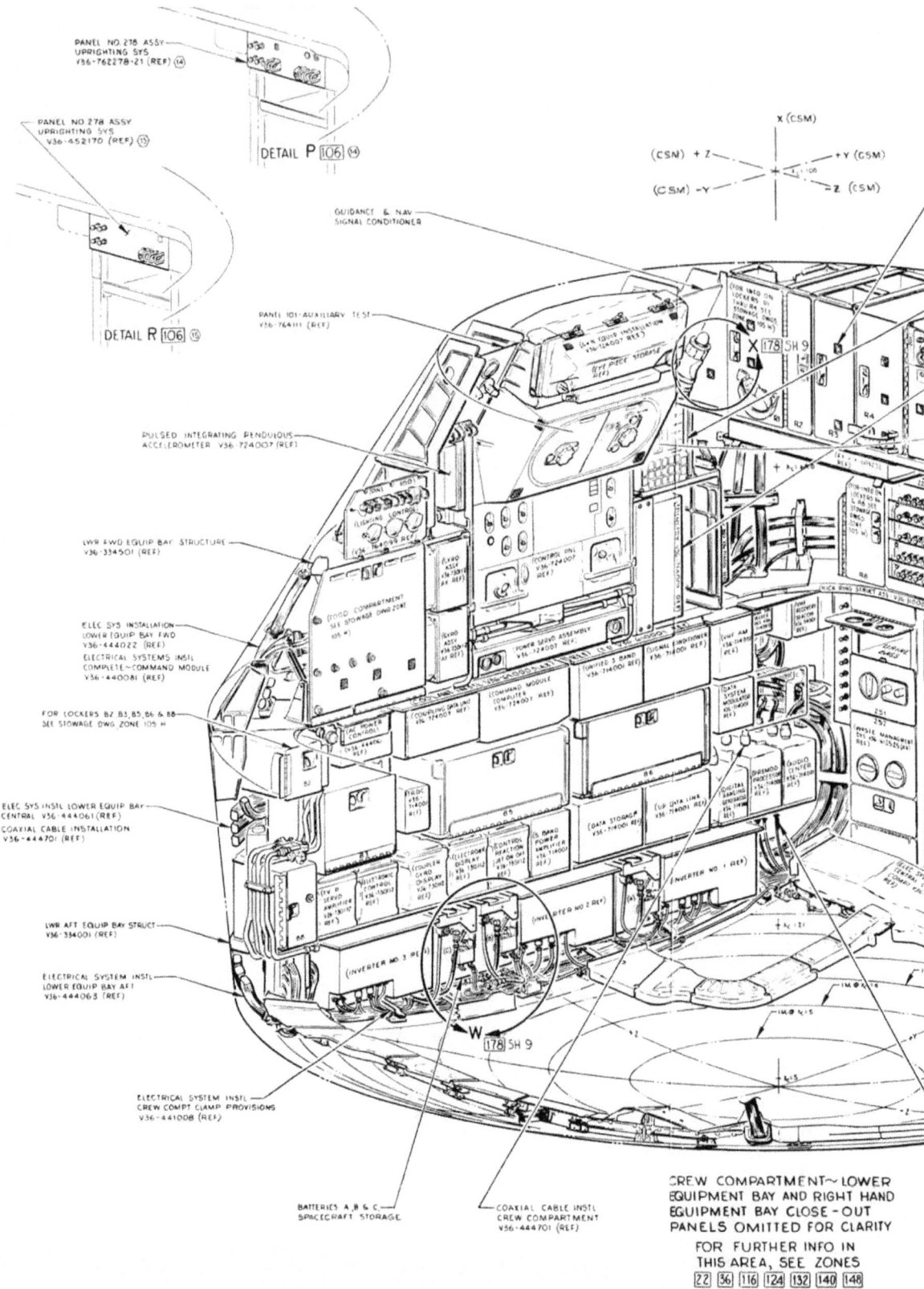

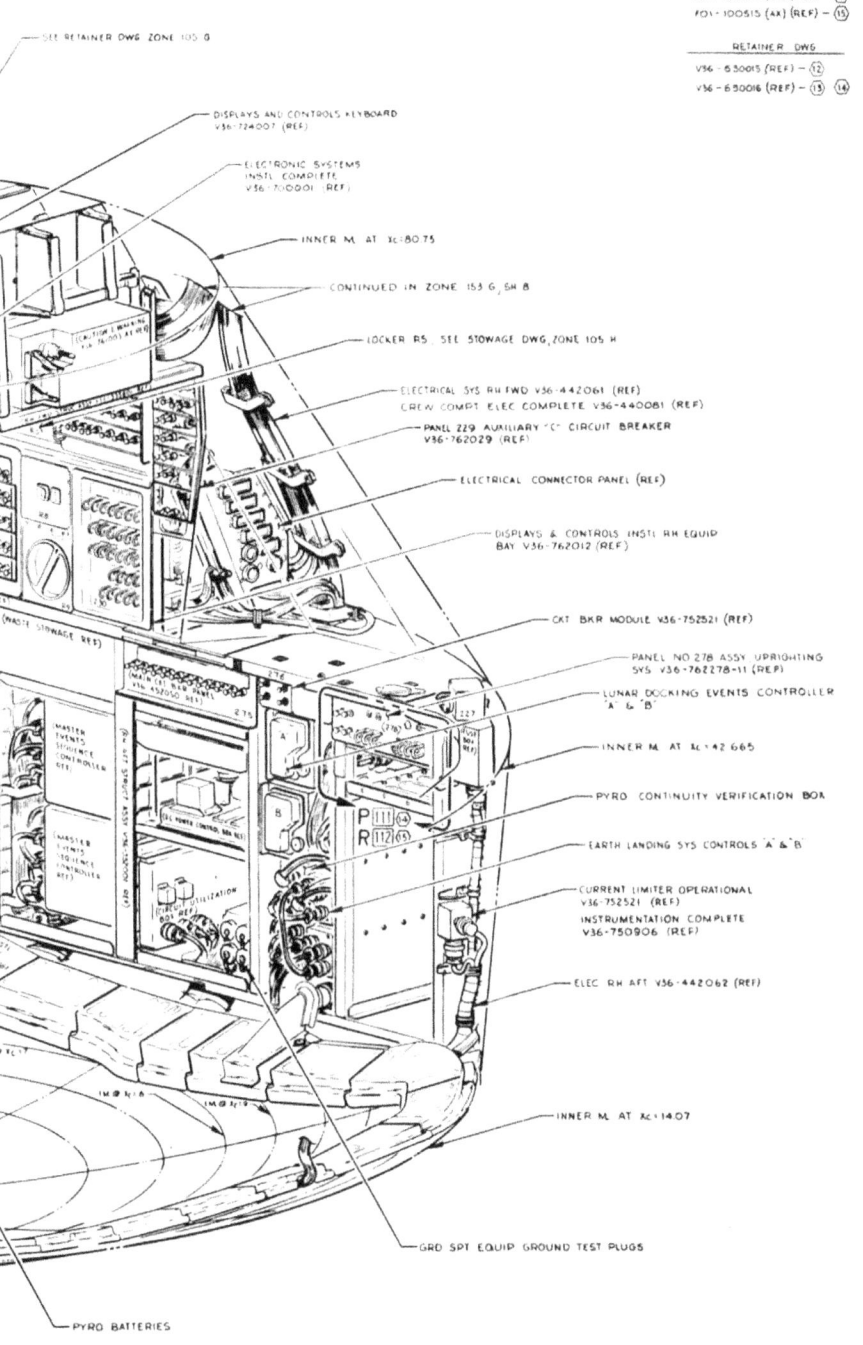

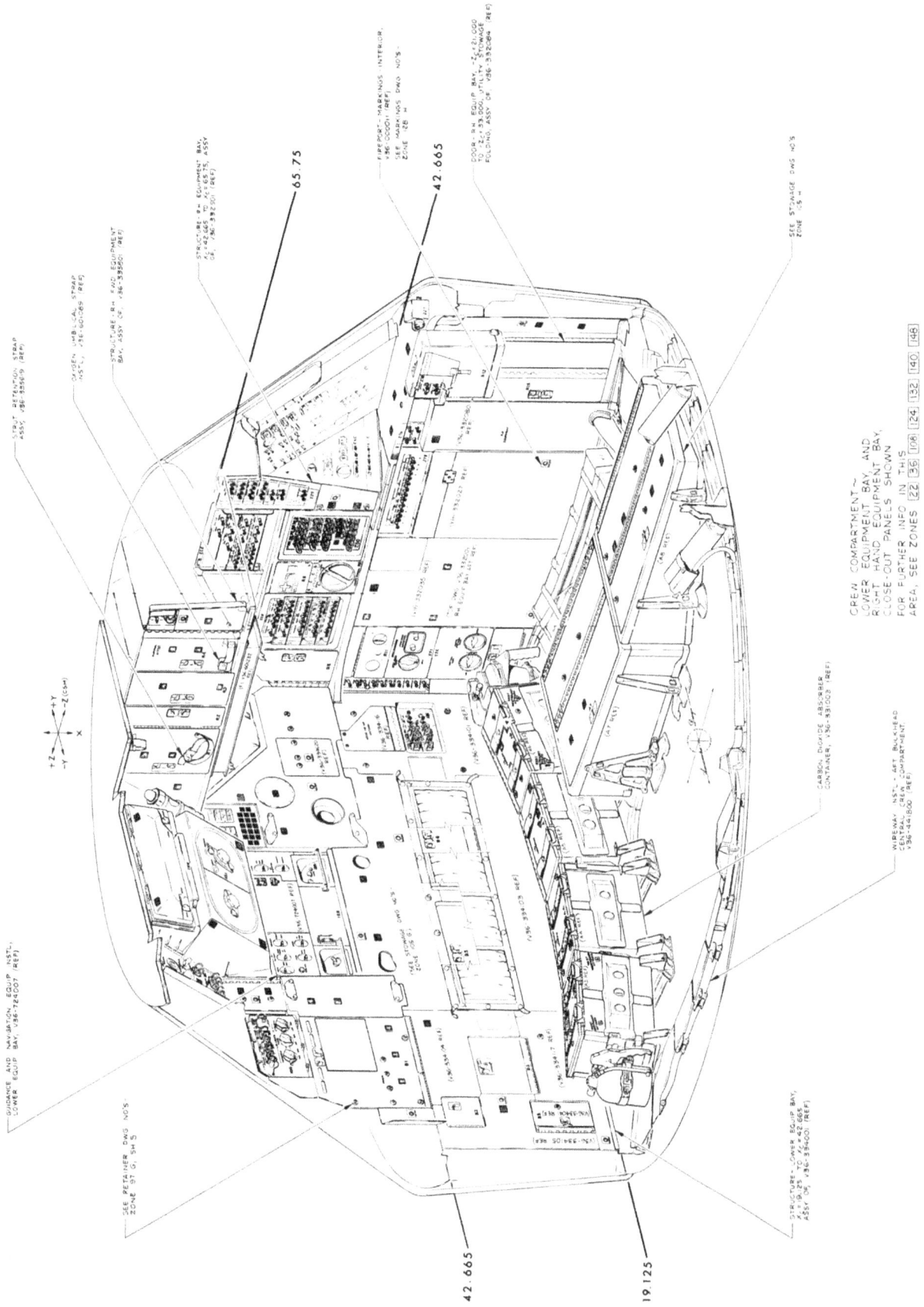

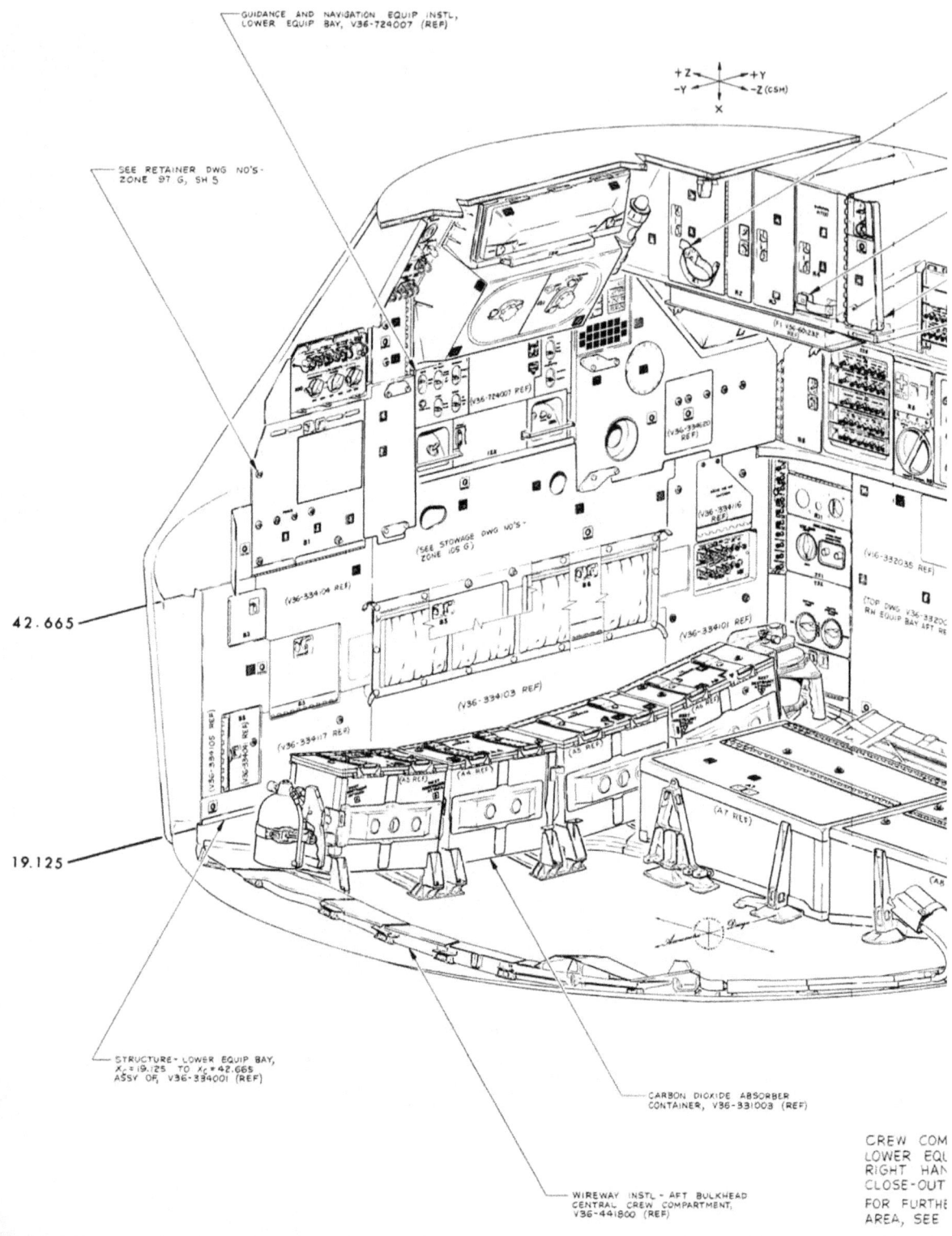

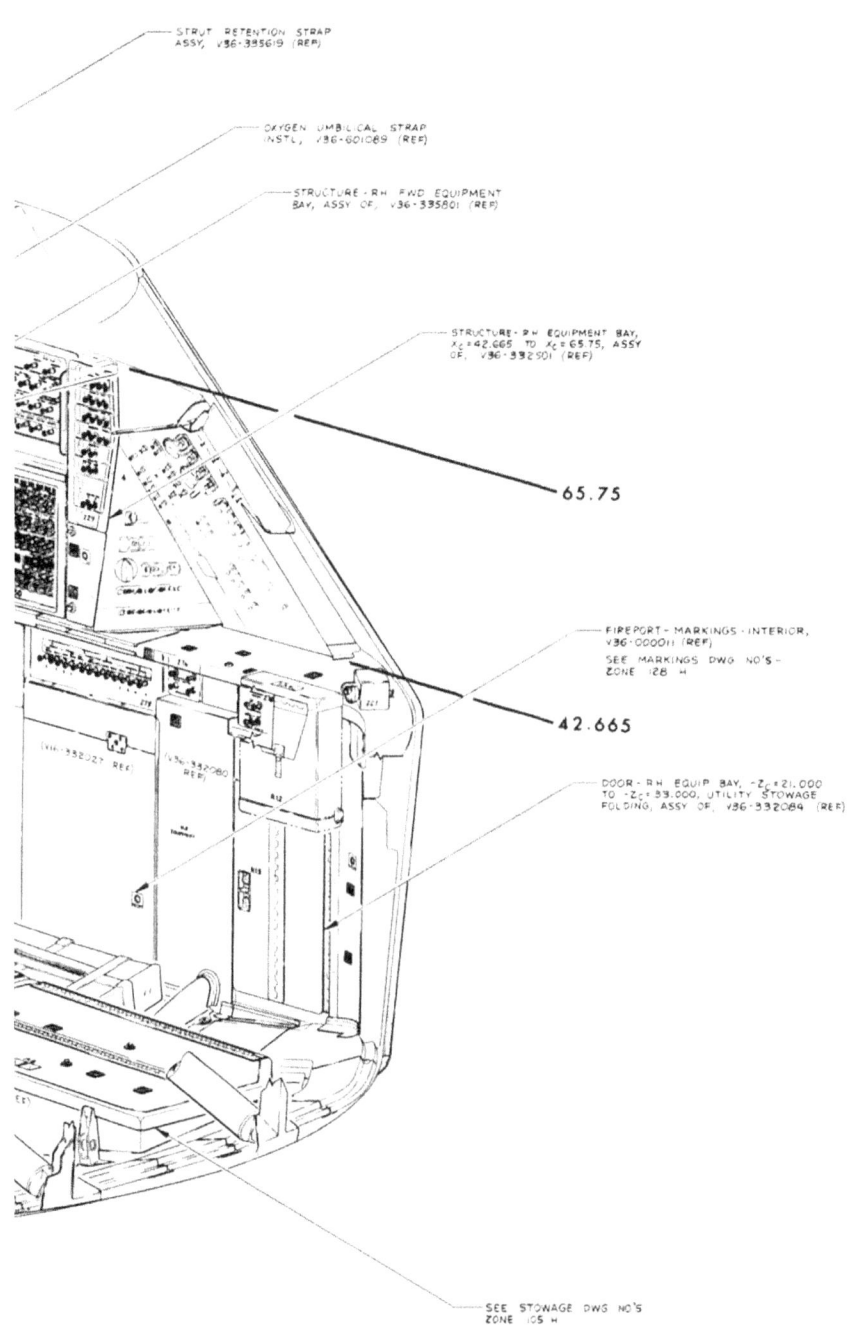

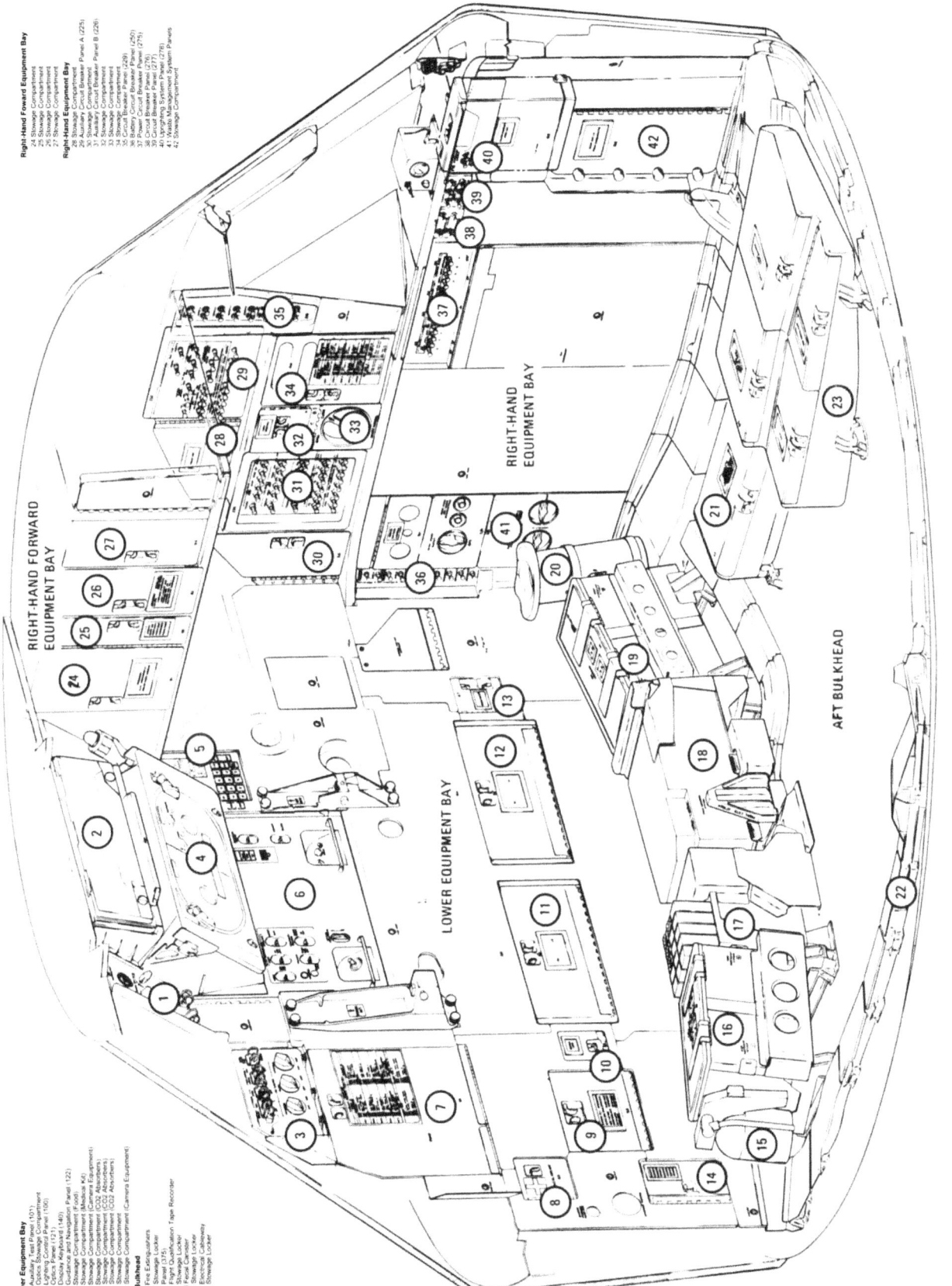

AFT BULKHEAD, LOWER, AND RIGHT-HAND EQUIPMENT BAYS

Lower Equipment Bay
1 - Auxiliary Test Panel (101)
2 - Optics Stowage Compartment
3 - Lighting Control Panel (106)
4 - Optics Panel (121)
5 - Display Keyboard (140)
6 - Guidance and Navigation Panel (122)
7 - Stowage Compartment (Food)
8 - Stowage Compartment (Medical Kit)
9 - Stowage Compartment (Camera Equipment)
10 - Auxiliary Stowage Compartment (CO2 Absorbers)
11 - Stowage Compartment (CO2 Absorbers)
12 - Stowage Compartment
13 - Stowage Compartment
14 - Stowage Compartment (Camera Equipment)

Aft Bulkhead
15 - Fire Extinguishers
16 - Stowage Locker
17 - Panel (315)
18 - Flight Qualification Tape Recorder
19 - Stowage Locker
20 - Fecal Canister
21 - Stowage Locker
22 - Electrical Cableway
23 - Stowage Locker

Right-Hand Forward Equipment Bay
24 - Stowage Compartment
25 - Stowage Compartment
26 - Stowage Compartment
27 - Stowage Compartment

Right-Hand Equipment Bay
28 - Stowage Compartment
29 - Auxiliary Circuit Breaker Panel A (225)
30 - Auxiliary Circuit Breaker Panel B (226)
31 - Stowage Compartment
32 - Stowage Compartment
33 - Stowage Compartment
34 - Stowage Compartment
35 - Circuit Breaker Panel (229)
36 - Battery Circuit Breaker Panel (250)
37 - Power Circuit Breaker Panel (275)
38 - Circuit Breaker Panel (276)
39 - Circuit Breaker Panel (277)
40 - Uprighting System Panel (278)
41 - Waste Management System Panels
42 - Stowage Compartment

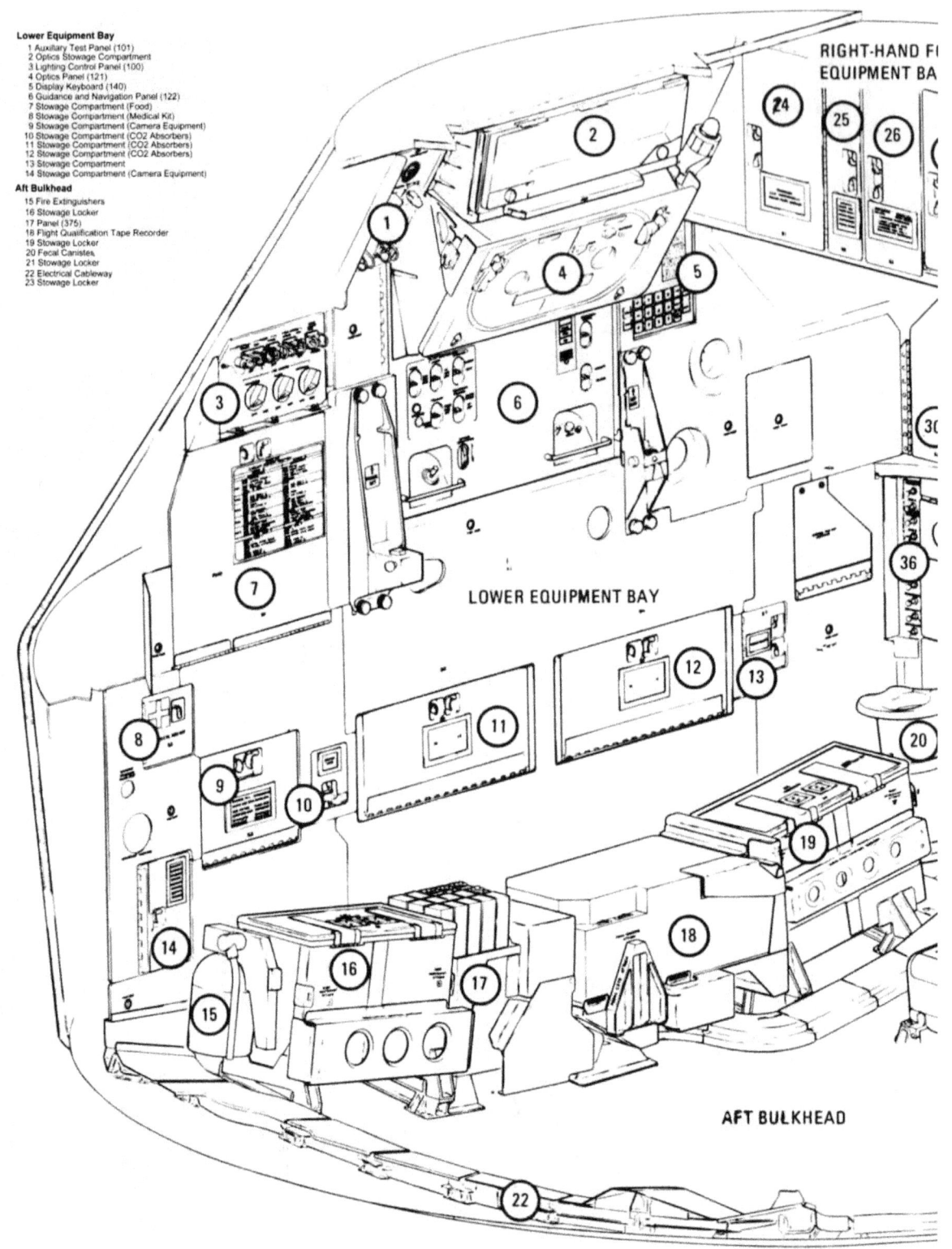

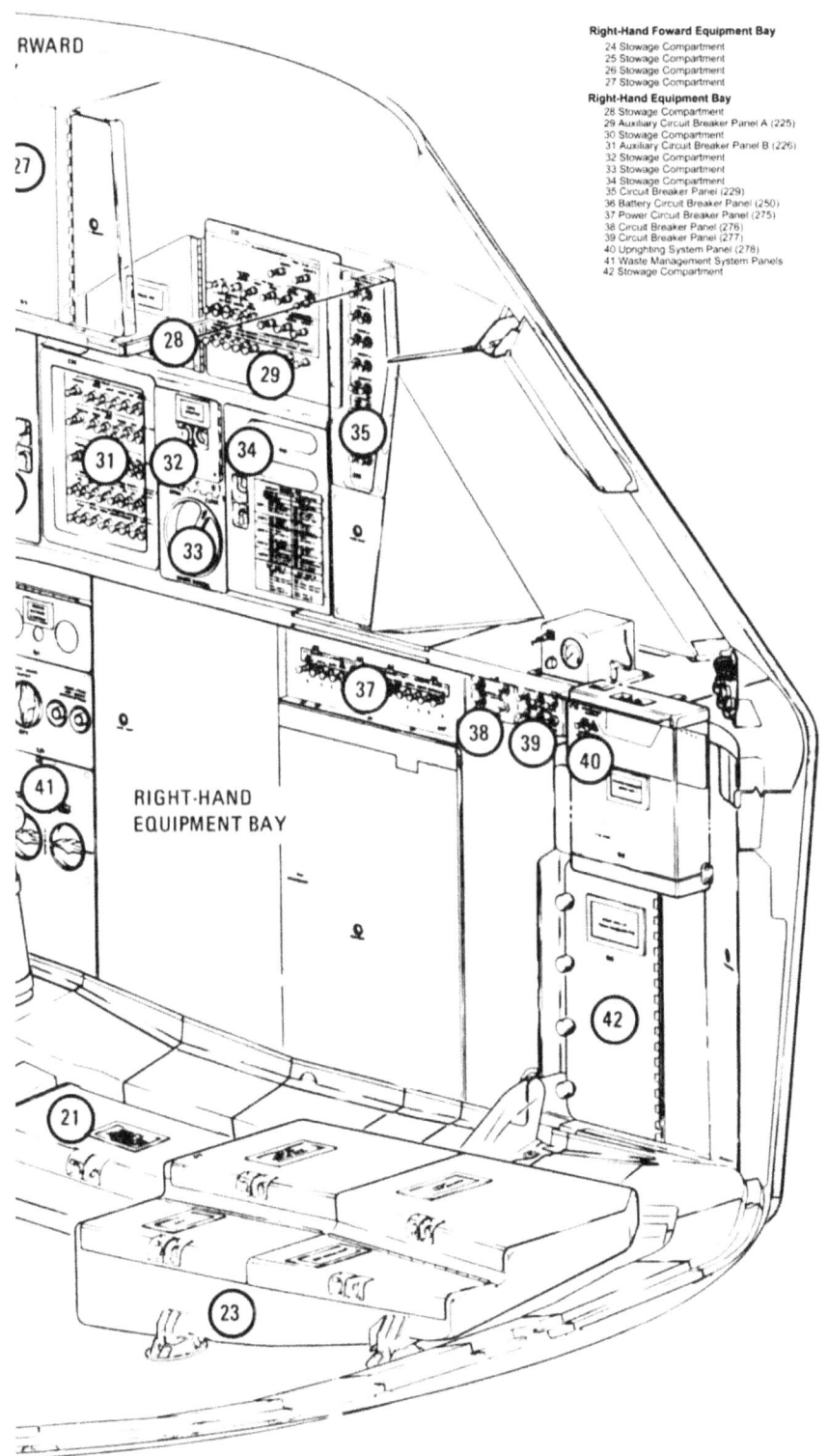

Right-Hand Foward Equipment Bay
24 Stowage Compartment
25 Stowage Compartment
26 Stowage Compartment
27 Stowage Compartment

Right-Hand Equipment Bay
28 Stowage Compartment
29 Auxiliary Circuit Breaker Panel A (225)
30 Stowage Compartment
31 Auxiliary Circuit Breaker Panel B (226)
32 Stowage Compartment
33 Stowage Compartment
34 Stowage Compartment
35 Circuit Breaker Panel (229)
36 Battery Circuit Breaker Panel (250)
37 Power Circuit Breaker Panel (275)
38 Circuit Breaker Panel (276)
39 Circuit Breaker Panel (277)
40 Uprighting System Panel (278)
41 Waste Management System Panels
42 Stowage Compartment

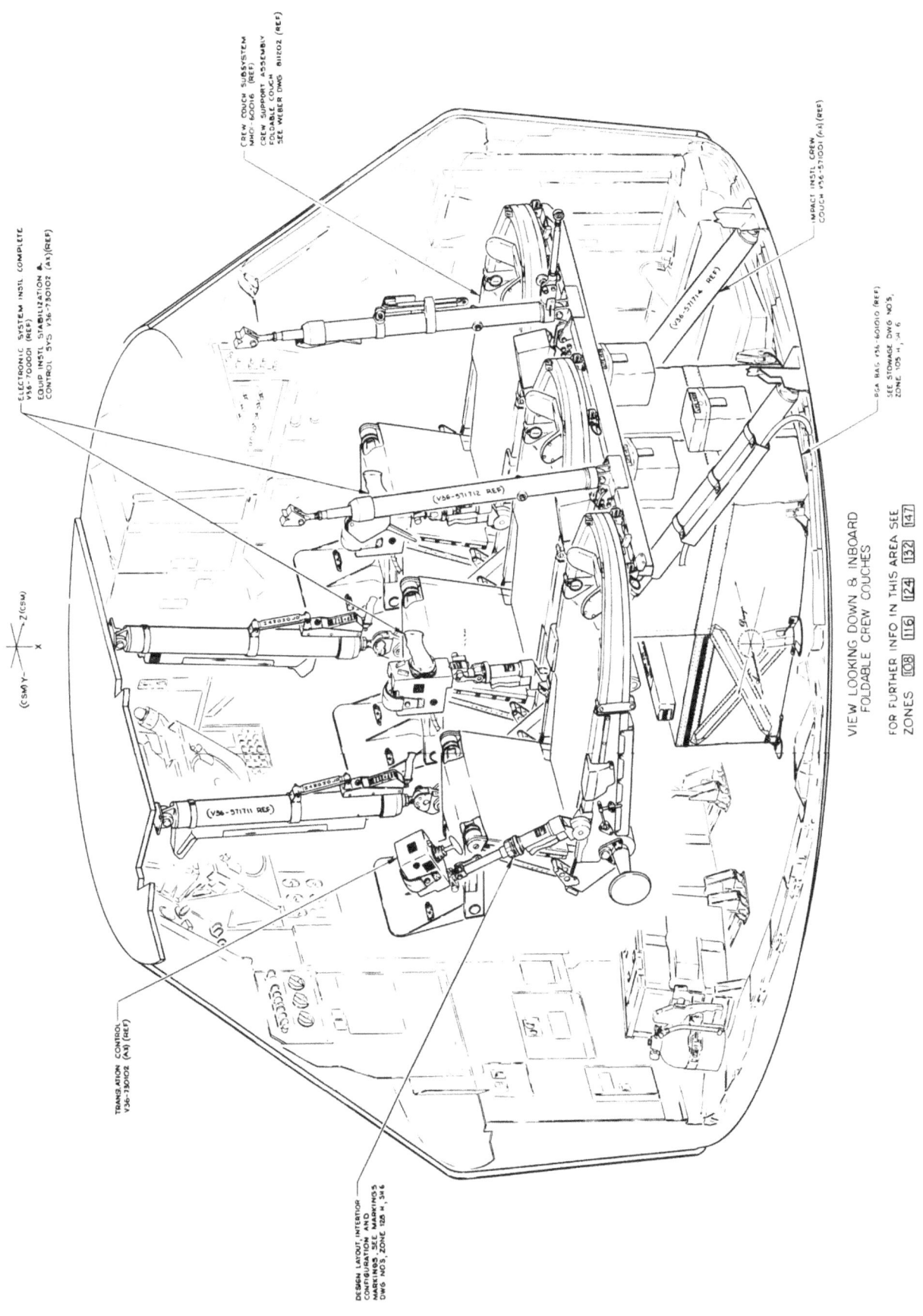

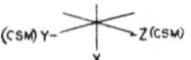

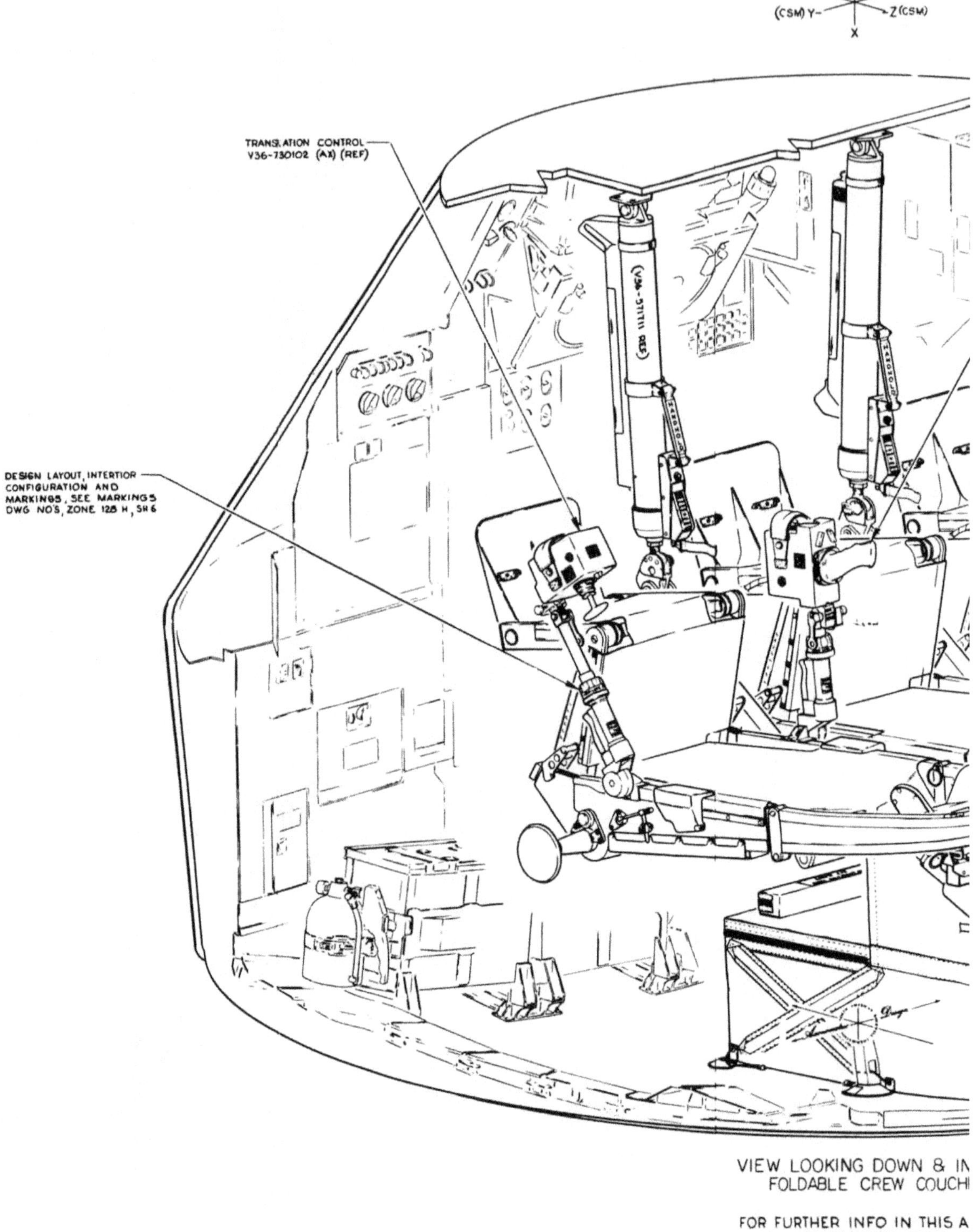
(CSM) Y-
Z (CSM)
X
TRANSLATION CONTROL
V36-730102 (AX) (REF)
DESIGN LAYOUT, INTERTIOR
CONFIGURATION AND
MARKINGS, SEE MARKINGS
DWG NO'S, ZONE 128 H, SH 6

VIEW LOOKING DOWN & IN
FOLDABLE CREW COUCH

FOR FURTHER INFO IN THIS A
ZONES 108 116 124

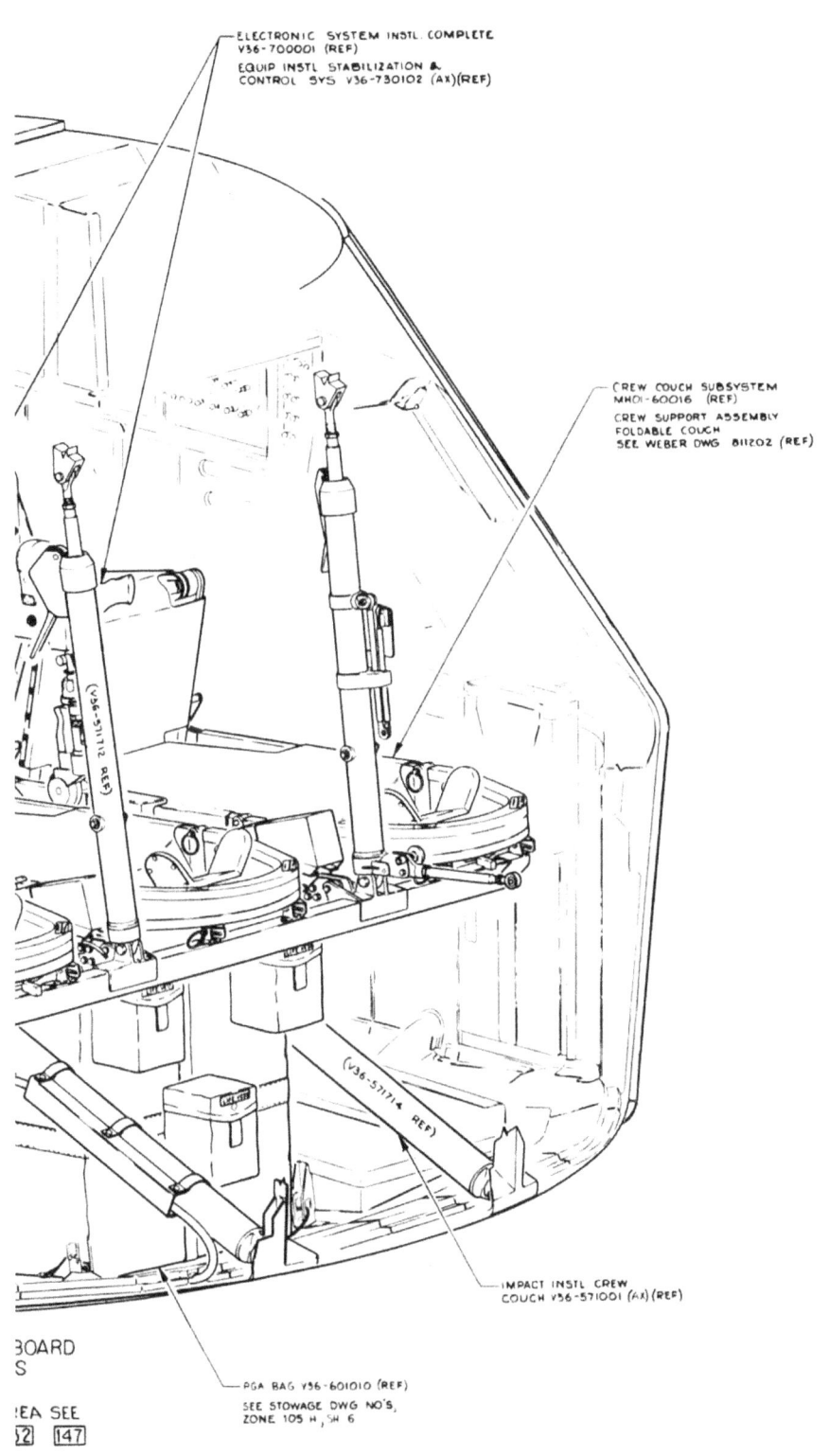

SERVICE MODULE

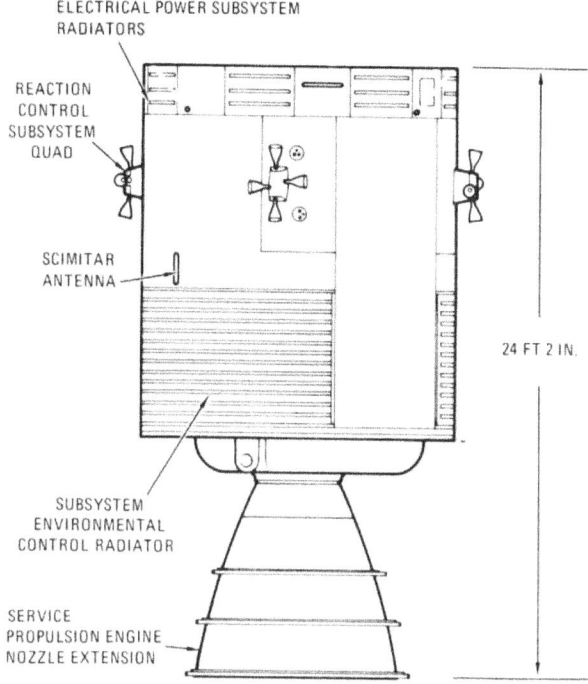
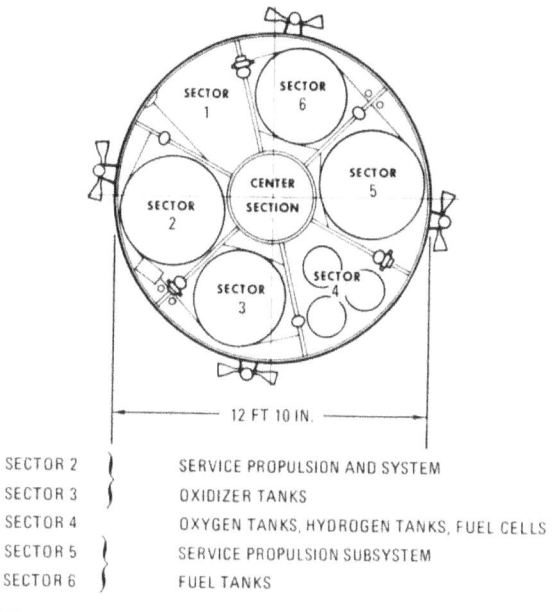

SECTOR 2 } SERVICE PROPULSION AND SYSTEM
SECTOR 3 } OXIDIZER TANKS
SECTOR 4   OXYGEN TANKS, HYDROGEN TANKS, FUEL CELLS
SECTOR 5 } SERVICE PROPULSION SUBSYSTEM
SECTOR 6 } FUEL TANKS

CENTER SECTION - SERVICE PROPULSION ENGINE AND HELIUM TANKS

Dimensions

Height	24 ft. 2 in.
Diameter	12 ft. 10 in.
Weight (loaded)	55,000 lb.
Weight (dry)	11,500 lb.

Propellant

SPS fuel	15,766 lb.
SPS oxidizer	25,208 lb.
RCS	1,362 lb.

Function

The service module contains the main spacecraft propulsion system and supplies most of the spacecraft's consumables (oxygen, water, propellant, hydrogen). It is not manned. The service module remains attached to the command module until just before entry, when it is jettisoned and is destroyed during entry.

Major Subsystems

Electrical power
Environmental control
Reaction control
Service propulsion
Telecommunications

Service module mated to command module at Kennedy Space Center

The service module is a cylindrical structure which serves as a storehouse of critical subsystems and supplies for almost the entire lunar mission. It is attached to the command module from launch until just before earth atmosphere entry.

The service module contains the spacecraft's main propulsion engine, which is used to brake the spacecraft and put it into orbit around the moon and to send it on the homeward journey from the moon. The engine also is used to correct the spacecraft's course on both the trips to and from the moon.

Besides the service propulsion engine and its propellant and helium tanks, the service module contains a major portion of the electrical power, environmental control, and reaction control subsystems, and a small portion of the communications subsystem.

It is strictly a servicing unit of the spacecraft, but it is more than twice as long and more than four times as heavy as the manned command module. About 75 percent of the service module's weight is in propellant for the service propulsion engine.

STRUCTURE

The service module is a relatively simple structure consisting of a center section or tunnel surrounded by six pie-shaped sectors.

The basic structural components are forward and aft (upper and lower) bulkheads, six radial beams, four sector honeycomb panels, four reaction control system honeycomb panels, an aft heat shield, and a fairing.

The radial beams are made of solid aluminum alloy which has been machined and chem-milled (metal removed by chemical action) to thicknesses varying between 2 inches and 0.018 inch, thus making a lightweight, efficient structure.

The forward and aft bulkheads cover the top and bottom of the module. Radial beam trusses extending above the forward bulkhead support and secure the command module. Three of these beams have compression pads and the other three have shear-compression pads and tension ties. Explosive charges in the center sections of these tension ties are used to separate the two modules.

An aft heat shield surrounds the service propulsion

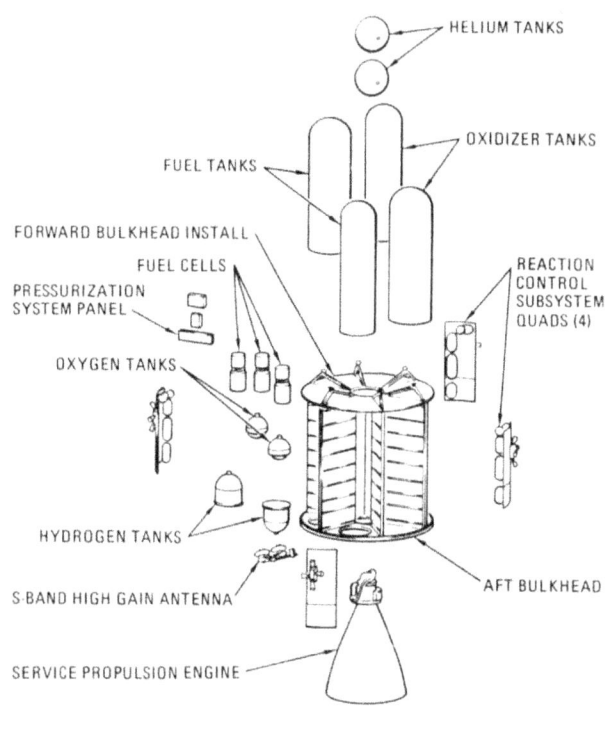

Main components of SM

engine to protect the service module from the engine's heat during thrusting. The gap between the command module and the forward bulkhead of the service module is closed off with a fairing which is ½-inch thick and 22 inches high. The fairing is composed of 16 pieces; eight electrical power subsystem radiators alternated with eight aluminum honeycomb panels.

The center section is circular and is 44 inches in diameter.

Maintenance doors around the exterior of the module provide access to equipment within each sector. These doors are designed for installation and checkout operations and are not used during space operations.

The sector and reaction control system panels are 1-inch thick and are made of aluminum honeycomb core between two aluminum face sheets. The sector panels are bolted to the radial beams. Radiators used to dissipate heat from the environmental control subsystem are bonded to the sector panels on opposite sides of the module. These radiators are each about 30 square feet in area.

P-67 *Technicians work on wiring and plumbing on "top deck" of service module*

SECTORS

The service module's six sectors are of three sizes, with two sectors each of the same size. The 360 degrees around the center section is divided among two 50-degree (Sectors 1 and 4), two 60-degree (Sectors 3 and 6), and two 70-degree (Sectors 2 and 5) compartments.

SECTOR 1

It is not currently planned to install any equipment in this sector. The space is thus available if any additional equipment needs to be added to the spacecraft for the lunar mission or if equipment is added for scientific experiments. Ballast may be stowed in the sector to maintain the service module's center of gravity if no equipment is added.

SECTOR 2

One of the two 70-degree sectors, contains part of a space radiator and a reaction control subsystem engine quad on its exterior panel, and the oxidizer sump tank, its plumbing, and the reaction control engine tanks and plumbing within the sector.

The oxidizer sump tank is the larger of the two tanks that hold the oxidizer (nitrogen tetroxide) for the service propulsion engine. A cylindrical tank made of titanium, it is 153.8 inches high (about 12 feet 9-¾ inches) and has a diameter of 51 inches (4 feet 3 inches). It holds 13,923 pounds of oxidizer. It is the tank from which oxidizer is fed to the engine. Feed lines connect the sump tank to the service propulsion engine and to the oxidizer storage tank.

SECTOR 3

Sector 3 is one of the 60-degree sectors, and contains the rest of the space radiator and a reaction control engine quad on its exterior panel, and the oxidizer storage tank and its plumbing within the sector.

The oxidizer storage tank is similar to the sump tank but not quite as large. It is 154.47 inches high (about 12 feet 10-½ inches) and has a diameter of 45 inches. It holds 11,284 pounds of oxidizer. Oxidizer is fed from it to the oxidizer sump tank in Sector 2.

SECTOR 4

Sector 4 is one of the 50-degree sectors and contains most of the electrical power subsystem equipment in the service module, including three fuel cell powerplants, two cryogenic oxygen and two cryogenic hydrogen tanks, and a power control relay box. A helium servicing panel also is located in this sector.

The three fuel cell powerplants are mounted on a shelf in the upper third of the sector. Each powerplant is 44 inches high, 22 inches in diameter, and weighs about 245 pounds. They supply most of the electrical power for the spacecraft as well as some of the drinking water.

The cryogenic (ultra low temperature) tanks supply oxygen to the environmental control subsystem and oxygen and hydrogen to the fuel cell powerplants. The tanks are spheres, with the oxygen tanks mounted side by side in the center of the sector and the hydrogen tanks mounted below them one on top of the other.

The oxygen tanks are made of Inconel (a nickel-steel alloy) and are a little over 26 inches in diameter. Each holds 326 pounds of oxygen in a semi-liquid, semi-gas state. Operating temperature of the tanks ranges from 300 degrees below zero to 80 above. Oxygen must be maintained at 297 degrees below zero to remain liquid.

The hydrogen tanks are made of titanium and are about 31-¾ inches in diameter. Each holds a little over 29 pounds of hydrogen. (Hydrogen is much lighter than oxygen, so that in vessels of the same volume the weight of the oxygen would be far greater.) The hydrogen also is in a semi-gas, semi-

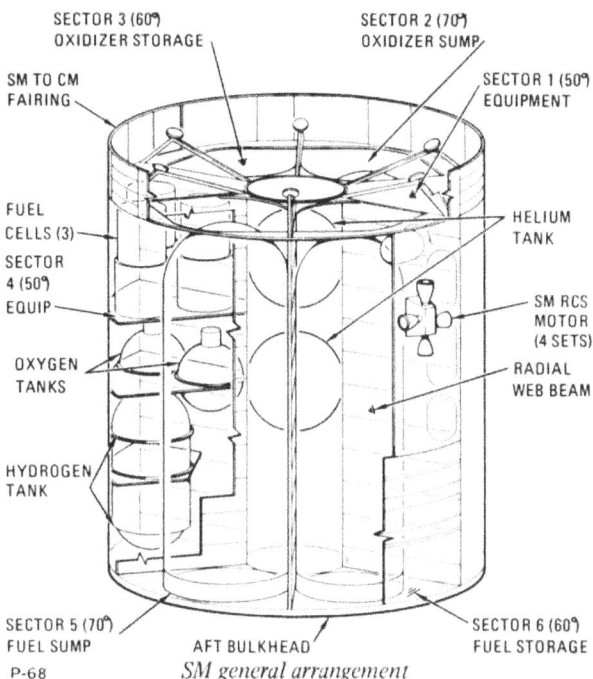

SM general arrangement

liquid stage, and its operating temperature ranges from 425 degrees below zero to 80 above. To remain liquid, hydrogen must be maintained at 423 degrees below zero.

The power control relay box operates in conjunction with the fuel cell powerplants to control the generation and distribution of electrical power.

SECTOR 5

This is the other 70-degree sector; it contains part of an environmental control radiator and a reaction control engine quad on the exterior panel, and the fuel sump tank within the sector.

The fuel sump tank occupies almost all of the space with the sector. It is a cylindrical titanium tank the same size as the oxidizer sump tank: 153.8 inches high (12 feet 9-¾ inches) and 51 inches in diameter. It holds 8,708 pounds of propellant (a 50-50 mixture of hydrazine and unsymmetrical dimethylhydrazine) for the service propulsion engine. It is the tank from which the fuel is fed to the engine; feed lines also connect it to the fuel storage tank.

SECTOR 6

The other 60-degree sector contains the rest of the space radiator and a reaction control engine

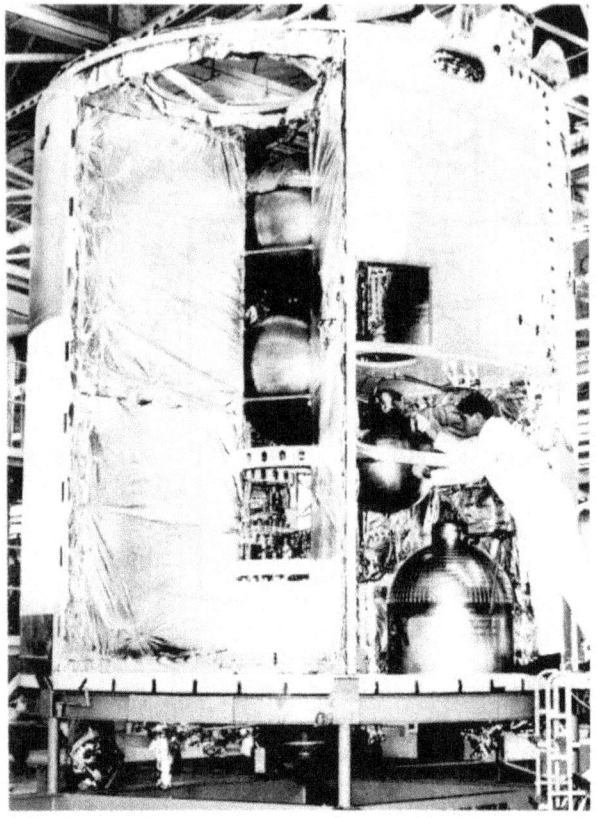

Technician finishes installation of cryogenic oxygen tank

quad on its exterior, and the fuel storage tank within the sector.

The fuel storage tank is the same size as the oxidizer storage tank: 154.47 inches high (about 12 feet 10-½ inches) and 45 inches in diameter. It holds 7,058 pounds of fuel. Fuel is fed from it to the fuel sump tank in Sector 5.

The sump tanks and the storage tanks for fuel and oxidizer are the same size; the difference in weight each contains is that the oxidizer is more than 50-percent heavier than the fuel.

CENTER SECTION

The center section or tunnel contains two helium tanks and the service propulsion engine.

The helium tanks are spherical vessels about 40 inches in diameter located one on top of the other in the upper half of the center section. Each contains 19.6 cubic feet of helium gas under a pressure of 3600 psi. This gas is used to pressurize the oxidizer and fuel tanks of the service propulsion subsystem. The pressure forces the propellant from one tank to another and through the feed lines to the engine.

The service propulsion engine is located in the lower half of the center section, with its nozzle extension skirt protruding more than 9 feet below the aft bulkhead of the module. The length of the engine including the skirt is 152.82 inches (about 12 feet 8 inches) and its weight is 650 pounds. This engine is used as a retrorocket to brake the spacecraft and put it into orbit around the moon, to supply the thrust for the return to earth from the moon, and for course corrections on the trips to and from the moon.

EXTERIOR

Located on the exterior of the service module are space radiators for both the environmental control and electrical power subsystems, reaction control subsystem engines, three antennas, umbilical connections, and several lights.

The environmental control subsystem space radiators are the larger ones and are located on the lower half of the service module on opposite sides. One is part of the panel covers for Sectors 2 and 3 and the other is part of the panel covers for Sectors 5 and 6. The radiators, each about 30 square feet, consist of five parallel primary tubes and four secondary tubes mounted horizontally and one series tube mounted vertically. The water-glycol coolant flows through these tubes to radiate to the cold of space the heat it has absorbed from the command module cabin and from operating electronic equipment.

The electrical power subsystem space radiators are located on the fairing at the top of the service module. Each of the eight radiator panels (which are alternated with eight aluminum honeycomb panels) contains three tubes which are used to radiate to space excess heat produced by the fuel cell powerplants. A separate radiation loop is used for each powerplant; that is, one of the tubes on each panel is connected to a specific powerplant.

The reaction control subsystem engines are located in four clusters of 90 degrees apart around the upper portion of service module. The clusters or quads are arranged in such a manner that the engines are on the outside of the panels and all the other

Technician prepares service propulsion engine for installation in service module

components are on the inside. The engines are mounted with two pointed up and down and two pointed to the sides in opposite directions. Components of the quad panels on the inside of the sectors include two oxidizer and two fuel tanks, a helium tank, and associated valves, regulators, and plumbing. Each quad package is eight feet long and nearly three feet wide.

The four antennas on the outside of the service module are the S-band high-gain antenna, mounted on the aft bulkhead; two VHF omni-directional antennas, mounted on opposite sides of the module near the top; and the rendezvous radar transponder antenna, mounted in the SM fairing. The S-band high-gain antenna, used for deep space communications, is composed of four 31-inch diameter reflectors surrounding an 11-inch square reflector. At launch it is folded down parallel to the service propulsion engine nozzle so that it fits within the spacecraft-LM adapter. After the CSM separates from the adapter the antenna is deployed at a right angle to the service module. The omnidirectional antennas, called scimitars because of their shape, are made of stainless steel and are approximately 13-½ inches long and only a hundredth of an inch thick.

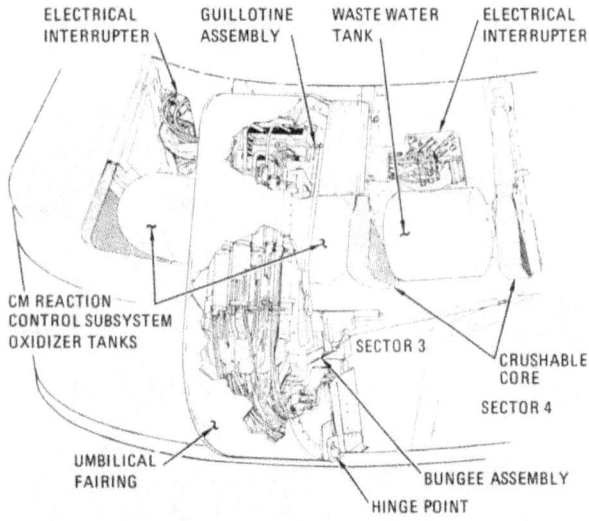

P-71 *CM-SM umbilical assembly*

The umbilicals consist of the main plumbing and wiring connections between the command module and service module, which are enclosed in a fairing (aluminum covering), and a "flyaway" umbilical which is connected to the launch tower. The latter supplies oxygen and nitrogen for cabin pressurization, water-glycol, electrical power from ground equipment, and purge gas.

Seven lights are mounted in the aluminum panels of the fairing. Four (one red, one green, and two amber) are used to aid the astronauts in docking, one is a floodlight which can be turned on to give astronauts visibility during extravehicular activities, one is a flashing beacon used to aid in rendezvous, and one is a spotlight used in rendezvous from 500 feet to docking with the lunar module.

SM-CM SEPARATION

Separation of the SM from the CM occurs shortly before entry. The sequence of events during separation is controlled automatically by two redundant service module jettison controllers, located on the forward bulkhead of the SM.

A number of events must occur in rapid sequence or at the same time for proper separation. These include physical separation of all the connections between the modules, transfer of electrical control, and firing of the service module's reaction control engines to increase the distance between the modules.

Before separation, the crewmen in the CM transfer electrical control to the CM reaction control subsystem (the SM reaction control subsystem is used for attitude maneuvers throughout the mission up to entry) so that they can pressurize it and check it out for the entry maneuvers.

Electrical control is then transferred back to the SM subsystem. Separation is started manually, by activation of either of two redundant switches on the main display console. These switches send signals to the SM jettison controllers. The controllers first send a signal to fire ordnance devices which activate the CM-SM electrical circuit interrupters; these interrupters deadface (cut off all power to) the electrical wires in the CM-SM umbilical.

A tenth of a second after the wires are deadfaced, the controllers send signals which fire ordnance devices to sever the physical connections between the modules. These connections are three tension ties and the umbilical. The tension ties are straps which hold the CM on three of the compression pads on the SM. Linear-shaped charges in each tension-tie assembly sever the tension ties to separate the modules. At the same time, explosive charges drive guillotines through the wiring and tubing in the umbilical.

Simultaneously with the firing of the ordnance devices, the controllers send signals which fire the SM reaction control engines. Roll engines are fired for 5 seconds to alter the SM's course from that of the CM, and the translation (thrust) engines are fired continuously until the propellant is depleted or fuel cell power is expended. These maneuvers carry the SM well away from the entry path of the CM.

The service module will enter the earth's atmosphere after separation and burn up.

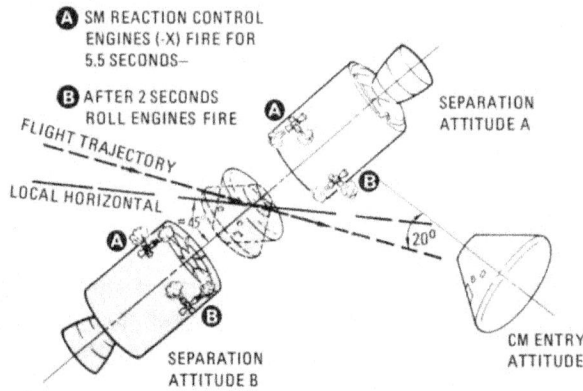

P-72 *Attitude of SM and CM during separation*

LUNAR MODULE

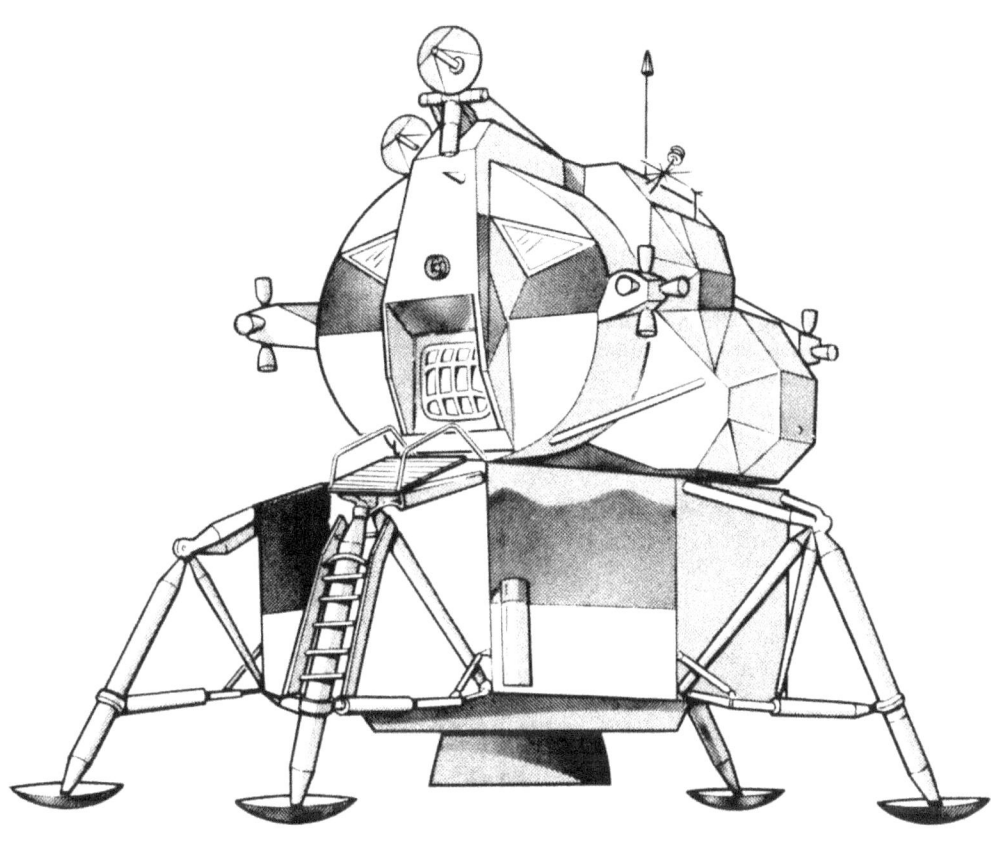

P-73

Dimensions

Height	22 ft 11 in. (with legs extended)
Diameter	31 ft (diagonally across landing gear)
Weight (with propellant and crew)	32,500 lb (approx.)
Weight (dry)	9,000 lb (approx.)
Pressurized volume	235 cu ft
Habitable volume	160 cu ft
Ascent stage	
Height	12 ft 4 in.
Diameter	14 ft 1 in.
Weight (dry)	4,850 lb (approx.)
Descent stage	
Height	10 ft 7 in.
Diameter	14 ft 1 in.
Weight (dry)	4,300 lb (approx.)
Propellant	
Ascent stage	5,170 lb tanked
Descent stage	17,880 lb tanked
RCS	605 lb tanked

Function

The lunar module carries two astronauts from lunar orbit to the surface of the moon; serves as living quarters and a base of operations on the moon, and returns the two men to the CSM in lunar orbit. The descent stage is left on the moon; the ascent stage is left in orbit around the moon.

The LM, built by Grumman Aircraft Engineering Corp., Bethpage, N.Y., is designed to operate for 48 hours while separated from the CM, with a maximum stay time of 35 hours on the moon. It consists of two main parts: a descent stage and an ascent stage. The former provides the means of landing on the moon, carries extra supplies, and serves as a launching platform for the liftoff from the moon. The latter contains the crew compartment in which the two astronauts will spend their time while not on the moon's surface, and the engines which will return the astronauts to the CM.

Information in this section was provided by Grumman Aircraft Engineering Corp. Complete details on the lunar module are contained in Grumman's Lunar Module News Reference.

ASCENT STAGE

The ascent stage houses the crew compartment, the ascent engine and its propellant tanks, and all the crew controls. It has essentially the same kind of subsystems found in the command and service modules, including propulsion, environmental control, communications, reaction control, and guidance and control. Overall height of the stage is 12 feet, 4 inches; overall width, with tankage, is 14 feet, 1 inch. Its weight, without propellant, is 4,850 pounds.

The ascent stage provides shelter and a base of operations for the two LM crewmen during their lunar stay. The crewmen use it to return to lunar orbit and rendezvous with the orbiting CSM. After the crewmen have transferred to the CM, the ascent stage is jettisoned and remains in orbit around the moon.

The primary structural components of the ascent stage are the crew compartment, the midsection, the aft equipment bay, and tank sections.

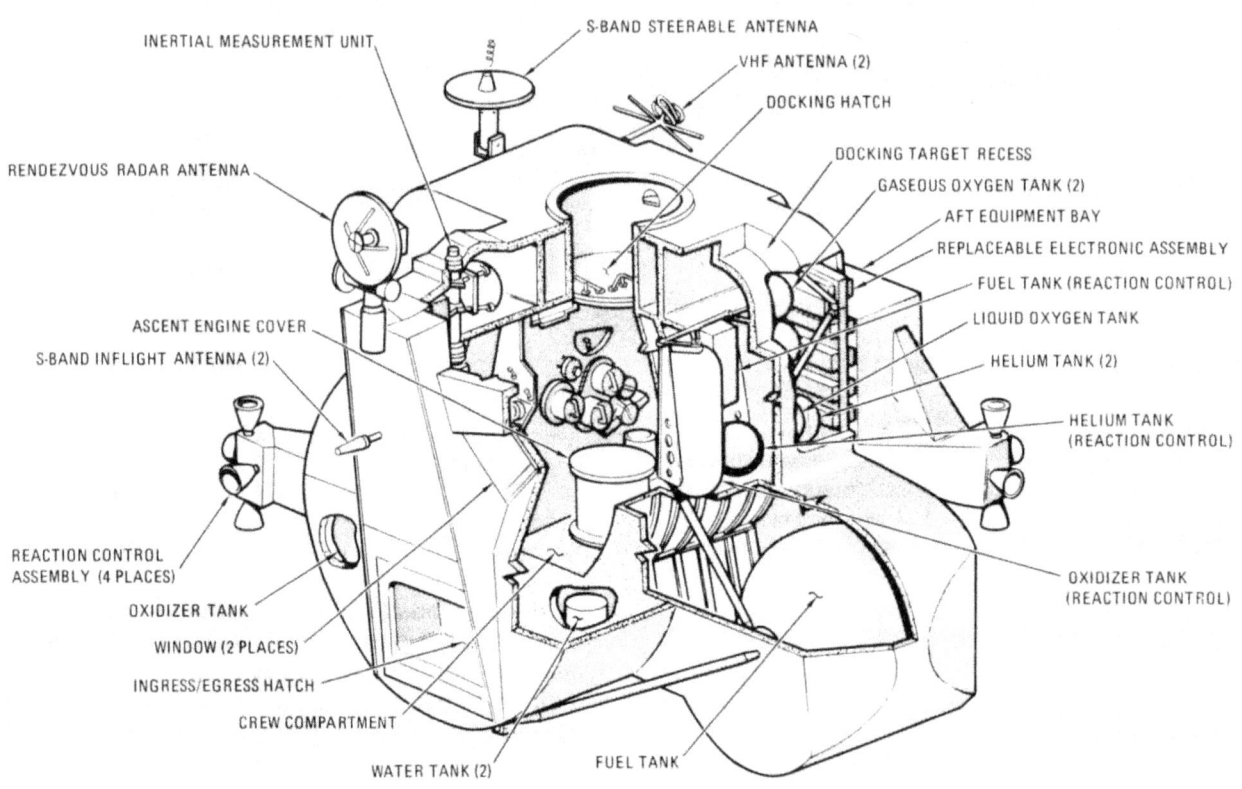

Ascent stage

P-74

The cabin is 92 inches in diameter and is made of welded aluminum alloy which is surrounded by a 3-inch-thick layer of insulating material. A thin outer skin of aluminum covers the insulation. The cabin is a pressurized shell in which the two astronauts will spend about two-thirds of their time during the lunar stay. The crew compartment is pressurized to 5 psi, its temperature is controlled at about 75 degrees F, and it has a 100-percent oxygen atmosphere. It contains the displays and controls that enable the astronauts to maneuver the module during descent, landing, lunar launch and ascent, and rendezvous and docking with the CSM. It also contains crew equipment, storage bays, and provisions for sleeping, eating, and waste management.

Astronaut stations are 44 inches apart and each man has a set of controllers and armrests but no seats. Total volume of the pressurized portion of the lunar module is 235 cubic feet; the habitable volume is about 160 cubic feet.

There are two hatches in the crew compartment, the docking hatch and the forward hatch. Both open inward. The round docking hatch (32 inches in diameter) is at the upper end of the docking tunnel and is used for the transfer of crewmen back and forth from the command module. The drogue portion of the docking subsystem is located outside the hatch. The forward hatch tunnel is beneath the center instrument console. The forward hatch is rectangular and is the one through which the astronauts will go to reach the lunar surface. Outside the forward hatch is a platform and a ladder mounted to the forward landing gear strut.

The crew compartment has three windows. The two triangular forward windows are approximately two square feet and canted down to the side to permit sideward and downward visibility. The third window, used for docking, is located on the left side of the cabin directly over the commander's position. It is about 5 inches wide and 12 inches long.

The midsection of the ascent stage is a smaller compartment directly behind the cabin; its floor is 18 inches above the crew compartment deck. Part of the midsection is pressurized. Ascent engine plumbing and valving extends above the deck. The midsection also contains the overhead docking tunnel (32 inches in diameter and 18 inches long), docking hatch, environmental control subsystem equipment, and stowage for equipment that must be accessible to the astronauts.

The aft equipment bay is unpressurized and is behind the midsection pressure-tight bulkhead. It contains an equipment rack with coldplates on which replaceable electronic assemblies are mounted. It also includes two oxygen tanks for the environmental control system, two helium tanks for ascent stage main propellant pressurization, and inverters and batteries for the electrical power subsystem.

The propellant tank sections are on either side of the midsection, outside the pressurized area. The tank sections contain ascent engine fuel and oxidizer tanks, and fuel, oxidizer, and helium tanks for the reaction control subsystem.

DESCENT STAGE

The descent stage is a modified octagonal shape. It is 10 feet, 7 inches high (with gear extended), 14 feet, 1 inch at its widest point, and has a diameter of 31 feet diagonally across the landing gear. It consists primarily of the descent engine and its propellant tanks, the landing gear assembly, batteries, a section to house scientific equipment for use on the moon, and extra oxygen, water, and helium tanks. The stage serves as a launching platform for the ascent stage and will remain on the moon. It is constructed of aluminum alloy chem-milled to reduce weight. (Chem-milling is a process of removing metal by chemical action.)

The descent engine provides the power for the complex maneuvers required to take the lunar module from orbit down to a soft landing on the moon. It is a throttleable, gimballed engine which provides from 1,050 to 9,710 pounds of thrust.

Four main propellant tanks (two oxidizer and two fuel) surround the engine. Such items as scientific equipment, the lunar surface antennas, four electrical power subsystem batteries, six portable life support system batteries, and tanks for helium, oxygen, and water are in bays adjacent to the propellant tanks.

The landing gear is of the cantilever type and consists of four legs connected to the outriggers. The legs extend from the front, rear, and sides of the LM. Each landing gear leg consists of a primary strut and footpad, a drive-out mechanism, two secondary struts, two downlock mechanisms, and a truss. Each strut has a shock-absorbing insert of crushable aluminum honeycomb material to soften the landing impact. The forward landing gear has a boarding ladder on the primary strut

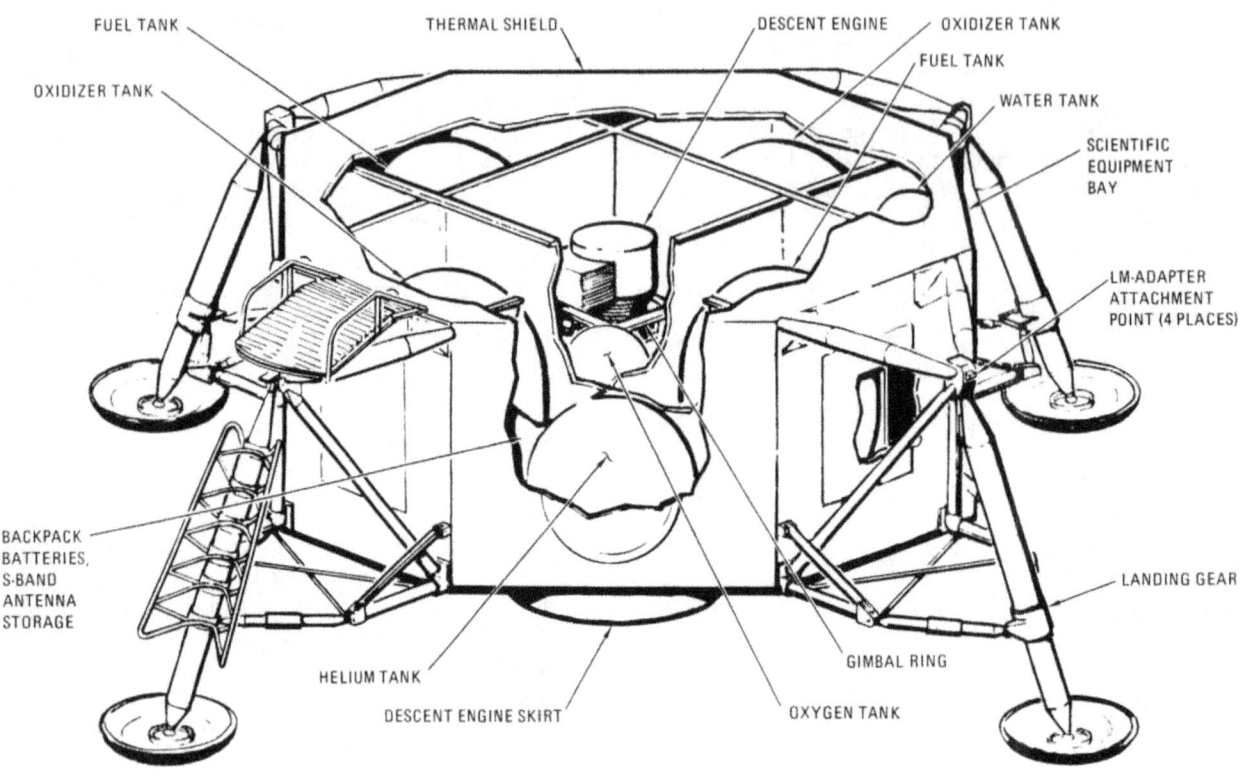

Descent stage

which the astronauts will climb to and from the forward hatch.

The landing gear is retracted until shortly after the astronauts enter the LM during lunar orbit. Extension is activated by a switch in the LM. The landing gear locks are then released by a mild explosive charge and springs in each drive-out mechanism extend the landing gear. The footpads, about 37 inches in diameter, are made of two layers of spun aluminum bonded to an aluminum honeycomb core.

SPACECRAFT-LM ADAPTER

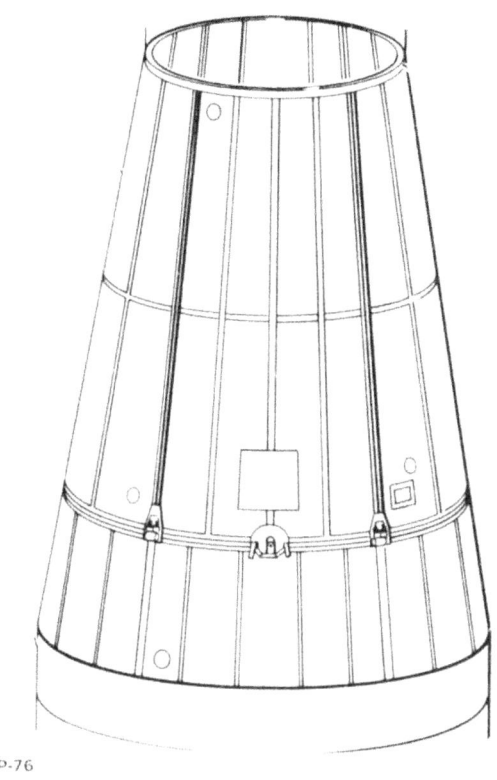

Dimensions

Height	28 ft
Diameter	12 ft 10 in. (top)
	21 ft 8 in. (bottom)
Weight	4050 lb (approx)
Volume	6700 cu ft
	(5000 cu ft usable)

Function

The spacecraft-LM adapter is an aluminum structure which protects the LM during launch and provides the structural attachment of the spacecraft to the launch vehicle.

The spacecraft-LM adapter (SLA) is constructed of eight 1.7-inch thick aluminum honeycomb panels which are arranged in two sets of four of equal size: the upper or forward panels, about 21 feet long, and the lower or aft panels, about 7 feet long. The exterior surface of the SLA is covered completely by a layer of cork 30/1000 of an inch thick. The cork helps insulate the LM from the heat generated by the spacecraft pushing through the atmosphere during boost.

The lunar module is attached to the SLA at four places around the joint between the upper and lower SLA panels. Besides the lunar module, the SLA encloses the nozzle of the service module's service propulsion engine (which extends down to the top of the LM) and an umbilical which houses connecting circuits between the launch vehicle and the spacecraft.

SLA-SM SEPARATION

The SLA and SM are joined by bolts through a flange that extends around the circumference of the two structures. The only other connection is an umbilical cable through which electrical power is supplied to the SLA. This power is used to trigger the separation devices.

Redundancy is provided in three areas to assure separation. The signals that initiate the ordinance are redundant; the detonators and cord trains are redundant, and the charges are "sympathetic"— that is, detonation on one charge will set off another.

The SLA and SM and the four upper panels of the SLA are separated by an explosive train which cuts through the metal connecting the structures.

The explosive train consists of 28 charge holders, each of which contains two strands of detonating cord, either one of which will sever the joint. The charge holders (aluminum strips to which the detonating cord is bonded) are mounted on the flange connecting the SM and SLA and on the splice plates (metal strips) which join the forward panels. Boosters (larger charges) are used at the ends of each charge holder and at crossover points to assure that the entire explosive train fires.

Although the explosive train fires like a fuse—that is, it travels from one point to another—it travels so fast that for practical purposes the entire train can be said to explode simultaneously.

P-77

Completed SLA in Tulsa before shipment

Two sets of thrusters—one pyrotechnic and one spring—are used in deploying and jettisoning the SLA's upper panels.

The four pyrotechnic thrusters are located at the top of the lower panels at the upper panel joints and are used to rotate the panels backwards. Each of these thrusters has two pistons, one acting on each panel, so that each panel has two pistons thrusting against it, one on each end.

The explosive train which separates the panels is routed through two pressure cartridges in each thruster assembly. Ignition of the pressure cartridges drives the pistons against the panels to begin deployment. Redundancy is provided because ignition of one pressure cartridge normally will sympathetically ignite the other.

The pyrotechnic thrusters apply only a small amount of impulse to the panels (for only 2 degrees of rotation), but this is enough to assure deployment. The speed (33 to 60 degrees per second of angular velocity) imparted by this thrust remains essentially constant. The panels are connected to the lower panels by two hinges. When the panels have rotated about 45 degrees, the hinges disengage and free the panels from the aft section of the SLA.

The spring thrusters are mounted on the outside of the upper panels. When the panel hinges disengage, the springs in the thruster push against the lower panels to propel the panels away from the vehicle. The opening speed and the spring thruster force are such that the panels will be pushed away from the vehicle at an angle of 110 degrees to the vehicle centerline and a speed of about 5-1/2 miles an hour. This assures that the panels will be headed away from the spacecraft.

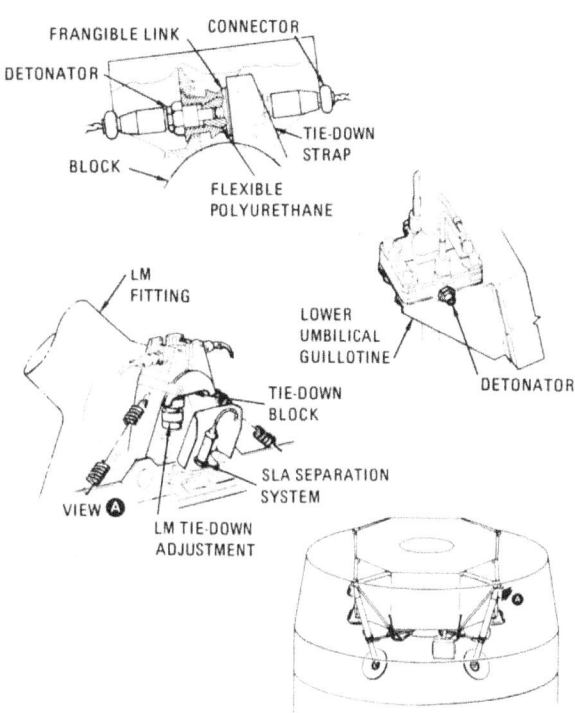

SLA-LM separation system

SLA-LM SEPARATION

Spring thrusters also are used to separate the LM from the SLA. After the CSM has docked with the LM, mild charges are fired to release the four connections which hold the LM to the SLA. Simultaneously, four spring thrusters mounted on the lower SLA panels push against the LM to separate the two vehicles.

The separation is controlled by two lunar module separation sequence controllers located inside the SLA near the attachment point to the instrument unit. The redundant controllers send signals which fire the charges that sever the connections and also fire a detonator to cut the LM-instrument unit umbilical. The detonator impels a guillotine blade which severs the umbilical wires.

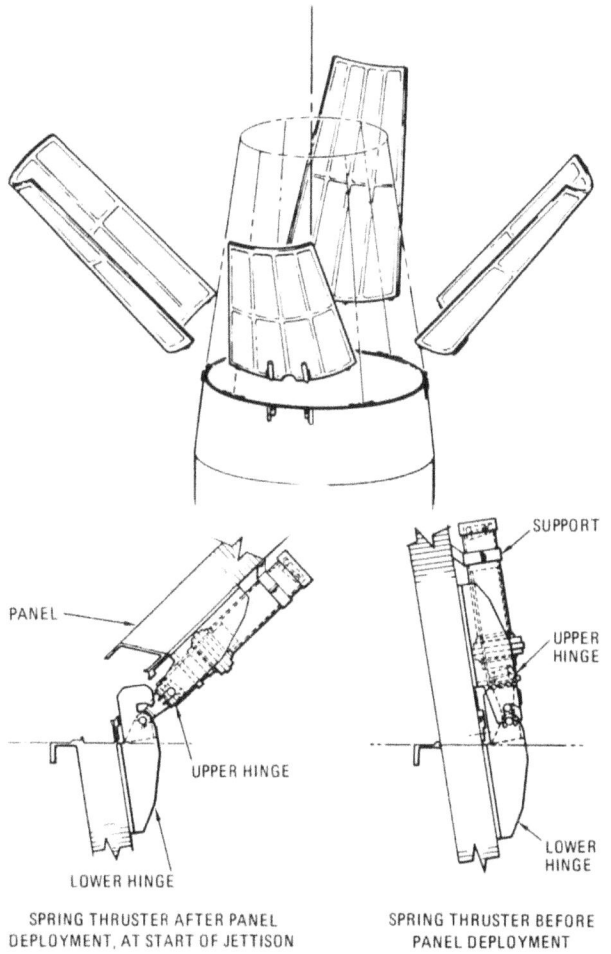

How SLA panels are jettisoned

Technicians at Tulsa put the finishing touches on SLA forward panel

CREW

Crew equipment includes all that provided for the protection, comfort, and assistance of the crew, as well as that for routine functions such as eating, sleeping, and cleansing. This section includes seven subsections: clothing, food and water, couches and restraints, hygiene equipment, operational aids, medical supplies, and survival equipment.

CLOTHING

The number of items a crewman wears varies during a mission. There are three basic conditions: unsuited, suited, and extravehicular. Unsuited, the crewman breathes the cabin oxygen and wears a bioinstrumentation harness, a communication soft hat, a constant wear garment, flight coveralls, and booties.

Under the space suit, the astronaut wears the bioinstrumentation harness, communications soft hat, and constant-wear garment. The extravehicular outfit, designed primarily for wear during lunar exploration, includes the bioinstrumentation harness, a fecal containment system, a liquid-cooled garment, the communication soft hat, the space suit, a portable life support system (backpack), an oxygen purge system, a thermal meteoroid garment, and an extravehicular visor. The space suit and the extravehicular equipment are described in the Space Suit section.

The bioinstrumentation harness has sensors, signal conditioners, a belt, and wire signal carriers. These monitor the crewman's physical condition and relay the information to the spacecraft's communications subsystem. This information is telemetered to the ground throughout the lunar mission. The sensors are attached to the crewman's skin and routed to signal conditioners in the biomedical belt. The belt is cloth, with four pockets, and snaps onto the constant wear garment. The sensors, which monitor heart beat and respiration, are silver chloride electrodes applied with paste and tape; at least four are worn by each astronaut.

The constant-wear garment is an undergarment for the suit and flight coveralls. Made of porous cloth, it is a one-piece short-sleeved garment with feet similar to long underwear. It is zippered from waist to neck and has openings front and rear.

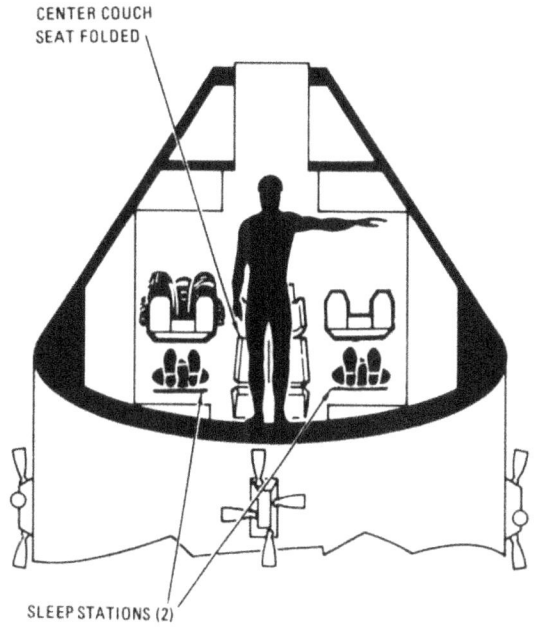
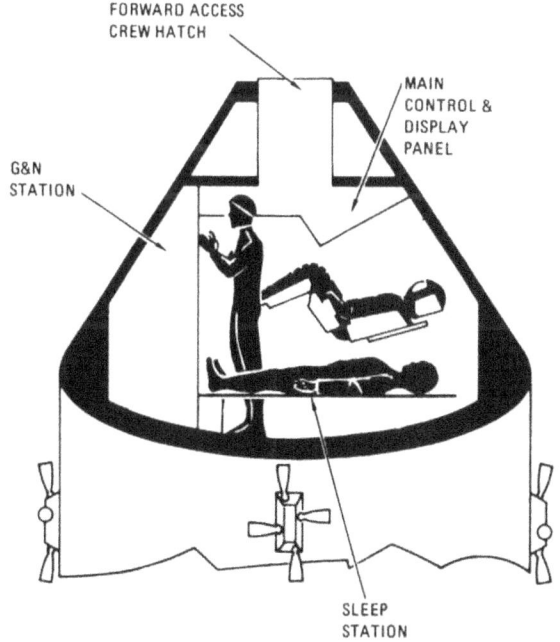

Crew stations

for personal hygiene. Pockets at the ankles, thighs, and chest hold passive radiation dosimeters. Spare garments are stowed on the aft bulkhead.

The flight coverall is the basic outer garment for

69

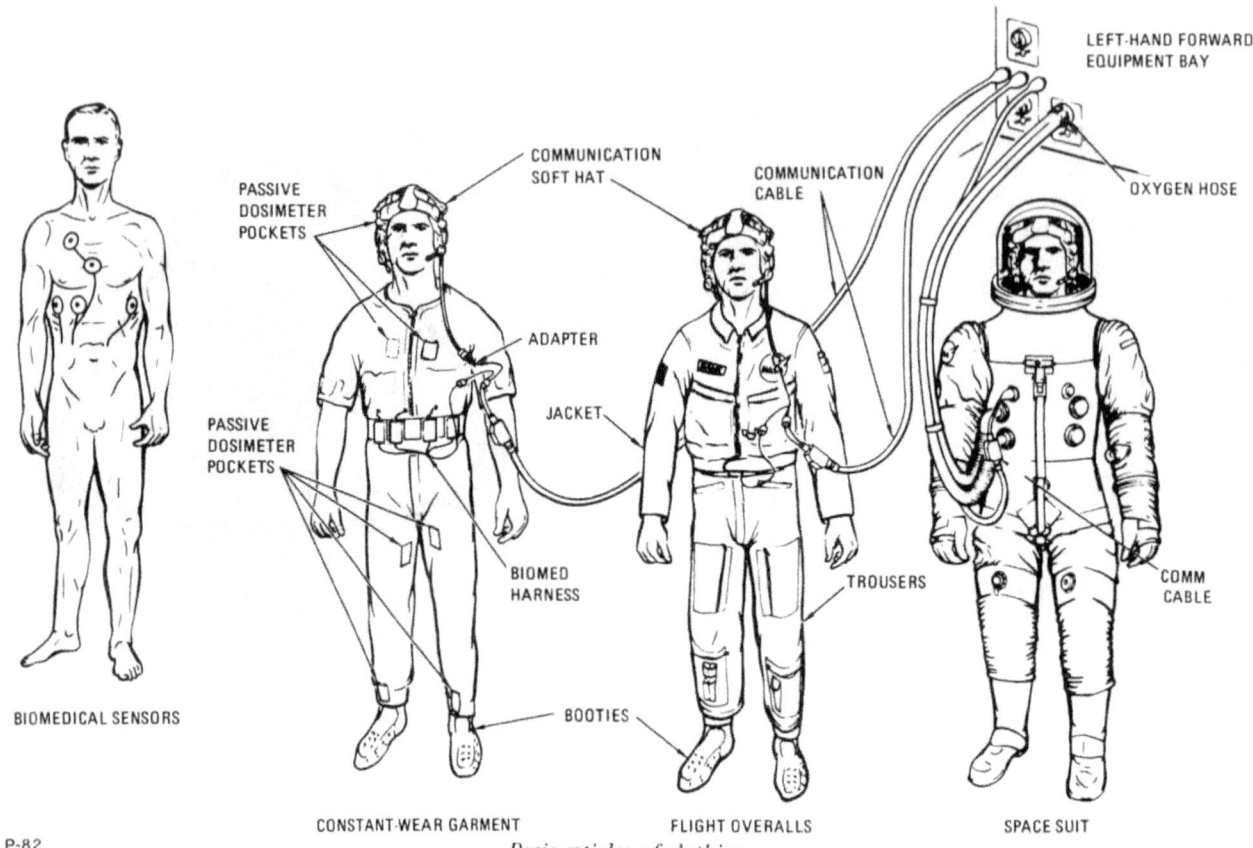

Basic articles of clothing

unsuited operations. It is a two-piece Beta cloth garment with pockets on the shins and thighs for personal equipment.

The communication soft hat is worn when suited. It has two earphones and two microphones, with voice tubes on two mounts that fit over the ears. Three straps adjust the hat for a snug fit and a chin strap is used to hold it on. A small pocket near the right temple holds a passive radiation dosimeter. An electrical cable runs from the hat to the communications cable. The lightweight headset is worn when crewmen are not in their suits.

Booties worn with the flight coveralls are made of Beta cloth with Velcro hook material bonded to the soles. During weightlessness, the Velcro hook engages Velcro pile patches attached to the floor to hold the crewmen in place.

FOOD AND WATER

Food supplies, furnished by NASA, are designed to supply each astronaut with a balanced diet of approximately **2800 calories** a day. The food is either freeze-dried or concentrated and is carried in vacuum-packaged plastic bags. Each bag of freeze-dried food has a one-way valve through which water is inserted and a second valve through which the food passes. Concentrated food is packaged in bite-size units and needs no reconstitution. Several bags are packaged together to make one meal bag. The meal bags have red, white, and blue dots to identify them for each crewman, as well as labels to identify them by day and meal.

The food is reconstituted by adding hot or cold water through the one-way valve. The astronaut kneads the bag for about 3 minutes. He then cuts the neck of the bag and squeezes the food into his mouth. After use, a germicide tablet attached to the bag is dropped through the mouthpiece; this prevents fermentation and gas. Empty bags are rolled as small as possible, banded, and returned to the food stowage drawer.

The two food stowage compartments (in the left-hand and lower equipment bays) have 5072 cubic inches of space, enough to store food for about 14 days. The food is prepared by the

Pillsbury Co. and packaged by the Whirlpool Corp.

Drinking water comes from the water chiller to two outlets: the water meter dispenser and the food preparation unit. The dispenser has an aluminum mounting bracket, a 72-inch coiled hose, and a dispensing valve unit in the form of a button-actuated pistol. The pistol barrel is placed in the mouth and the button pushed for each half-ounce of water. The meter records the amount of water drunk. A valve is provided to shut off the system in case the dispenser develops a leak or malfunction.

Food preparation water is dispensed from a unit which has hot (150°F) and cold (50°F) water. The water is dispensed in 1-ounce amounts by two syringe-like valves and a nozzle. The nozzle cover is removed and the food bag valve pushed over the nozzle. The syringe valve (either hot or cold) then is pulled as many times as needed.

Cold water comes directly to the unit from the water chiller. Hot water is accumulated in a 38-ounce tank which contains three heaters that keep the water at 150°.

COUCHES AND RESTRAINTS

The astronauts spend much of their time in their couches. These are individually adjustable units made of hollow steel tubing and covered with a heavy, fireproof fiberglass cloth (Armalon). The couches (produced by Weber Aircraft Division of Walter Kidde and Co., Burbank, Calif.) rest on a head beam and two side stabilizer beams supported by eight attenuator struts (two each for the Y and Z axes and four for the X axis) which absorb the impact of landing.

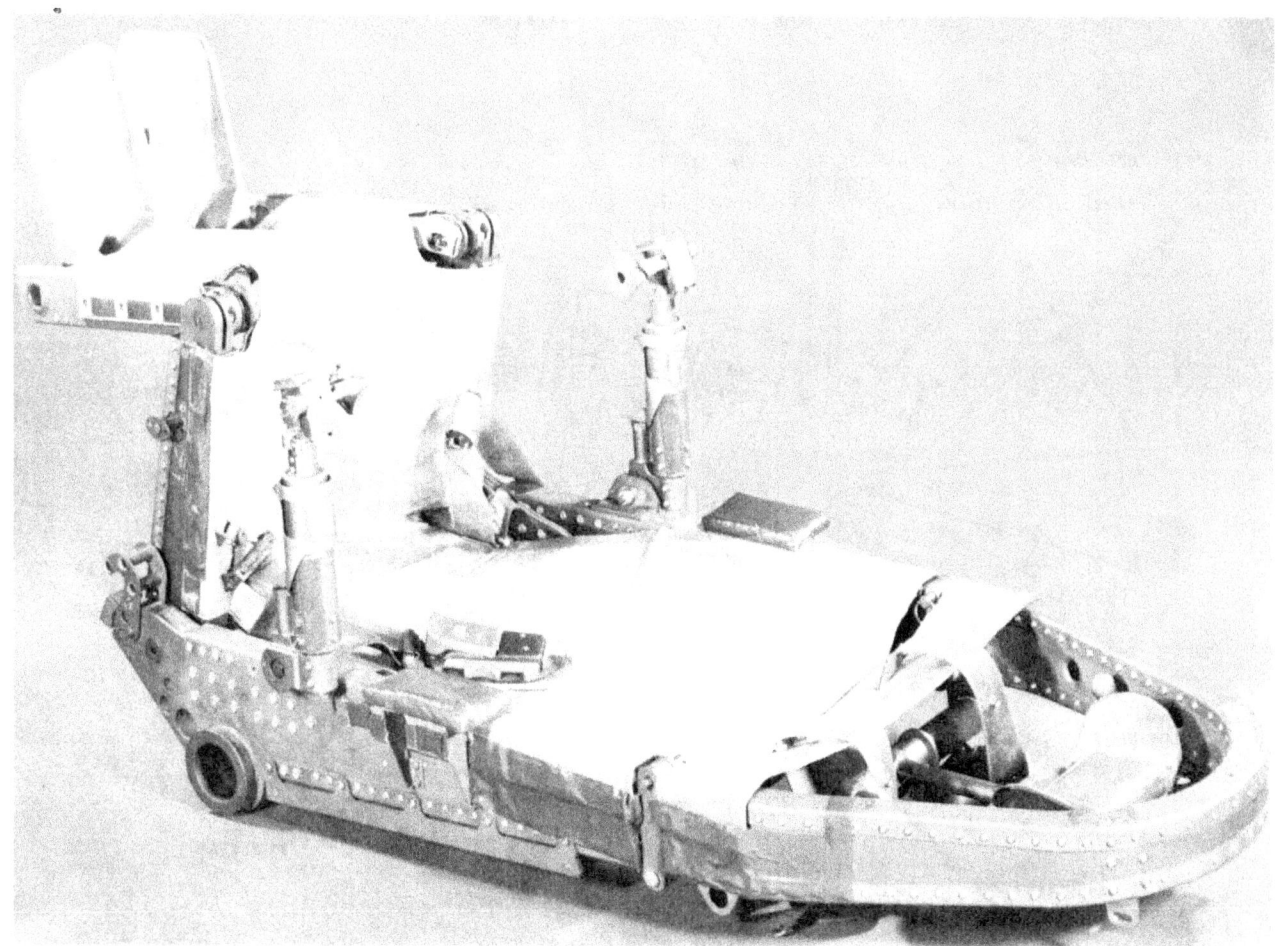

Couch

The couches — called foldable — support the crewmen during acceleration and deceleration, position the crewmen at their duty stations, and provide support for translation and rotation hand controls, lights, and other equipment. A lap belt and shoulder straps are attached to the couches.

The couches can be folded or adjusted into a number of seat positions. The one used most is the 85-degree position assumed for launch, orbit entry, and landing. The 170-degree (flat-out) position is used primarily for the center couch, so that crewmen can move into the lower equipment bay. The armrests on either side of the center couch can be folded footward so the astronauts from the two outside couches can slide over easily. The hip pan of the center couch can be disconnected and the couch pivoted around the head beam and laid on the aft bulkhead (floor) of the CM. This provides room for the astronauts to stand and easier access to the side hatch for extravehicular activity.

The three couches are basically the same. The head rest can be moved 6-1/2 inches up and down to adjust for crewman height. Two armrests are attached to the back pan of the left couch and two armrests to the right couch. The center couch has no armrests. The translation and rotation controls can be mounted to any of the four armrests. A support at the end of each armrest rotates 100 degrees to provide proper tilt for the controls. The couch seat pan and leg pan are formed of framing and cloth, and the foot pan is all steel. The foot pan contains a boot restraint device which engages the boot heel and holds it in place.

The couch restraint harness consists of a lap belt

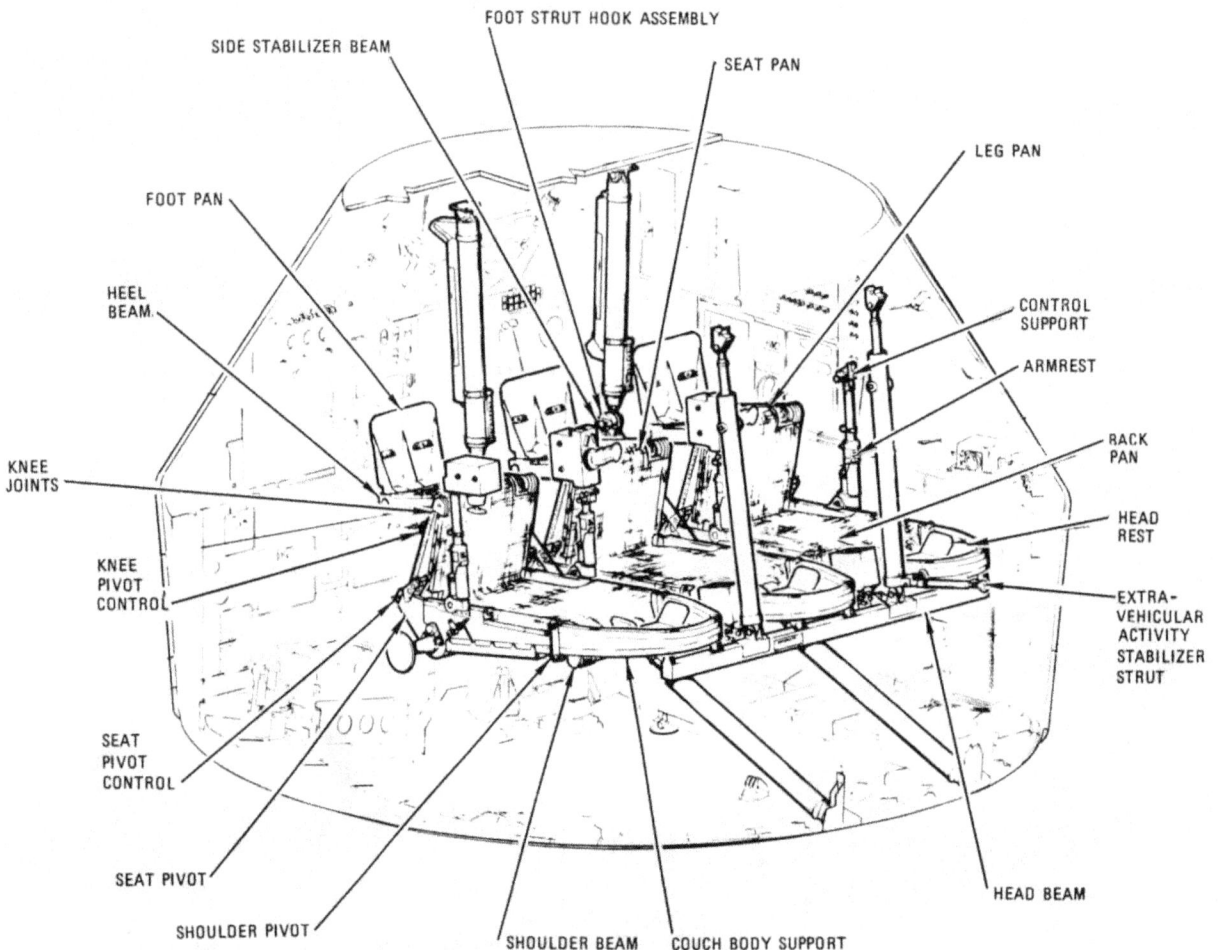

Foldable couch

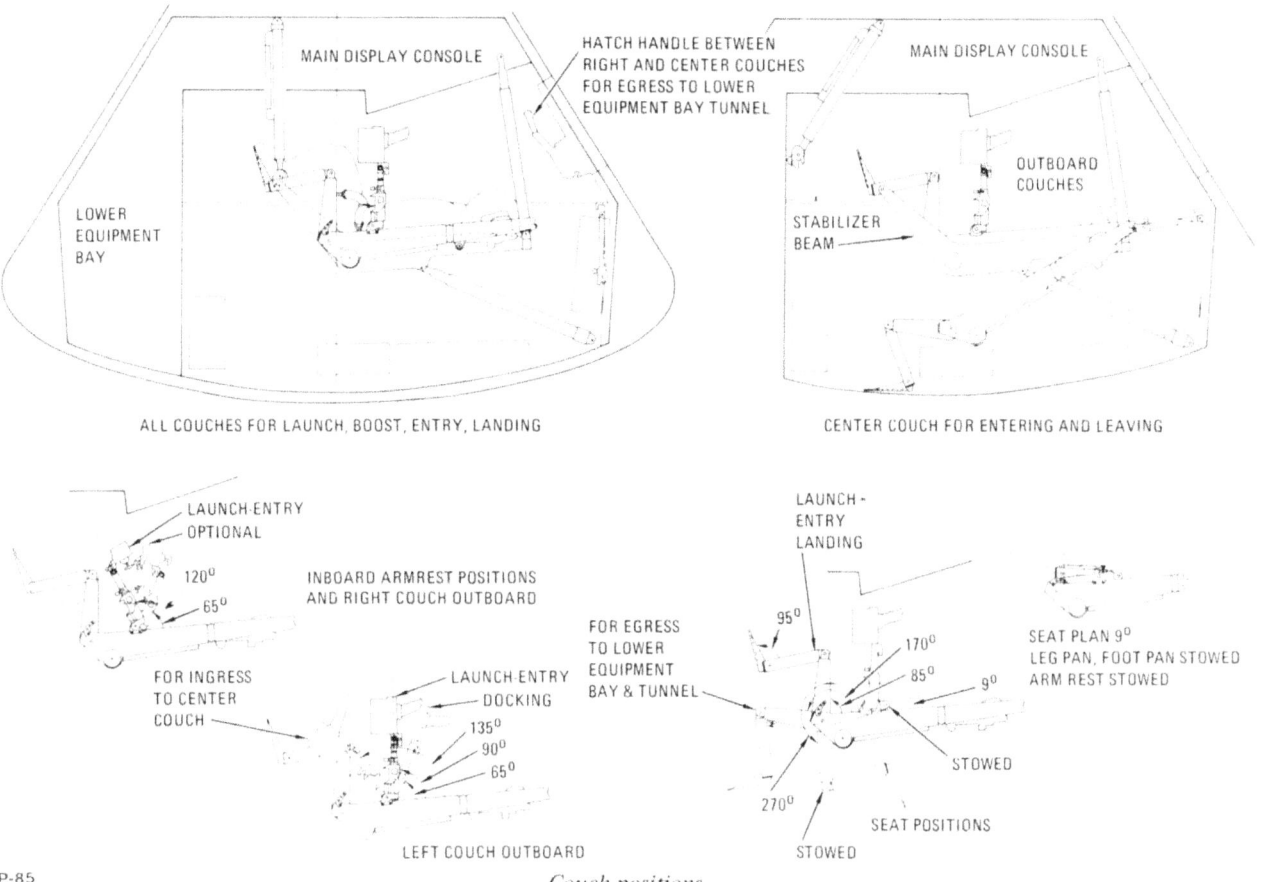

Couch positions

and two shoulder straps which connect to the lap belt at the buckle. The shoulder straps connect to the shoulder beam of the couch. The lap belt buckle is a lever-operated, three-point release mechanism. By pulling a lever, the shoulder straps and right-lap belt strap will be released. The strap ends and buckle have button snaps which are fastened to mating snaps on the controllers and struts to prevent them from floating during zero gravity.

Other restraints in the CM include handholds, a hand bar, hand straps and patches of Velcro which hold crewmen when they wear sandals.

Two aluminum handholds are by the side windows close to the main display console. The hand bar on the main display console near the side hatch helps crewmen move through the hatch and move couches to the locking position. The hand bar can be stowed or extended.

The hand straps, made of Fluorel attached by brackets at each end, serve as a maneuvering aid during weightlessness. There are five hand straps behind the main display console and one on the left-hand equipment bay.

The astronauts sleep in bags under the left and right couches with the head toward the hatch. The two sleeping bags are lightweight Beta fabric 64 inches long, with zipper openings for the torso and 7-inch-diameter neck openings. They are supported by two longitudinal straps that attach to storage boxes in the lower equipment bay and to the CM inner structure. The astronauts sleep in the bags when unsuited and on top of the bags when they have the space suit on.

HYGIENE EQUIPMENT

Hygiene equipment includes wet and dry cloths for cleaning, towels, a toothbrush, and the waste management system. The cloths are 4 inches square and come in sealed plastic bags packaged with the food. The wet cloths are saturated with a germicide and water. Twelve-inch square towels similar to a washcloth are stowed under the left couch.

The urine subsystem contains urine collection bags, a 100-inch flexible hose (capable of reaching the crewman in a couch) with a 3/8-inch suit urine valve quick-disconnect, and controls. The urine collection bag connects to the flexible hose at one end and to a hold-on cuff at the other. The rubber cuff is rolled onto the penis. The flexible hose can withstand a 5-psia differential pressure.

The fecal subsystem has 30 bag assemblies and tissue dispensers. The bag assemblies have inner and outer fecal-emesis bags with pouches containing a germicide and a skin-cleaning towel. The rim of the inner bag is covered with a cement-like material covered with a thin plastic. For use, the plastic is peeled off and the bag "pasted" to the buttocks. After use, the germicide is inserted in the inner bag and the latter is sealed in the outer bag and kneaded. The sealed bags are stowed in the waste disposal compartment; a split membrane trap prevents the bags from floating through the door into the cabin when the door is open. Tissue dispensers contain tissue (Kleenex) for wiping and are stored in a container under the center couch.

OPERATIONAL AIDS

These include data files, tools, workshelf, cameras, fire extinguishers, oxygen masks, and waste bags.

The crew has a data file containing checklists, manuals, and charts needed for operation of the spacecraft. At launch the data file is stored in a compartment in the right-hand forward equipment bay. Data books used by the LM pilot are stored in

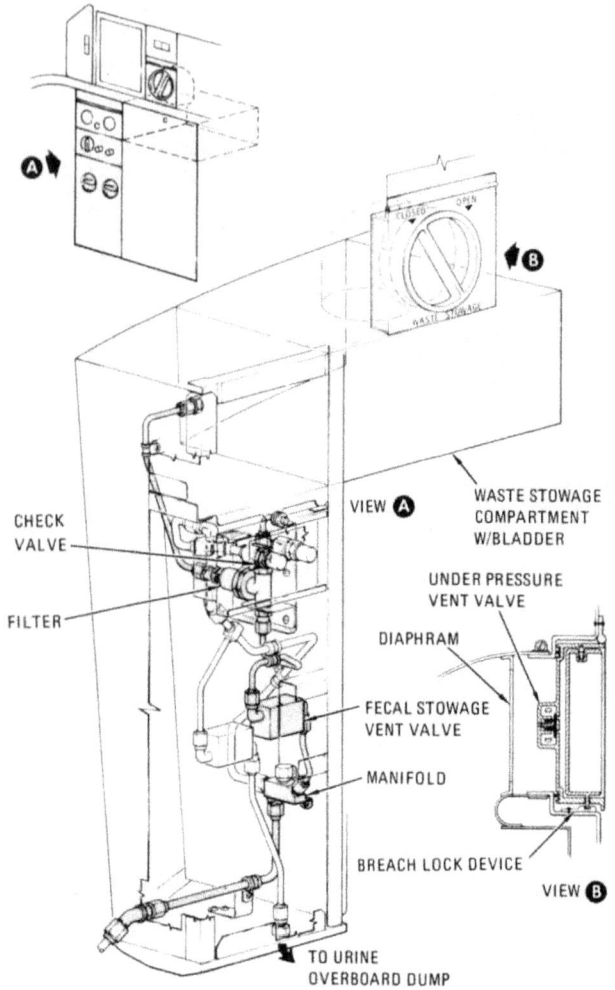

Waste stowage vent system

The waste management system controls and disposes of waste solids, liquids, and gases. The major portion of the system is in the right-hand equipment bay. The system stores feces, removes odors, dumps urine overboard, and removes urine from the space suit.

Urine, oxygen, and fecal odors, as well as emergency relief of fluids from the CM batteries and excess water from the water system, are routed overboard through the water/urine dump line. A small (0.055-inch) nozzle restricts gas flow to 0.4 cubic feet per minute and liquid flow to 1 pound per minute. The limited gas flow prevents excessive loss of cabin atmosphere during fecal canister use. The liquid flow is restricted to prevent the formation of ice in the nozzle. Redundant heaters at the nozzle also help prevent ice.

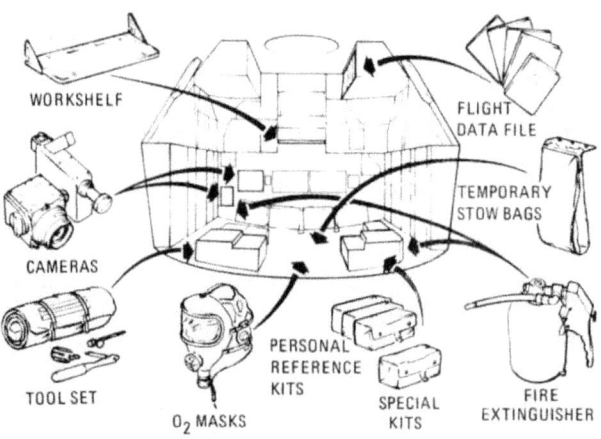

Operational aids

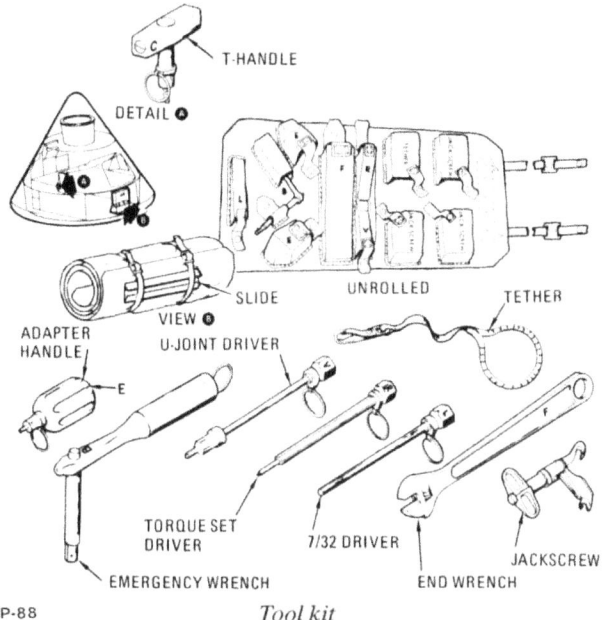

Tool kit

mately 30 seconds at high pressure. The expelling agent is Freon. Safety features prevent sparking and over-heating. Fire ports are located at various panels so that the extinguisher's nozzle can be inserted to put out a fire behind the panel. The fire extinguishers are produced by Southwest Research Co., San Antonio, Tex.

a container at his right side during flight. The files contain such things as the flight plan, mission log, landmark maps, star charts, and subsystem data.

The tool set holder is a synthetic cloth pouch which contains pockets for a number of tools. It rolls for stowage (on the aft bulkhead) and has snaps on the back so it can be attached to the CM structure. Among the tools in the set are an adjustable wrench, an adapter handle, hatch securing tools, two drivers, and a 20-inch tether. Placards throughout the CM indicate which tools are to be used and the direction of rotation.

The astronauts will have still and movie cameras in the CM. The movie camera is a 16mm Maurer sequence camera that operates at 6 frames per second, 1 frame per second, single frame, and time exposure. Accessories include 18 and 5mm lenses and a right-angle mirror. The camera is powered by the spacecraft's dc system and can be mounted to the CM structure. The still camera is a 70mm Hasselblad. It is hand-held and manually operated, and has a ring sight and an 80mm lens. An exposure meter and a spotmeter also are provided.

The CM has one fire extinguisher, located adjacent to the left-hand and lower equipment bays. The extinguisher weighs about 8 pounds, is about 10 inches high, and has a 7-inch nozzle and handle. The tank body is a stainless steel cylinder with a dome. The extinguishing agent is an aqueous gel expelled in 2 cubic feet of foam for approxi-

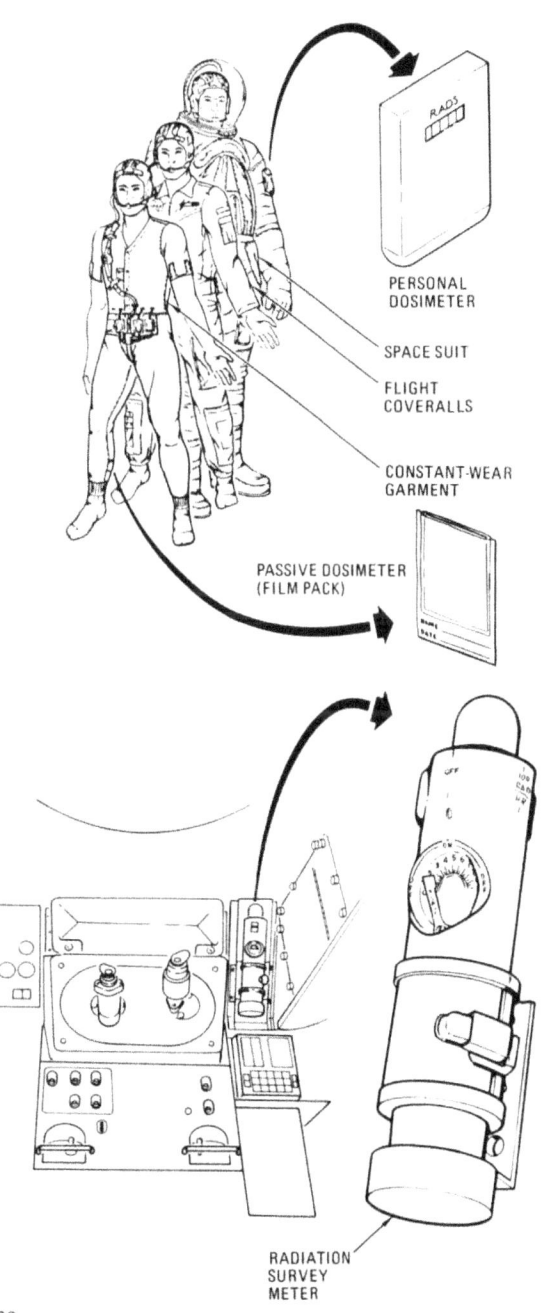

Radiation measuring and monitoring units

Oxygen masks are provided for each astronaut in case of smoke, toxic gas, or other hostile atmosphere in the cabin while the astronauts are out of their suits. The masks are a modified commercial type with headstraps. Oxygen is supplied through a flexible hose from the emergency oxygen/repressurization unit in the upper equipment bay. The masks are stowed in a cloth bag in the aft equipment bay below the emergency oxygen unit.

MEDICAL SUPPLIES

Medical equipment aboard the command module includes monitoring devices and emergency supplies.

The sensors attached to each crewman are the principal monitoring devices. These sensors are connected to signal conditioners which fit into pockets on the bioinstrumentation belt. The signal conditioners (somewhat smaller than a cigarette package and weighing about 2 ounces) amplify the low-level signals from the sensors, and transmit them to an electrocardiograph and an impedance pneumograph (which measures respiration rate).

A number of devices are used to monitor radiation level. These include passive dosimeters, personal radiation dosimeters, a radiation survey meter, a Van Allen Belt dosimeter, and a nuclear particle detection system.

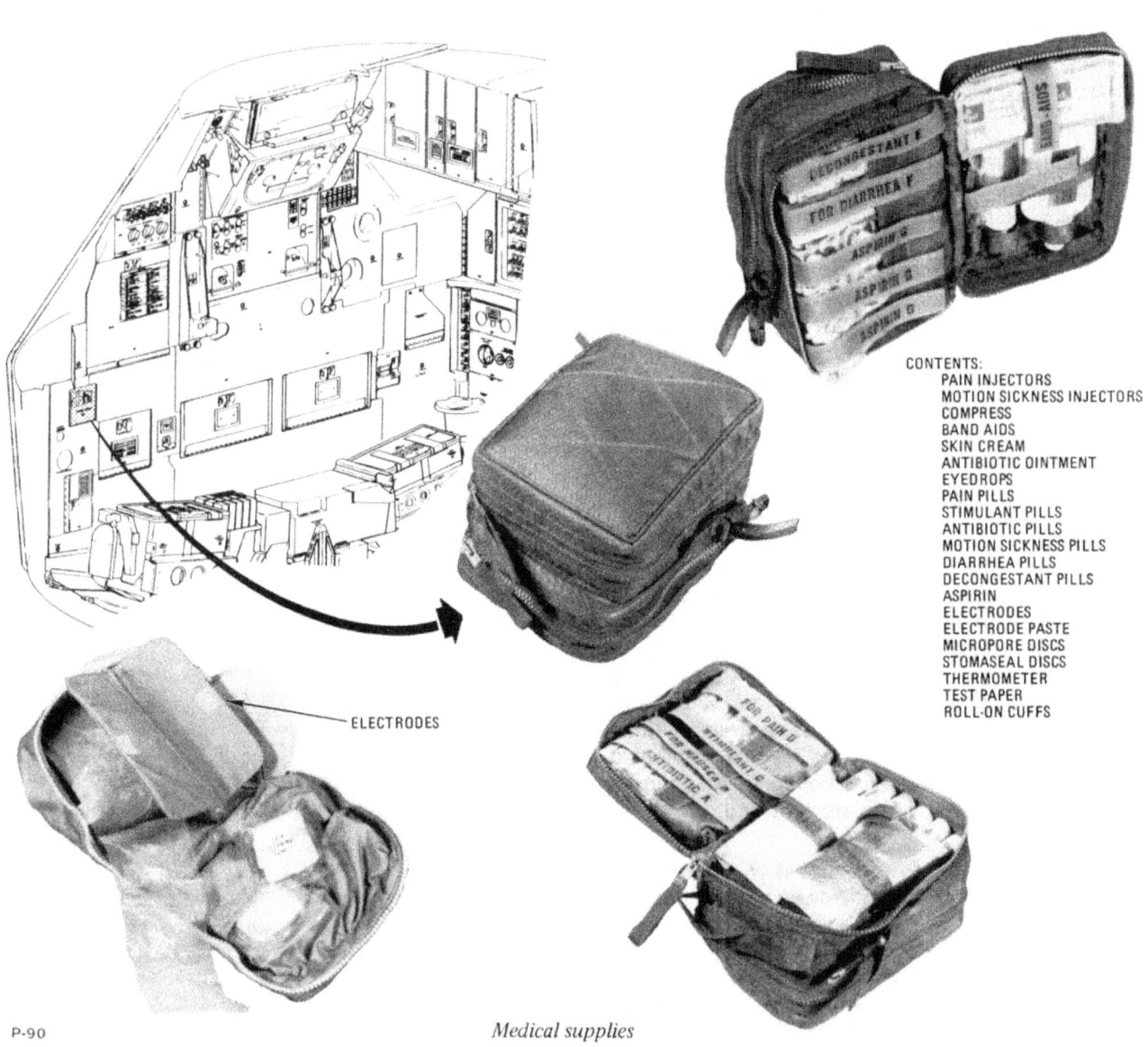

CONTENTS:
PAIN INJECTORS
MOTION SICKNESS INJECTORS
COMPRESS
BAND AIDS
SKIN CREAM
ANTIBIOTIC OINTMENT
EYEDROPS
PAIN PILLS
STIMULANT PILLS
ANTIBIOTIC PILLS
MOTION SICKNESS PILLS
DIARRHEA PILLS
DECONGESTANT PILLS
ASPIRIN
ELECTRODES
ELECTRODE PASTE
MICROPORE DISCS
STOMASEAL DISCS
THERMOMETER
TEST PAPER
ROLL-ON CUFFS

Medical supplies

Each crewman wears four passive dosimeters (film packs) in pockets at the temple, chest, thigh, and ankle. These register total radiation dosage and are processed on the ground after the mission. Each crewman also wears a personal radiation dosimeter which is battery-powered and about the size of a cigarette package; it has a readout which indicates the total dosage received during the mission.

The radiation survey meter determines the magnitude of the immediate radiation field; it is flashlight-like about 10 inches long and 2 inches in diameter. It is battery-operated and is clamped to a bracket near the guidance and navigation station in the lower equipment bay.

The Van Allen Belt dosimeter measures dose rates to the skin and blood-forming organs (depth dose measurement). It consists of two individual dosimeters (skin and depth) which have ionization chambers as sensors; the measurements of the sensors are telemetered to the ground. This is mounted in the structure of the CM near the hatch.

The nuclear particle detection system (produced by Philco Corp's Western Development Laboratory, Palo Alto, Calif.) measures proton and alpha particle rates and telemeters the information to the ground. It is normally located in the area between the command and service modules that is enclosed by the SM-CM fairing.

Medical supplies are contained in an emergency medical kit, about 7 by 5 by 5 inches, which is stored in the lower equipment bay. It contains oral drugs and pills (pain capsules, stimulant, antibiotic, motion sickness, diarrhea, decongestant, and aspirin), injectable drugs (for pain and motion sickness), bandages, topical agents (first aid cream, sun cream, and an antibiotic ointment), and eye drops.

SURVIVAL EQUIPMENT

Survival equipment, intended for use in an emergency after earth landing, is stowed in two rucksacks in the right-hand forward equipment bay. The rucksacks are 18 inches long, 6 inches wide, and 6 inches deep and weigh about 54 pounds.

One of the rucksacks contains a three-man rubber life raft with an inflation assembly,

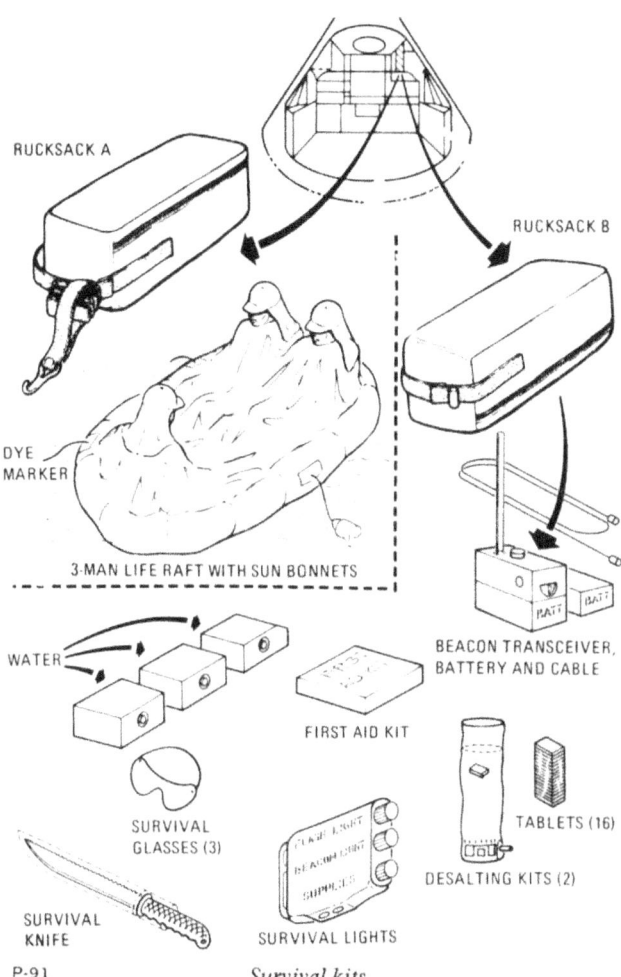

Survival kits

carbon-dioxide cylinder, a sea anchor, dye marker, and a sunbonnet for each crewman.

The other rucksack contains a beacon transceiver, survival lights, desalter kits, machete, sun-glasses, water cans, and a medical kit.

The UHF beacon transceiver (manufactured by Sperry Phoenix Co.) is a hand-held battery-powered radio tuned permanently to a VHF frequency of 243 megacycles. The receiver-transmitter and battery pack form a watertight assembly about 8 by 4-1/2 by 3 inches; a tapered, flexible steel tape antenna can be extended to 11-1/2 inches. The transceiver unit can be used for voice communications through a speaker and microphone or as a beacon, in which case it will transmit an intermittent signal for up to 24 hours. A spare battery and a spacecraft connector cable are provided.

The survival lights are contained in a waterproof, three-in-one device. The unit has a flashlight, a strobe light for night signaling, and a waterproof compartment containing fish hook, line, a sparky kit (striker and pith balls), needle and thread, and a whistle. The top of the unit is a compass with a folding signal mirror on one side.

The two desalter kits contain a process bag, tablets, and bag repair tape. The bag is plastic with a filter at the bottom and holds about a pint. Sea water is put into the bag and a tablet added; after about an hour the water can be drunk through a valve on the bottom.

The machete (protected with a cloth sheath) is very thin with a razor edge on one side and a saw edge on the other. The three sunglasses are polarized plastic sheet with a gold coating that reflects heat and radio waves. The three water cans are aluminum, hold a little more than a half-gallon each, and have a drinking valve.

The survival medical kit contains the same type of supplies as the emergency medical kit: 6 bandages, 6 injectors, 30 tablets, and one tube of all-purpose ointment.

EQUIPMENT

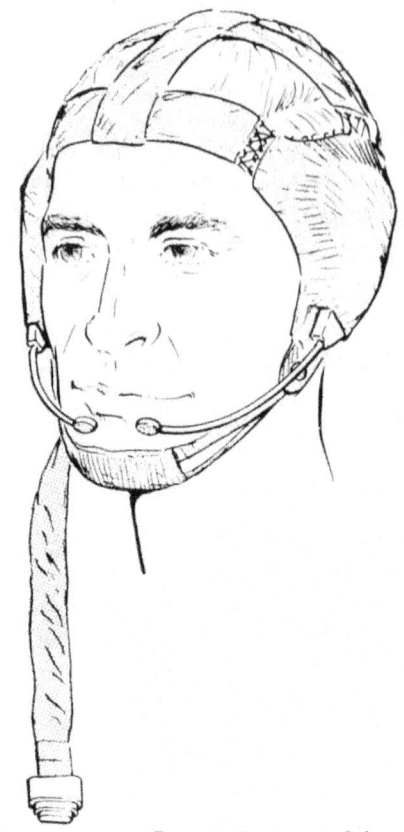

P-92 *Communications soft hat*

Bioinstrumentation Harness — Sensors are attached to bodies of each crewman with paste and tape to monitor heartbeat (electrocardiogram) and respiration (impedance pneumograph). The assembly includes three signal conditioners (cigarette package size) and accessories located in the medical accessories kit.

Cameras — There are a 16-mm Maurer sequence camera and a 70-mm Hasselblad camera. Accessories for the Maurer are an 18-mm lens, a right angle mirror, power cable and film magazines. The camera weighs about 1.9 pounds and each magazine, 1/2 pound. Ten to 15 magazines are carried. It is electrically operated on 28 volts dc. Normally, it is mounted on the right rendezvous window frame. There are two electrical outlets in the cabin from which it can be used as a hand-held camera to take sequential pictures of crew activity. It is also used to show docking and rendezvous movies and any other sequential operation. It can be operated at speeds of six or one frame per second, or for test, single frame or time exposure. The 70-mm Hasselblad is hand-held and has an 80-mm Zeiss planar lens and a ring sight. The camera weighs 1.9 pounds and each of five cassettes weighs 1.59 pounds. Each cassette has 150 exposures. This camera will be used for high-resolution photography. It will verify landmark tracking; record third-stage Saturn separation; photograph disturbed weather regions (hurricanes, typhoons etc.) debris collection on spacecraft windows, other equipment in space, the lunar module during rendezvous and docking, the terrain of the moon, and act as a back up to the 16-mm camera. The cameras are stowed in the lower equipment bay.

Communications Soft Hat — It has two earphones and two microphones, with voice tubes, on two mounts that fit over the ears. It is made of cloth and plastic. There are three straps attached to the mounts with laces for individual fitting. A chin strap secures it to the head. A small pocket on the inside near the right temple holds a passive dosimeter film packet. An electrical cable with a 21-socket connector will connect to the constant-wear garment adapter or the space suit. The hat is worn at all times for the purpose of communications.

<u>Constant Wear Garment</u> (David Clark Co.) — This is a 13-ounce undergarment for the space suit. It is a porous cotton cloth, one-piece garment similar to long underwear. It has a zipper from the waist to the neck and openings for urination and defecation. There are snaps at the midsection to attach the biomedical belt with signal conditioners, and pockets for film packet passive dosimeters at the ankles, thighs, and chest. The garment also has integral socks. Each crewman will wear one garment and will have another garment stowed in a locker on the aft bulkhead bay.

<u>Couches</u> (Weber Aircraft, Burbank, Calif.) — The three crew couches (left, center, and right) are made of steel framing and tubing and covered with Armalon, a heavy glass fiber cloth impregnated with Teflon. The couches are clamped together and weigh 280 pounds. They are suspended and supported by eight attenuator struts, and each couch is adjustable. The backpan of each couch is 32 by 22 inches and is concave. Headrests can be moved 6-1/4 inches up and down to adjust for crewman height. The hiprest and upper legrest form the seat of the seatpan. The seatpan also has a footrest, which engages the boot heel and holds it in place. The couch seats are adjustable to a number of positions, but the ones used most are 85 degrees (for launch, orbit, entry, and landing), 170 degrees (when crewmen want to get to lower equipment bay), and 11 degrees (when crewmen want to get to right, left, and lower equipment

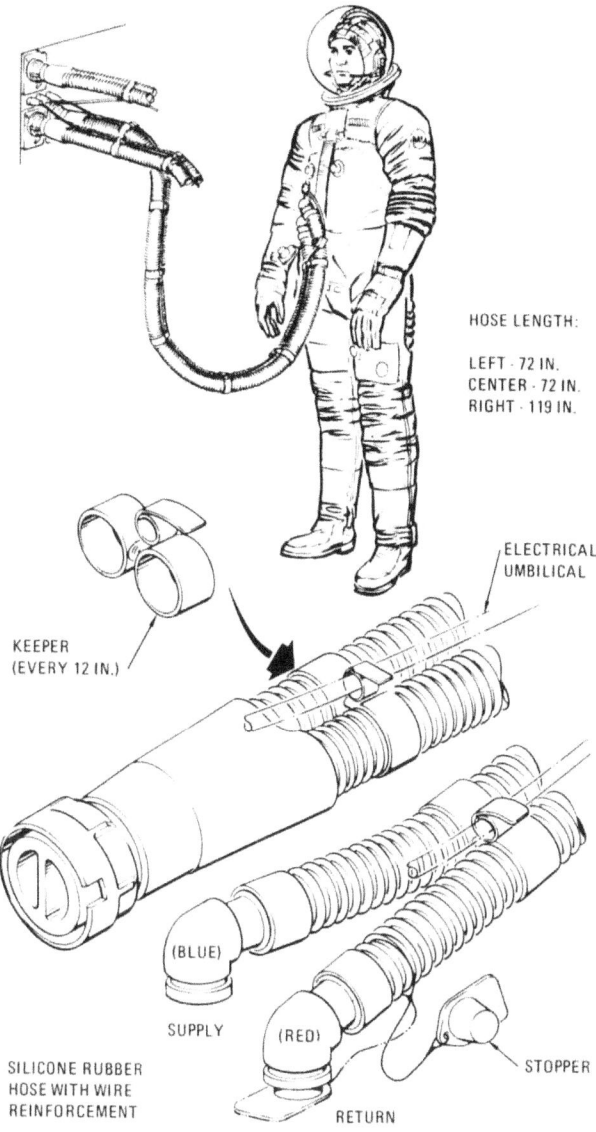

Oxygen hose assembly

bay stowage areas). It can also be adjusted to 270 degrees for lower equipment accessibility. Positions are indicative of the angle of the seatpan (which is movable) to the backrest.

<u>Crewman Umbilical Assembly</u> — Cable and hose assembly is connected to the space suit — the cable for electrical power for communications equipment and the hose for oxygen. There are three oxygen hoses made of Fluorel, or silicon rubber in a fiberglass cloth sock. Two of the hoses are 72 inches long with a diameter of 1-1/4 inches and weigh 5.3 pounds each. The third, used for transfer to the lunar module, is

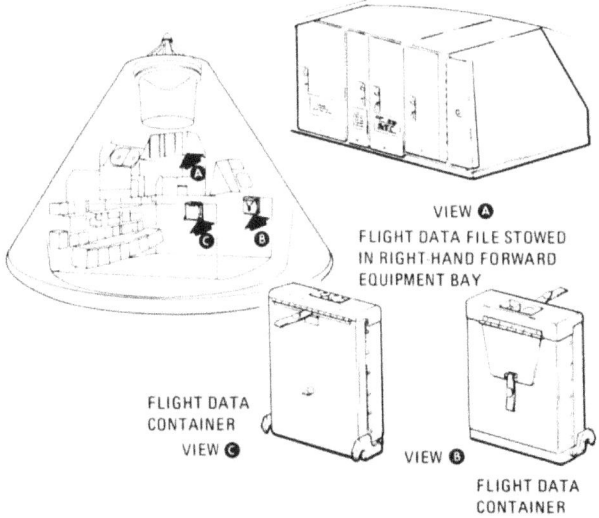

Flight data file

119 inches long with a diameter of 1-1/4 inches and weighs 8.2 pounds. The communications cables are Fluorel-covered and consist of a cable and control head. Two are 74 inches long with a 3/8-inch diameter and weigh 2.5 pounds each. Two others, one of which is a spare, are 121 inches long with a 3/8-inch diameter and weigh 3.5 pounds each.

Data Files — There are 20 volumes, containing all reference data for the mission, including checklists, light plans, photo logs, star charts, lunar and earth landmark maps, orbital map, malfunction procedures, and crew logs. They weigh a total of 20 pounds and are stored in the right-hand forward equipment stowage compartment.

Emergency Oxygen Mask and Hose (Darling Co.) — Each crewman has a mask to wear in case of smoke, toxic gas, or other hostile environment when he is in shirtsleeves. Total weight for masks and hoses is 4 pounds. The masks are stowed in a Beta cloth bag on the aft bulkhead below the emergency oxygen/repressurization unit.

Fire Extinguisher (Southwest Research Co.) — There is one extinguisher stowed in the lower equipment bay beside a CO_2 absorber container. It weighs about 8 pounds, is 10 inches high, and has a 7-inch flexible nozzle and a handle both of which are insulated. It is a stainless steel cylinder with dome containing 2 cubic feet of aqueous gel foam at 6 pounds per square inch. It is activated by pulling a pin and pressing a button.

Flight Coveralls — This is a two-piece Beta Cloth garment with a pair of booties. The coveralls and booties weigh 3 pounds. The coveralls are worn over the constant wear garment and provide additional warmth as well as stowage for miscellaneous personal equipment while in a shirtsleeve environment. The booties are cloth fabric boots with snaps at the ankles. They have Velcro patches at the balls of the feet.

Medical Accessories Kit — Beta cloth bag, about 7 by 5 by 5 inches, has two zippers for access to oral drugs and pills, injectable drugs, dressings, creams and ointments, and bioinstrumentation accessories. The kit weighs 2.5 pounds and is in the lower equipment bay.

Nuclear Particle Detection System — This consists of a detector assembly in the form of a telescope

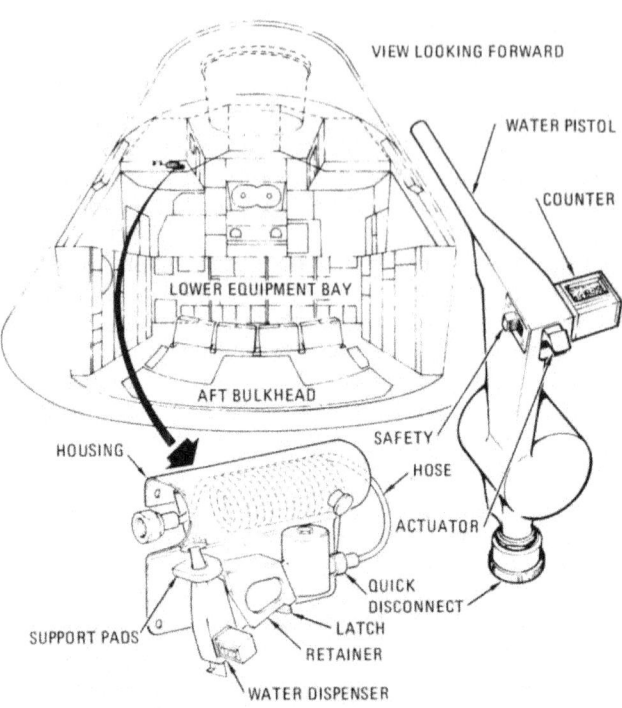

P-95 *Water metering dispenser assembly*

and a signal analyzer assembly. The pulse rates from the detector assembly at which particles enter the various energy intervals are converted to d-c voltage levels by ratemeters in the signal analyzer; the outputs of the ratemeters are then telemetered to ground. The system measures proton and alpha particle rates in four proton and three alpha differential energy bands and one integral proton energy band. The unit is in the adapter section between the command and service modules and is mounted on the forward bulkhead of the service module.

Passive Dosimeters — Each crewman wears three of the radiation-measuring devices, which are processed after recovery to determine total dosage received. They are in form film packs and are worn in constant wear garment pockets at the chest to simulate a 5-centimeter depth to measure bone marrow dosage and at the thigh and ankle to measure skin dosage. Each weighs about half an ounce.

Personal Dosimeters — Each crewman wears a battery-powered radiation dosimeter about the size of a cigarette package in the pocket of a sleeve of the space suit or flight coveralls. Its readout dial will indicate in radians accumulated dosage of radiation received by the crewman during the mission. The dosimeter weighs half a pound.

Personal Hygiene — Each crewman has a toothbrush, 4-by-4-inch wet and dry cleansing cloths, and a tube of ingestible toothpaste, items that are packaged with the food. The wet cloths are saturated with a germicide and water. Towels, 12 by 12 inches and similar to washcloths, will be packaged in containers and stowed in an aft bulkhead locker.

Pilots Preference Kits — Each crewman has an 8 by 4 by 2-inch Beta cloth bag for personal items. The bags are kept in an aft bulkhead stowage locker.

Radiation Survey Meter — It is a 1.58-pound cylinder 10 inches long and 2 inches in diameter. It measures the magnitude of the immediate radiation field. It has an on-off switch, a direct readout dial calibrated in radians per hour, and is battery operated. The meter is clamped in a bracket mounted on the guidance and navigation signal conditioning panel. It is transferred to the lunar module during crew transfer.

Restraints — Gravity-load restraints are three harnesses (one for each crewman) to hold crewmen in their couches; two handholds, and a hand bar. Each restraint harness has a lap belt and two shoulder straps connected at a lap belt buckle. The lap belt buckle is a lever-operated, three-point release mechanism. The strap ends and buckle have snap fasteners and they can be fastened to mating snaps on the couch and struts when not in use. The handholds are strong aluminum handles, one on each longeron by the side windows. The hand bar is near the side hatch and can be stowed or extended. A crewman can hold it with two hands for ingress or egress from the command module side hatch. Zero-gravity restraints are five hand straps behind the main display console, one on the left-hand equipment bay, and one on each lower x-x axis strut.

Sighting Aids — Each of the command module's five windows has an aluminum sheet window shade held on by wing latches. They are .20-inch thick with .250-inch frames and are stowed in a stowage bag in the upper equipment bay. There are two mirror subsystems: internal and external. Each couch has a 4-by-6-inch metal internal viewing mirror assembly consisting of a mounting base, a two-segmented arm, and a mirror. The mirrors for the left and right astronauts are

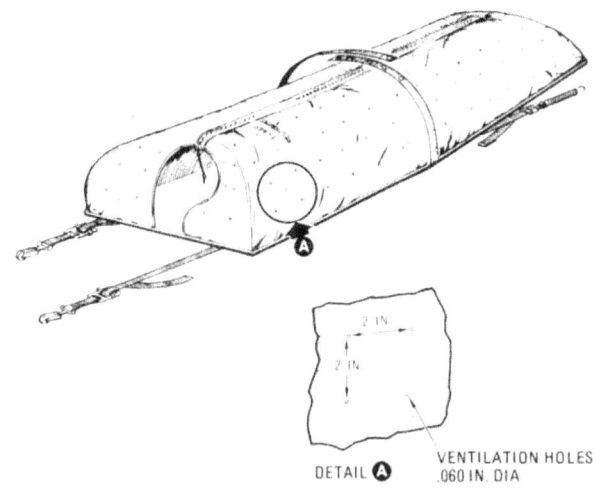

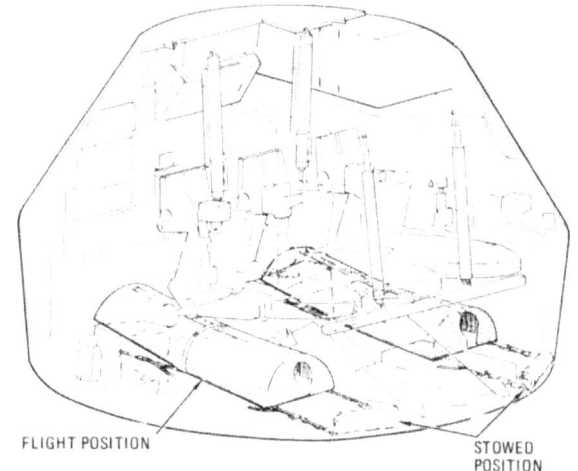

P-96 *Sleep station restraints*

mounted on the side of the lighting and audio control console above the side viewing window; they have folding joints. The center astronaut's mirror is mounted on the right-hand upper x-x axis attenuator strut. External viewing mirrors, which also are metal, consist of an upper mirror assembly mounted on the side wall near the upper rim on the right rendezvous window frame and a lower mirror assembly mounted on the right rendezvous window housing near the lower rim of the window frame. A mirror assembly is a mirror and a bracket.

Sleeping Bags — There are two Beta-cloth sleeping bags weighing 2.5 pounds each. They are 64 inches long with a 7-inch diameter neck opening when the bag is zippered closed. There is a torso zipper opening. The bags have holes for ventilation every 4 inches. The bags are under the left and right couches for sleeping. At launch, they are rolled and strapped against the upper side wall and aft bulkhead.

Survival Kit — The kit is stowed in the right-hand forward equipment bay in two rucksacks, which are Beta cloth zippered bags. Both rucksacks are 18 by 6 by 6 inches. One, which weighs 34.8 pounds, contains a VHF beacon-transceiver which is a hand-held, battery-powered radio fixed-tuned to a VHF frequency of 243 megacycles; it is manufactured by Sperry Phoenix Co. The receiver-transmitter and battery pack assemblies mate to form a water-tight 8-by-4-1/2-by-3-inch unit. The antenna is an 11-1/2-inch tapered flexible steel tape ending in a coaxial RF connector and is normally stored in a retaining spool and clipped on top of the radio. The rucksack also contains a spare battery and spacecraft connector cable; a survival light unit containing a flashlight, a strobe light for night signaling, a fish hook and line, a sparky kit (striker and pith balls), needle and thread, whistle, compass, and a folding signal mirror; plastic desalter bag with a filter at the bottom; desalter tablets; desalter bag repair tape; one machete protected with a cloth sheath (the knife has a razor edge and the back edge is a saw); three pair of sunglasses that are a polarized plastic sheet with Sierra Coat III — a gold coating that reflects heat and radio waves — to protect eyes against the sun and glare; three 5-pound capacity aluminum water cans; and medical kit with 6 Band-Aids, 6 injectors, 30 tablets, and one tube of all-purpose ointment. The other rucksack weighs 17.8 pounds and contains a three-man life raft with lanyard with an inflation assembly and CO_2 cylinder, a sea anchor, a dye marker kit, and three sunbonnets.

Temporary Stowage Bags — Each crewman has a 3-foot-by-l-foot-by-3-inch bag with inner and outer pocket. It is made of Beta cloth (fiberglass cloth). The inner bag is for temporary stowage of small items; the outer bag is for dry, uncontaminated waste. The bags are stowed in an upper equipment bay locker.

Tools — A rolled pouch contains an emergency wrench, two adapter handles, a crescent wrench, a 4-inch torque set driver, three jack screws, and a 20-inch tether. Each tool has a tether ring, and each is designated by a letter of the alphabet, except the three jack screws. The pouch and

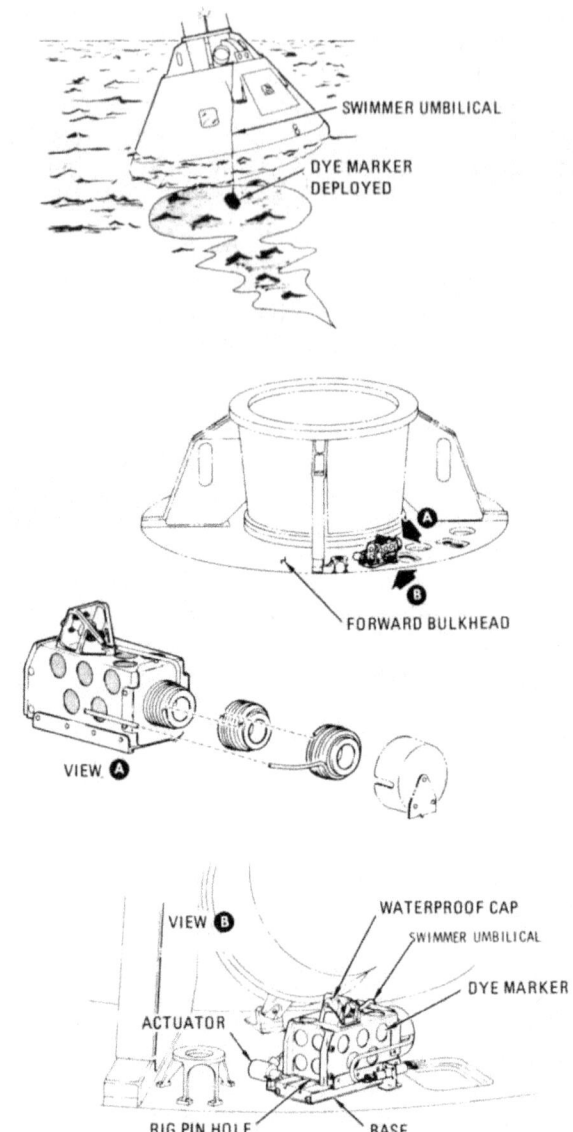

Dye marker and swimmer umbilical

tools weigh 2.19 pounds. They are stowed in a locker in the aft bulkhead.

Van Allen Belt Dosimeter — The meter is 2-1/2 by 8 by 3 inches and weighs 6 pounds. It measures dose rates to the skin and blood-forming organs. It has two dosimeters (skin and depth), which have ionization chambers as sensors. The meter is mounted in the command module between longeron No. 4 and the hatch. It has three telemetry channels and is powered by 28 volts dc.

DISPLAYS AND CONTROLS

There are hundreds of controls and displays located in the cabin of the Apollo command module. A majority of these are on the main display console, which faces the three crew couches and extends on both sides of them. The console is nearly seven feet long and three feet high, with the two wings each about three feet wide and two feet deep.

The console is the heart of the command module: on it are the switches, dials, meters, circuit breakers, and other controls and displays through which the three-man crew will control the spacecraft and monitor its performance. Crew members can see and operate controls on the console while in their restraint harnesses.

Other displays and controls are placed throughout the cabin in the various equipment bays and on the crew couches. In general, these are controls and displays that do not need frequent attention or are used during parts of the mission when crewmen can be out of the couches. Most of the guidance and navigation equipment is in the lower equipment bay, at the foot of the center couch. This equipment, including the sextant and telescope, is operated by an astronaut standing and using a simple restraint system. The non-time-critical controls of the environmental control system are located in the left-hand equipment bays, while all the controls of the waste management system are on a panel in the right-hand equipment bay. The rotation and translation controllers used for attitude, thrust vector, and translation maneuvers are located on the arms of two crew couches. In addition, a rotation controller can be mounted at the navigation position in the lower equipment bay.

The main display console has been arranged to provide for the expected duties of crew members. These duties fall into the categories of commander, CM pilot, and LM pilot, occupying the left, center, and right couches, respectively. The CM pilot, in the center couch, also acts as the principal navigator.

While each astronaut has a primary responsibility, each Apollo crewman also must know all the controls and displays in the spacecraft. During a mission each might at some time take over the duties of the other crewmen: during sleep or rest periods, while other crewmen are occupied with

- LAUNCH VEHICLE EMERGENCY DETECTION
- FLIGHT ATTITUDE
- MISSION SEQUENCE
- VELOCITY CHANGE MONITOR
- ENTRY MONITOR
- PROPELLANT GAUGING
- ENVIRONMENT CONTROL
- COMMUNICATIONS CONTROL
- POWER DISTRIBUTION
- CAUTION & WARNING

Main display console

experiments, and, of course, during an emergency.

Flight controls are located on the left-center and left side of the main display console, opposite the commander. These include controls for such subsystems as stabilization and control, propulsion, crew safety, earth landing, and emergency detection. One of two guidance and navigation computer panels also is located here, as are velocity, attitude, and altitude indicators.

The astronaut in the center couch (CM pilot) faces the center of the console, and thus can reach many of the flight controls, as well as the system controls on the right side of the console. Displays and controls directly opposite him include reaction control propellant management, caution and warning, environmental control and cryogenic storage subsystems.

The right-hand (LM pilot's) couch faces the right-center and right side of the console. Communications, electrical control, data storage, and fuel cell subsystem components are located here, as well as service propulsion of subsystem propellant management.

All controls have been designed so they can be operated by astronauts wearing gloves. The controls are predominantly of four basic types: toggle switches, rotary switches with click-stops (detents), thumbwheels, and push buttons. Critical switches are guarded so that they cannot be thrown inadvertently. In addition, some critical controls have locks that must be released before they can be operated.

In any mission, the Apollo crewmen will spend a great deal of their time manipulating controls and monitoring displays on the main display console. Crew duties, broken down by mission phase, are determined by NASA and compiled into a checklist for each astronaut. These checklists are part of the flight data file for the mission. This file consists of ten documents divided among three packages. The two smaller packages, called data file bags, each contain two documents and are attached to the outer sides of the left and right couches at about shoulder height. The other six documents are kept in a fiberglass container stowed in the lower equipment bay.

The data file bag on the left couch contains the commander's checklist and the mission flight plan. The bag on the right couch contains the LM pilot's checklist and the mission log, which is used as a backup to the voice recorder log. The data file container includes the CM pilot's checklist, landmark maps, star charts, orbital maps, an experiment checklist, and spacecraft subsystem data.

Despite the man-hours spent in mission simulators, it would be difficult if not impossible for the astronauts to remember all the procedures required for long-duration mission. The checklists contain the detailed procedures for each phase of the mission.

Among the checklists carried by the astronauts are those for subsystem management. These are compilations of procedures that are common to more than one phase of a mission. These procedures involve system monitoring, periodic checks, and unique functions of the service propulsion, reaction control, electrical power, environmental control, and caution and warning subsystems.

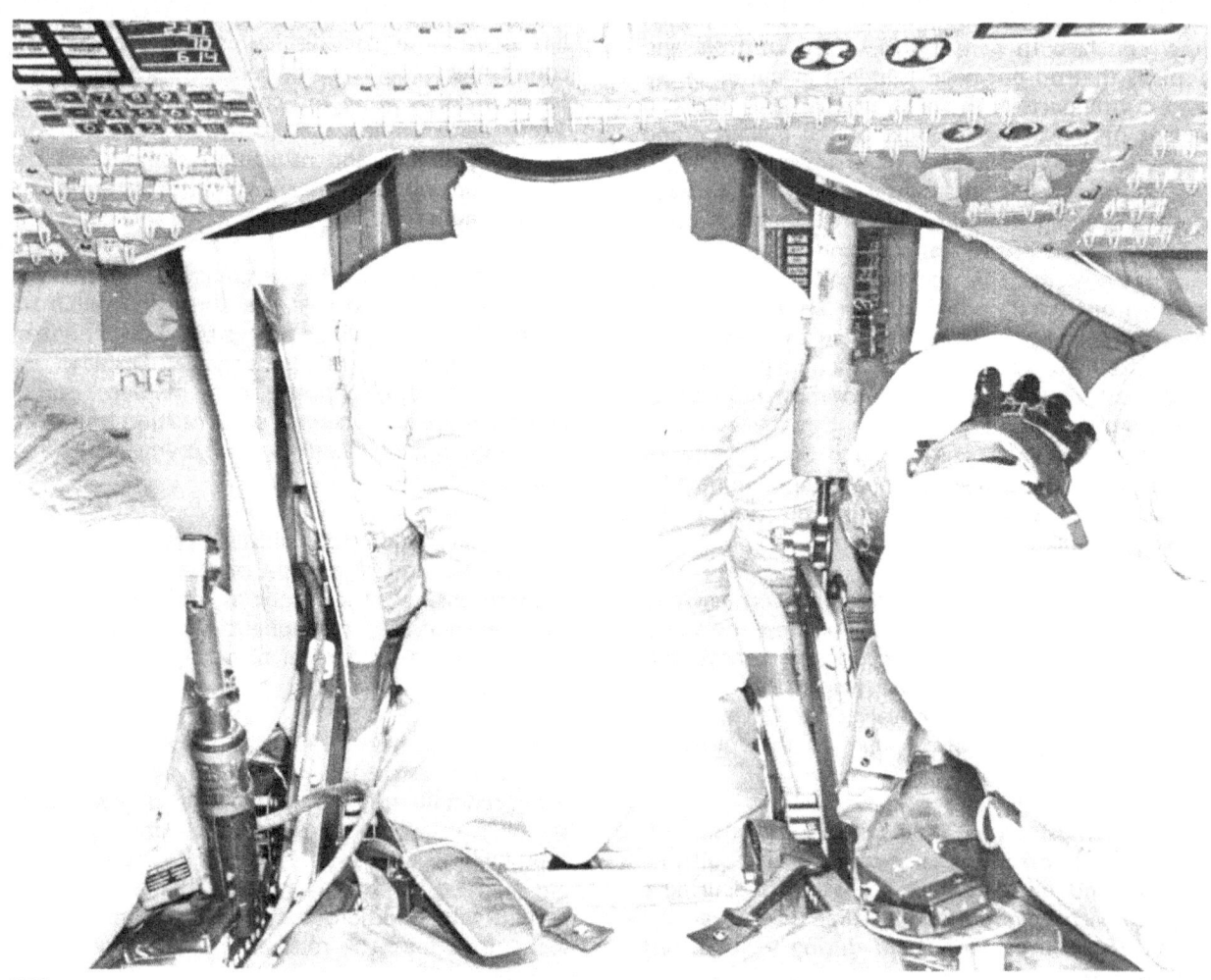

Astronaut stands at navigation station in lower equipment bay

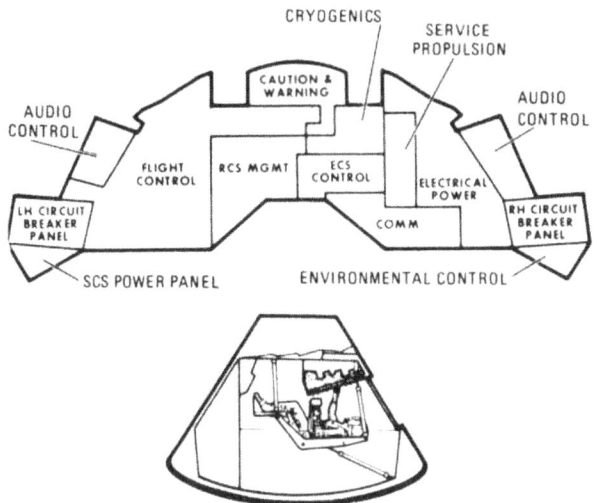

Grouping of controls and displays

Periodic checks are performed from every hour to every 24 hours, depending on the subsystem, throughout a mission. In addition, other checks or tests are performed at specific times or events, such as the service propulsion subsystem tests before and after every velocity change.

CAUTION AND WARNING SYSTEM

Critical conditions of most spacecraft systems are monitored by a caution and warning system. A malfunction or out-of-tolerance condition results in illumination of a status light that identifies the abnormality. It also activates the master alarm circuit, which illuminates two master alarm lights on the main display console and one in the lower equipment bay and sends an alarm tone to the astronauts' headsets. The master alarm lights and tone continue until a crewman resets the master alarm circuit. This can be done before the crewmen deal with the problem indicated. The caution and warning system also contains equipment to sense its own malfunctions.

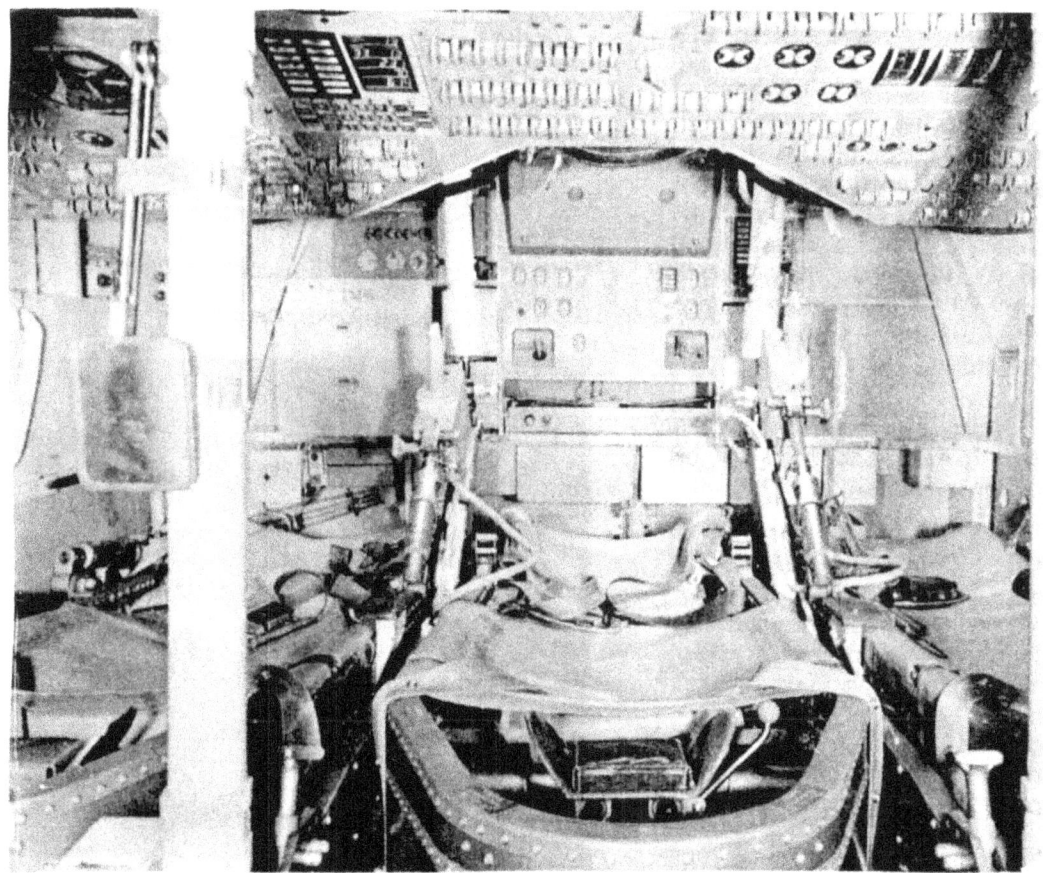

Panels and crew couches viewed through command module hatch

DETAILED DESCRIPTION

The major portion of the main display console is composed of nine panels, three in the center section and three in each of its two wings. These panels are numbered 1 through 9. Because of the multitude and complexity of the displays and controls on the three center panels, they have been arbitrarily divided into zones for easier reference. The zone markings do not appear on the actual console.

Panel 1A

Contains entry displays.

Altimeter	Indicates altitude of CM to 60,000 feet.
Entry monitor panel displays	ΔV/EMS SET switch Increases or decreases ΔV/RANGE display; also used to move velocity scroll.
FUNCTION switch	Used to select operation: EMS TEST positions check proper operation of displays; RNG SET enables moving RANGE display to initial condition; Vo SET enables moving VELOCITY scroll to initial condition; ENTRY sets panel for entry displays; ΔV TEST checks operation of thrust displays; ΔV SET enables moving of ΔV display to initial condition; ΔV is position for service propulsion engine thrust monitoring.
GTA switch	Provides bias signal in ground tests to nullify earth gravity; off during mission.
MODE switch	On STBY, inhibits all operations except ΔV SET, RNG SET and Vo SET; on NORMAL, activates function selected; on BACKUP/VHF RNG, activates displays for manual entry or thrust vector control, and enables VHF ranging information to be displayed on ΔV/ranging display.
RANGE/ΔV display	Shows either range to go (in nautical miles) or ΔV remaining

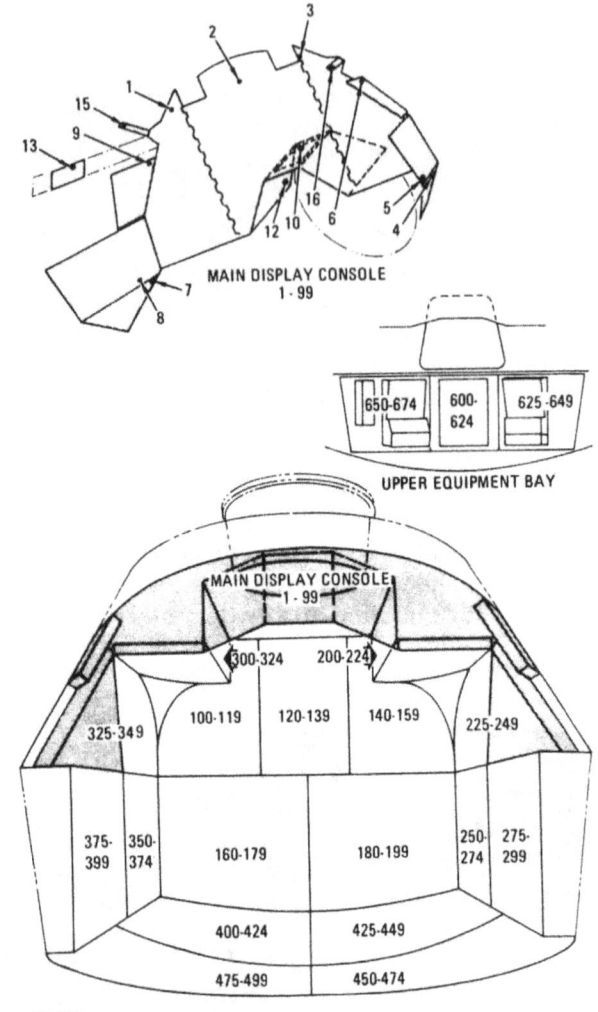

Panel numbering system

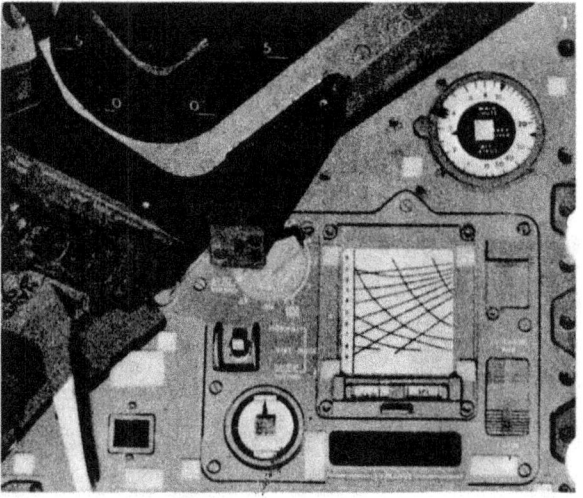

Panel 1A

	(in feet per second), depending on position of FUNCTION switch; signals automatic cutoff of service propulsion engine in stabilization and control configuration; provides readout of LM-CSM ranging during rendezvous.
SPS THRUST light	When lit, indicates thrust-on command to service propulsion engine (operation depends on MODE and FUNCTION switches).
0.05G light	When lit, indicates deceleration greater than 0.05G.
VELOCITY indicator	Shows acceleration (G scribe) versus velocity (velocity scroll) during entry.
Roll Attitude Indicator	Pointer shows stability axis roll attitude (lift vector orientation) and should be directed toward lamp that is lit; top lamp lights 10 seconds after 0.05G light to indicate deceleration equal to or greater than 0.262G; bottom lamp lights 10 seconds after 0.05G light to indicate deceleration equal to or less than 0.262G.
MASTER ALARM light	Glows red to alert crewmen to malfunction or out-of-tolerance condition; at the same time, an alarm tone is sent to headsets of each crewman and a light is illuminated on caution and warning panel to indicate the malfunction; pushing MASTER ALARM button extinguishes light and resets alarm circuit. The light may be inhibited during boost.

Panel 1B

Contains displays and manual controls for boost, entry, and abort.

ABORT light	Glows red to alert crew that abort has been requested by the ground; backup to voice communications.
Digital event timer	Starts automatically at liftoff and is automatically reset to zero if abort is initiated; controlled by switches on Panel 1D.
Launch Vehicle lights	LV RATE light glows red if launch vehicle angular rates are excessive, which can automatically initiate abort; LV GUID light glows red to indicate failure in launch vehicle guidance system, which can result in abort; SII SEP light glows white when staging is initiated and goes out when interstage skirt drops away from S-II after ignition; LV ENGINES lights (arranged in same pattern as on vehicle—1, 2, 3, and 4 are outboard and 5 is inboard) glow yellow before ignition and go out when an engine reaches 90 percent of thrust; No. 1 light also operated for third-stage engine; two lights glowing (two engines below 90 percent thrust) can initiate automatic abort during first-stage burning.
Backup switches	LIFTOFF light glows white at liftoff and goes out before first stage separation; NO AUTO ABORT portion glows red at liftoff if either of the emergency detection system automatic abort systems has not been enabled. LES MOTOR FIRE is backup switch to fire launch escape subsystem motor; CANARD DEPLOY is backup to deploy canards on launch escape subsystem during abort; CSM/LV SEP is switch for normal CSM-launch vehicle separation after ascent; APEX COVER JETT is backup to deploy the apex cover (forward heat shield) during descent; DROGUE DEPLOY and MAIN DEPLOY are backup switches to deploy drogue and main parachutes during descent; CM RCS He DUMP is backup for starting reaction control subsystem helium purge. The backup switches are all guarded to prevent inadvertent operation.

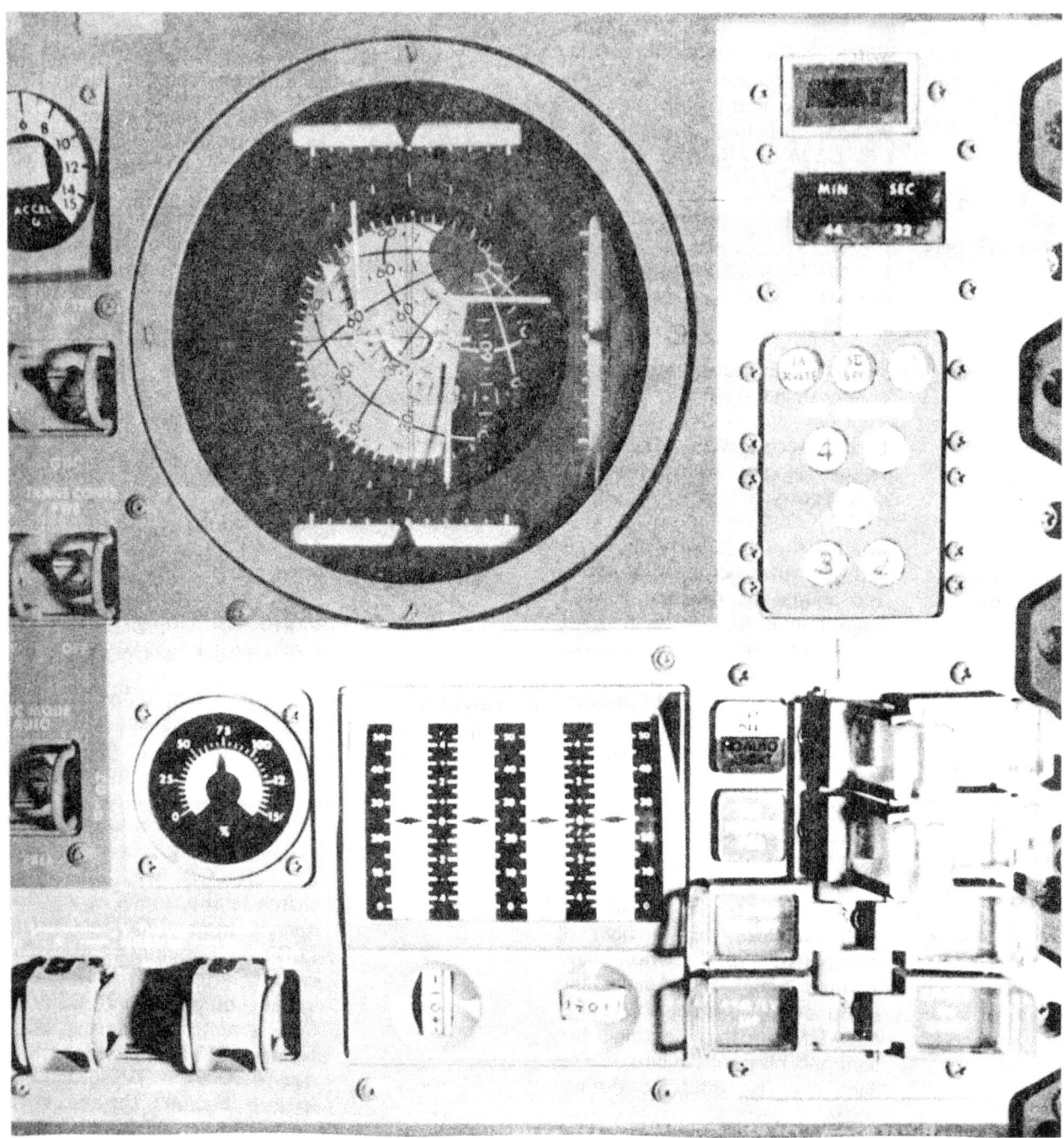

Panel 1B

LV TANK PRESS indicators	Shows pressure in the fuel tank of the second stage (S-II) and pressure in oxidizer and fuel tanks of the third stage (S-IVB) on left, center, and right scales, or position of service propulsion engine pitch and yaw gimbals on other two scales. Display is controlled by switch on Panel 1D; tank pressures are monitored during boost, gimbal position during rest of mission.	
SPS GIMBAL thumbwheels	Used to set position of service propulsion engine pitch and yaw gimbals in SCS mode.	
LV/SPS Pc	Shows either off-axis pressure	

measured by Q-ball atop launch escape subsystem or service propulsion engine combustion chamber pressure, depending on position of switch on Panel 1D; switch is in former position only during early phases of boost.

Panel 1C

Contains flight director attitude indicator No. 1.

Rate indicators are at top (roll), right (pitch), and bottom (yaw). Spacecraft body rates are indicated by the displacement of the arrows from the center and correspond to the direction the rotation control must be moved to correct the displacement.

The attitude error needles (on the outside rim of the ball adjacent to the rate indicators) show the difference between the actual and the desired spacecraft attitude. These also correspond to the direction the rotation control must be moved to reduce the error to zero.

The ball shows spacecraft orientation with respect to a selected inertial reference. The ball rotates about three independent axes which correspond to pitch, yaw, and roll. Pitch attitude is shown by great semicircles; the semicircle shown under the inverted wing symbol is inertial pitch at time of readout. Yaw attitude is shown by minor circles; the minor circle under the inverted wing is the inertial yaw at the time of readout. Roll attitude is shown by small arrowhead on outer scale of ball.

Panel 1D

Consists of switches and indicators for flight control of the spacecraft.

ACCEL G	Meter shows G forces along spacecraft's X axis.
CMC ATT	On IMU (inertial measurement unit), allows attitude transformations in gyro display coupler and motor excitation to either or both attitude balls. GDC (gyro display coupler) was planned as backup but is not used.
FDAI switches	SCALE switch selects scale (degrees of error and degrees per second of rate) on three attitude indicators. SELECT switch selects sources for attitude balls: 1/2 sends signals to both balls, 2 sends signals just to FDAI No. 2, and 1 sends signals just to No. 1. SOURCE switch selects source of signals to balls (if SELECT switch is on either 1 or 2 position); CMC is computer, ATT SET is based on position of ATT SET switch, GDC is gyro display coupler.
ATT SET	On IMU (inertial measurement unit), signals from unit are sent to attitude set panel; on GDC, signals from gyro display coupler are sent to panel.
MANUAL ATTITUDE switches	ACCEL CMD position for all three switches disables automatic control of reaction control engines and prepares for manual control; RATE CMD is normal position for automatic control; MIN IMP position prepared electronics for minimum-impulse engine firing.

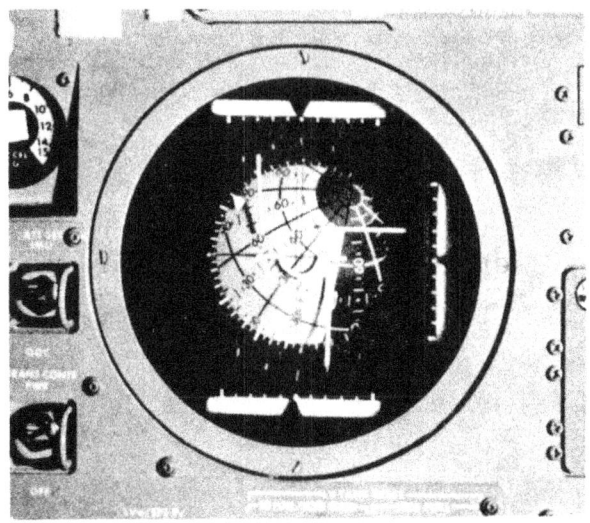

Panel 1C

LIMIT CYCLE switch	When OFF inhibits pseudo-rate feedback in electronic control assembly.	TRANS CONTR switch	When on PWR, applies power to translational hand controller.
ATT BEADBAND switch	Selects additional 4 degrees of deadband in attitude control electronics (deadband is the area in which no response will be provoked, like the play in a car steering wheel).	ROT CONT PWR switches	NORMAL switches apply power to rotational hand controllers for normal operation; DIRECT switches apply power to rotational hand controllers for direct mode of operation.
RATE switch	Selects high or low rate and attitude deadbands and proportional rate command capability in electronic control assembly.	SC CONT switch	ON CMC, provides computer control; on SCS, provides stabilization and control subsystem control.
		CMC MODE switch	Selects computer mode of operation.

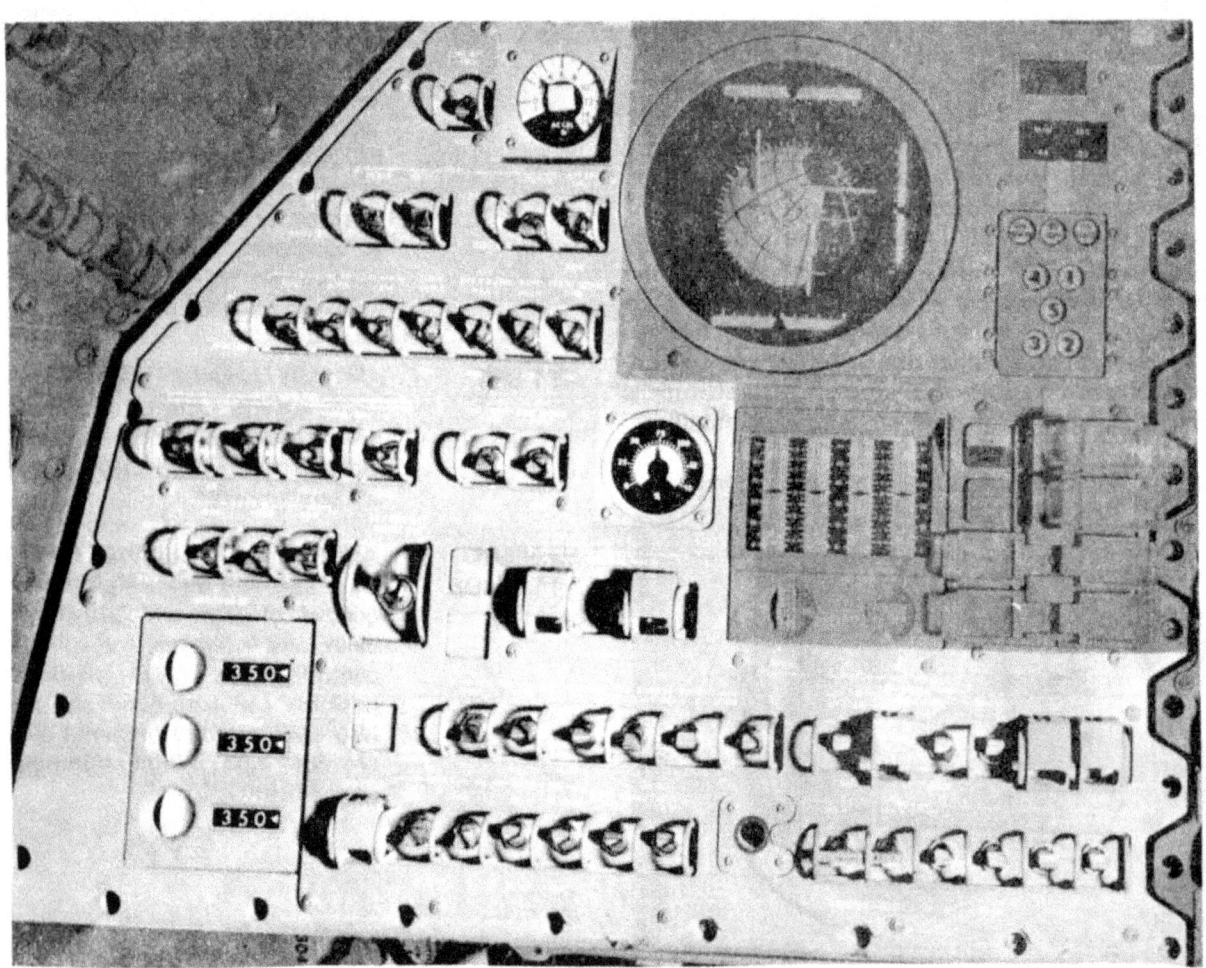

Panel 1D

BMAGMODE switches	Routes signals from body-mounted attitude gyros to electronics from gyro assembly No. 1 (RATE 1) or No. 2 (RATE 2), or uses signals from both assemblies for different functions (ATT 1/RATE 2).
SPS THRUST DIRECT ON switch	On DIRECT ON, energizes circuitry which permits service propulsion engine firing; on NORMAL, de-energizes circuits.
DIRECT ULLAGE pushbutton	Backup switch for ullage maneuver before service propulsion engine firing; translation control normally provides for ullage maneuver.
THRUST ON pushbutton	Signals circuitry to fire service propulsion engine.
ΔV THRUST switches	On NORMAL, applies power to injector pre-valves A or B, signals computer that engine is ready to fire, and supplies power to permit thrusting. Two switches are guarded to the OFF position.
ATTITUDE SET panel	Thumbwheels and indicators allow crewman to insert a desired attitude reference into stabilization and control subsystem.
GDC ALIGN pushbutton	When pressed, sends attitude information dialed on ATTITUDE SET panel to gyro display coupler.
SCS TVC switches	On AUTO, provides automatic control of thrust vector; on RATE CMD, provides for manual control of thrust vector; on ACCEL CMD, inhibits body-mounted attitude gyro signals.
SPS GIMBAL MOTORS switches	Switches for primary (1) and secondary (2) gimbal motors; normally down (off), they are moved up to start the motors and switch is spring-loaded to return to center (on) position. Gimbal motors move service propulsion engine direction of thrust; position can be set manually by thumbwheels on Panel 1B.
ΔV CG switch	On LM/CSM when lunar module is attached; on CSM when lunar module no longer attached.
ELS switches	On LOGIC, provides power to arm earth landing subsystem circuitry; switch is guarded until entry or abort. Second switch is normally on AUTO for automatic control of landing subsystem; set to MAN only after drogue parachute deployment during an abort in first minute of boost.
CM PROPELLANT JETT switches	These control dumping of CM reaction control subsystem propellant. LOGIC switch activates DUMP switch and prepares circuits for automatic dumping of propellant. Guarded DUMP switch starts process of burning propellant during descent. PURGE switch sends helium through CM lines and reaction control engines after propellant has been burned.
IMU CAGE switch	Guarded switch which when up cages the inertial measurement unit platform with all three gyros at 0 degrees.
ENTRY switches	EMS ROLL switch sends signal to gyro display coupler to drive roll attitude indicator on entry monitor display. 0.05G switch provides signals which couple roll and yaw rates and sum them for display on roll stability indicator and attitude ball.
LV/SPS IND switches	Controls what is displayed on indicators on Panel 1B. Left switch connects Q ball ( ) or service propulsion engine sensor (Pc) to LV/SPS Pc indicator. Right switch provides for display of launch vehicle tank pressures (SII/SIVB) or gimbal position

(GPI) on the LV FUEL TANK PRESS indicator.

TVC GMBL DRIVE switches — Allows crew to select a specific channel or automatic control for thrust vector control. AUTO is automatic; 1 applies commands only to primary channel of gimbal actuator; 2 applies commands to secondary channel.

EVENT TIMER switches — These control digital event timer on Panel 1B. RESET position sets timer to zero; center position causes counter to count up; DOWN position causes it to count down. START switch starts timer, STOP halts it, center position has no function. MIN and SEC switches turn counter by tens or ones (units) and can be turned up or down depending on position of RESET/DOWN switch.

Panel 2A

Contains caution and warning system controls and displays, docking switches, and mission timer switches and display.

Caution & Warning lights — These lights illuminate to indicate a malfunction or out-of-tolerance condition in a system or component. The MASTER ALARM lights and audio signal are activated when one of these caution and warning lights come on. Most of the 37 lights are yellow; five are red. The lights and what they indicate are: BMAG 1 and BMAG 2 (temperature too high in gyro assembly 1 or 1); PITCH GMBL 1 or 2 and YAW GMBL 1 or 2 (overcurrent in drive motor of

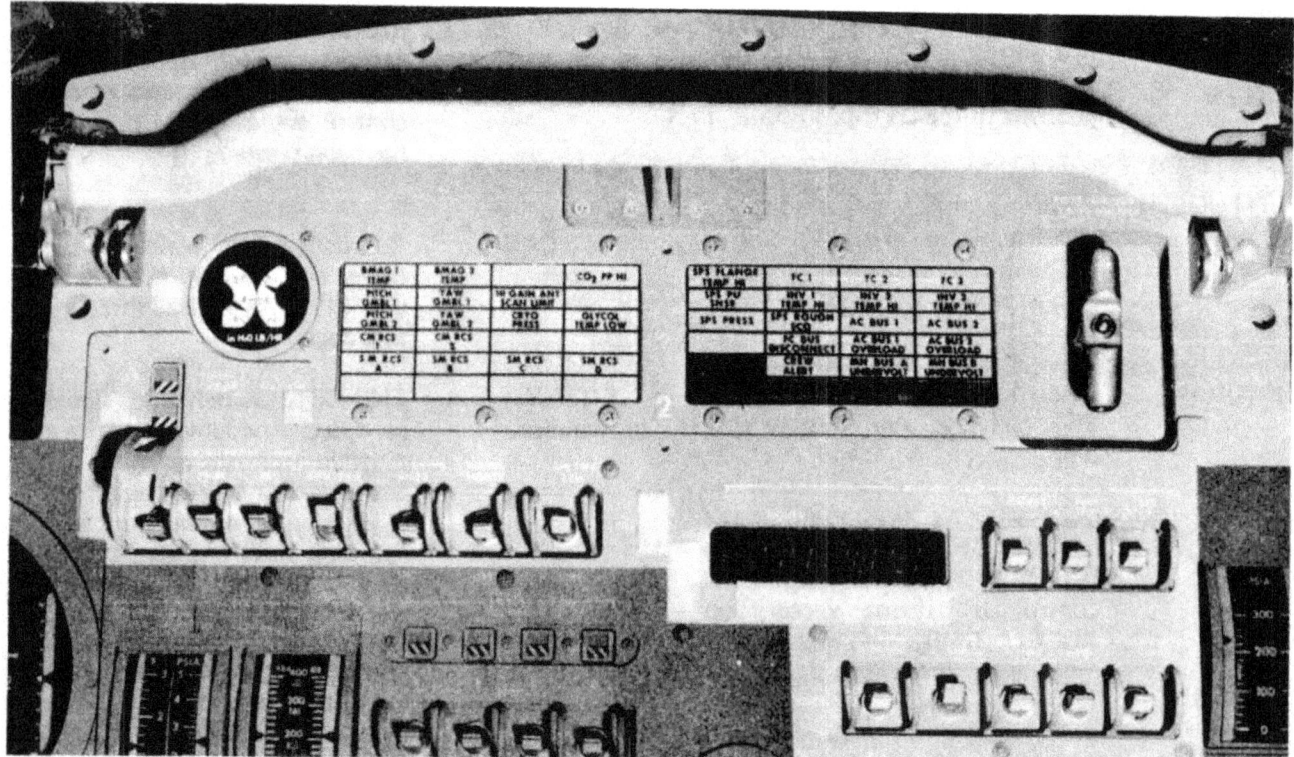

Panel 2A

gimbal actuator); CM RCS 1 or 2 (over- or under-pressure in CM reaction control subsystem propellant tanks); SM RCS A, B, C, or D (pressure or temperature too high or too low in SM reaction control subsystem quad); HI GAIN ANT SCAN LIMIT (scan limit exceeded on S-band high-gain antenna); CRYO PRESS (under- or over-pressure in cryogenic hydrogen or oxygen tanks of electrical power subsystem); CO_2 PP HI (carbon dioxide partial pressure is high); GLYCOL TEMP LOW (temperature of water glycol from radiator is low); SPS FLANGE TEMP HI (temperature too high in service propulsion engine combustion chamber flange); CMC (red) (loss of power or failure of any one of a number of components in computer); ISS (red) (failure in one of key units in inertial subsystem); FC 1, 2, or 3 (any one of a number of temperature, pressure, or flow conditions in a fuel cell); INV 1 (2 or 3) TEMP HI (temperature too high in an inverter); AC BUS 1 or 2 (under- or over-voltage in ac electrical bus); FC BUS DISCONNECT (a fuel cell has automatically disconnected from dc main bus); AC BUS 1 (or 2) OVERLOAD (ac bus overloaded); CREW ALERT (activated by ground and can only be extinguished by ground); MN BUS A (or B) UNDERVOLT (dc bus undervoltage); C/W (red) (power supply voltage to caution and warning system is out of range); O_2 FLOW HI (red) (oxygen flow rate too high;

SUIT COMPRESSOR (red) (suit compressor differential pressure low).

PROBE EXTD/RETR indicators	Striped lines indicate probe is in motion; gray indicates movement has stopped.
DOCKING PROBE switches	Guarded left-hand switch on EXTD/REL extends probe and releases latches. Two RETRACT switches (primary and secondary) retract probe.
LIGHTS switches	EXTERIOR switches for RUN/EVA and rendezvous or spotlights. TUNNEL switch turns on CM tunnel lights.
LM PWR switch	Connects CM power to heater circuits in LM; on RESET, disconnects CM power and connects heater circuits to LM power.
MISSION TIMER indicator	Shows time in hours, minutes, and seconds.
MISSION TIMER switches	Sets timer in hours, minutes, and seconds; switches can change timer display by tens or units (ones) but only counts up (it is necessary to go all the way to zero to get lower numbers).
MSN TIMER switch	Starts, stops, and resets the timer.
CAUTION/ WARNING switches	Left-hand switch is set to NORMAL for most of mission; on BOOST, Master Alarm light on Panel 1A is disabled to prevent confusion with Abort red light near it; on ACK, caution and warning lights on panel 2 only are disabled. CSM/CM switch selects

systems to to be monitored; CSM is normal, CM is used after separation so that only CM systems are monitored. POWER switch selects power source for caution and warning system; LAMP TEST tests lamps of system; 1 position tests left-hand lamps, 2 position tests right-hand lamps.

Panel 2B

Contains flight director attitude indicator No. 2. Identical in operation to No. 1. See Panel 1C for description of displays.

Panel 2C

Contains display and keyboard for command module computer. Group of status lights at top left include UPLINK ACTY (computer is receiving data from ground); TEMP (temperature of stable platform is out of tolerance); NO ATT (inertial subsystem is not in a mode to provide attitude reference); GIMBAL LOCK (middle gimbal angle exceeds 70 degrees); STBY (computer subsystem is on standby); PROG (computer requests additional information to complete program); KEY REL (internal program needs keyboard circuits to continue program); RESTART (computer has gone into restart program); OPR ERR (error has been made on keyboard); TRACKER (failure of one of the optical coupling data units).

The displays include COMP ACTY light (computer is calculating); PROG, NOUN, and VERB lights, which show the program, noun, and verb being used; and Registers 1, 2, and 3, which provide information to the crew.

The keyboard (at bottom) is the means by which the crew operates the computer. VERB key sets computer to accept next two digits as verb code. NOUN key sets computer to accept next two digits as noun code. CLR key blanks the register being loaded. STBY key puts computer in standby; when pressed again computer resumes normal operation. KEY RLSE releases the displays from keyboard control so computer can display internal program information. ENTR key informs the computer that the data has all been loaded and tells it to execute the desired function. RSET key extinguishes the status lamps controlled by the computer.

Panel 2D

Contains key switches for abort, boost, and entry.

EDS switch — On AUTO, prepares the emergency detection system for automatic abort; the system is part of the launch vehicle and operates only during boost.

CSM/LM FINAL SEP switches — Guarded switches spring-loaded to the down position. Pushing up jettisons the docking ring and probe; this also jettisons the LM ascent stage in lunar orbit.

CM/SM SEP switches — Guarded redundant switches spring-loaded to the down position. When pushed up, send signal to master events sequence controller to start CM-SM separation procedure.

S-IVB/LM SEP switch — Guarded switch used to separate LM from S-IVB after transposition and docking.

ABORT SYSTEM switches — PRPLNT DUMP switch is placed on AUTO during first minute of boost to enable rapid dumping of reaction control subsystem propellant in case of an abort; RCS CMD position used afterwards to enable dumping sequence by firing unused propellant through reaction control engines. 2 ENG OUT switch on AUTO activates

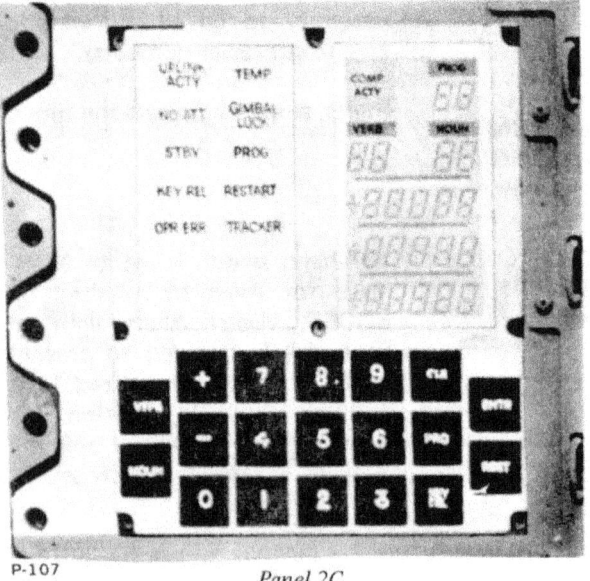

Panel 2C

the emergency detection system for an automatic abort if two booster engines fail; operates only during boost. LV RATES on AUTO activates the emergency detection system for automatic abort if launch vehicle rates are excessive; operates only during boost. Guarded redundant TWR JETT switches start jettisoning of launch escape tower; tower normally is jettisoned automatically shortly after first stage separation.

LAUNCH VEHICLE switches

GUIDANCE switch selects means of guidance during boost; this is normally on IU (instrument unit), allowing Saturn guidance system to control boost; CMC (command module computer) position will disconnect Saturn guidance and CM computer will control flight. S-II/S-IVB guarded switch initiates separation of the Saturn V second stage (S-II) from the third stage (S-IVB); this is backup to jettison second stage in emergency. XLUNAR switch will allow or inhibit translunar injection.

MAIN RELEASE switch

Guarded switch which is electrically activated when CM descends to 10,000 feet and main parachutes are deployed. Spring-loaded to down position, it is used only after splashdown, to release main parachutes from command module.

Panel 2E

Contains displays and controls for managing reaction control subsystem, plus two switches of telecommunications subsystem.

SM RCS, CM RCS indicators

Four meters show various conditions in SM and CM reaction control subsystems; display is controlled by RCS INDICATORS switch. For SM subsystem, indicators show TMP PKG (temperature of the SM quad selected), He PRESS (helium tank pressure in quad selected), SEC FUEL (secondary tank pressure in quad selected), and PRPLNT QTY (propellant quantity in quad selected) or He TK TEMP (helium supply temperature of quad selected). CM subsystem indicators show He TEMP (helium tank temperature of system selected), He PRESS (helium tank pressure of system selected), MANF PRESS (regulated helium pressure in the manifold of the system selected).

SM RCS HELIUM switches and indicators

Two groups each with four switches, with each switch having an indicator window above it. Switches are identical, spring-loaded to return to center position. OPEN position opens isolation valve in helium pressure system, CLOSE position closes valve. HELIUM 1 switches control valves in one of redundant systems, HELIUM 2 switches control valves in other redundant system. Striped-line display on indicators means valve is closed; gray display means valve is open.

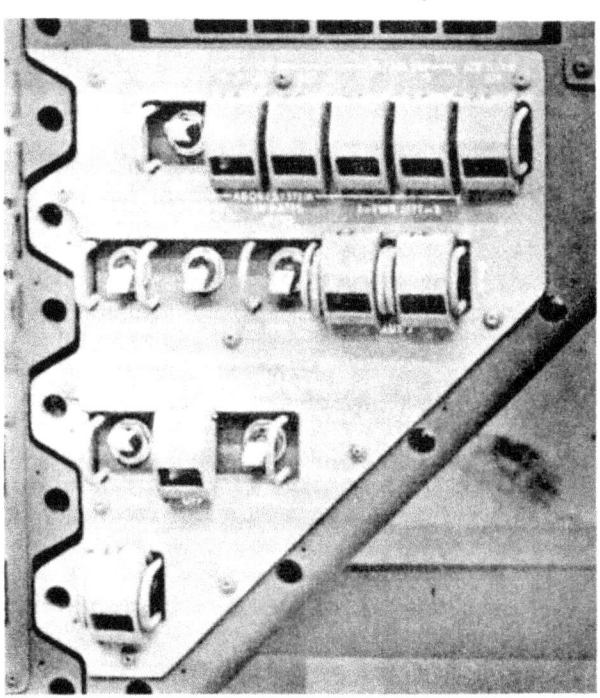

Panel 2D

Panel 2E

RCS INDICATORS switch	Selects portion of reaction control subsystem to be displayed on indicators. CM1 and 2 are command module Systems 1 and 2; SM A, B, C, and D are service module quads.	SM RCS IND switch	Controls what is displayed on right-hand reaction control subsystem indicator when SM quads are selected by RCS INDICATORS switch. He TK TEMP is helium supply temperature of quad selected and PRPLNT QTY is propellant quantity in quad selected.
UPTLM switches	On ACCEPT, enables telemetry from the ground to go to the instrument unit (IU) or CM computer (CM) the up-data link equipment. On BLOCK, blocks the message.	SM RCS HEATERS switches	Four identical switches apply power to primary (high-temperature) or secondary (low-
CM RCS PRESS switch	Guarded switch which when pushed up activates the two helium isolation valves in CM reaction control Systems 1 and 2.		

	temperature) thermoswitches in SM reaction control subsystem quads. Thermoswitches automatically turn heaters on and off.
SM RCS PRPLNT switches and indicators	Two groups of four identical switches each, with each switch having an indicator above it. Each switch controls two isolation valves (one fuel and one oxidizer) in each SM reaction control subsystem quad. Switches are spring-loaded to center position. Indicators show whether valves are closed (striped lines) or open (gray).
CM RCS PRPLNT switches and indicators	Two identical switches which control fuel and oxidizer isolation valves in CM reaction control System 1 and 2. Pushing switch up opens valves, down closes them; switch is spring-loaded to return to center position. Indicators show whether valve is closed (striped lines) or open (gray).
RCS switches	CMD switch, when on, enables automatic operation of reaction jet engine control assembly (electronics used by the stabilization and control subsystem to fire reaction engines for attitude control). OFF position disables the assembly. Switch is spring-loaded to the center position. TRNFR switch is backup for automatic transfer of reaction control between SM and CM subsystems.

Panel 2F

Contains displays and controls for management of the environmental control subsystem.

CRYOGENIC TANKS indicators	Four meters which show pressure and quantity in the cryogenic (ultra low-temperature) oxygen and hydrogen tanks in the SM. PRESSURE indicators show pressure in each of the two hydrogen and each of the two oxygen tanks; QUANTITY indicators shows percentage of fluid remaining. O_2 (oxygen) pressure indicator also can be used to show pressure in environmental control subsystem surge tank; display is controlled by O_2 PRESS IND switch below indicator.
CABIN FAN switches	Used to turn on cabin air fans for cooling; both normally are used at same time.
H_2 HEATERS switches	Used to control redundant heaters in two cryogenic hydrogen tanks. AUTO position enables automatic operation of heaters by pressure switches in tanks; ON position bypasses automatic switches. Center position is off.
O_2 HEATERS switches	Identical to H_2 HEATERS switches, except for redundant heaters in cryogenic oxygen tanks.
O_2 PRESS IND switch	Controls what is shown on O_2 1 indicator above it. TANK 1 position shows pressure in cryogenic oxygen tank 1 on indicator; SURGE TANK shows pressure in oxygen surge tank on indicator.
H_2 FANS switches	Used to control redundant fan motors in two cryogenic hydrogen tanks. AUTO position enables automatic operation of fan motors by pressure switches in tanks; ON position bypasses automatic switches. Center position is off.
O_2 FANS switches	Identical to H_2 FANS switches except for redundant fan motors in cryogenic oxygen tanks.
ECS INDICATORS switch	Selects primary or secondary glycol loop data for display on four indicators. The four are the three circular indicators immediately to the right of the switch, and the right-hand circular indicator below them.

ECS RADIATOR TEMP indicators — PRIM/SEC display shows temperature of water glycol entering primary or secondary loop sections of space radiators (depending on position of ECS INDICATORS switch). PRIM display shows temperature of water glycol leaving primary loop space radiators. SEC display shows temperature of water glycol leaving secondary loop space radiators.

GLY EVAP indicator — TEMP (right-hand side of center indicator) shows temperature of water glycol at outlet of water glycol evaporator. STEAM PRESS (left-hand side of third indicator) shows pressure of steam discharge from evaporators. Either primary or secondary glycol loop can be selected by ECS INDICATORS switch.

GLY DISCH PRESS — Display (right-hand side of third indicator) shows output pressure of either primary or secondary loop water glycol pumps depending on ECS INDICATORS switch.

TEMP indicators — SUIT indicator shows temperature of suit circuit atmosphere; CABIN shows temperature of cabin atmosphere.

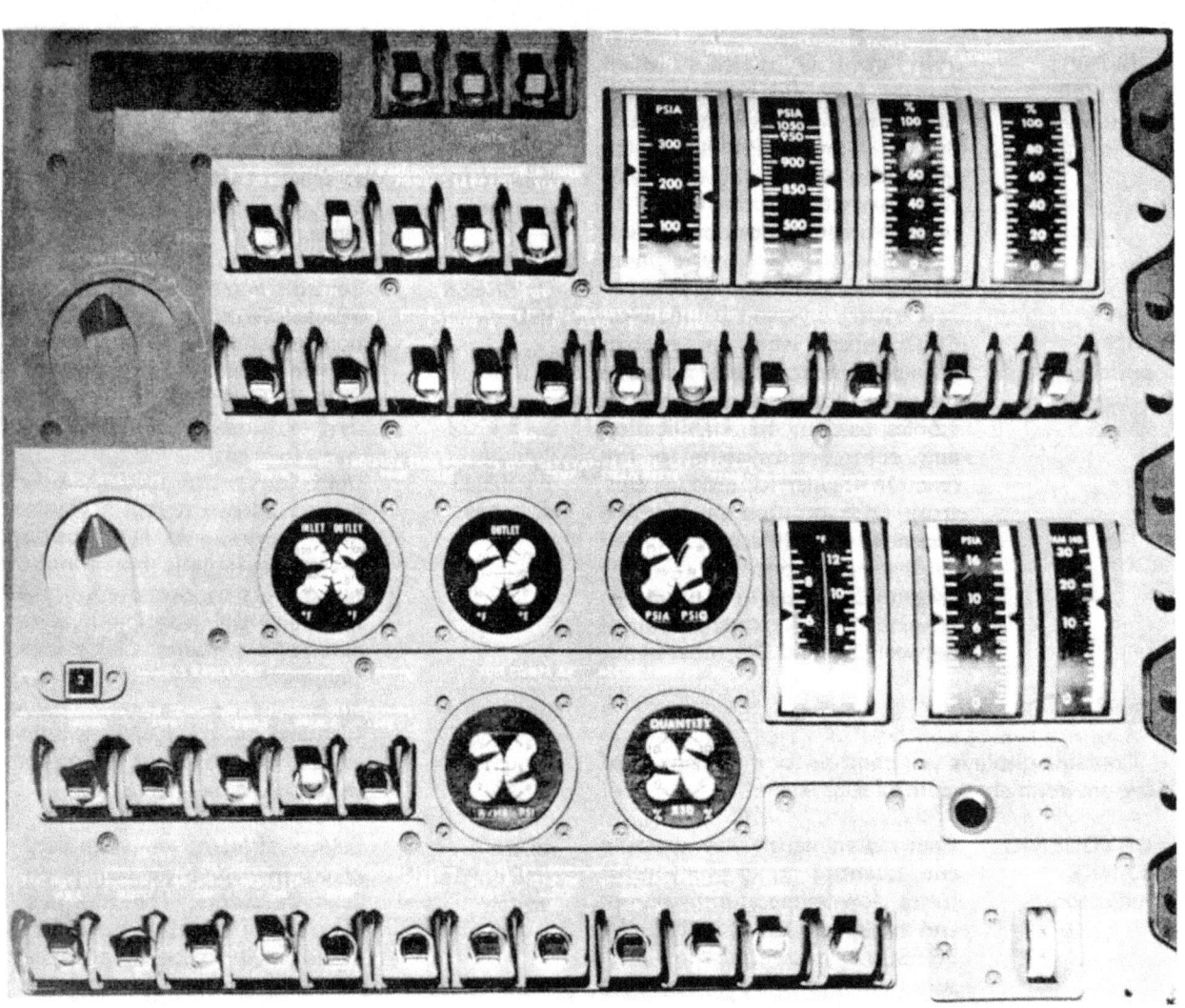

P-110

86L

PRESS indicators	SUIT indicator shows pressure of suit circuit atmosphere; CABIN shows pressure of cabin atmosphere.	POT H_2O HTR	Selects either dc Main Bus A or dc Main Bus B to supply power to the heater in the potable water tank. Center position cuts off power to the heater.
PART PRESS CO_2 indicator	Shows partial pressure of carbon dioxide in suit circuit atmosphere.	SUIT CIRCUIT switches	H_2O ACCUM switches control cycling of water accumulator valve. AUTO 1 and 2 positions on left-hand switch enables automatic operation of No. 1 and No. 2 accumulator valves; center position activates right-hand switch for manual control of valves (ON 1 is accumulator valve 1, ON 2 is accumulator valve 2, center position is off). HEAT EXCH switch controls flow of water-glycol through heat suit exchanger (ON) or around it (BYPASS); the center position is off.
ECS RADIATORS	Indicator shows gray, indicating flow proportioning control no. 1 is in operation, or the figure 2, indicating flow proportioning control No. 2 is in operation. The FLOW CONTROL switches are used to select flow proportioning control to space radiators; AUTO is automatic control, 1 is No. 1 control, and 2 is No. 2 control. PWR applies power to the AUTO switch, MAN SEL MODE applies power to the MAN SEL switch. MAN SEL switch selects radiator No. 1 (closes radiator No. 2 isolation valves), radiator No. 2 (closes radiator No. 1 isolation valves), or closes isolation valves to both radiators (center position). HEATER switches select No. 1 or No. 2 heater controllers or the secondary glycol loop heater controller, or turns them off.		
		SEC COOLANT LOOP switches	EVAP switch enables activation of the secondary water-glycol evaporator temperature control. PUMP switch supplies power to the secondary water-glycol loop pump from either ac Bus 1 or ac Bus 2, or turns off the pump (center position).
O_2 FLOW indicator	Shows total oxygen flow rate from main pressure regulator.	H_2O QTY IND switch	Controls what is displayed on H_2O indicator; POT shows quantity in potable water tank on indicator, WASTE shows quantity in waste water tank on indicator.
ΔP indicator	Shows pressure differential between suit and cabin.	GLYCOL EVAP switches	TEMP IN provides either automatic or manual control of temperature of coolant entering evaporator by mixing hot and cold water-glycol. (Manual control is achieved by operating a valve in the left-hand equipment bay.) The STEAM PRESS switches provide automatic or manual regulation of pressure in the evaporator steam duct; MAN position of left-hand activates right-hand switch, which is used to increase
ACCUM indicator	Shows quantity of water-glycol in either primary or secondary glycol loop accumulators, depending on ECS INDICATORS switch.		
H_2O indicator	Shows quantity of water in potable water tank or waste water tank, depending on H_2O QTY IND switch.		

or decrease steam duct pressure. H₂O FLOW switch provides automatic or manual control of valve which regulates flow of water into evaporator; Auto is automatic control, ON is manual control (valve is opened only while switch is held down), and center position is off.

CABIN TEMP switch and thumbwheel — The switch provides for either automatic or manual control of cabin temperature. AUTO position activates temperature control unit which regulates temperature by controlling flow of water-glycol through cabin heat exchanger. Manual control (used only if automatic control fails) is achieved through operation of valve in left-hand forward equipment bay. Thumbwheel is used to increase (up) or decrease the cabin temperature maintained by the automatic control unit.

Panel 2G

Contains displays and controls for operation of the S-band high-gain antenna.

TRACK switch — On AUTO, the antenna automatically points toward a ground station; on MAN, the antenna points in the direction set by the PITCH and YAW POSITION dials; on REACQ, the antenna automatically scans (moves) to acquire a ground station.

BEAMWIDTH switch — Sets the antenna for wide, medium or narrow beamwidth.

PITCH indicator — Shows position in degrees of the antenna in the pitch plane.

Panel 2G

PITCH POSITION switch — Allows manual positioning of the antenna in the pitch plane.

S-BAND ANT indicator — Shows strength of S-band signal after lock-up with ground. Display is related to decibels so that full scale would be about -60 db.

YAW indicator — Shows position in degrees of the antenna in the yaw plane.

YAW POSITION switch — Allows manual positioning of the antenna in the yaw plane.

POWER switch — On POWER, provides electrical power to antenna equipment; on STBY, provides heater power to boom components only; on OFF, cuts off all power to antenna.

SERVO ELEC switch — Selects one of two redundant electronic servoassemblies in high-gain antenna.

Panel 3A

Contains displays and control for management of the service propulsion subsystem and a VHF antenna control.

VHF ANTENNA — Connects one of three VHF antennas into telecommunications subsystem. SM LEFT is the left

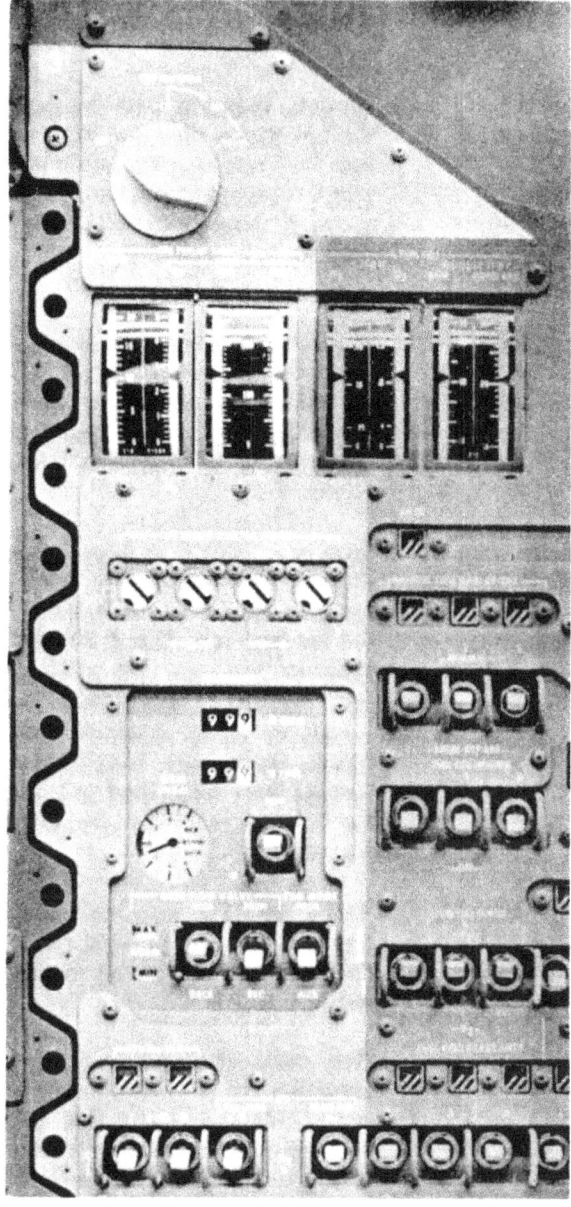

Panel 3A

	(He), fuel (FUEL), and oxidizer (OXID) tanks. The helium tank pressure indicator also can be used to show pressure in the nitrogen storage tank of either of the redundant engine pneumatic valve control system, depending on the position of the PRESS IND switch at the bottom of panel.
SPS ENGINE INJECTOR VALVES indicators	Four identical indicators which show whether service propulsion engine main propellant valves are open or closed. Each indicator shows status of pair of valves (one fuel and one oxidizer); pointer to left is closed, to right is open.
SPS QUANTITY indicator	Two digital counters which show percentage of oxidizer (% OXID) remaining in the oxidizer tanks and percentage of fuel (% FUEL) remaining in the fuel tanks.
OXID UNBAL indicator	Shows unbalance of remaining propellant. Each of six graduations on scale represent 100 pounds of propellant unbalance. Upper half of scale shows increased oxidizer flow required to achieve proper balance; lower half shows decreased flow required.
TEST SWITCH	Used to check proper functioning of SPS QUANTITY and OXID UNBAL indicators. Spring-loaded to center position; pushing up sends simulated signal to sensing system, causing indicators to function; pushing down sends simulated signal in reverse polarity.
OXID FLOW VALVE switches and indicators	Upper indicator shows word MAX when propellant utilization valve is in increased oxidizer flow rate position, gray when it is not. Lower indicator shows word MIN when valve is in decreased oxidizer flow rate position, gray when it is not. Switches control

scimitar VHF antenna on the service module; SM RIGHT is the right scimitar antenna. RECY is the VHF recovery antenna on the command module.

SPS PRPLNT TANKS indicators	Shows temperature and pressure of service propulsion subsystem tanks. TEMP provides constant monitoring of temperature in propellant tanks. PRESS indicators show pressure in the helium

860

	increase (INCR), decrease (DECR), or normal (center position) oxidizer flow by positioning propellant utilization valve. PRIM means signal is applied to primary servo amplifier and SEC is secondary servo amplifier. Switches are used to bring fuel and oxidizer into proper ratio (1.6 oxidizer to 1 fuel) by controlling flow of oxidizer.
PUG MODE switch	Controls source of propellant quantity indicating and warning displays; PRIM is primary propellant quantity sensing system, AUX is auxiliary sensing system, and NORM is both sensing systems together for warning but only primary being displayed.
SPS He VLV switches and indicators	Two identical switches used to apply power to helium isolation valve solenoid; on AUTO, valve opening and closing is controlled automatically; on ON, valves are opened; center position is off. Indicators show striped lines when valves are closed, gray when valves are open.
SPS switches	LINE HTRS switch applies power to heating systems in engine feed lines and bipropellant valves; center position is off. PRESS IND switch determines whether pressure in helium tank (He) or in one of nitrogen storage tank engine control valve systems (N_2A or N_2B) is displayed on indicator at top of panel.

Panel 3B

Contains displays and controls for management of the electrical power subsystem and a second master alarm light.

FUEL CELL indicators	FLOW meters show flow rate of hydrogen (H_2) and oxygen (O_2) into a fuel cell selected by the FUEL CELL INDICATOR switch. MODULE TEMP meters shows skin (SKIN) or condenser exhaust (COND EXH) temperature of the fuel cell selected.
pH HI indicator	Window shows striped line when the pH factor of water from the fuel cell selected by the FUEL CELL INDICATOR switch is above 9, shows gray when factor is below 9.
FC RAD TEMP LOW indicator	Shows striped lines when water-glycol temperature at radiator outlet of selected fuel cell is -30°F or below; shows gray when temperature at outlet is above -30°F.
FUEL CELL RADIATORS switches and indicators	Switches control radiator panels used by each fuel cell; NORMAL position allows use of full radiator for fuel cell, EMER BYPASS position bypasses 3/8 of radiator. Switch is springloaded to return to center position. Indicators above each switch show striped lines when part of radiator is being bypassed, gray when full radiator is in use.
FUEL CELL HEATERS switches	Three identical switches control heaters in fuel cells; when switches are up heaters operate automatically.
MASTER ALARM light	Red light illuminates to alert crewmen to a malfunction or out-of-tolerance condition; tone also sounds in crewmen's headsets. Pushing button shuts off light and audio tone and resets caution and warning system. A second Master Alarm light is on Panel 1A.
FUEL CELL INDICATOR switch	Controls which fuel cell will be connected to indicators.
FUEL CELL MAIN BUS A and MAIN BUS B switches and indicators	Two groups of four switches and three indicators each, one group for Main Bus A and one for Main Bus B. Switches for each fuel cell are pushed up to connect cell

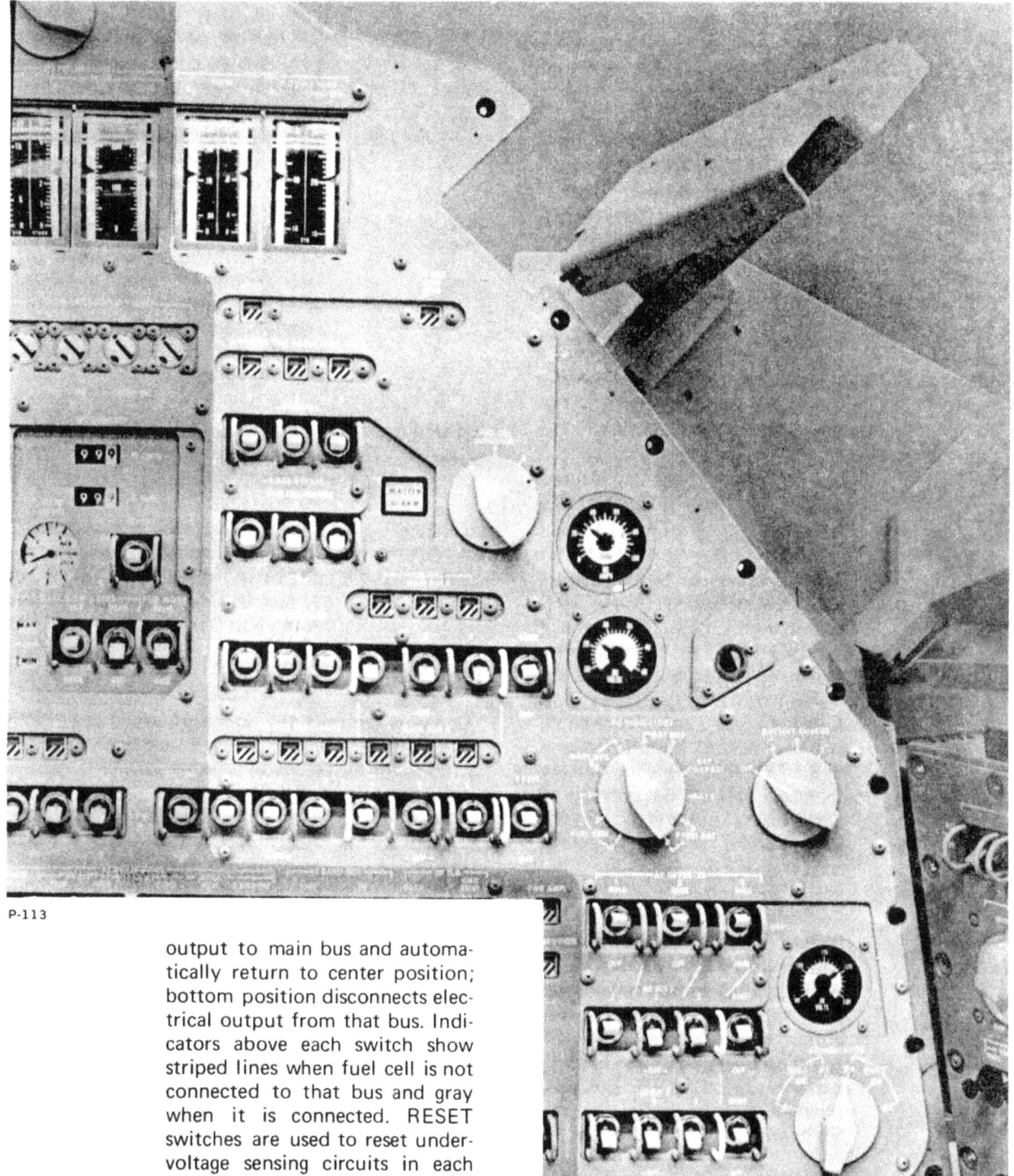

Panel 3B

output to main bus and automatically return to center position; bottom position disconnects electrical output from that bus. Indicators above each switch show striped lines when fuel cell is not connected to that bus and gray when it is connected. RESET switches are used to reset undervoltage sensing circuits in each main bus after a voltage drop (below 26.25 volts) that has illuminated the caution light.

FUEL CELL PURGE switches	These control purging of the fuel cells. H_2 position (up) opens valve on hydrogen side of

	selected fuel cell to purge impurities from hydrogen electrodes; O_2 position acts in same manner for oxygen side. Center position is off.
FUEL CELL REACTANTS switches and indicators	Control flow of reactants (hydrogen and oxygen) to selected fuel cell. ON opens valve to permit hydrogen and oxygen flow; switch automatically returns to center position. Indicators show striped lines if reactant flow has been cut off, gray if flow is normal.
DC AMPS meter	Indicates dc current of source selected by DC INDICATORS switch.
DC VOLTS meter	Indicates dc voltage of source selected by DC INDICATORS switch.
DC INDICATORS switch	Selects source to be monitored on DC AMPS and DC VOLTS meters. FUEL CELL 1, 2, and 3 are the three fuel cell powerplants; MAIN BUS A and B are the two dc main buses; BAT BUS A and B are the buses from entry and post-landing batteries A and B; BAT CHARGER is the battery charger; BAT C is entry and post-landing battery C; PYRO BAT A and B are the pyrotechnic batteries.
BATTERY CHARGE switch	Controls ac and dc power to the battery charger and selects the battery to be charged. Positions A, B, and C routes ac and dc power to the charger and routes output power of the charger to the entry and post-landing battery selected. OFF position disconnects electrical power to the charger.
AC INVERTER switches	Nine switches in one group of three and two groups of four. Top three switches control dc power to the three inverters, which convert dc to ac power. Inverter 1 can receive power only from dc Main Bus A, Inverter 2 can receive power only from dc Main Bus B, and Inverter 3 can receive power from either dc main bus. The two groups of four switches control the output of the inverters; in upper group applies each inverter's output to ac Bus 1 and the lower group applies the output to ac Bus 2. RESET switches at the right of each group are used to reset sensing circuits after over- or undervoltage has been sensed (which activates caution and warning lights).
AC VOLTS meter	Indicates ac voltage of source selected by AC INDICATOR switch.
AC INDICATOR switch	Selects source for display on AC VOLTS meter. BUS 1 sources are phases (ϕ) A, B, and C of ac Bus 1, and BUS 2 sources are Phases A, B, and C of Bus 2.

Panel 3C

Contains controls for operation and management of the telecommunications subsystem and two switches for the electrical power subsystem.

S-BAND NORMAL switches	The XPNDR switch activates the No. 1 (PRIM) or No. 2 (SEC) transponder of the unified S-band equipment, or switches both transponders off (center position). The PWR AMPL switches select either the No. 1 (PRIM) or No. 2 (SEC) S-band power amplifier (left switch) and the high-level (HIGH) or low-level (LOW) mode of operation; the center position selects the bypass mode for the transponder selected by the XPNDR switch. The MODE switches select what will be transmitted on the S-band equipment: VOICE, RELAY (relays VHF-AM voice from LM or voice and data from an extravehicular astronaut) PCM (pulse-code modulation

86R

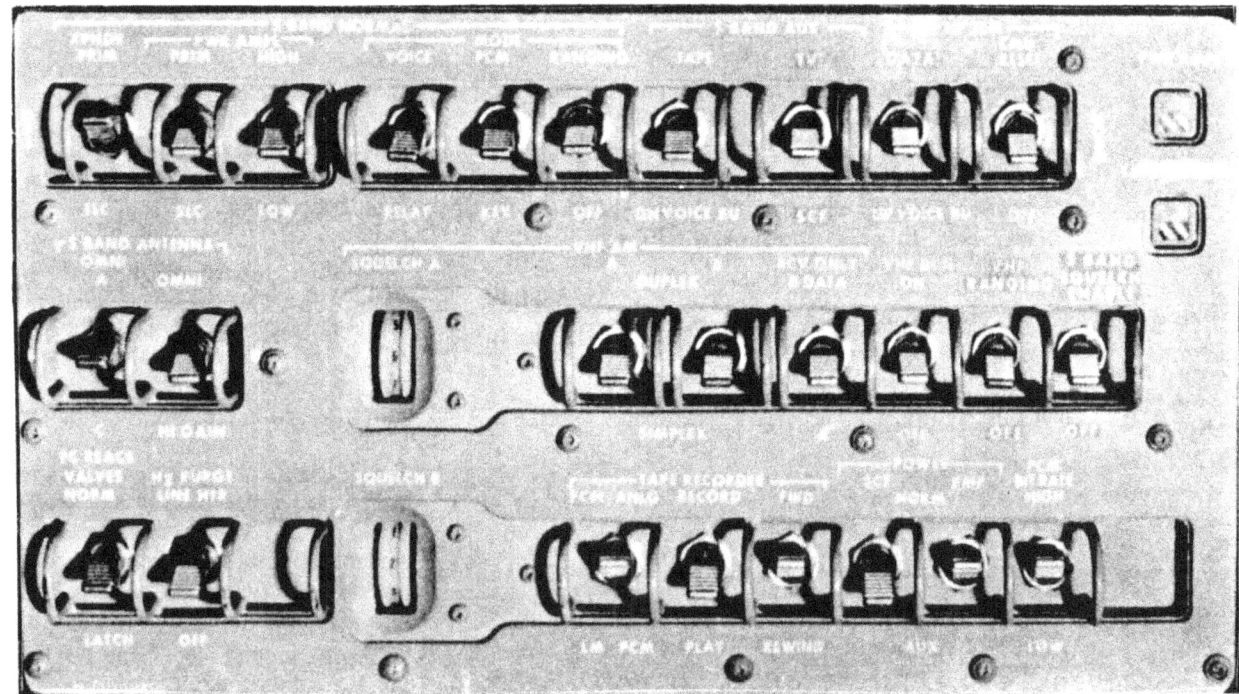

Panel 3C

S-BAND AUX switches	data), KEY (emergency key), and RANGING (retransmits ranging signals). These activate FM transmitter and the power amplifier not selected by the PWR AMPL switch and connect equipment to the transmitter: tape playback function selected by left-hand TAPE RECORDER switch, TV is television camera, SCI is scientific instruments. DWN VOICE BU position selects pulse-modulated baseband voice mode of transponder selected by XPNDR switch.
UP TLM switches	These control mode of telemetry from ground. DATA is normal position for receiving voice and data from ground; UP VOICE BU (backup) switches voice transmission from the ground to another carrier; NORM applies power to the up-data link equipment CSM RESET resets all of the real-time command relays except one (switch is spring-loaded to return to the NORM position); OFF cuts off power to the up-data link quipment.
PWR AMPL indicator	Shows striped lines when either S-band power amplifier is activated, gray when both are off.
TAPE MOTION indicator	Shows striped lines when data storage equipment tape is in motion, gray when inactive.
S-BAND ANTENNA switches	These select one of four flush-mounted S-band antennas on command module for transmission and reception. Left-hand switch is for selection of omnidirectional antennas A, B, or C; right-hand switch in up (OMNI) position activates the left-hand switch and in down position selects omni-directional antenna D. In HIGH-GAIN selects high-gain antenna on service module and disables A, B, C or D omni-directional antennas.
VHF-AM switches and thumbwheels	The A and B switches select either the DUPLEX (different frequencies for transmission and re-

86S

	ception) or SIMPLEX modes; the B modes are backup. The RCV ONLY switch selects a received only; DATA B position selects receiver for low-bit-rate PCM data; A position is used to select receiver for monitoring during recovery. SQUELCH thumbwheels are used to adjust minimum RF (radio frequency) levels.
VHF BCN switch	Used to activate the VHF beacon equipment for recovery.
VHF RANGING	Used to activate VHF ranging generator.
FC REACS VALVES switch	Latch position applies holding voltage to fuel cell reactant valves to prevent inadvertent closing of them during launch, ascent, and orbital insertion. NORM switch disconnects the holding voltage.
H₂ PURGE LINE HTR switch	Applies power to redundant heater system in fuel cell hydrogen purge line to prevent freezing during hydrogen purge.
TAPE RECORDER switches	These control operation of data storage (tape recorder) equipment. Center switch turns the recorder on (RECORD), sets it to play back (PLAY), or turns it off (center position). The right-hand switch controls the direction of the tape: forward (TWD) or backward (REWIND); the center position is off. The left-hand switch selects what will be played back: PCM/ANLG is CSM pulse-code modulated data, CSM and LM voice, and three analog channels of scientific information; LM/PCM is LM pulse-code modulated data recorded by the CSM data storage equipment.
TLM INPUTS	These control what is transmitted to ground. The PCM switch on HIGH selects the normal pulse-code modulation data mode and the normal recording speed; LOW selects the narrow-bit-rate mode and a slow recording speed.

pilot (center couch), and RIGHT the LM pilot (right couch).

Panel 4

Contains ac power switches and circuit breakers for the environmental control, telecommunications, and service propulsion subsystems.

SPS GAUGING switch	Applies ac power to service propulsion subsystem quantity gauging system. AC1 is ac Bus 1, AC 2 is ac Bus 2, and center position cuts off all power to gauging system.
TELCOM switches	Two identical switches which apply power from ac Bus 1 or ac Bus 2 to telecommunications subsystem circuit breakers. Group 1 is essential equipment; Group 2 is non-essential telecommunications equipment (for example, data storage equipment, and voice recorder).
ECS GLYCOL PUMPS switch	Selects source of ac power for environmental control subsystem primary loop water-glycol pump. Only one pump can be operated at a time; the other is a backup.
SUIT COMPRESSOR switches	These select source of ac power to motors in suit compressors No. 1 and No. 2. They may be operated simultaneously

Panel 4

CIRCUIT BREAKERS	Circuit breakers are for the suit compressors and water-glycol pumps of the environmental control subsystem. The circuit breakers are the push-pull reset type.

Panel 5

Contains controls for the electrical power subsystem and interior CM lighting and circuit breakers for instruments and the electrical power and environmental control subsystems.

FUEL CELL PUMP switches	These select source of ac power to parallel pump motors in each of the three fuel cell powerplants; one pump drives the hydrogen circulating pump and water separation centrifuge, the other drives the water-glycol circulating pump.
G/N PWR switch	Selects source of ac power for guidance and navigation lighting.
MAIN BUS TIE switches	These control connection of batteries to main dc buses. BAT A/C position connects battery bus A to dc Main Bus A and battery bus C to dc Main Bus B and disconnects BATTERY CHARGER switch on Panel 3B from battery bus A. BAT B/C position connects battery bus B to dc main bus B and battery bus C to Main Bus A and disconnects BATTERY CHARGER switch from battery bus B and battery C. OFF positions disconnect batteries and main buses and connect the respective batteries to the BATTERY CHARGE switch. AUTO positions connects respective batteries to main buses automatically at CM-SM separation.
BAT CHGR switch	Selects source of ac power for the battery charger.
NONESS BUS switch	Selects source of dc power (MNA is dc Main Bus A and MNB is dc Main Bus B) to non-essential equipment bus, or cuts off power to non-essential equipment.
INTERIOR LIGHTS	These control lights in CM cabin. INTEGRAL rheostat controls

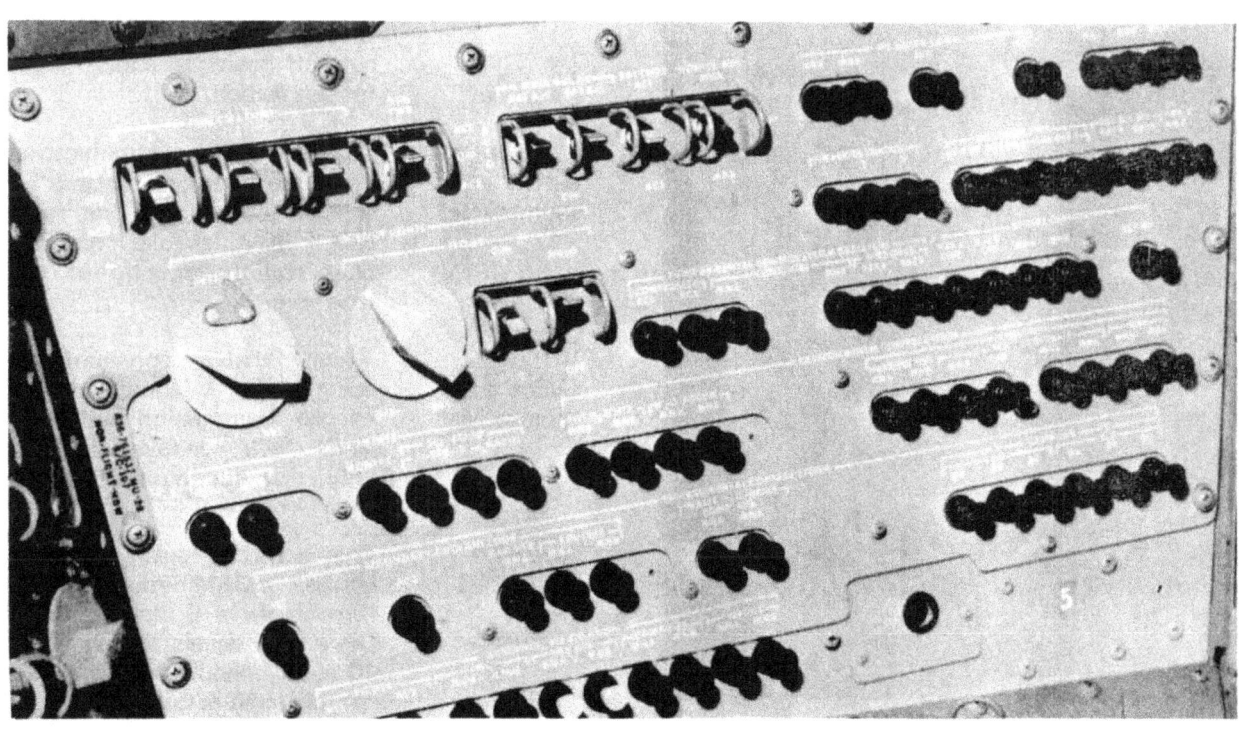

Panel 5

brightness or turns off electroluminescent or turns off floodlights in pilot's area. DIM switch selects primary (1) or secondary (2) floodlights. FIXED switch turns on lights not controlled by rheostat.

Panel 6

Contains audio center controls for the LM pilot (right couch). Identical controls are located on Panel 9 for the commander (left couch) and Panel 10 for the CM pilot (center couch). Panel 10 is mounted at right angles to the bottom of Panel 2.

MODE INTERCOM and thumbwheel	The switch selects the method of communications. INTERCOM/PTT position provides hot mike and VOX (voice-operated relay) operation for the intercom and push-to-talk (PTT) operation for RF (radio frequency) transmission; PTT position provides push-to-talk operation only for intercom and RF transmission; VOX position provides voice-operated relay operation only for intercom and RF transmission. Thumbwheel adjusts sensitivity of voice-operated relay.
PAD COMM switch and thumbwheel	On T/R, enables headsets to receive and transmit over hard lines (cable) with ground. On RCV, crewman can receive only over the circuits. The thumbwheel adjusts the volume of the reception.
S-BAND switch and thumbwheel	These enable communications using S-band equipment. T/R is transmit and receive, RCV is receive only, and the thumbwheel controls volume.
MASTER VOLUME thumbwheel	Adjusts volume from the earphone amplifier to the earphone.
POWER switch	On AUDIO/TONE, provides primary power to the audio center module and enables the crew alarm to be heard in LM pilot's headset. On AUDIO, the crew alarm tone will not sound in the LM pilot's headset. Off position cuts off power to audio module for this station.
INTERCOM switch and thumbwheel	These enable communications using the intercom system. T/R enables transmission and reception, RCV enables reception only, and the thumbwheel controls volume.
VHF AM switch and thumbwheel	These enable communication using VHF AM equipment. T/R enables transmission and reception, RCV enables reception only, and the thumbwheel controls volume.
AUDIO CONTROL switch	This controls routing of audio signals. NORM routes signals through Panel 6 module; BACK UP routes signals through Panel 10 audio module in case Panel 6 module malfunctions.

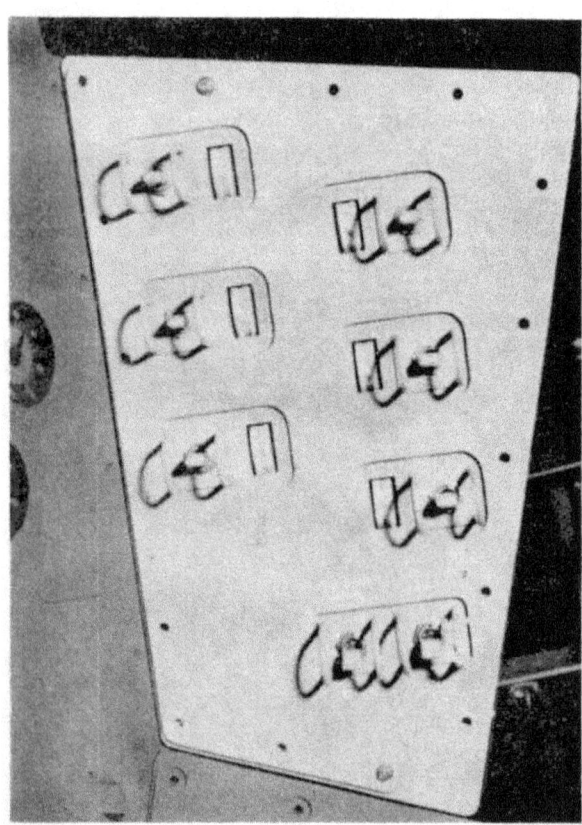

Panel 6

SUIT POWER switch — When up, applies power to the two microphones and the biomedical preamplifiers worn by the pilot.

Panel 7

Contains controls to supply power for elements of the stabilization and control subsystem.

EDS POWER switch — Supplies power from all three entry batteries to the emergency detection system; system operates only during boost and is turned off at all other times.

TVC SERVO POWER switches — These control ac power source to servo channels in thrust vector position servo amplifier and dc power source to servo clutches through the amplifier. Switch 1 controls the primary channels and clutches and Switch 2 the secondary channels and clutches. AC1/MNA is ac Bus 1 and dc Main Bus A, and AC2/MNB is ac Bus 2 and dc Main Bus B.

FDAI POWER switch — Supplies power that drives electronics for flight director attitude indicator No. 1 (1 position) or No. 2 (2 position), or both at the same time (BOTH).

SCS HAND CONTROL switch — Supplies power for operation of rotation and translation controls. Position 1 is for Rotation control No. 1, position 2 for rotation control No. 2, and BOTH for both at the same time. All three positions supply power for operation of translation control.

SCS ELECTRONICS switch — Supplies power to electronics control assembly (ECA) or to gyro display couplers and electronics control assembly (GDC/ECA).

BMAG POWER swtiches — Supplies power to gyro assemblies which contain body-mounted attitude gyros. Switch 1 is for Gyro Assembly No. 1 and Switch 2 for Gyro Assembly No. 2. WARN UP position supplies power to electronics and heaters in assemblies, ON position is for operation of assemblies.

DIRECT O_2 valve — Turned counterclockwise (to left) to open valve for flow of oxygen directly into the suit circuit.

Panel 8

Contains CM lighting, float bag, and sequential events control switches and circuit breakers for stabilization and control, reaction control, service propulsion, and other subsystems.

INTERIOR LIGHTS switches — These control lights in CM cabin. NUMERICS rheostat controls brightness and turns off the numerics (flashing numbers) on the two computer keyboards, the entry monitor system, and the timers. The FLOOD rheostat controls brightness and turns off floodlights in the commander's (left-hand) area. INTEGRAL

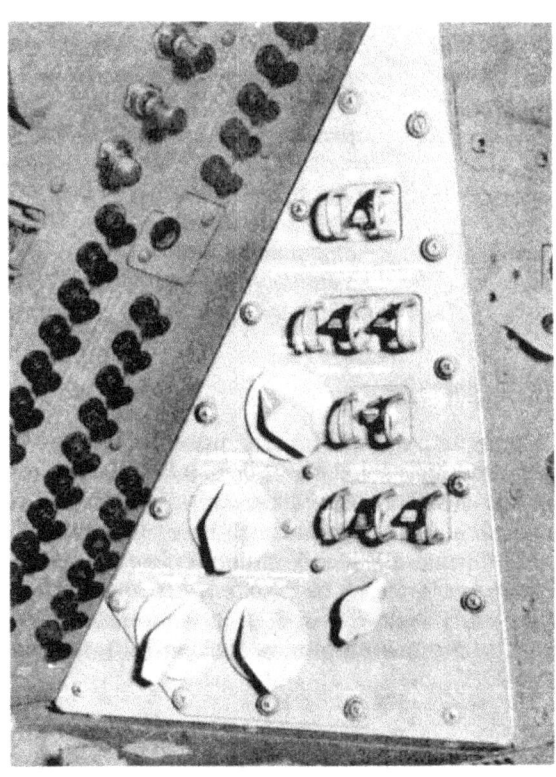

Panel 7

Panel 8

	theostat controls brightness and turns off electroluminescent lights on panels in commander's area (Panels 1, 7, 8, 9, and part of Panel 2).
FLOOD switches	DIM switch selects the primary (1) or secondary (2) floodlights in commander's area. FIXED position of right-hand switch turns on lamps not controlled by FLOOD rheostat; POST LDG position connects post-landing battery bus to lights on commander's couch and left-hand area.
FLOAT BAG switches	Three identical switches, one for each of three float bags of CM uprighting system. FILL positions start compressors which inflate bags, and VENT positions turn off compressors and open vent lines to bags. These lever lock-type switches remain in the VENT position until splashdown.
SEQ EVENTS CONTROL SYSTEM switches	Two LOGIC switches supply power to master events sequence controllers. PYRO ARM switches supply power to controllers for pyrotechnic devices; a key-operated lock and guard assembly is placed over the switches to hold them in the SAFE position until just before launch, when commander unlocks and removes assembly. All four SEQ EVENTS switches are lever lock-type.

Panels 9 and 10

Contain communications controls identical to those described for Panel 6. The BACKUP position of the AUDIO CONTROL switch routes the commander's audio signals through Panel 6 and the CM pilot's through Panel 9. Although containing identical controls, Panel 10 is square rather than triangular as are Panels 6 and 9. It is mounted at right angles to the main display console at the bottom of Panel 2.

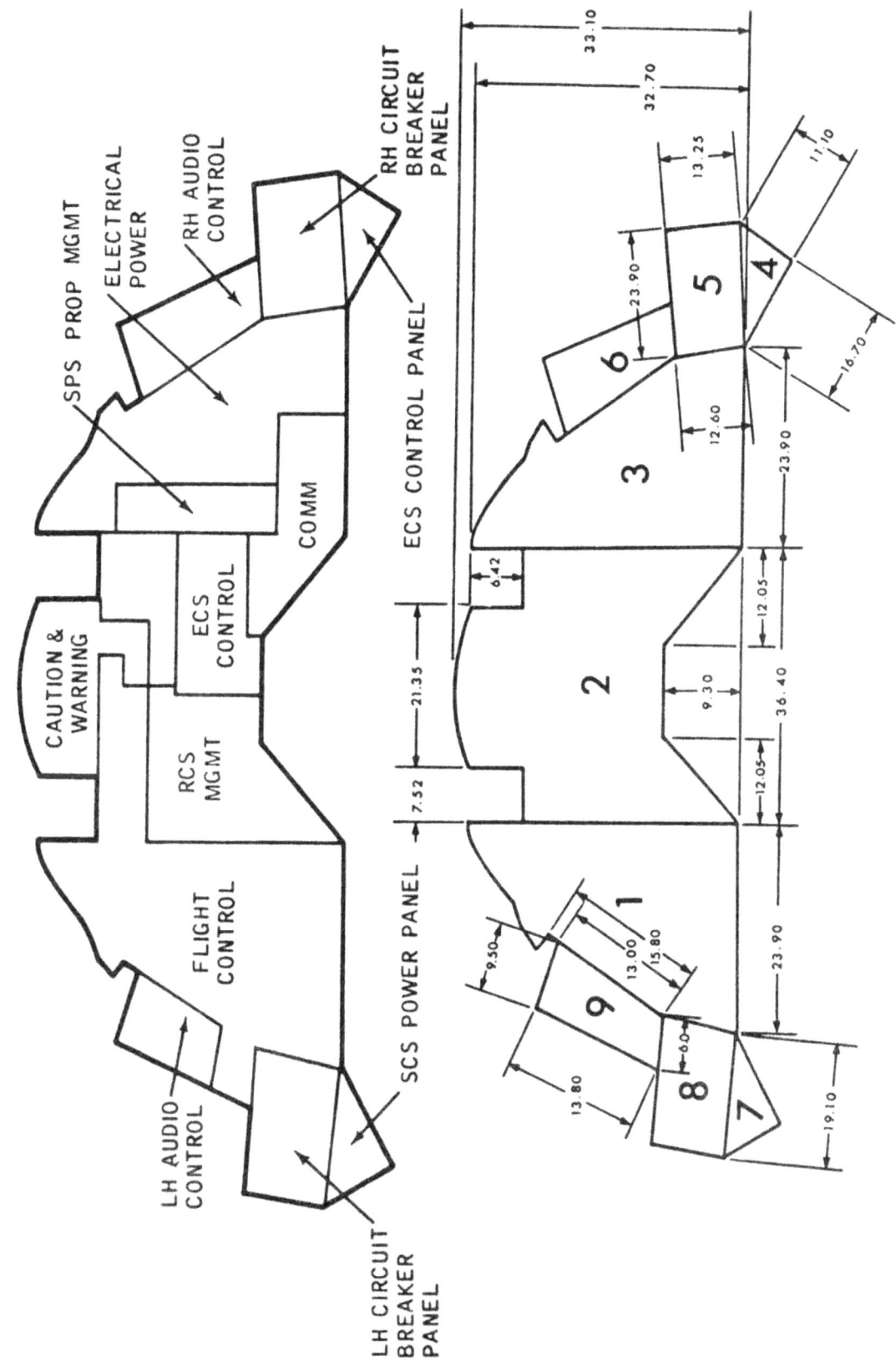

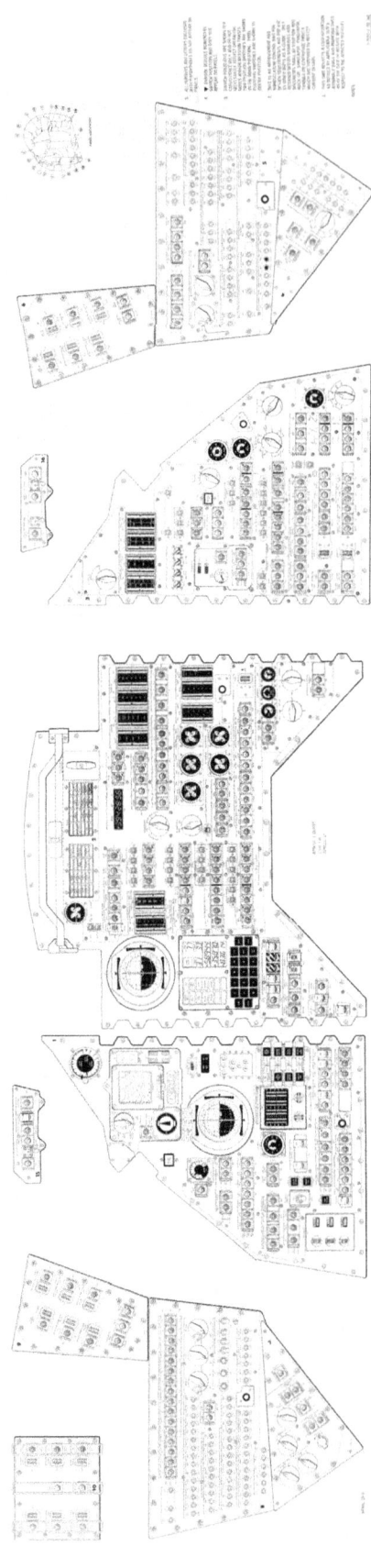

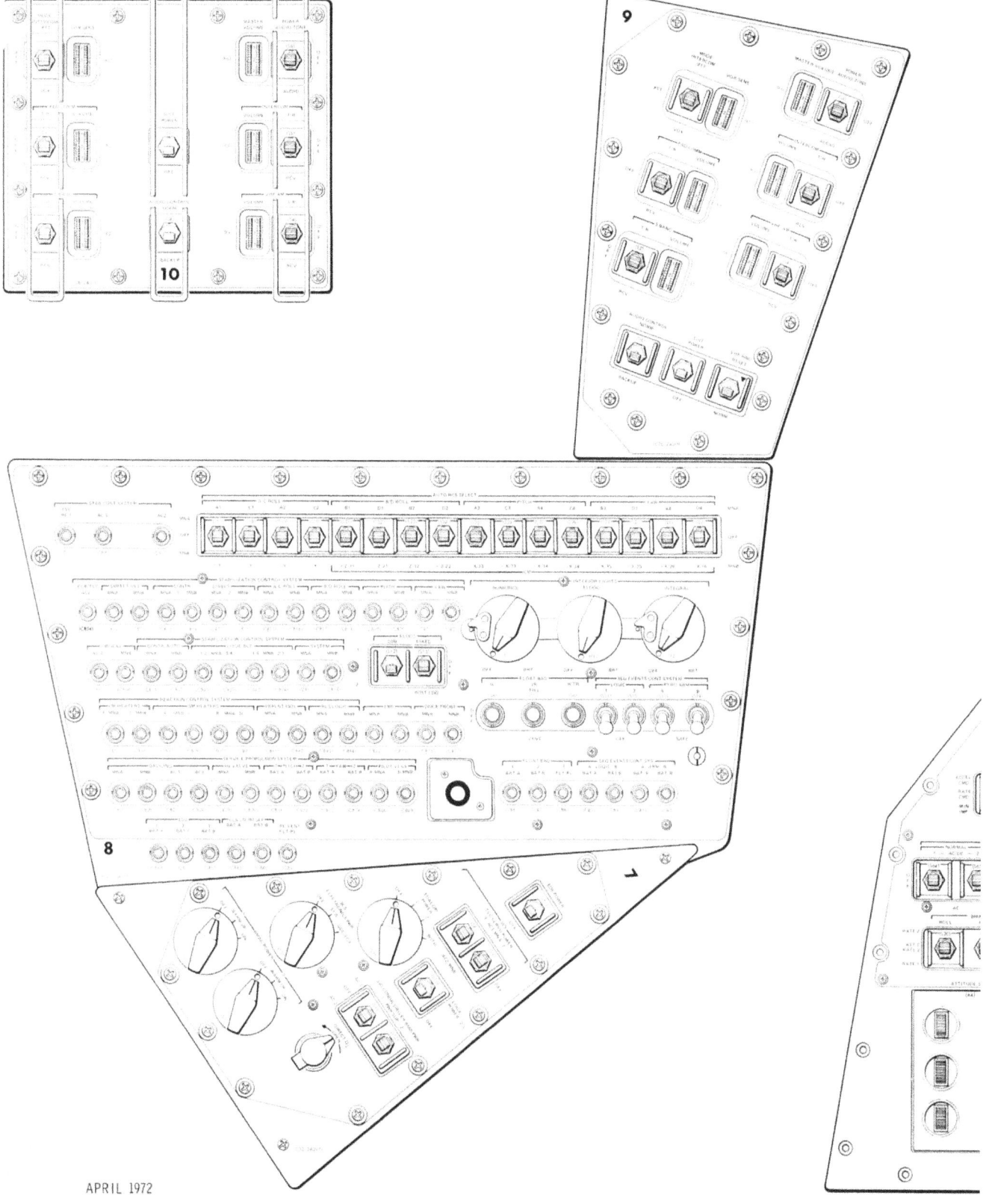

APRIL 1972

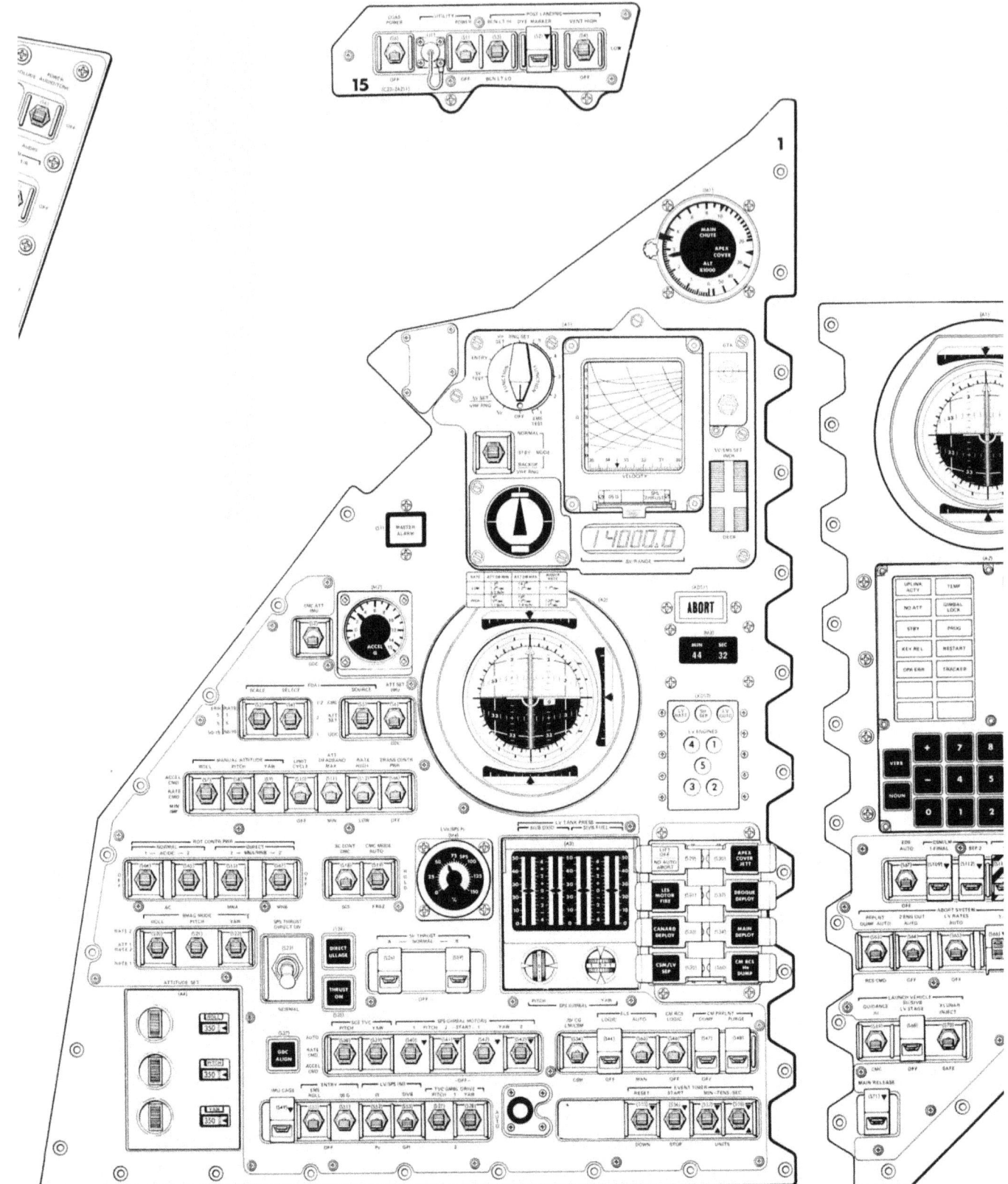

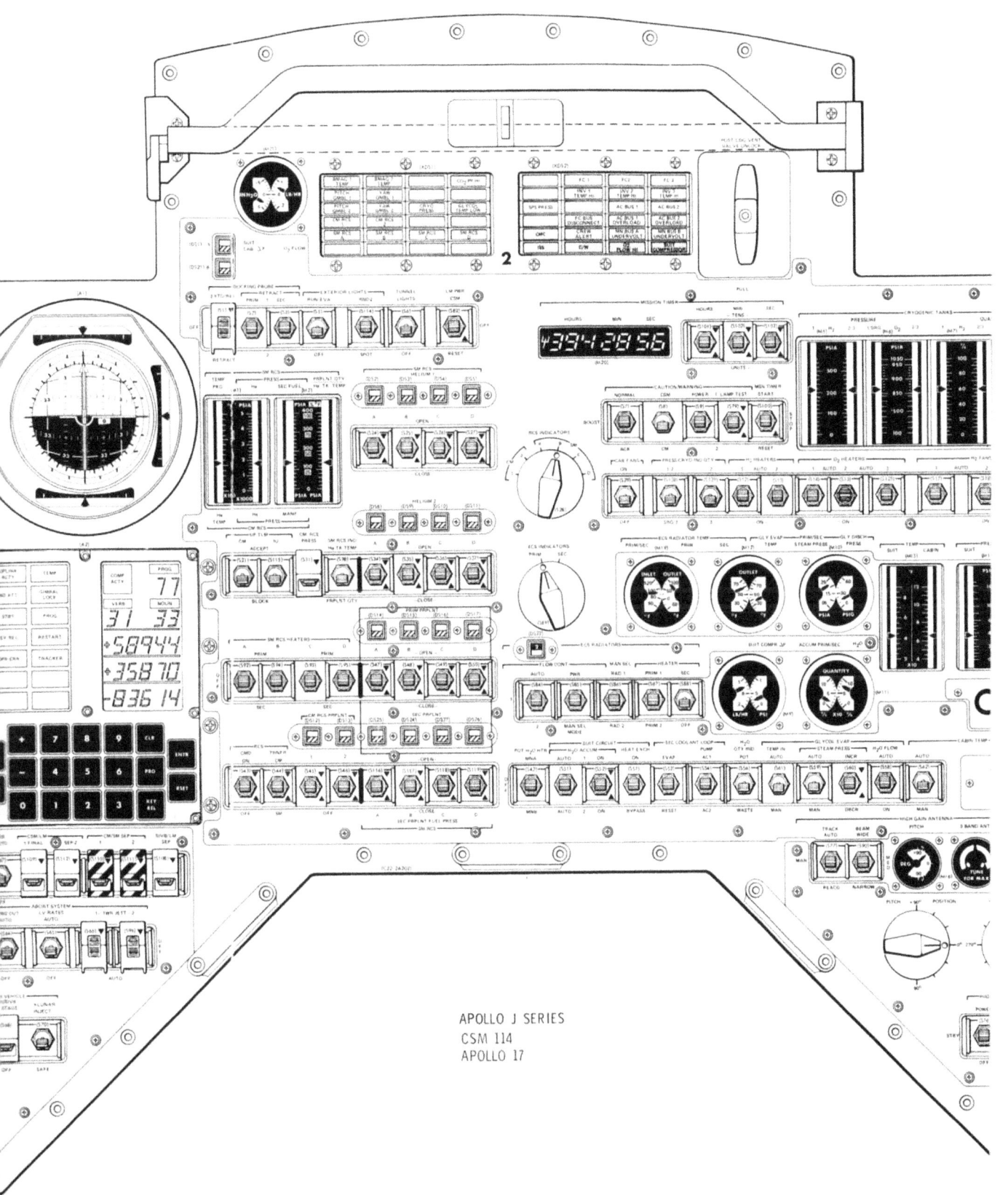

APOLLO J SERIES
CSM 114
APOLLO 17

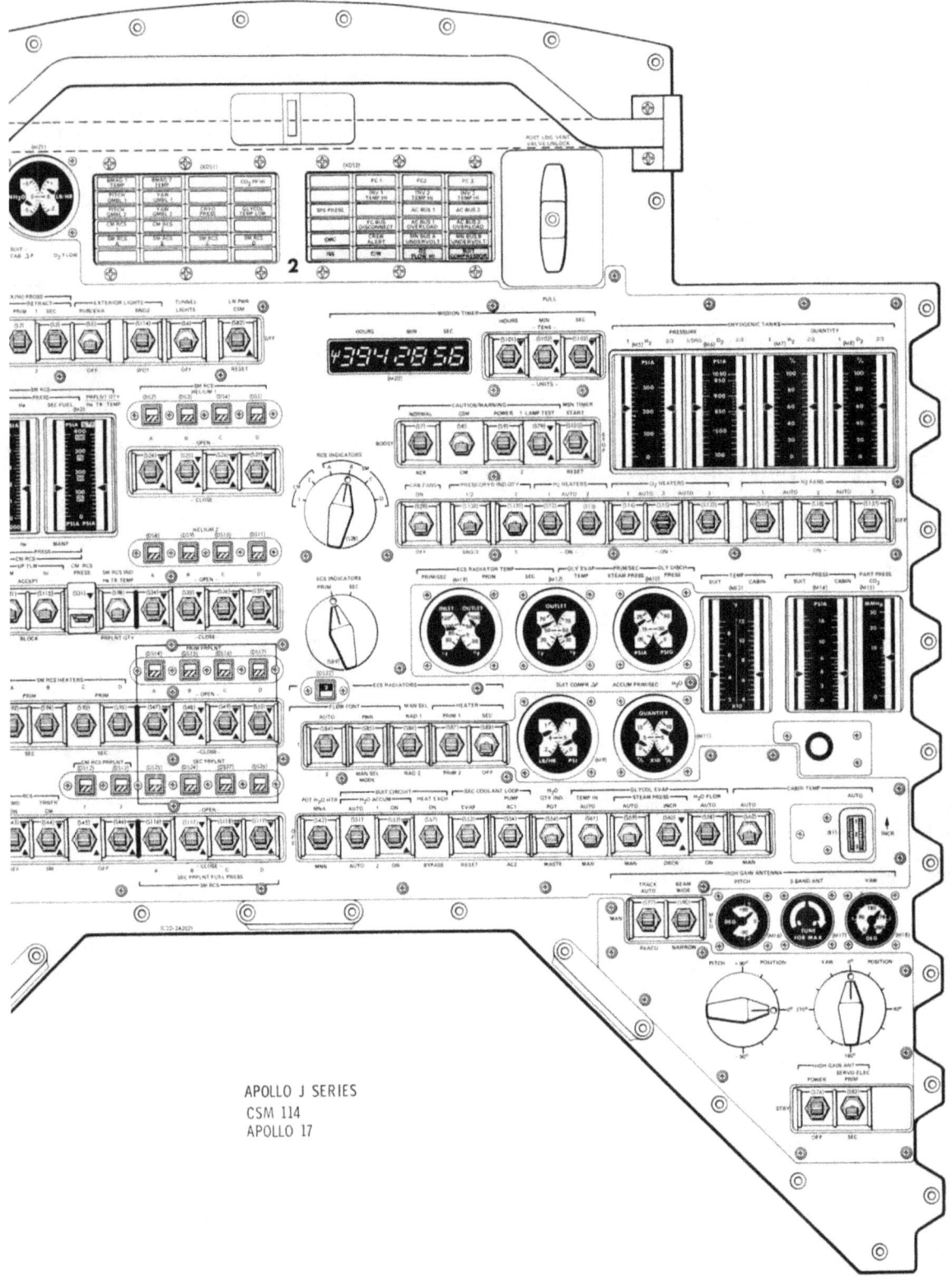

APOLLO J SERIES
CSM 114
APOLLO 17

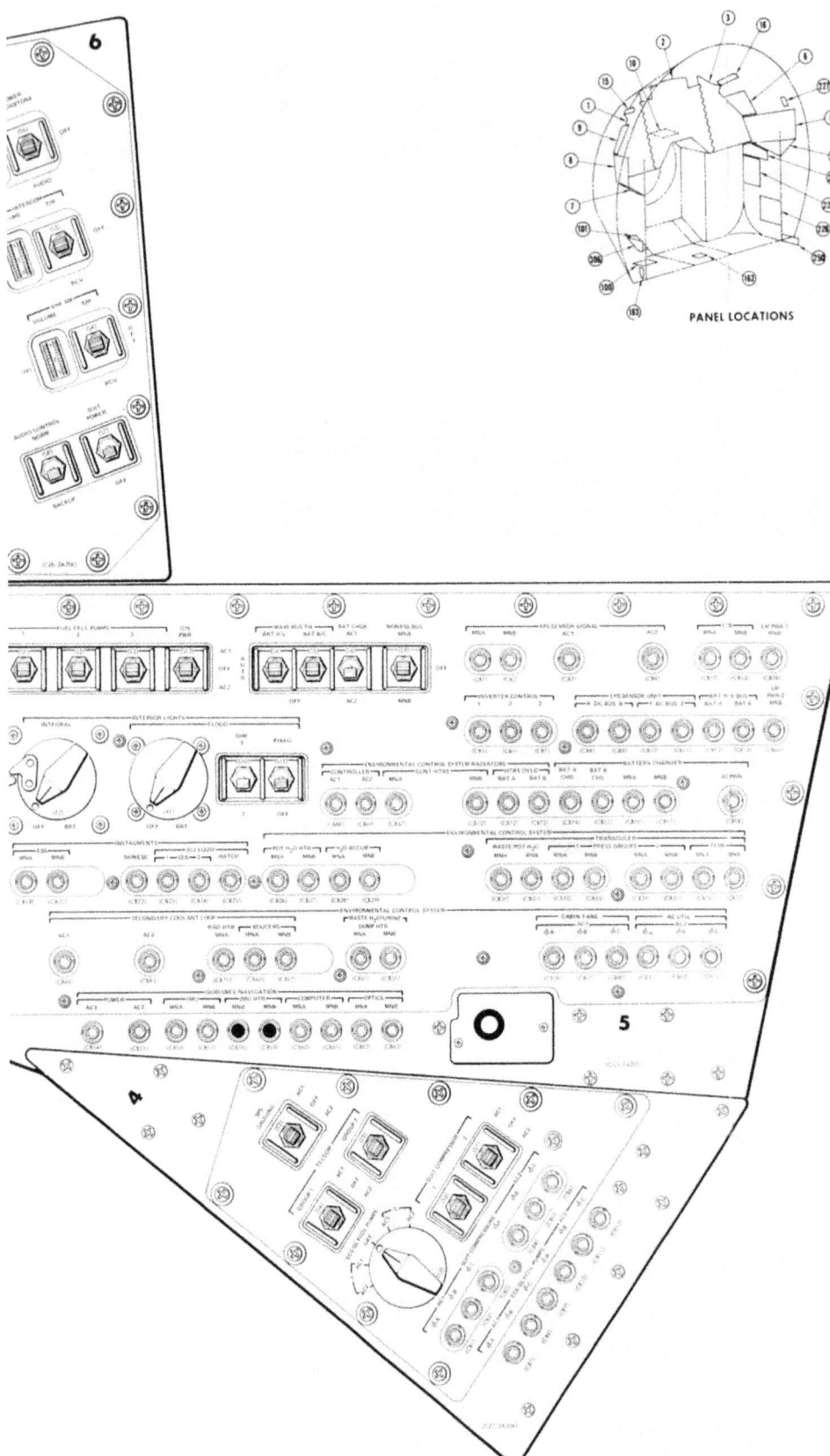

PANEL LOCATIONS

5. ALL NUMBERS AND LETTERS ENCLOSED WITH PARENTHESIS DO NOT APPEAR ON PANELS.

4. ▼ SYMBOL DENOTES MOMENTARY SWITCH POSITION AND DOES NOT APPEAR ON PANELS.

3. SWITCH POSITIONS ARE SHOWN FOR CONVENIENCE ONLY AND DO NOT NECESSARILY REFLECT OPERATING MODES AND/OR MISSION PHASES. TWO POSITION SWITCHES ARE SHOWN IN THE DOWN POSITION. THREE POSITION SWITCHES ARE SHOWN IN CENTER POSITION.

2. THIS IS AN ARRANGEMENT AND NOMENCLATURE CONTROL DWG FOR DESIGN REQUIREMENTS AND FOR USE BY OTHER DEPTS AS A GUIDE. ONLY RELEASED DESIGN DRAWINGS AND DOCUMENTS SHALL BE USED FOR MFG, MOCK-UP, SIMULATOR, EVALUATOR, TRAINER OR CENTRIFUGE PANELS WHICH ARE INTENDED TO REFLECT CURRENT DESIGN.

1. THIS DWG REFLECTS A CONFIGURATION AS DEFINED BY APPLICABLE SCD'S, ASSEMBLY DWGS AND MARKING DWGS AS OF THE DATE OF RELEASE WITH RESPECT TO THE AFFECTED VEHICLES.

NOTES:

J-MDC-1 TO 16G
CSM LOGISTICS TRAINING

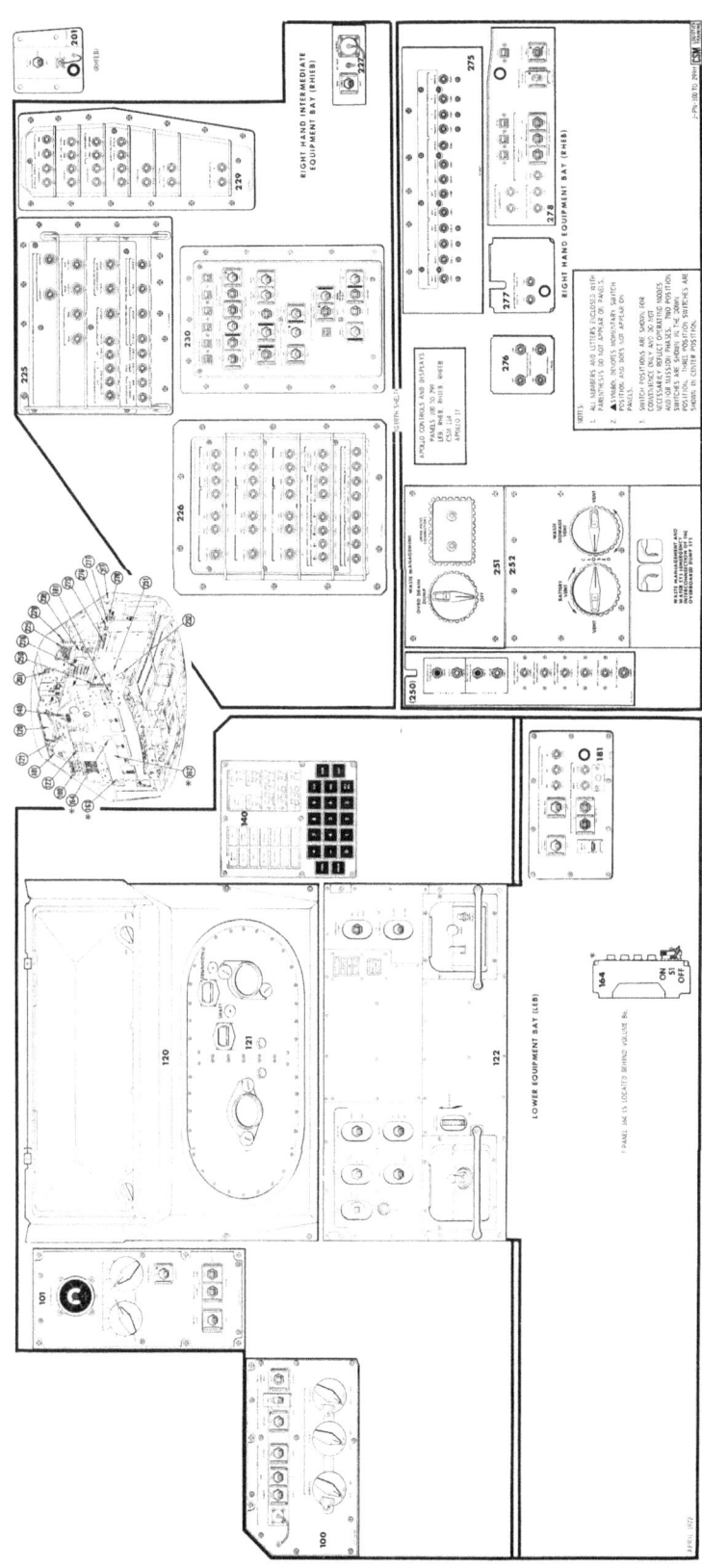

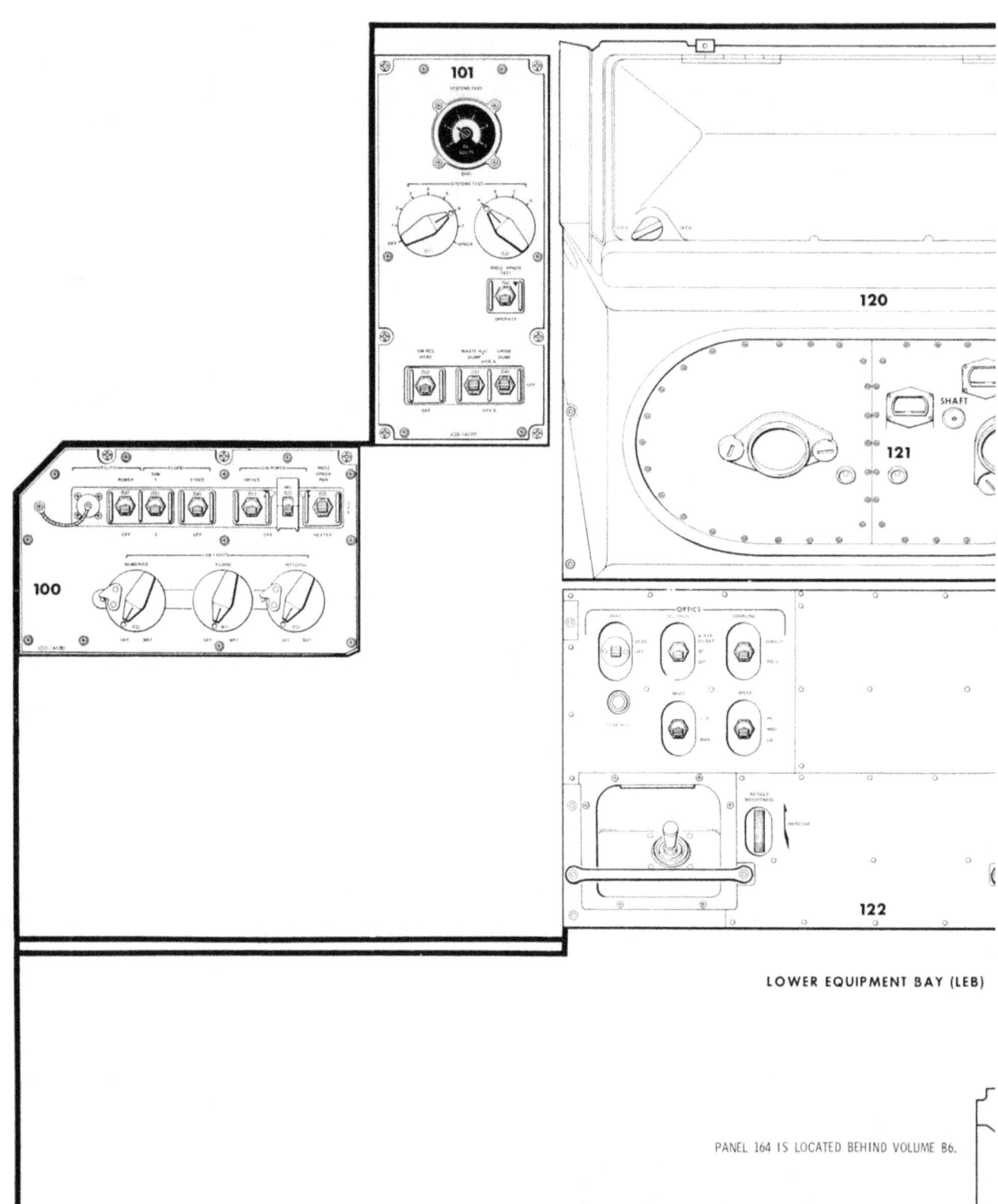

LOWER EQUIPMENT BAY (LEB)

PANEL 164 IS LOCATED BEHIND VOLUME B6.

APRIL 1972

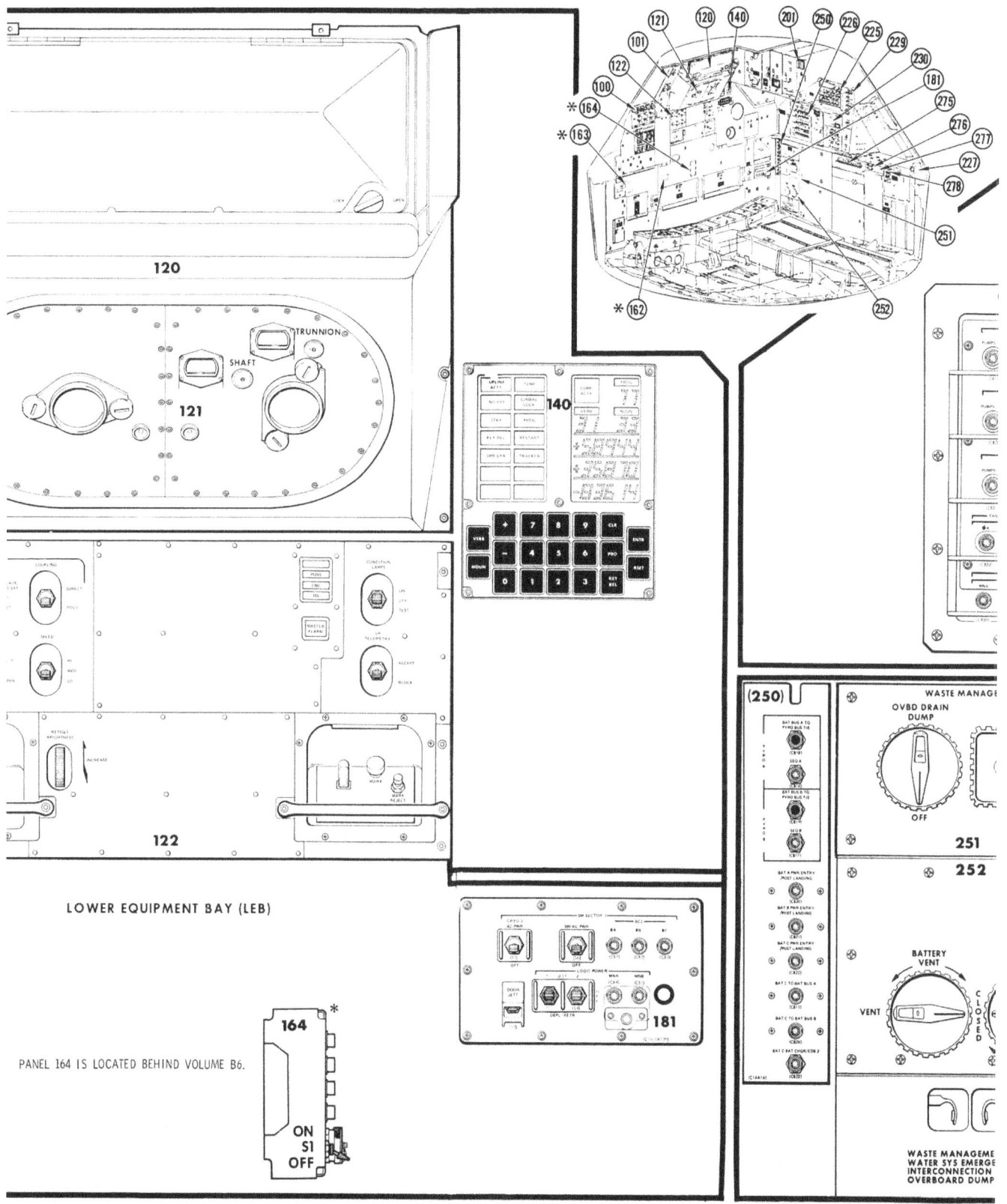

LOWER EQUIPMENT BAY (LEB)

PANEL 164 IS LOCATED BEHIND VOLUME B6.

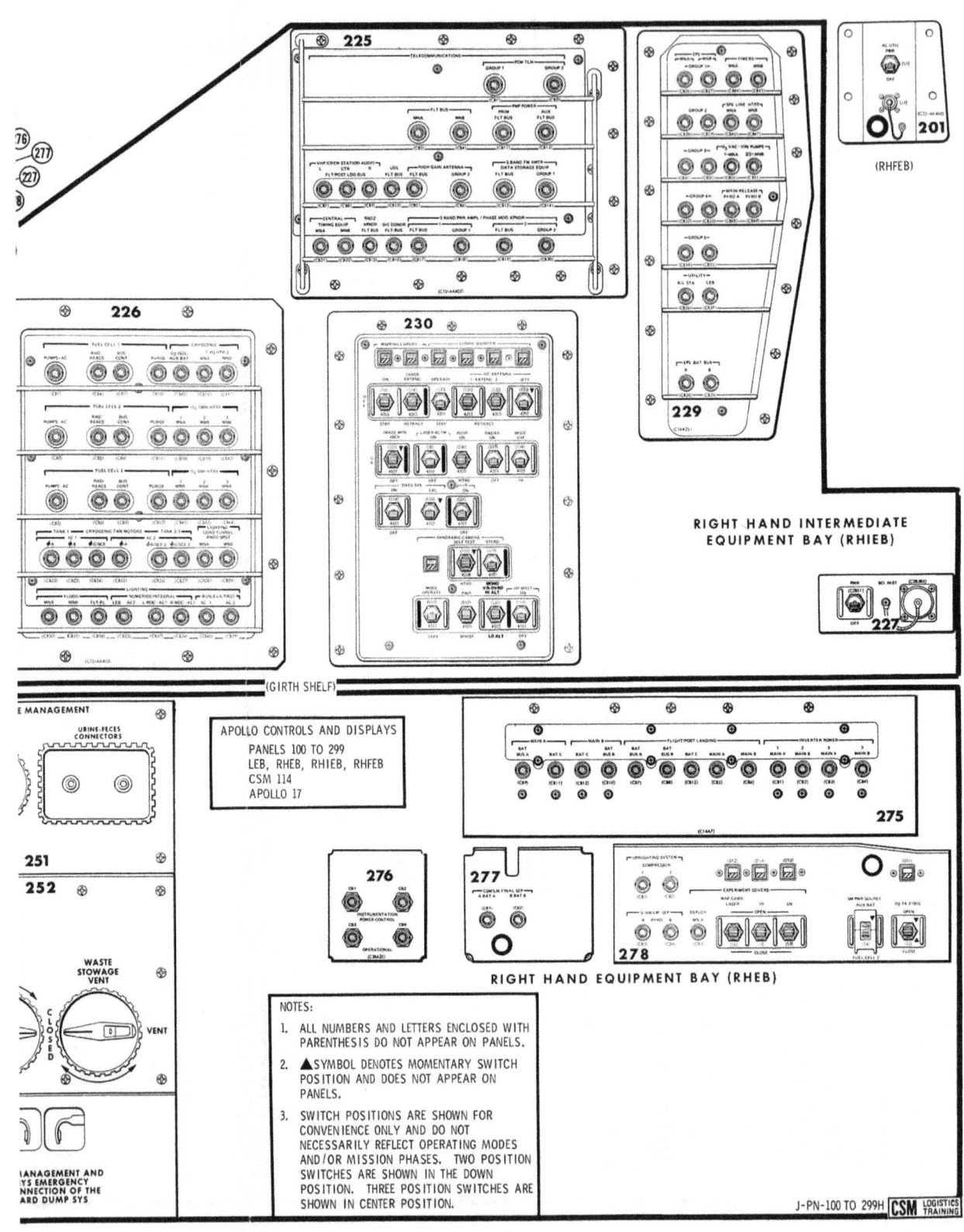

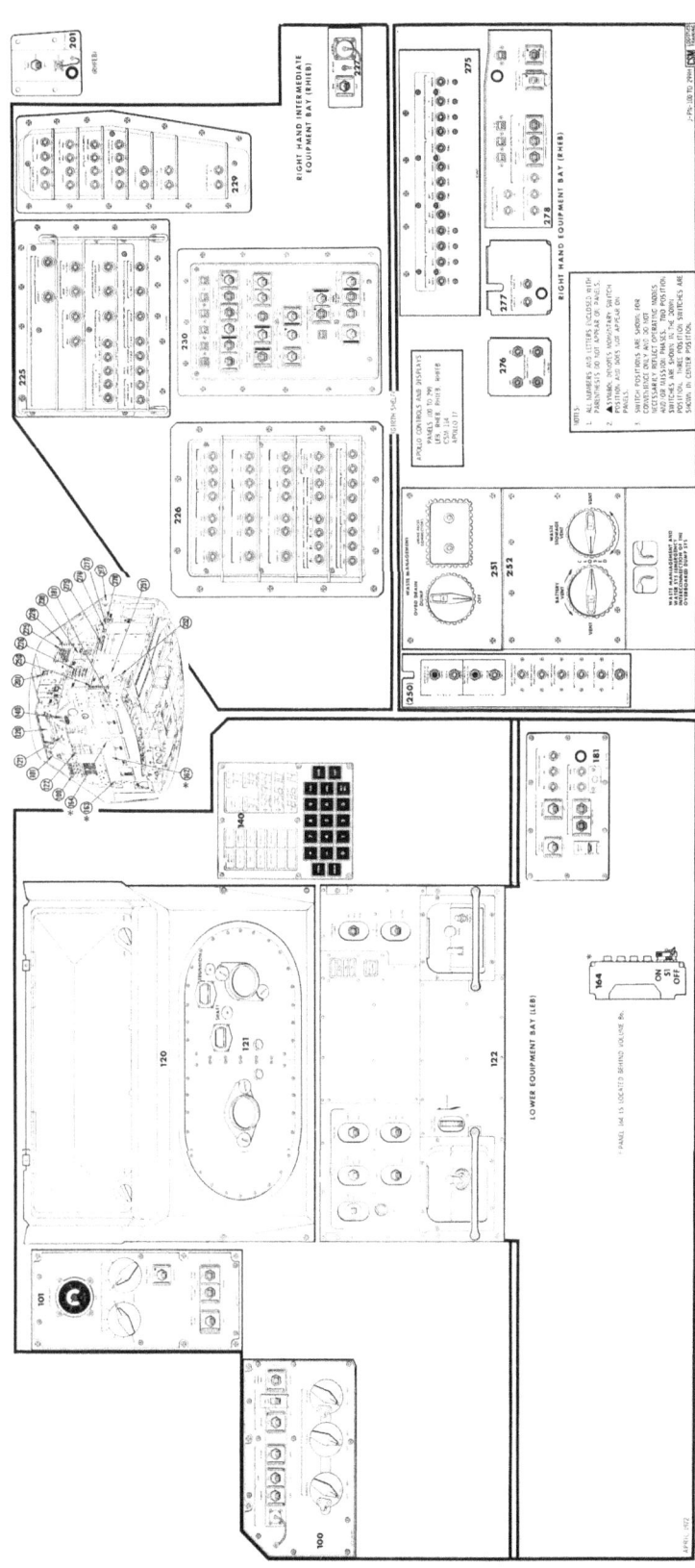

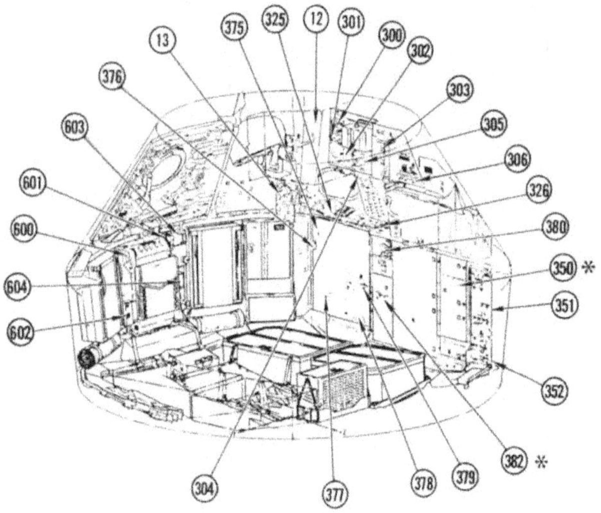

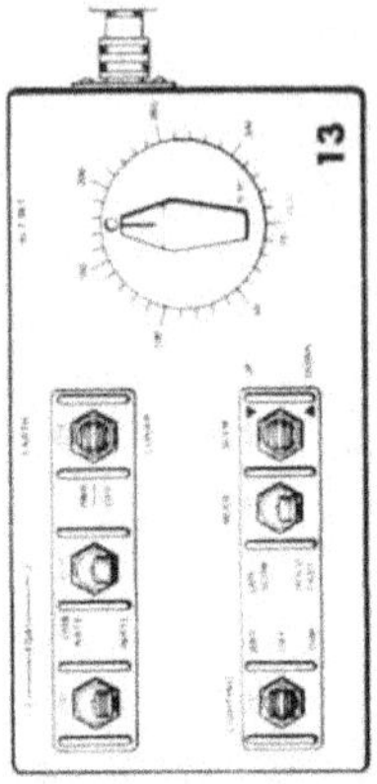

MDC (REAR TUNNEL)

LEFT HAND INTERMEDIATE
EQUIPMENT BAY (LHIEB)

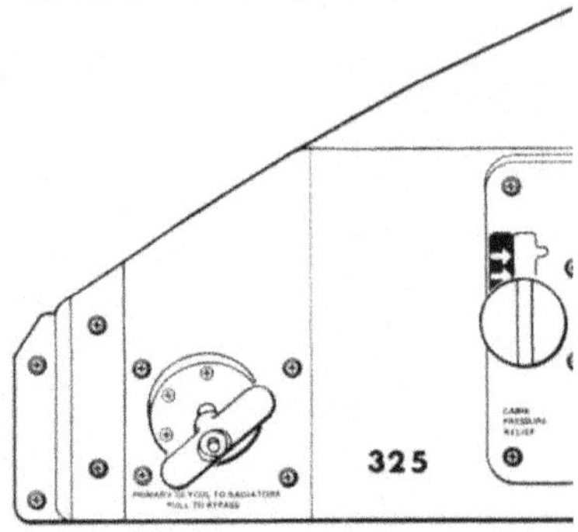

(GIRTH SHELF)

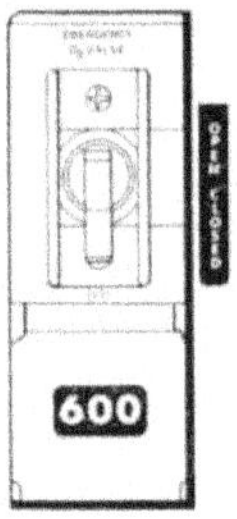

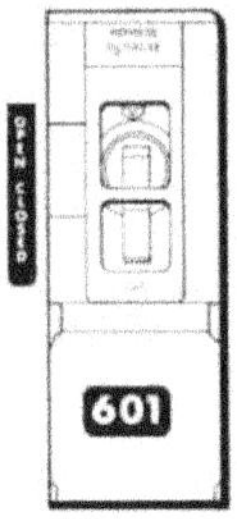

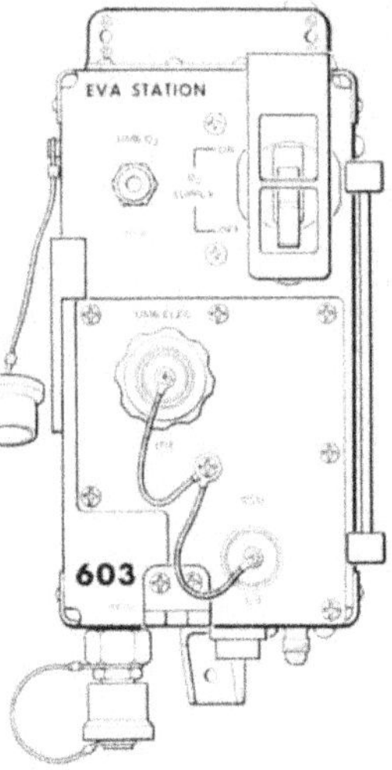

UPPER EQUIPMENT BAY (UEB)

NOTES:

1. ALL NUMBERS AND LETTERS ENCLOSED WITH PARENTHESIS DO NOT APPEAR ON PANELS.

2. ▼ SYMBOL DENOTES MOMENTARY SWITCH POSITION AND DOES NOT APPEAR ON PANELS.

3. SWITCH POSITIONS ARE SHOWN FOR CONVENIENCE ONLY AND DO NOT NECESSARILY REFLECT OPERATING MODES AND/OR MISSION PHASES. TWO POSITION SWITCHES ARE SHOWN IN THE DOWN POSITION. THREE POSITION SWITCHES ARE SHOWN IN CENTER POSITION.

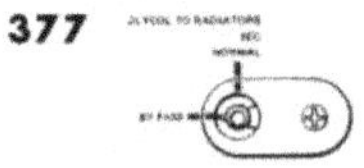

APRIL 1972

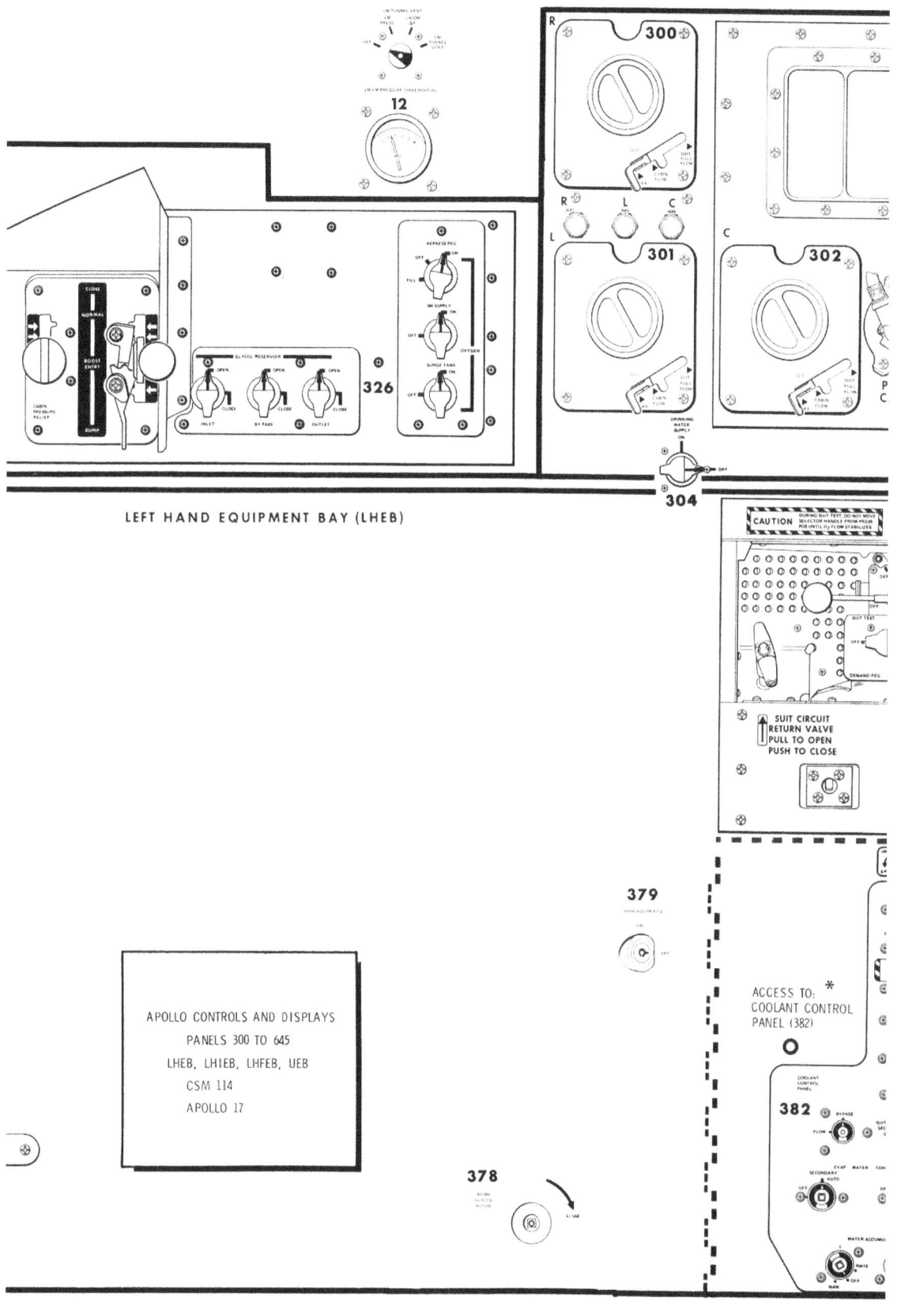

LEFT HAND EQUIPMENT BAY (LHEB)

APOLLO CONTROLS AND DISPLAYS
PANELS 300 TO 645
LHEB, LHIEB, LHFEB, UEB
CSM 114
APOLLO 17

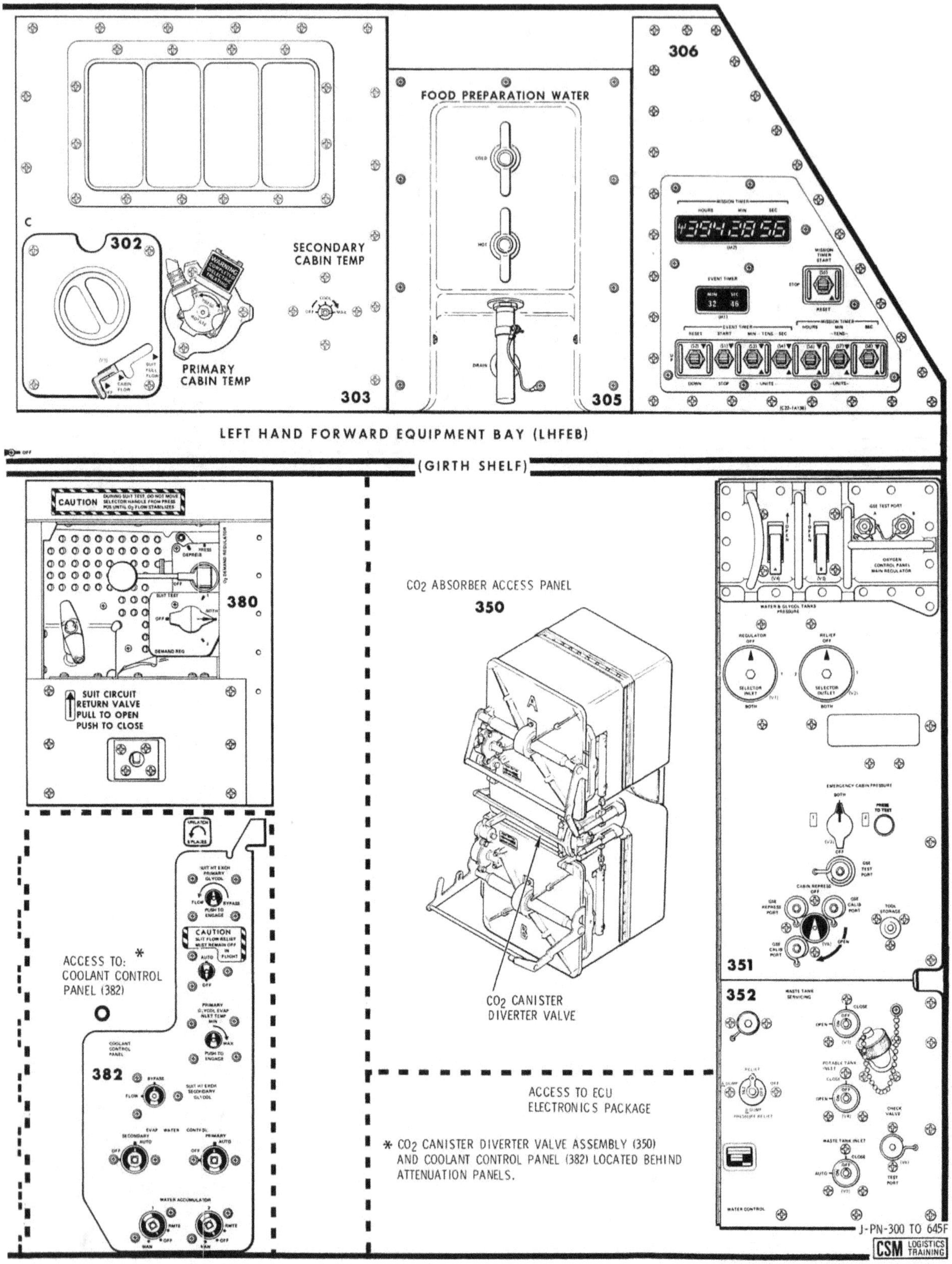

LEFT HAND FORWARD EQUIPMENT BAY (LHFEB)

(GIRTH SHELF)

- 302
- 303 PRIMARY CABIN TEMP / SECONDARY CABIN TEMP
- 305 FOOD PREPARATION WATER
- 306 (Mission Timer / Event Timer)
- 380 (Suit Circuit Return Valve – PULL TO OPEN / PUSH TO CLOSE)
- 350 CO₂ ABSORBER ACCESS PANEL — CO₂ CANISTER DIVERTER VALVE
- 351
- 352
- 382 ACCESS TO: COOLANT CONTROL PANEL (382)

ACCESS TO ECU ELECTRONICS PACKAGE

* CO₂ CANISTER DIVERTER VALVE ASSEMBLY (350) AND COOLANT CONTROL PANEL (382) LOCATED BEHIND ATTENUATION PANELS.

J-PN-300 TO 645F

CSM LOGISTICS TRAINING

DOCKING

The docking subsystem provides the means to connect and disconnect the lunar module and the command module. It is used twice during a normal lunar mission: at the beginning of the translunar flight when the CM docks with the LM, and in lunar orbit when the ascent stage of the lunar module docks with the CM.

Docking is achieved by maneuvering one of the modules close enough to the other so that a probe on the CM engages a drogue on the LM. In the first docking, the CM is the maneuvering vehicle and the LM the passive one; in the second docking the LM ascent stage is active and the CM passive.

The probe, drogue, tension tie, and a docking ring are the principal components of the docking subsystem. Each module also contains a docking pressure hatch and a tunnel through which the astronauts will transfer from one vehicle to the other.

The docking maneuvers are controlled by the commander through short bursts of the reaction control engines on the active vehicle. He is aided in maneuvering his craft by the crewman alignment sight, an optical device something like the range finder of a camera which is mounted at a rendezvous window.

Before the docking maneuvers begin a crewman in the CM activates a switch which extends the probe. When the probe comes into contact with the drogue, it is guided into the socket at the bottom of the drogue. Three capture latches in the probe head then hold the two modules together.

A crewman then activates the probe retraction device (a nitrogen pressure system located in the probe) which automatically pulls the LM and CM together. At contact 12 latches mounted on the CM docking ring are automatically activated to form a pressure-tight seal between the two modules.

A CM crewman then equalizes the pressure between the LM and CM tunnels through a valve provided for the purpose and removes the CM's docking hatch. After removing the hatch, he first checks all the docking ring latches to make sure they are engaged (locking of any 3 of the 12 latches is sufficient to assure a pressure-tight seal and allow entry into tunnel), and manually locks those that aren't. He then connects the two electrical umbilicals (stowed in the LM) to the CM ring connectors, providing power from the CM to the LM. The probe and drogue are then removed and passed down into the CM along with the hatch. Finally, he operates the valve in the LM docking hatch to equalize the

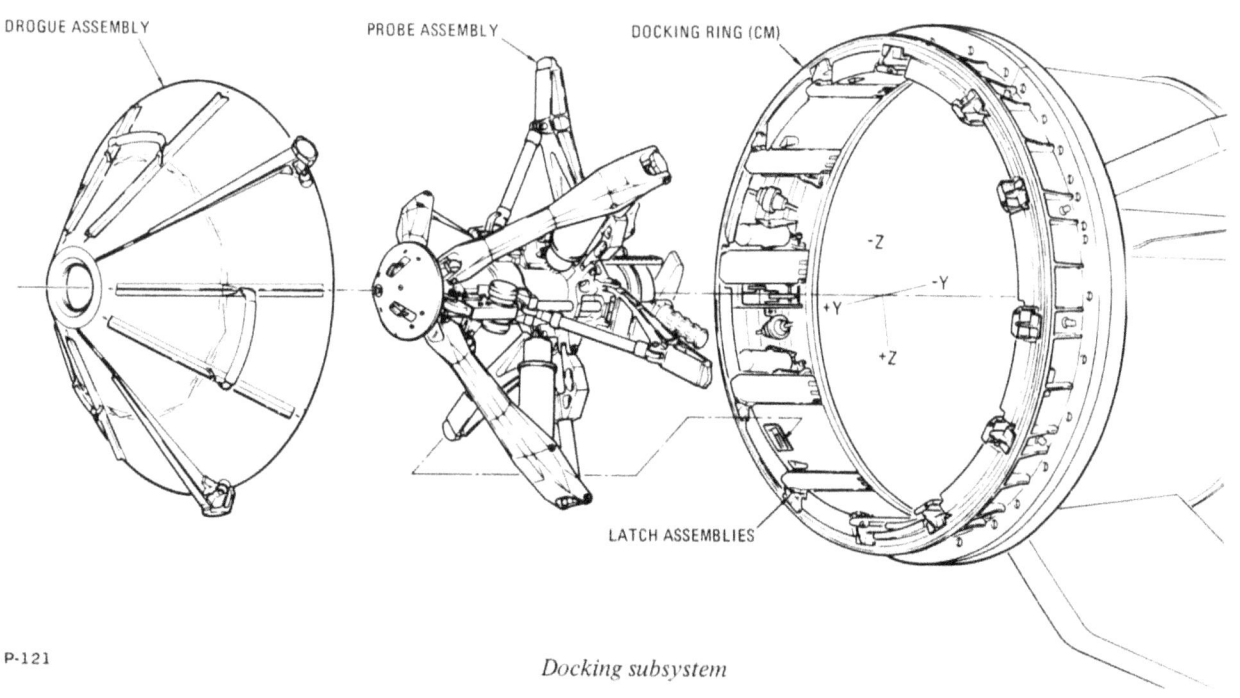

Docking subsystem

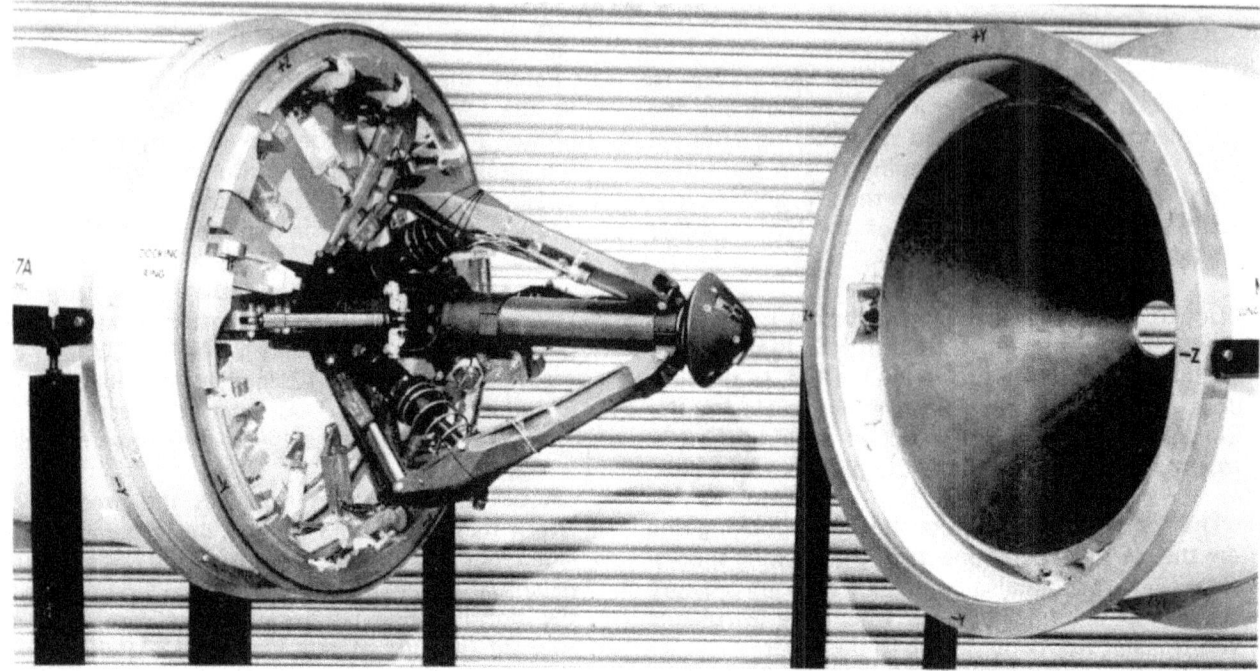

Mockup of docking subsystem

pressure in the LM. The LM hatch can then be opened; this is hinged to swing inward into the LM crew compartment.

In the first docking, the probe and drogue will not be removed, nor will the LM hatch be opened. The CM crewman will remove the CM hatch, check the locking latches, connect the umbilicals, and replace the CM hatch.

Once in orbit around the moon, the passage between the two modules is opened, the hatch, probe and drogue are removed, and two astronauts transfer into the LM. The remaining CM crewman then passes the drogue back to an LM crewman and it is re-installed in the tunnel. The CM crewman re-installs the probe and disconnects the LM/CM umbilicals for stowage in the tunnel area, manually cocks all the 12 ring latches, and then closes and seals both hatches. The two vehicles are separated for lunar descent through remote electrical release of latches on the probe assembly.

In the second docking, the LM is the maneuvering or active vehicle, but the operations are otherwise similar. The CM crewman must remove the CM hatch, check the latches, and remove the probe into the CM. An LM crewman opens the LM hatch and removes the drogue to open the tunnel.

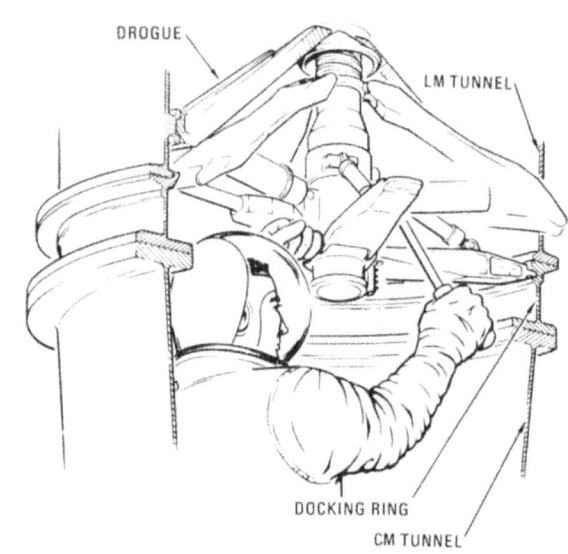

Operation of ratchet assembly

After all equipment has been transferred from the LM to the CM, the probe and drogue are stowed in the LM, since they are no longer needed. Both hatches are then replaced and sealed and the modules are ready for separation.

In this case, separation is accomplished by firing an explosive train located around the circumference

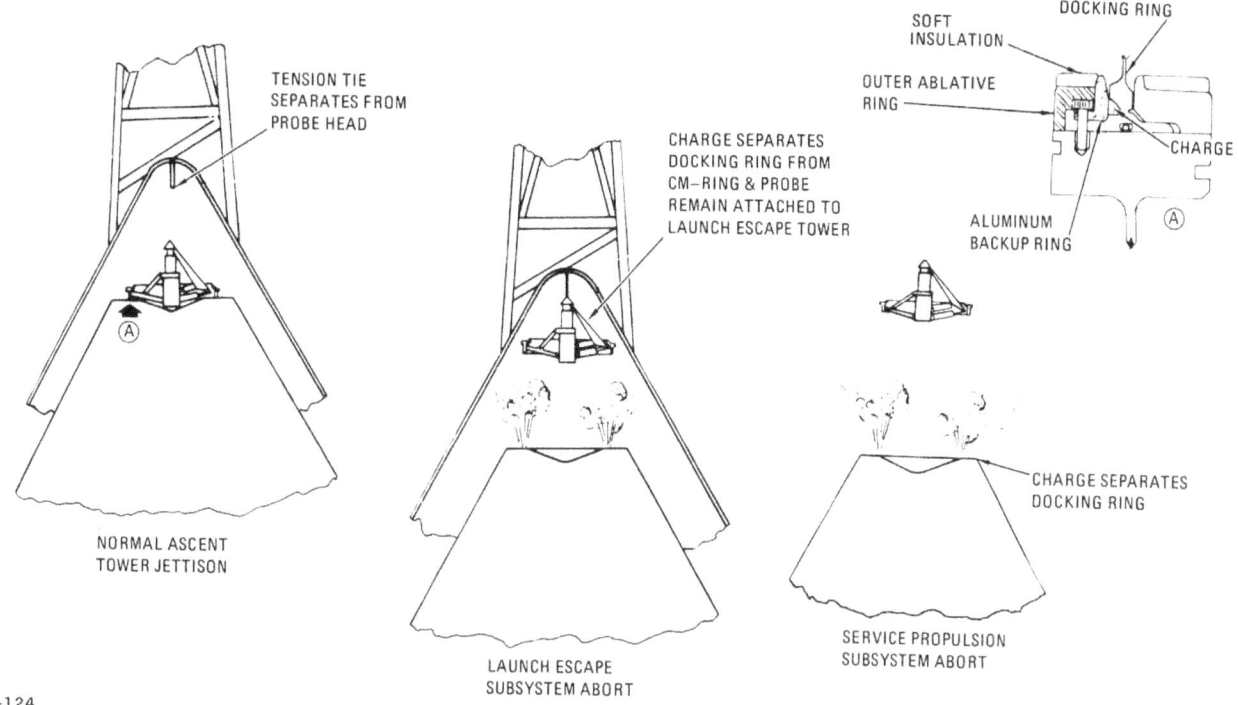

Modes of separating probe

of the docking ring. This separates the entire docking ring from the CM and provides enough impetus to separate the modules. The docking ring remains attached to the LM ascent stage, which remains in orbit around the moon.

EQUIPMENT

Probe—Consists of aluminum inner and outer cylinders sized to allow a maximum of 10 inches of travel of the inner cylinder, and a probe head gimbal-mounted on the inner cylinder. The probe head is self-centering and houses the three capture latches. The probe is mounted at three points to the docking ring by a support structure attached to the outer cylinder and is designed to fold so that it can be removed from either LM or CM side. Its components include pitch arms and tension linkages, shock attenuators, extension latch assembly, capture latches, ratchet assembly, a retraction system, and probe umbilicals.

Pitch Arms and Tension Linkages—These make contact with the drogue surface for CM/LM alignment and shock attenuation; the contacting surfaces of the pitch arms are contoured to match the curvature of the drogue. The tension links transmit loads to the shock attenuators.

Shock Attenuators—These are piston, variable orifice, fluid displacement units hermetically sealed with a metal bellows. They absorb the shock of impact during docking.

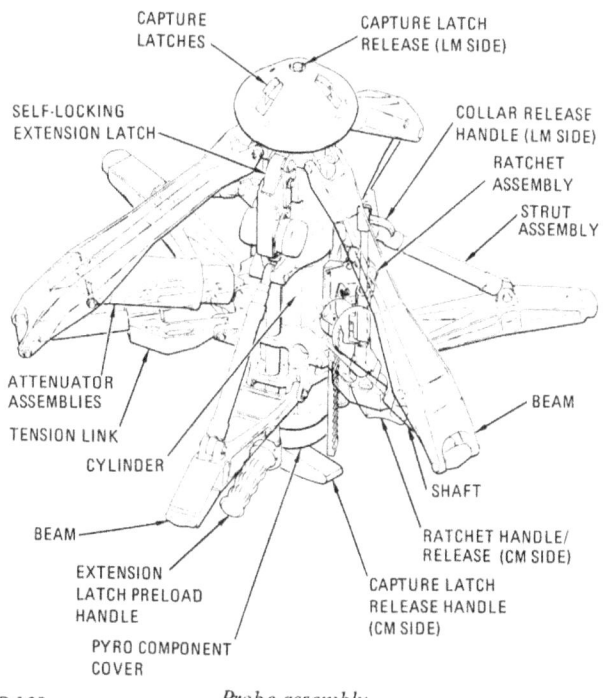

Probe assembly

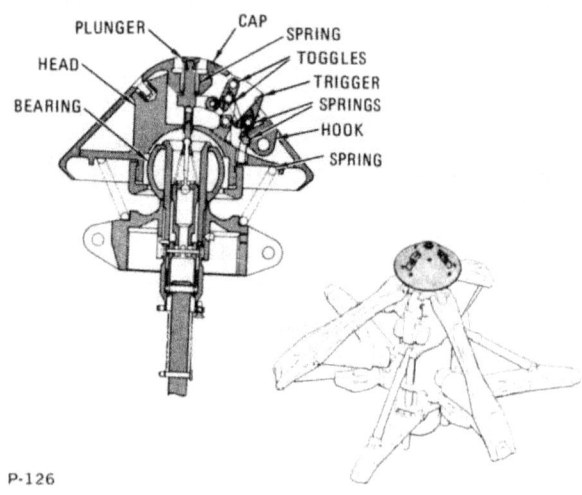

Capture latch assembly

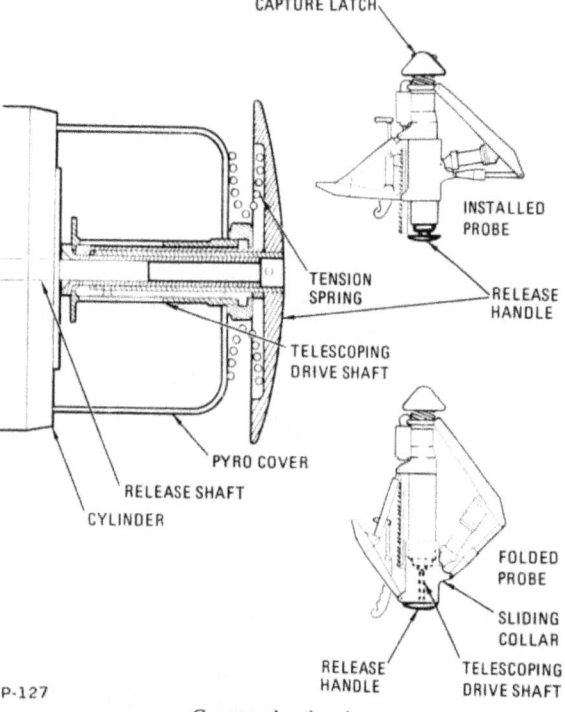

Capture latch release

released remotely from the CM or by a manual release handle from the CM side. They can also be released from the LM side by depressing the center button in the probe head.

Ratchet Assembly—This mechanism provides a hold for handling the probe, and performs the ratcheting operation to install the probe support arms. A handle is provided from either the CM or LM side to unlock the ratchet sliding collar for folding and removing the probe.

Retraction System—This is a cold gas system pressurized from four hermetically sealed nitrogen bottles located inside the probe body. When the gas pressure is released it compresses a piston with sufficient force to draw the LM to the CM, compress the interface seal, and engage the locking latches. The retraction system is activated automatically by capture latch engagement into the drogue opening or manually by a CM crewman.

Probe Umbilicals— The microdot connectors and harness assemblies are provided for probe instrumentation and power. The connectors are installed on the side of the docking ring and can be mated and demated from either the CM or LM side. During probe removal and stowing, the umbilical connectors are attached to receptacles on the probe support beams.

Drogue Assembly—The drogue consists of an internal conical surface facing the CM, a support structure and mounting provisions, and a locking mechanism that prevents it from turning during the docking maneuvers. It is made of aluminum honeycomb with aluminum support beams. The drogue may be unlocked and removed from either side.

Docking Ring—This is an aluminum structure bolted to the CM tunnel just forward of the top hatch. It contains seals and the shaped charge for final separation. It also serves as the mounting point for the probe and docking latches. The docking ring must withstand all loads from docking and from course corrections, and must maintain proper alignment of the docked vehicles. The ring also contains a covered passageway for the electrical harnesses and connections for attaching the umbilicals between the modules.

Extension Latch Assembly—This engages and retains the probe in a fully retracted position after docking. It is released remotely from the CM to allow probe extension.

Capture Latches—These are mounted in the probe head and engage automatically when the probe head centers and bottoms in the drogue. Engagement of these latches operates a switch on the probe which initiates automatic operation of the retraction mechanism. The capture latches can be

Docking Latches—The 12 docking latches are spaced an equal distance around the inner peri-

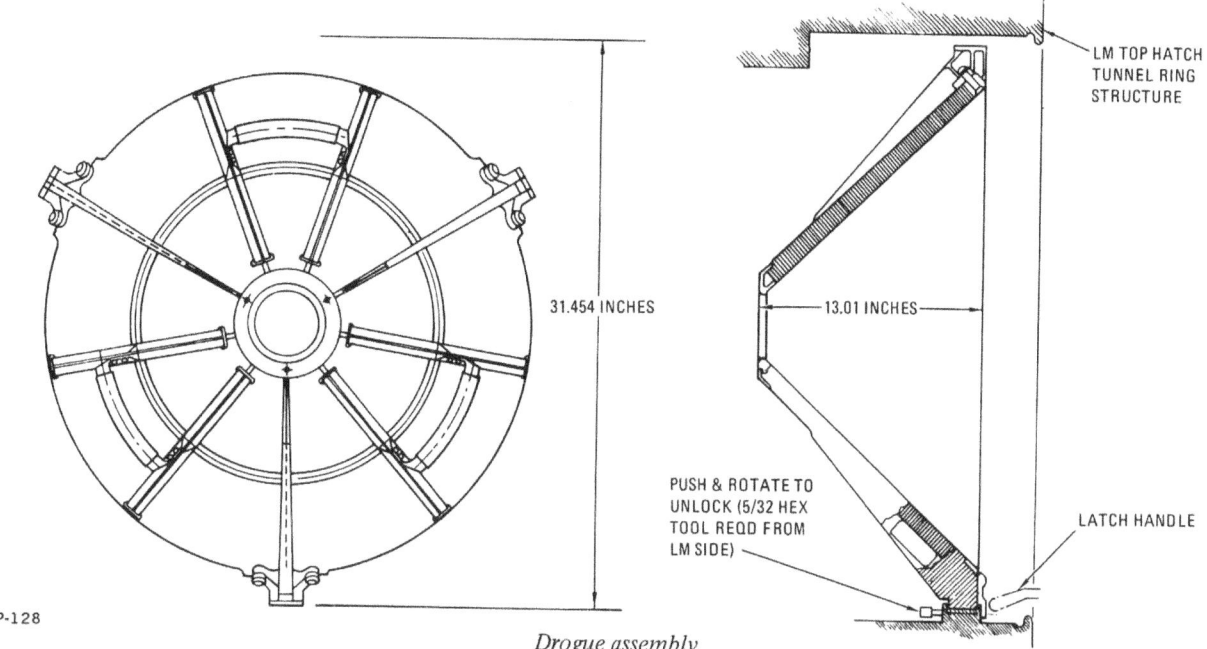

Drogue assembly

phery of the docking ring. The latches automatically seek and engage the back surface of the LM docking flange. The latch trigger mechanism is activated by contact with the flange. The latch will retract its hook, seat it on the back of the flange, and compress the docking seals. The latches are released manually by a crewman pulling the handle two times. This relieves the load from the hook and cocks the mechanism for the next docking engagement.

Docking Seals—These seals are round and hollow and made of a silicon material; they compress when the two modules come together to form a pressure-tight seal.

Umbilicals—Two electrical umbilicals are attached to stowage connectors on the LM tunnel wall so that they are clear of the probe and drogue supports during docking. The umbilicals can be reached from the CM tunnel and are connected to the receptacles on the CM docking ring before LM withdrawal.

Tunnel Hatches—The hatch at the forward end of the CM can be opened from either side (by a handle on the CM side and with a tool from the LM side), but can be removed only into the CM. The hatch contains a pressure equalization valve which can be operated from either side to equalize the pressure in the tunnel and LM before removing the hatch. The LM docking hatch is located at the lower end of the LM tunnel and is not removable; it swings down into the LM crew compartment. It also can be opened from either side, and has a pressure dump valve. Although the valve's primary function is to release pressure from the LM cabin, it can also be used to equalize pressure in the tunnel.

Passive Tension Tie—A tension-type bolt arrangement connecting the probe and the boost protective cover. During normal flight, the attaching pins are sheared in the probe head, leaving the probe intact for docking. During an abort, a pyrotechnic charge separates the docking ring, allowing the launch escape tower to take the ring and probe with it.

Pyrotechnic Charge—This is a mild detonating fuse located around the periphery of the docking ring. When fired, the charge separates the ring structure between the forward heat shield and the probe mounting, leaving all docking hardware with the LM. The charge is initiated by a signal from the CM. The charge normally is fired for final separation of the LM ascent stage, just before the CSM is injected on the transearth flight. It also would be fired during an abort, and the docking hardware would be carried away by the launch escape subsystem.

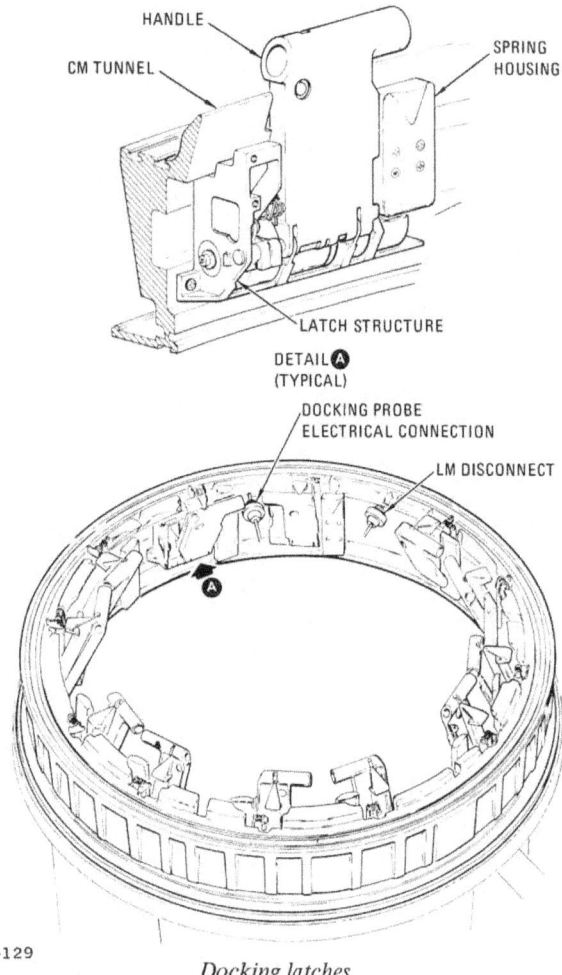

Docking latches

Single-eye projection keeps the system small and eliminates apparent parallax of the projected image caused by focusing and by convergence when sighting during the final docking phase. Sideward motion of the astronaut's head during sighting will not affect the system.

Docking Targets—These are mounted in a window of the CM and LM to aid in docking maneuvers. The LM active docking target is mounted in the right rendezvous window of the CM. Its base is 8 inches in diameter and contains green electroluminescent lamps and a black stripe pattern on the first. An upright support at a right angle to base contains a red incandescent lamp. The docking target in the LM is similar but about twice the size.

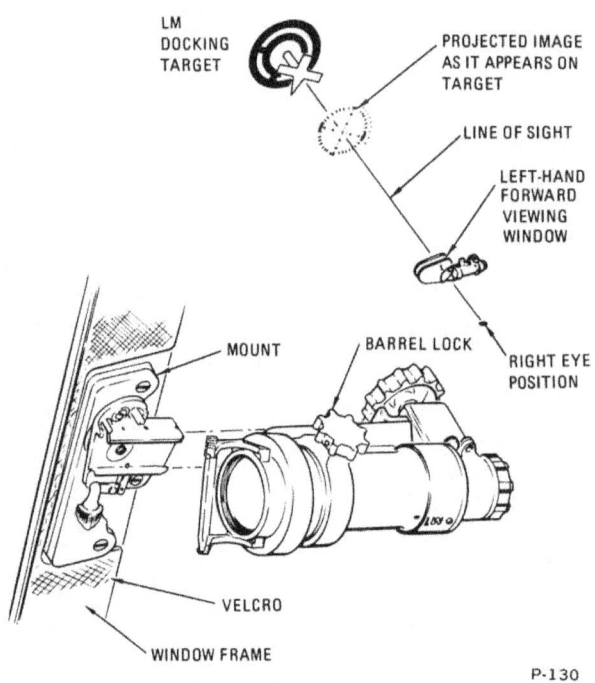

Crewman optical alignment sight

Crewman Optical Alignment Sight—This is used to help astronauts align the maneuvering module with the passive one. It is a collimator-type device that provides the astronaut with a fixed line-of-sight reference image. When viewed through the rendezvous window, the image appears to be the same distance away as the target. It is about 8 inches long, can be mounted near either rendezvous window, and has a control to permit adjustment for proper brightness against any background lighting conditions. The operator controls firing of the reaction control engines so as to keep the projected image superimposed on the passive module target. Range and rate of closure are determined by the size relationship between the image and the target. The adjustable light source projects the image from within its housing to the crewman's right eye by means of a beam splitter or combining glass. Though the image is projected to the right eye only, the crewman sights with both eyes open.

EARTH LANDING

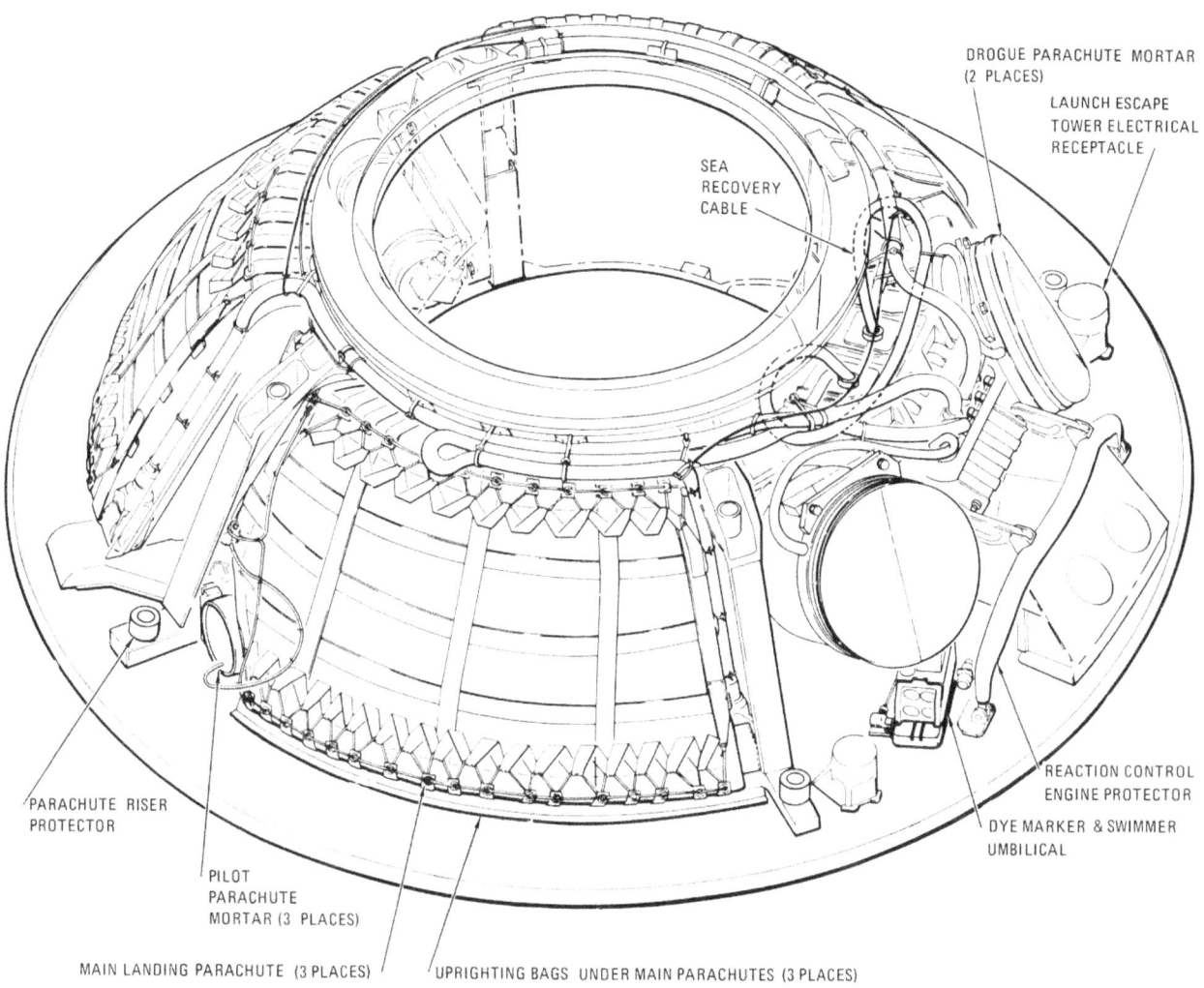

Location of main components of earth landing subsystem

The earth landing subsystem provides a safe landing for the astronauts and the command module. Several recovery aids which are activated after splashdown are part of the subsystem. Operation of the subsystem normally is automatic, timed and activated by the sequential control system. There are, however, backup manual controls for astronaut operation.

For normal entry, the subsystem operation begins with jettison of the forward heat shield when the CM descends to about 24,000 feet. About 1-1/2 seconds later the two drogue parachutes are deployed to orient the module properly and to provide initial deceleration. At about 10,000 feet, the drogue parachutes are released and the three pilot parachutes are deployed; these pull the main parachutes from the forward section of the CM. The main parachutes hold the CM at an angle of 27-1/2 degrees so the module will hit the water on its "toe," which will produce water penetration at the least impact load. If one of the main parachutes fails to open, the remaining two will be able to land the CM safely.

This sequence of operations differs slightly for an abort. In that case, the forward heat shield is jettisoned 0.4 second after jettisoning of the launch escape tower. The drogue chutes are then deployed 1.6 seconds later. For low-altitude aborts the main

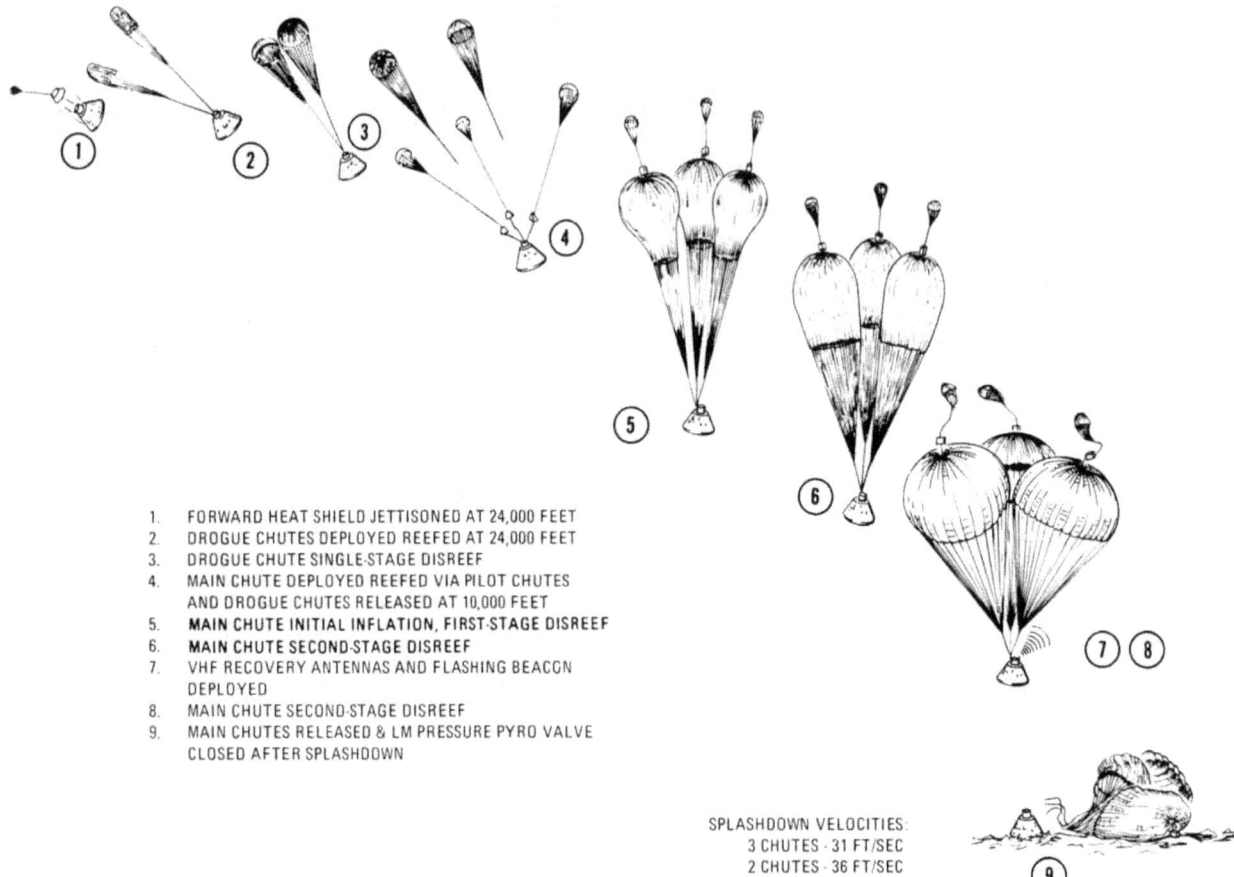

1. FORWARD HEAT SHIELD JETTISONED AT 24,000 FEET
2. DROGUE CHUTES DEPLOYED REEFED AT 24,000 FEET
3. DROGUE CHUTE SINGLE-STAGE DISREEF
4. MAIN CHUTE DEPLOYED REEFED VIA PILOT CHUTES AND DROGUE CHUTES RELEASED AT 10,000 FEET
5. **MAIN CHUTE INITIAL INFLATION, FIRST-STAGE DISREEF**
6. **MAIN CHUTE SECOND-STAGE DISREEF**
7. VHF RECOVERY ANTENNAS AND FLASHING BEACON DEPLOYED
8. MAIN CHUTE SECOND-STAGE DISREEF
9. MAIN CHUTES RELEASED & LM PRESSURE PYRO VALVE CLOSED AFTER SPLASHDOWN

SPLASHDOWN VELOCITIES:
3 CHUTES - 31 FT/SEC
2 CHUTES - 36 FT/SEC

P-132

Normal sequence of operation for earth landing subsystem

parachutes are automatically deployed 12 seconds later, or between zero and 12 seconds by the astronauts. For high-altitude aborts, the main parachutes are deployed in the same manner as normal entry.

After splashdown, the main parachutes are released and the recovery aid subsystem is set in operation by the crew. The subsystem consists of an uprighting system, swimmer's umbilical cable, a sea dye marker, a flashing beacon, and a VHF beacon transmitter. A sea recovery sling of steel cable also is provided to lift the CM aboard a recovery ship.

The two VHF recovery antennas are located in the forward compartment with the parachutes. They are deployed automatically 8 seconds after the main parachutes. One of them is connected to the beacon transmitter, which emits a 2-second signal every 5 seconds to aid recovery forces in locating the CM. The other is connected to the VHF/AM transmitter and receiver to provide voice communications between the crew and recovery forces.

Automatic operation of the earth landing subsystem is provided by the event sequencing system located in the right-hand equipment bay of the command module. The system contains the barometric pressure switches, time delays, and relays necessary to control automatically the jettisoning of the heat shield and the deployment of the parachutes.

The parachute subsystem is produced by the Northrop Corporation, Ventura Division, Newbury Park, Calif.

EQUIPMENT

Drogue Parachutes (Northrop-Ventura, Newbury Park, Calif.) — Two white nylon conical-ribbon parachutes with canopy diameters of 16.5 feet. They are deployed at 23,000 feet to orient and slow the spacecraft from 300 miles an hour to 175 miles an hour so that the main parachutes can be safely deployed. They are 65 feet above the command module.

<u>Pilot Parachutes</u> (Northrop-Ventura) — Three white nylon ring-slot parachutes with canopy diameters of 7.2 feet. They deploy the main parachutes and are 58 feet above the main parachutes.

<u>Main Parachutes</u> (Northrop-Ventura) — Three orange-and-white-striped ringsail parachutes with canopy diameters of 83.5 feet. Each weighs 127 pounds counting canopy, risers, and deployment bag. They are deployed at 10,000 feet to reduce the speed of the spacecraft from 175 miles an hour to 22 miles an hour when it enters the water. The parachutes are 120 feet above the command module.

<u>Reefing Line Cutters</u> (Northrop-Ventura) — Each assembly is about the size of a fountain pen and consists of slow-burning powder, small explosive charge, blade, and hole through which the reefing line passes. Burning powder sets off a charge, which drives the blade into the reefing line, severing it and allowing full inflation of the parachutes.

<u>Sequence Controller</u> — Two 3 by 3 by 6-inch boxes each containing four barometric pressure switches, four time delays, and four relays to sequence and

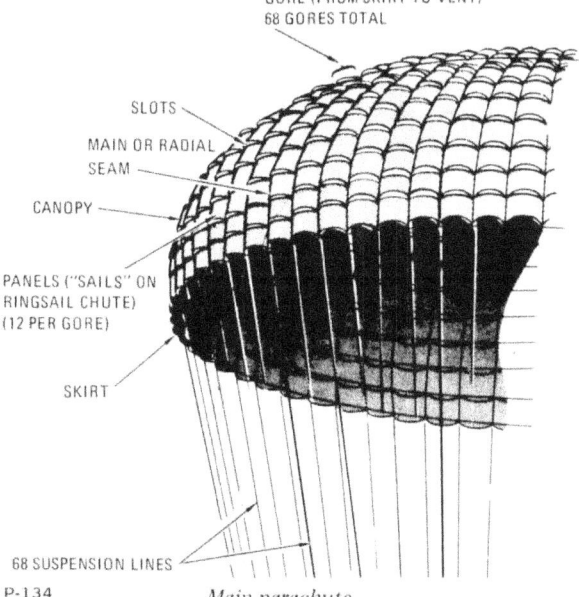

Main parachute

control the deployment and release of parachutes automatically. Each box weighs less than four pounds. They are in the forward compartment of the command module.

<u>Pyro Continuity Verification Box</u> — A 2-1/2 by 8 by 10-inch box containing relays and fusistors located in the forward compartment of the command module. It provides an accessible point within the command module to verify the continuity of the pyrotechnic device firing circuits.

<u>Uprighting System</u> — Three inflatable bags in the forward compartment of the command module

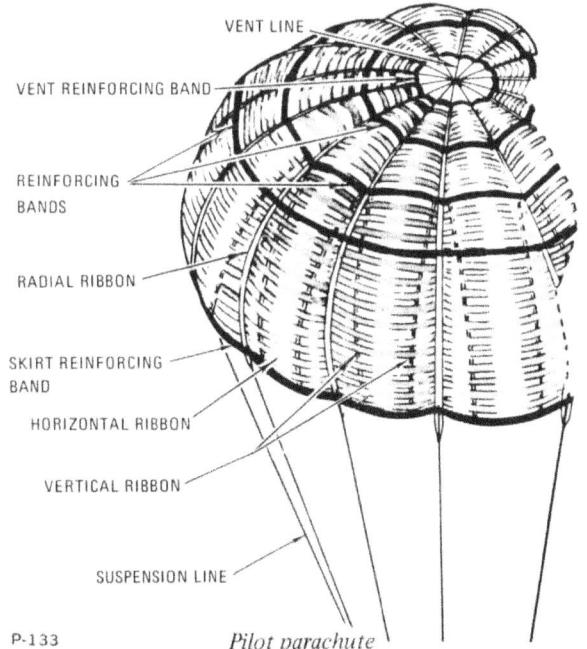

Pilot parachute

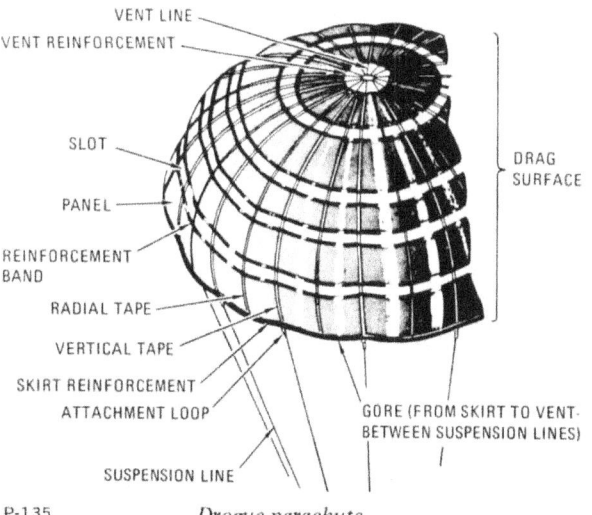

Drogue parachute

and two air compressors in the aft compartment. The compressors are manually initiated to provide air to the bags through tubes. Each bag has a volume of 22 cubic feet. If the command module turns apex down after landing, the air bags are inflated to right the spacecraft.

Sea Dye Marker — Powdered fluorescein dye packed in a 3 by 3 by 6-inch metal container. When the marker is deployed, dye colors the sea yellow-green around the command module. The marker is in the forward compartment of the command module. It is released manually and is connected to the command module by a 12-foot tether. The dye lasts approximately 12 hours.

Flashing Beacon — Flashing self-powered strobe light to aid in recovery of crew and command module. Eight seconds after main parachutes are deployed, the beacon is automatically extended from the forward compartment of the command module. The light is turned on manually. The arm is one-foot long.

Swimmer's Umbilical — This is the 12-foot dye marker tether. A recovery frogman can connect his communications equipment to the end of the tether to talk to the command module crew.

Automatic and manual control circuits for jettisoning the forward heat shield are included in the integrated master events sequence controllers, the earth landing sequence controllers, and the lunar docking events controllers.

The shield is jettisoned by the use of a thruster mechanism and a drag parachute. When gas pressure is generated by the pressure cartridges, two pistons are forced apart, breaking a tension tie which connects the shield to the forward compartment structure. The lower piston is forced against a stop and the upper piston is forced out of its cylinder. The piston rod ends are fastened to fittings on the shield, which is thrusted away from the CM. Two of the thruster assemblies have breeches and pressure cartridges; plumbing connects the breeches to thrusters mounted on diametrically opposite CM structural members.

The mortar-deployed parachute drags the heat shield out of the area of negative air pressure following the CM and prevents recontact with the CM. Lanyard-actuated switches are used to fire mortar pressure cartridges.

The forward heat shield must be jettisoned before earth landing equipment may be used. Drogue and pilot parachutes are mortar-deployed to assure that they are ejected beyond the turbulent air around and following the CM. An engine protector bar prevents damage to CM reaction control engines by the drogue cables. The drogue cables (risers) are protected from damage by spring-loaded covers over the launch escape tower attachment studs.

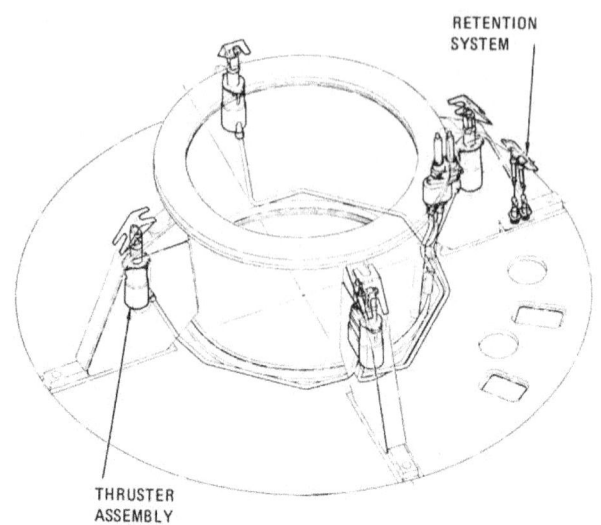

Forward heat shield retention and thruster system

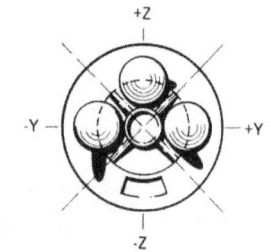

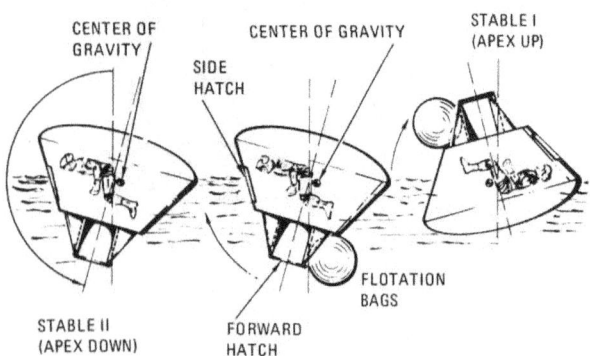

CM uprighting system

The sea recovery cable loop will spring into position after the parachutes have been deployed. Three uprighting bags are installed under the main parachutes. A switch is provided for the crew to deploy the sea dye marker and swimmer umbilical any time after landing.

Eight parachutes are used in the earth landing subsystem: two drogue, three pilot, and three main. The drogue parachutes are the conical ribbon type and are 16.5 feet in diameter. The pilot parachutes are the ring slot type and are 7.2 feet in diameter. The main parachutes are the ring sail type and are 83.5 feet in diameter.

The drogue and main parachutes are deployed reefed for 10 seconds. The reefing lines run through rings which are sewn to the inside of the parachute skirts and reefing line cutters. When the suspension lines pull taut, a lanyard pulls the sear release from the reefing line cutter and a pyrotechnic time-delay is started. When the delay train has burned through, a propellant is ignited, driving a cutter through the reefing line.

Each of the drogue parachutes has two reefing lines with two cutters per line to prevent disreefing in case one reefing line cutter fires prematurely. Each of the three main parachutes has three reefing lines with two cutters per line. Two of the three reefing lines are severed after 6 seconds, allowing the main parachutes to open slightly wider than when deployed. The remaining reefing line is cut 4 seconds later, or 10 seconds after deployment. At this time the parachutes inflate fully.

Reefing line cutters also are used in the deployment of the two VHF antennas and the flashing beacon light during descent. These recovery devices are retained by spring-loaded devices which are secured with parachute rigging cord. The cord is passed through reefing line cutters which are activated by action of the main parachute risers.

Redundant channels in the sequential events system command the sequence of operations of the subsystem.

Activation of the earth landing subsystem switch ("ELS Logic") on the main display console closes logic power circuits to redundant transistorized switches in the master events sequence controllers. Each of these solid-state switches requires two conditions to close; one is closure of the logic

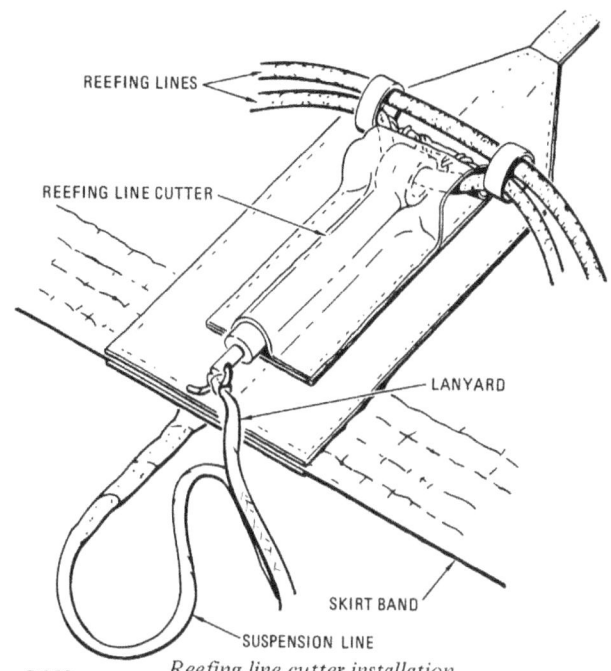

Reefing line cutter installation

power circuits, the other is closure of the 24,000-foot barometric switches. Assuming that the subsystem is set for automatic operation, closure of the barometric switches activates the earth landing subsystem controller and begins the landing sequence.

The barometric switches are devices which use air pressure to trigger a switch at a set altitude. The dorgue baroswitches are set to close at the normal pressure for 24,000 feet; because air pressure varies with meteorological conditions, however, the switches may close a little above or a little below that altitude.

In addition to activating the earth landing subsystem controllers, closure of the 24,000-foot baroswitches energizes the 24,000-foot lockup relay in the controllers. This establishes logic power holding circuits which bypass the baroswitches. When the controllers are activated a signal is relayed to the unlatching coils of the reaction jet engine control to disable the automatic firing of the reaction control engines.

The first function of the earth landing subsystem controllers is to jettison the forward heat shield. After a delay of 0.4 second, the shield thrusters are fired and the lanyard-actuated switches used to deploy the drag parachute are armed. The lanyard, which is attached to the forward heat shield, pulls

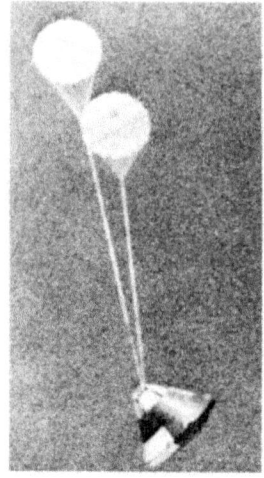

Parachute sequence: drogues (top left) open at 24,000 feet, pilots pull out main chutes (bottom left), and main chutes open fully

holding pins from the switches which, because of spring loading, close circuits to energize relays to fire the drag parachute mortar.

Two seconds after the forward heat shield is jettisoned, relays close to send signals that fire the drogue mortars and deploy the drogue parachutes. When the CM has descended to 10,000 feet, another set of baroswitches closes to energize relays which fire the mortars to deploy the pilot parachutes. The main parachutes are deployed by the pilot chutes.

After splashdown, a crewman activates a switch on the main display console which fires ordnance devices to drive a chisel-type cutter through the main parachute risers, releasing the chutes.

ELECTRICAL POWER

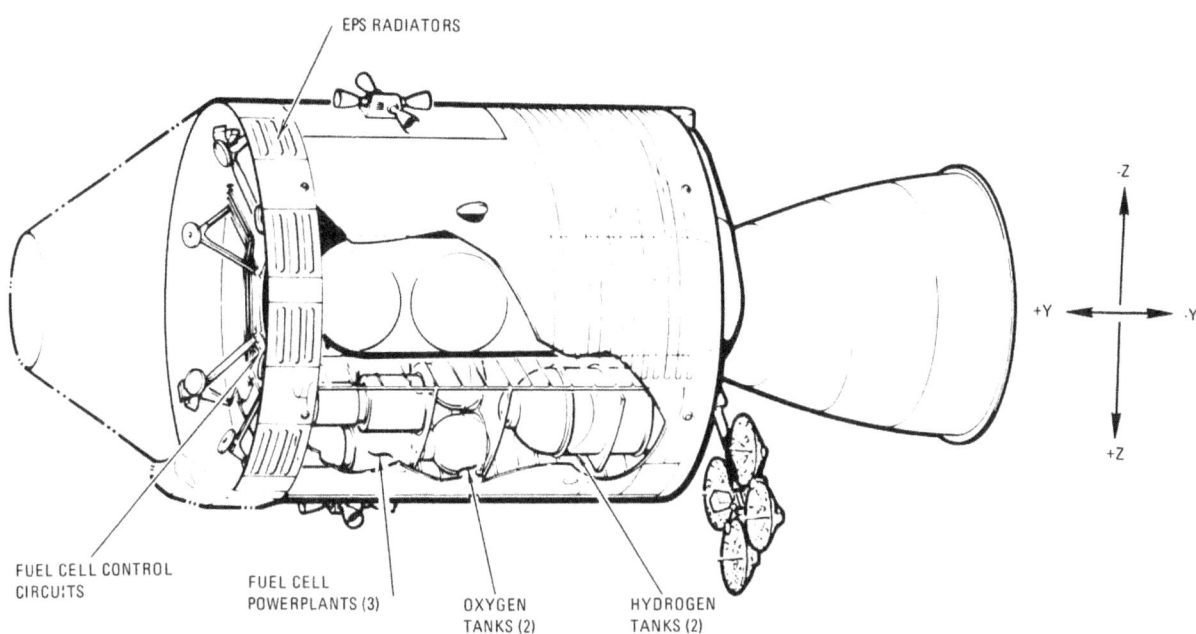

Location of major electrical power subsystem equipment

The electrical power subsystem provides electrical energy sources, power generation and control, power conversion and conditioning, and power distribution to the spacecraft throughout the mission. For checkout before launch, dc electrical power is supplied by ground support equipment. The electrical power subsystem furnishes drinking water to the astronauts as a byproduct of its fuel cell powerplants, and includes the cryogenic gas storage system.

Each of the three fuel cell powerplants (produced by United Aircraft Corp.'s Pratt & Whitney Aircraft Division, Hartford, Conn.) consists of 31 cells connected in series. Each cell consists of a hydrogen compartment, an oxygen compartment, and two electrodes (conductors) — one hydrogen and one oxygen. The electrolyte (substance through which ions are conducted) is a mixture of approximately 72 percent potassium hydroxide and approximately 28 percent water and provides a constant conduction path between electrodes. The hydrogen electrode is nickel and the oxygen electrode is nickel and nickel oxide.

The reactants (hydrogen and oxygen) are supplied to the cell under regulated pressure (referenced to a nitrogen gas supply which also is used to pressurize the powerplants). Chemical reaction produces electricity, water, and heat, with the reactants being consumed in proportion to the electrical load. The byproducts—water and heat—are used to maintain the drinking water supply and to keep the electrolyte at the proper operating temperature. Excess heat is rejected to space through the space radiators. The fuel cell powerplants are located in Sector 4 of the service module.

Three silver oxide-zinc storage batteries supply power to the CM during entry and after landing, provide power for sequence controllers, and supplement the fuel cells during periods of peak power demand. These batteries are located in the lower equipment bay of the CM. A battery charger located in the same bay re-charges the batteries after each use and assures that they will be fully charged before entry. These batteries are produced by Eagle Picher Co., Joplin, Mo.

Two other silver oxide-zinc batteries, independent of and completely isolated from the rest of the dc power system, are used to supply power for explosive devices. These batteries are not recharged. They are produced by the Electric Storage Battery Co., Raleigh, N.C.

The cryogenic (ultra low temperature) gas storage system (produced by Beech Aircraft Corp., Boulder, Colo.) supplies the hydrogen and oxygen used in the fuel cell powerplants, as well as the oxygen used in the environmental control subsystem. The system consists of storage tanks and associated valves, switches, lines, and other plumbing. The hydrogen and oxygen are stored in a semi-gas, semi-liquid state; by the time they reach the fuel cells, however, they have warmed considerably and are in a gaseous state. The system is located in Sector 4 of the service module beneath the fuel cell powerplants.

Three solid-state inverters, located in the lower equipment bay of the CM, supply the ac power for the spacecraft. These inverters are devices which convert dc electrical power into ac. Both the fuel cell powerplants and batteries, the two electrical power sources in the spacecraft, produce dc power.

The inverters operate from the two 28-volt dc main buses (connecting circuits) to supply 115/120-volt, 400-cycle, 3-phase ac power to two ac buses. Normally two inverters are used; however, one inverter can supply all primary ac electrical power needed by the spacecraft. If one inverter fails, a crewman can switch in the standby. Two inverters cannot be paralleled (hooked up together).

The inverters are produced by Westinghouse Electric's Aerospace Electrical Division, Lima, Ohio.

EQUIPMENT

Oxygen Tanks (Beech Aircraft Corp., Boulder, Colo.) — Two spherical dewar-type tanks made of Inconel (nickel-steel alloy) in Sector 4 of the service module store oxygen for production of power by fuel cells, for command module pressurization, and for metabolic consumption. Outer diameter of each is 26.55 inches and wall thickness is 0.020 inch. Tanks with accessories are 36.39 inches tall. Each tank has an inner vessel with a diameter of 25.06 inches and wall thickness of 0.061 inch. Rupture pressure of the tanks is 1530 pounds per square inch. Insulation between the inner and outer shells is fiberglass, paper mating, and aluminum foil. In addition, a pump maintains a vacuum between the inner and outer vessels. Each tank weighs 79-1/2 pounds, has a volume of 4.73 cubic feet, and a capacity of 320 pounds — 210 pounds for fuel cells and 110 pounds for environmental control. Each tank has a repressurization probe with two heaters and two fans to keep the tank pressurized and a capacitive probe which measures the amount of oxygen. A resistance element measures temperature.

Hydrogen Tanks (Beech) — Two spherical dewar-type titanium tanks located in Sector 4 of the service module contain the hydrogen that powers the fuel cell. Outer diameter of each is 31.80 inches and wall thickness is 0.033 inch. Each tank is 31.9 inches tall and has an inner vessel 28.24 inches in diameter with a wall thickness of 0.046 inch. Rupture pressure is 450 pounds per square inch. Unlike the oxygen tank, which has insulation between the inner and outer shells, the hydrogen tanks have a vapor-cooled shield suspended in a vacuum as a heat barrier. Each tank weighs 69

Fuel cell powerplant

P-141

pounds, has a volume of 6.75 cubic feet, and a capacity of 28 pounds of usable fluid. Similar to the oxygen tanks, they contain repressurization and capacitive probes, a pump to maintain the vacuum, and temperature transducers.

Batteries (Eagle Picher Co., Joplin, Mo.) — Three silver oxide-zinc storage batteries are in the command module lower equipment bay. Each has 20 cells with potassium hydroxide and water as an electrolyte. Battery cases are plastic, coated with fiberglass epoxy, and are vented overboard for outgassing. Each is 6-7/8 by 5-3/4 inches and weighs 28 pounds. The batteries are rated at 40 ampere hours, providing a high power-to-weight ratio. Open circuit voltage is 37.2 volts. The battery characteristics are such that a minimum of 27 volts can be maintained until the battery is depleted. The batteries have been shock-tested to 80-g impact. The batteries provide all CM power during entry and after landing. They also supplement fuel cells during major thrusting maneuvers and provide power for the sequence system and fuel cell and inverter control circuits.

Battery Charger — The constant-voltage, current-limited charger is 4 by 6 by 6 inches and weighs 4.3 pounds. The current is limited to 2.8 amperes so as not to overheat the batteries. It has an operating life of more than 1,000 hours. The charger is located near the entry and postlanding batteries.

Pyrotechnic Batteries (Electric Storage Battery Co., Raleigh, N.C.) — Each of the two silver oxide-zinc batteries in the lower equipment bay has 20 cells with potassium hydroxide and water as an electrolyte. The cases are plastic. There is a relief valve venting arrangement for outgassing. Each is 2-3/4 by 3 by 6-3/4 inches and is rated at 0.75 ampere hours with an open-circuit voltage of 37.2 volts and a 20-volt minimum underrated load. They power mild explosive devices for CM-SM separation, parachute deployment and separation, Saturn third stage separation, launch escape tower separation, and other functions.

Fuel Cell Powerplants (United Aircraft Corporation's Pratt & Whitney Aircraft Division, Hartford, Conn.) — Three fuel cell powerplants, each 44 inches high, 22 inches in diameter, and weighing 245 pounds, are located in Sector 4 of the SM. They are mainly constructed of titanium, stainless steel, and nickel. They are rated at 27 to 31 volts under normal loads. There are 31 separate cells in a stack, each producing 1 volt, with potassium hydroxide and water as electrolyte. Each cell consists of a hydrogen and an oxygen

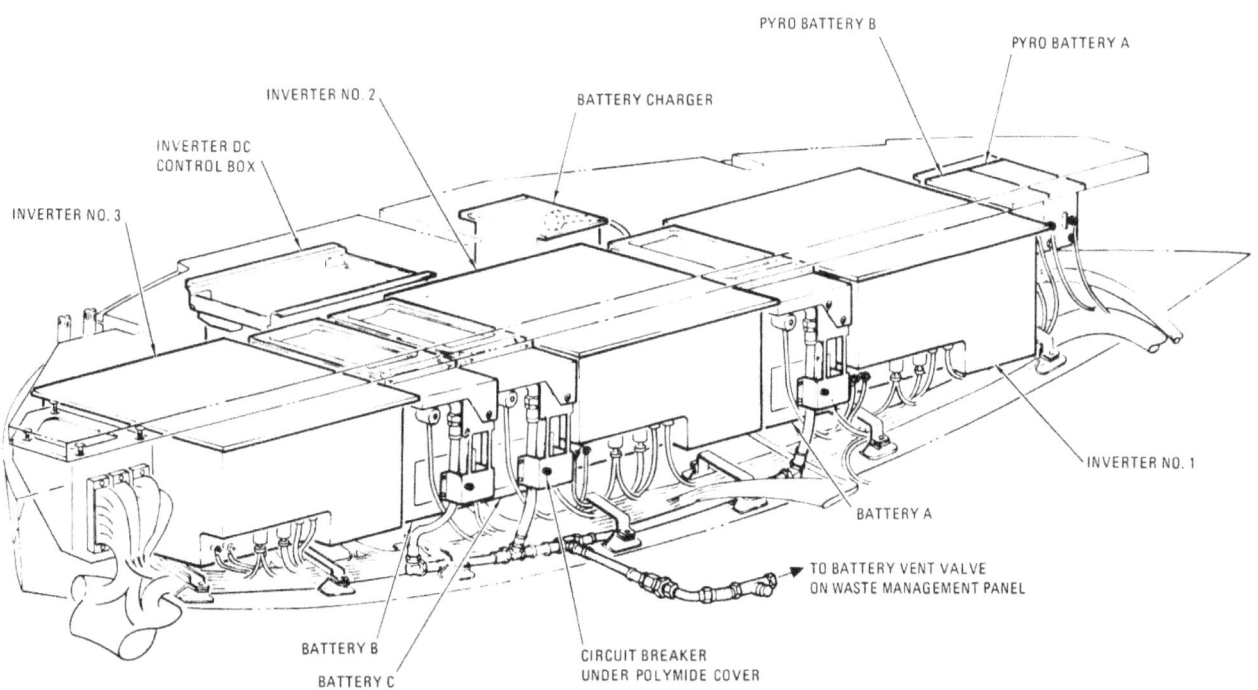

Electrical power subsystem components in CM lower equipment bay

electrode, a hydrogen and an oxygen gas compartment, and the electrolyte. Each gas reacts independently to produce a flow of electrons. The fuel cells are nonregenerative. They are normally operated at 400 degrees F with limits of 385 and 500 degrees. Water-glycol is used for temperature control. The fuel cells use hydrogen, oxygen, and nitrogen under regulated pressure to produce power and, as a by-product, water.

<u>Inverters</u> (Westinghouse Electric's Aerospace Electrical Division, Lima, Ohio) — Three solid-state inverters are in the lower equipment bay. Each is contained in an aluminum enclosure and coldplated with a water-glycol loop. The inverters weigh 53 pounds each and are 14-3/4 by 15 by 5 inches. They produce 1250 volt-amperes each. They convert 28-volt dc to 115-volt ac, 3 phase, 400 Hertz. They are designed to compensate for input and output voltage variations. Two of the three inverters are in constant use. They provide alternating current for fuel cell pumps, environmental control system glycol pumps, space suit compressors, and other circuitry.

DETAILED DESCRIPTION

CRYOGENIC STORAGE

The cryogenic storage subsystem supplies oxygen and hydrogen to the fuel cell powerplants and oxygen to the environmental control subsystem and for initial pressurization of the lunar module. Each of the two tanks of hydrogen and oxygen holds enough fluid to assure a safe return from the furthest point of the mission. The cryogenic tanks

Command module wire harness is first assembled on this mockup stand

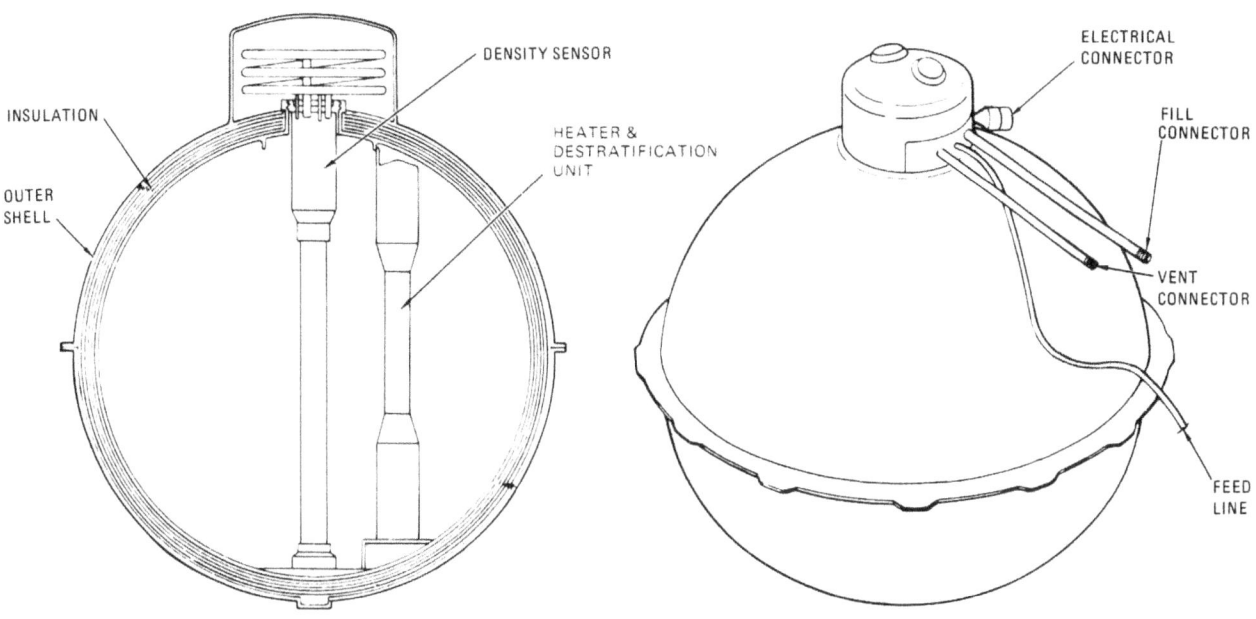

Cryogenic tank (hydrogen)

are pressurized by internal heaters after filling is complete.

Two parallel dc heaters in each tank supply heat necessary to maintain pressure. Two parallel 3-phase ac circulating fans circulate the fluid over the heating elements to maintain a uniform density and decrease the probability of thermal stratification. Relief valves provide overpressure relief and check valves provide tank isolation. A malfunctioning fuel cell powerplant can be isolated by a shutoff valve. Filters extract particles from the flowing fluid to protect components. Pressure transducers and temperature probes indicate the thermodynamic state of the fluid and capacitive quantity probes indicate the amount of fluid in the tanks.

The systems can be repressurized automatically or manually. The automatic mode is designed to give a single-phase reactant flow into the fuel cell and feed lines at design pressures. The heaters and fans are automatically controlled through pressure and motor switches. As pressure decreases, the pressure switch in each tank closes to energize the motor switch, closing contacts in the heater and fan circuits. Both tanks have to decrease in pressure before heater and fan circuits are energized. When either tank reaches the upper operating pressure limit, its pressure switch opens, again energizing the motor switch and opening the heater and fan circuits to both tanks. The oxygen tank circuits are energized at 865 psia minimum and de-energized at 935 psia maximum. The hydrogen circuits energize at 225 psia minimum and de-energize at 260 psia maximum.

When the systems reach the point where heater and fan cycling is at a minimum (due to a reduced heat requirement), the heat leak of the tank is sufficient to maintain proper pressures provided flow is within proper values. The minimum heat requirement region for oxygen starts at approximately 40-percent quantity in the tanks and ends at approximately 25-percent quantity. Between these tank quantities, minimum heater and fan cycling will occur under normal usage. The heat needed for pressurization at quantities below 25 percent starts to increase until at the 5-percent level practically continuous heater and fan operation is required. In the hydrogen system, the quantity levels for minimum heater and fan cycling are between approximately 53 and 33 percent, with continuous operation occurring at approximately 1 percent.

The oxygen system heaters and fans can sustain proper pressures for 30 minutes at a total flow of 10.4 pounds per hour (5.2 pounds per hour per tank). The hydrogen system heaters and fans can sustain proper pressures at a total flow of 1.02 pounds per hour (0.51 pound per hour per tank).

Manual repressurization supplies power directly to the heaters and fans through the control switches.

It can be used in case of automatic control failure, heater failure, or fan failure.

Tank pressure and quantity are monitored on meters located on the main display console. The caution and warning system will activate on alarm when oxygen pressure in either tank exceeds 950 psia or falls below 800 psia or when the hydrogen system pressure exceeds 270 psia or drops below 220 psia. Pressure, quantity, and reactant temperature of each tank are telemetered to MSFN.

Oxygen relief valves vent at a pressure between 983 and 1010 psig and reseat at 965 psig. Hydrogen relief valves vent at a pressure between 273 and 285 psig and reseat at 268 psig. Relief opening of the relief valves will be prevented if possible to minimize the probability of improper reseating, resulting in eventual depletion of one tank.

Overpressurization may be prevented in two ways. The first is to disable the heater and fan circuits when tank quantities reach approximately 55 percent, allowing pressure in the tanks to decrease. This provides wider range for eventual pressure increase during minimum-value operation. This method retains the maximum amount of fluid for spacecraft use. The second method is to perform an unscheduled fuel cell purge to deplete tank pressure.

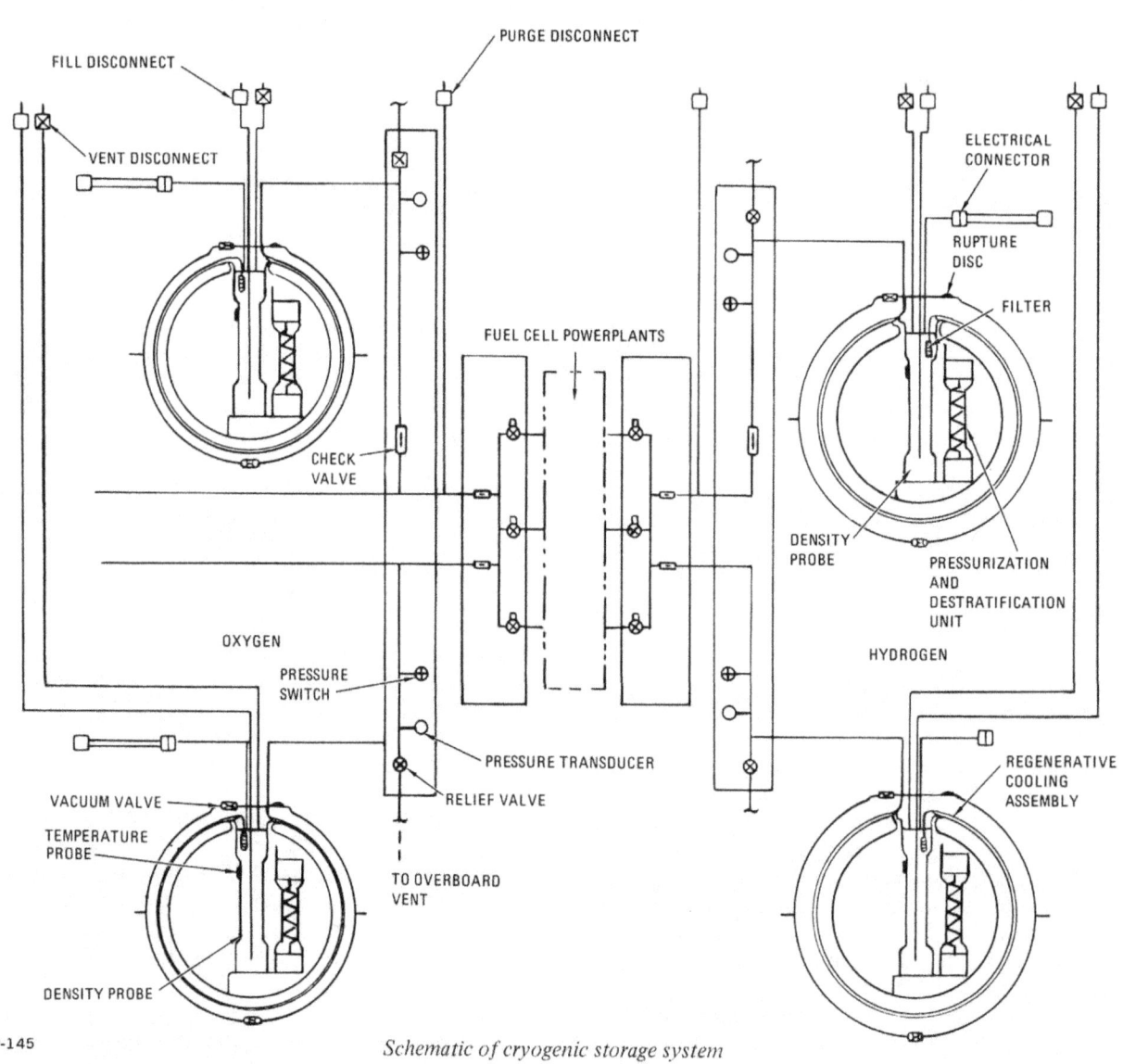

Schematic of cryogenic storage system

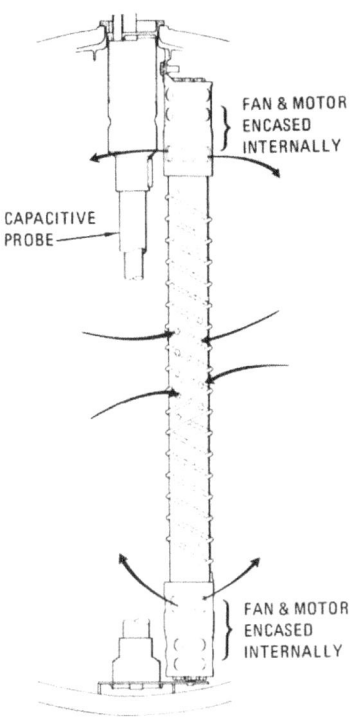

P-146

Cryogenic tank pressure and quantity measurement devices

The reactant tanks have vacuum-ion pumps which function as ion traps to maintain the vacuum between the inner and outer shells.

BATTERIES

The five silver oxide-zinc storage batteries of the electrical power subsystem are located in the lower equipment bay of the CM.

Three rechargeable entry and post-landing batteries (A, B, and C) power the CM systems after CM-SM separation. Before separation, the batteries provide a secondary source of power while the fuel cells are the primary source. They supplement fuel cell power during peak load periods (velocity change maneuvers), provide power during emergency operations (failure of two fuel cells), and provide power for power subsystem control circuitry (relays, indicators, etc.) and sequencer logic. They can also be used to power pyro circuits.

Each entry and post-landing battery consists of 20 silver oxide-zinc cells connected in series. The cells are individually encased in plastic containers which contain relief valves that open at 35 ± 5 psig, venting during an overpressure into the battery case. Each battery case is vented overboard through a common manifold and the urine/water dump line. The vent line prevents battery-generated gas from entering the crew compartment.

In the event a battery case fractures, the vent is closed. The battery manifold pressure is monitored on the meter and when it approaches CM pressure the vent valve is opened to prevent the gas going into the cabin. Battery manifold pressure can be used as an indication of urine/waste water dump line plugging.

Each battery delivers a minimum of 40 ampere-hours at a current output of 35 amps for 15 minutes and a subsequent output of 2 amps, or at a current output of 25 amps for 30 minutes and a subsequent output of 2 amps. At Apollo mission loads, each battery can provide 50 ampere-hours.

Open circuit voltage is 37.2 volts. Since sustained battery loads are extremely light (2 to 3 watts), voltages very close to open circuit voltage will be indicated on the spacecraft voltmeter, except when the main bus tie switches have been activated to tie the battery output to the main dc buses. Normally only batteries A and B will be connected to the main dc buses. Battery C is isolated during the pre-launch period and provides a backup for main dc bus power. The two-battery configuration provides more efficient use of fuel cell power during peak power loads and decreases overall battery recharge time.

The two pyrotechnic batteries supply power to activate ordnance devices in the spacecraft. The

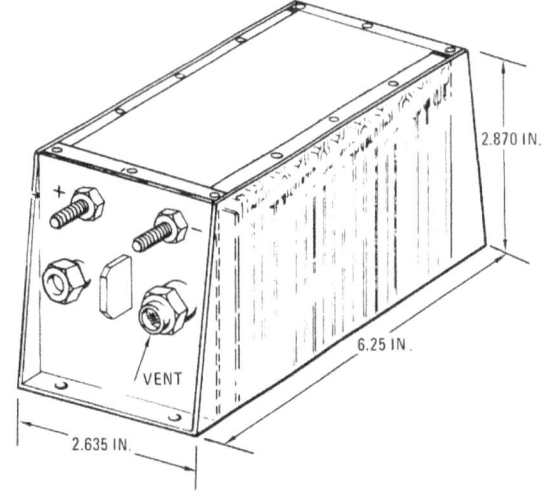

P-147

Pyrotechnic battery

pyrotechnic batteries are isolated from the rest of the electrical power system to prevent the high-power surges in the pyrotechnic system from affecting it and to assure source power when required. These batteries are not recharged in flight. The entry and post-landing batteries can be used as a redundant source of power for initiating pyro circuits if either pyro battery fails.

FUEL CELL POWERPLANTS

Each of the three Bacon-type fuel cell powerplants is individually coupled to a heat rejection (radiator) system, the hydrogen and oxygen cryogenic storage systems, a water storage system, and a power distribution system.

The powerplants generate dc power on demand through an exothermic chemical reaction. A byproduct of this chemical reaction is water, which is fed to a potable water storage tank in the CM where it is used for astronaut consumption and for cooling purposes in the environmental control subsystem. The amount of water produced is proportional to the ampere-hours.

Each powerplant consists of 31 single cells connected in series and enclosed in a metal pressure jacket. The water separation, reactant control, and heat transfer components are mounted in a compact accessory section attached directly above the pressure jacket.

Powerplant temperature is controlled by the primary (hydrogen) and secondary (glycol) loops. The hydrogen pump, providing continuous circulation of hydrogen in the primary loop, withdraws water vapor and heat from the stack of cells. The primary bypass valve regulates flow through the hydrogen

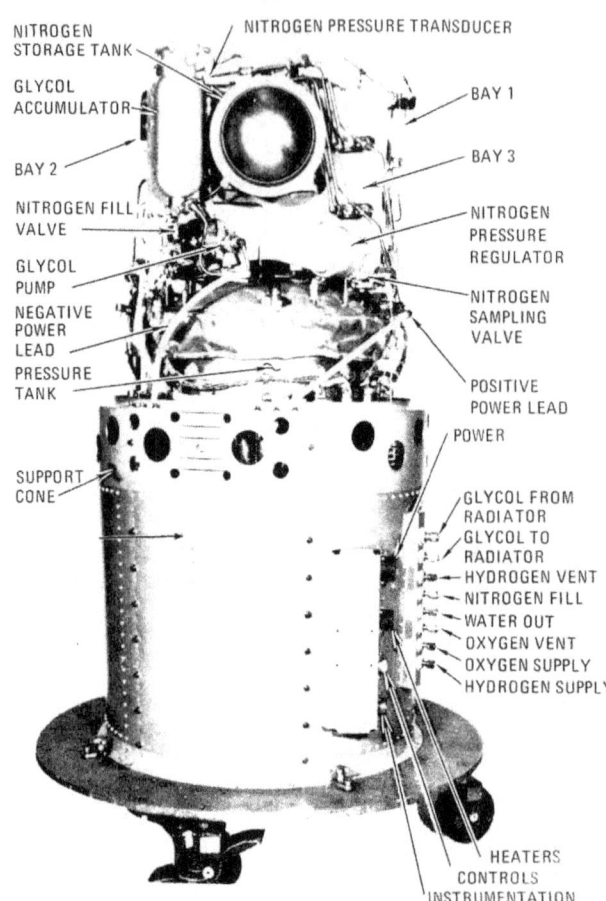

Fuel cell module accessories

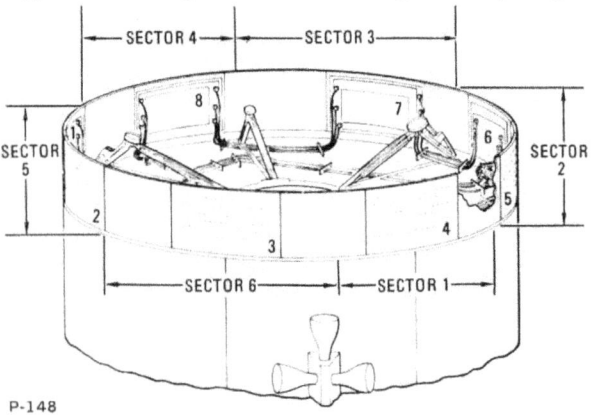

Location of electrical power subsystem radiators

regenerator to impart exhaust heat to the incoming hydrogen gas as required to maintain the proper cell temperature. The exhaust gas flows to the condenser where waste heat is transferred to the glycol, the resultant temperature decrease liquifying some of the water vapor. The motor-driven centrifugal water separator extracts the liquid and feeds it to the potable water tank in the CM. The temperature of the hydrogen-water vapor exiting from the condenser is controlled by a bypass valve which regulates flow through a secondary regenerator to a control condenser exhaust within desired limits. The cool gas is then pumped back to the fuel cell through the primary regenerator by a motor-driven vane pump, which also compensates for pressure losses due to water extraction and cooling. Waste heat, transferred to the glycol in the condenser, is transported to the radiators located on the fairing between the CM and SM, where it is radiated into space. Radiator area is sized to reject the waste heat resulting from operation in the normal power range. If an emergency arises in which an extremely low power level is required, individual controls can bypass three of the eight radiator panels for each powerplant. This area reduction improves the margin for radiator freezing which could result from the lack of sufficient waste heat to maintain adequate glycol temperature. This is not a normal procedure and is considered irreversible due to freezing of the bypassed panels.

Reactant valves provide the connection between the powerplants and the cryogenic system. They are opened during pre-launch fuel cell startup and closed only after a powerplant malfunction necessitating its isolation from the cryogenic system. Before launch, a valve switch is operated to apply a holding voltage to the open solenoid of the hydrogen and oxygen reactant valves of the three powerplants. This voltage is required only during boost to prevent inadvertent closure due to the effects of high vibration. The reactant valves cannot be closed with this holding voltage applied. After earth orbit insertion, the holding voltage is removed and three circuit breakers are opened to prevent valve closure through inadvertent activation of the reactant valve switches.

Nitrogen is stored in each powerplant at 1500 psia and regulated to a pressure of 53 psia. Output of the regulator pressurizes the electrolyte in each cell through a diaphragm arrangement, the coolant loop through an accumulator, and is coupled to the oxygen and hydrogen regulators as a reference pressure.

Cryogenic oxygen, supplied to the powerplants at 900 ± 35 psia, absorbs heat in the lines, absorbs additional heat in the fuel cell powerplant reactant preheater, and reaches the oxygen regulator in a gaseous form at temperatures above $0^\circ F$. The differential oxygen regulator reduces pressure to 9.5 psia above the nitrogen reference, thus supplying it to the fuel cell stack at 62.5 psia. Within the porous oxygen electrodes, the oxygen reacts with the water in the electrolyte and the electrons

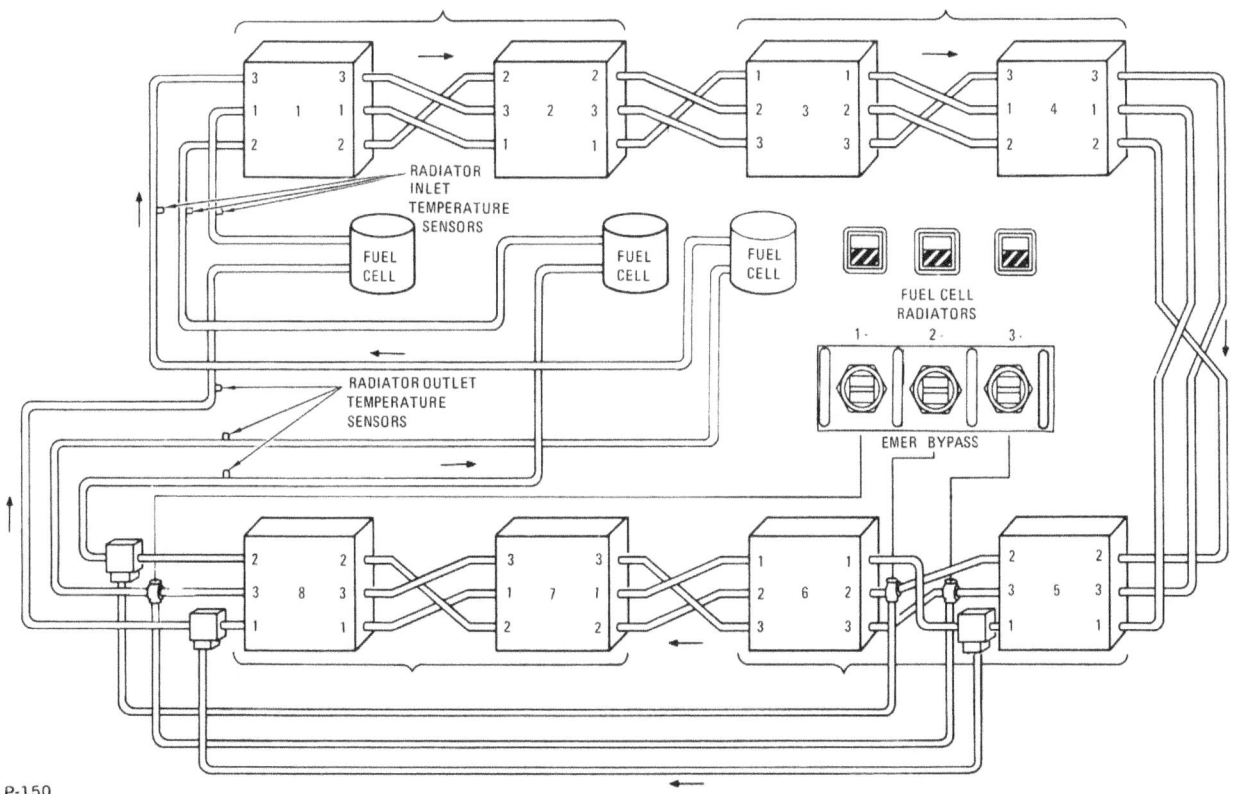

Flow and control of electrical power subsystem radiators

provided by the external circuit to produce hydroxyl ions.

Cryogenic hydrogen, supplied to the powerplants at 245 (+15, -20) psia, is heated in the same manner as the oxygen. The differential hydrogen regulator reduces the pressure to 8.5 psia above the reference nitrogen, thus supplying it in a gaseous form to the fuel cells at 61.5 psia. The hydrogen reacts in the porous hydrogen electrodes with the hydroxyl ions in the electrolyte to produce electrons, water vapor, and heat. The nickel electrodes act as a catalyst in the reaction. The water vapor and heat are withdrawn by the circulation of hydrogen gas in the primary loop and the electrons are supplied to the load.

Each of the 31 cells comprising a powerplant contains electrolyte which on initial fill consists of approximately 83 percent potassium hydroxide (KOH) and 17 percent water by weight. The powerplant is initially conditioned to increase the water ratio, and during normal operation, water content will vary between 23 and 28 percent. At this ratio, the electrolyte has a critical temperature of $360°F$. Powerplant electrochemical reaction becomes effective at the critical temperature. The powerplants are heated above the critical temperature by ground support equipment. A load on the powerplant of approximately 563 watts is required to maintain it above the normal minimum operating temperature of $385°F$. The automatic in-line heater circuit will maintain powerplant temperature in this range with smaller loads applied.

Purging is a function of power demand and gas purity. Oxygen purging requires 2 minutes and hydrogen purging 80 seconds. The purge frequency is determined by the mission power profile and gas purity as sampled after spacecraft tank fill. A degradation purge can be performed if powerplant current output decreases approximately 3 to 5

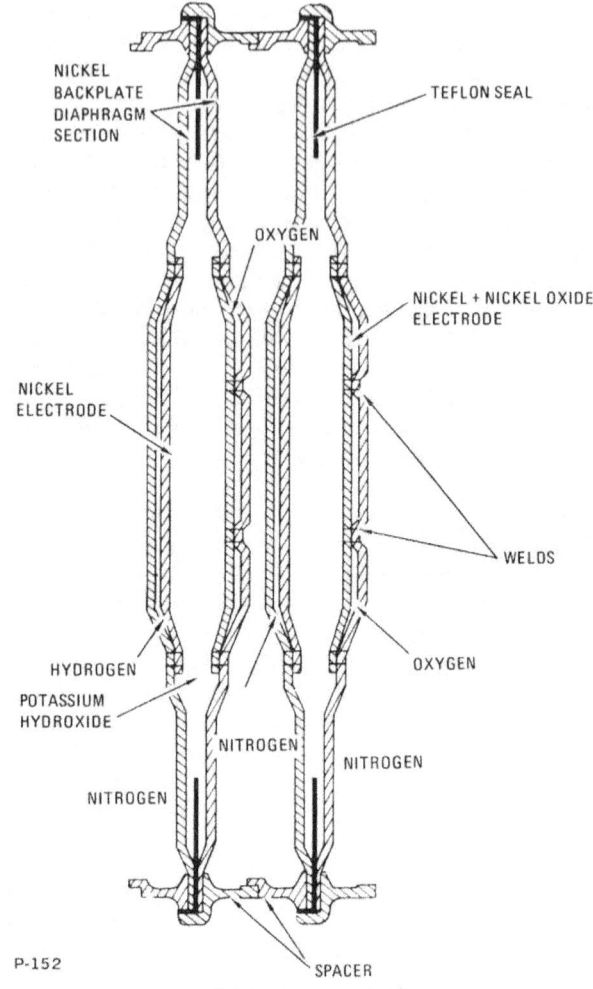

Cutaway view of cell

amps during sustained operation. The oxygen purge has more effect during this type of purge, although it would be followed by a hydrogen purge if recovery to normal were not realized. If the performance degradation were due to powerplant electrolyte flooding, which would be indicated by activation of the pH high indicator, purging would not be performed due to the possibility of increasing the flooding.

The application and removal of fuel cell loads causes the terminal voltage to decrease and increase, respectively. A decrease in terminal voltage, resulting from an increased load, is followed by a gradual increase in fuel cell skin temperature which causes an increase in terminal voltage. Conversely, an increase in terminal voltage, resulting from a decreased load, is followed by a gradual decrease in fuel cell skin temperature which causes a decrease in terminal voltage. This performance change with

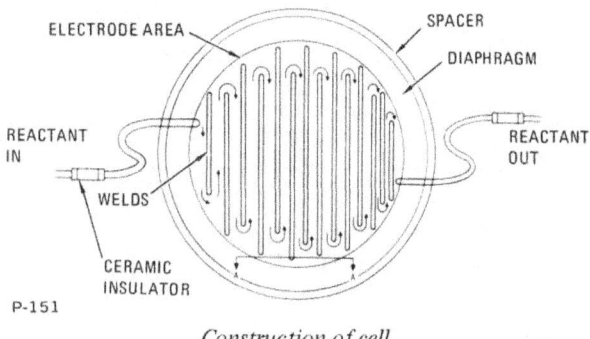

Construction of cell

temperature is regulated by the primary regenerator bypass valve and provides the capability of operating over an increased power range within voltage regulation limits.

The range in which the terminal voltage is permitted to vary is determined by the high and low voltage input design limits of the components being powered. For most components the limits are 30 volts dc and 25 volts dc. To remain within these design limits, the dc bus voltage must be maintained between 31.0 and 26.2 volts dc. Bus voltage is maintained within prescribed limits during high power requirements by the use of entry and post-landing batteries.

Spacecraft systems are powered up in one continuous sequence providing the main bus voltage does not decrease below 26.2 volts. If bus voltage

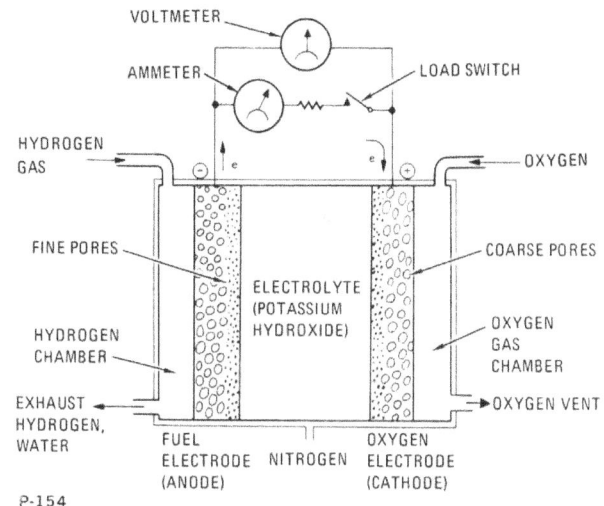

Electrochemical flow in fuel cell

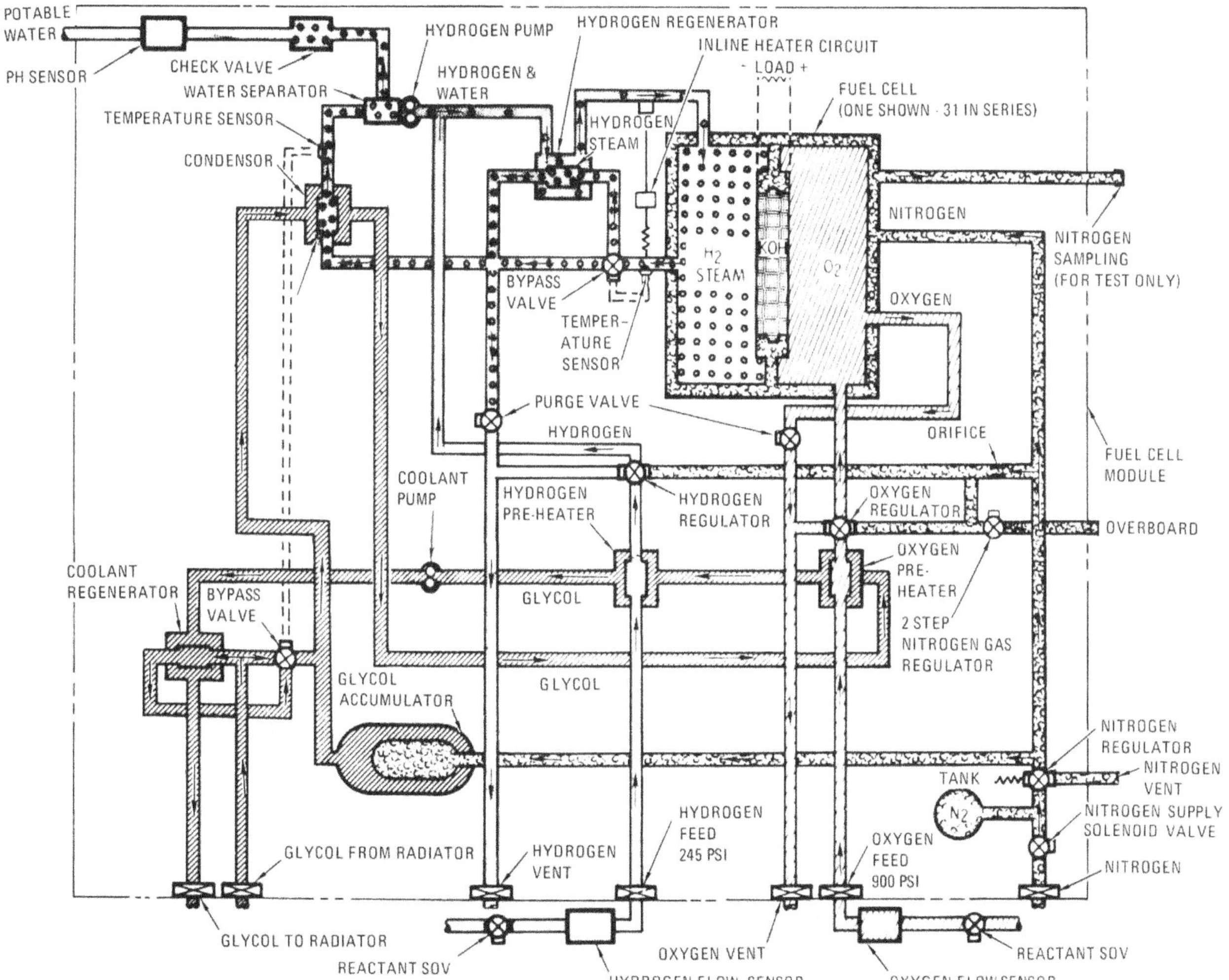

Schematic of fuel cell module

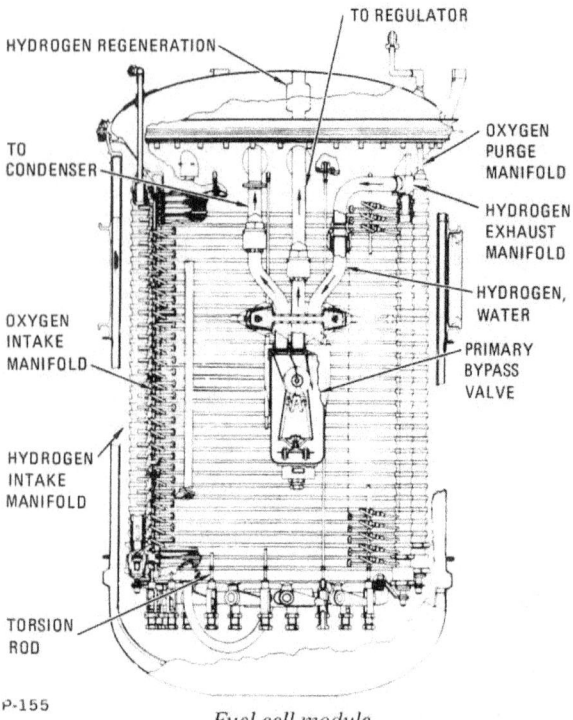

Fuel cell module

decreases to this value the power-up sequence can be interrupted for the time required for fuel cell temperatures to increase with the resultant voltage increase, or the batteries can be connected to the main buses reducing the fuel cell load. In most cases, powering up can be performed in one continuous sequence; however, when starting from an extremely low spacecraft load, it is probable that a power-up interruption or batteries will be required. The greatest load increase occurs while powering up for a velocity change maneuver.

Spacecraft systems are powered down in one continuous sequence providing the main bus voltage does not increase above 31.0 volts. In powering down from relatively high spacecraft load levels, the sequence may have to be interrupted for the time required for fuel cell temperature, and thus bus voltage, to decrease.

If a powerplant fails it is disconnected from the main dc buses and the in-line heater circuit is deactivated. Before disconnecting a fuel cell, if a single inverter is being used, the two remaining powerplants are connected to each main dc bus to enhance load sharing since bus loads are unbalanced. If two inverters are being used, each of the remaining powerplants is connected to a separate main dc bus for bus isolation, since bus loads are relatively equal.

INVERTERS

Each inverter is composed of an oscillator, an eight-stage digital countdown section, a dc line filter, two silicon-controlled rectifiers, a magnetic amplifier, a buck-boost amplifier, a demodulator, two dc filters, an eight-stage power inversion section, a harmonic neutralization transformer, an ac output filter, current sensing transformers, a Zener diode reference bridge, a low-voltage control, and an overcurrent trip circuit. The inverter normally uses a 6.4 kiloHertz square wave synchronizing signal from the central timing equipment which maintains inverter output at 400 Hertz. If this external signal is completely lost, the free running oscillator within the inverter will provide pulses that will maintain inverter output within ±7 Hertz. The internal oscillator is normally synchronized by the external pulse.

The 6.4 kiloHertz square wave provided by central timing equipment is applied through the internal oscillator to the eight-stage digital countdown section. The oscillator has two divider circuits which provide a 1600-pulse per second to the magnetic amplifier.

The eight-stage digital countdown section produces eight 400-cycle square waves, each mutually displaced one pulse-time from the preceding and following wave. One pulse-time is 156 microseconds and represents 22.5 electrical degrees. The eight square waves are applied to the eight-stage power inversion section.

The eight-stage power inversion section, fed by a controlled voltage from the buck-boost amplifier, amplifies the eight 400-Hertz, square waves produced by the eight-stage digital countdown section. The amplified square waves, still mutually displaced 22.5 electrical degrees, are next applied to the harmonic neutralization transformer.

The harmonic neutralization section consists of 31 transformer windings on one core. This section accepts the 400-Hertz square-wave output of the eight-stage power inversion section and transforms it into a 3-phase, 400-Hertz 115-volt signal. The manner in which these transformers are wound on a single core produces flux cancellation which eliminates all harmonics up to and including the fifteenth of the fundamental frequency. The 22.5-degree displacement of the square waves provides a means of electrically rotating the square wave excited primary windings around the 3-phase,

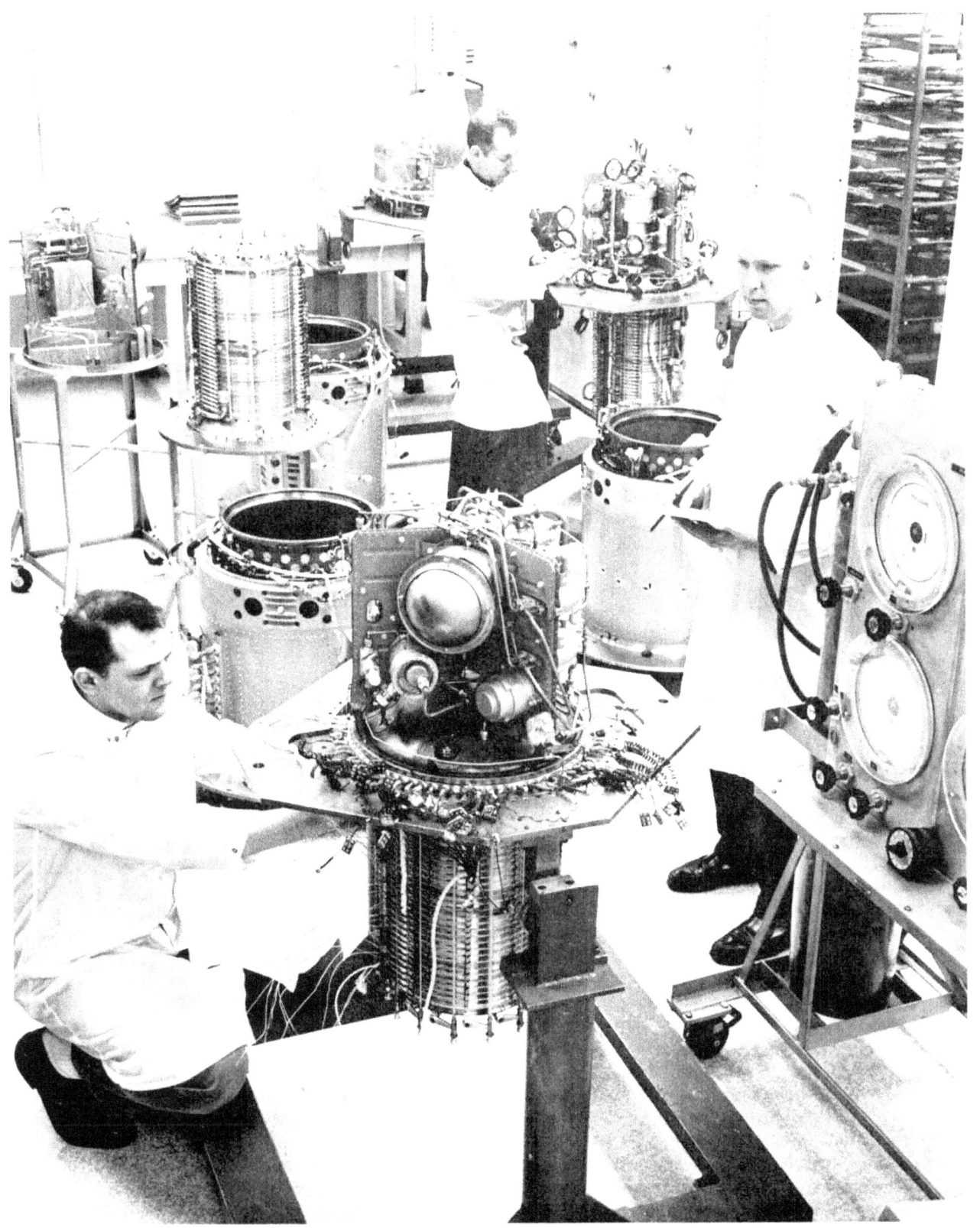

Pratt & Whitney technicians assemble fuel cell powerplants at plant in Hartford, Conn.

wye-connected secondary windings, thus producing the 3-phase 400 cycle sine wave output. This 115-volt signal is then applied to the ac output filter.

The ac output filter eliminates the remaining higher harmonics. Since the lower harmonics were eliminated by the harmonic neutralization transformer, the size and weight of this output filter is reduced. Circuitry in this filter also produces a rectified signal which is applied to the Zener diode reference bridge for voltage regulation. The amplitude of this signal is a function of the amplitude of ac output voltage. After filtering, the 3-phase, 115-volt ac 400-hertz sine wave is applied to the ac buses through individual phase current-sensing transformers.

The current-sensing transformers produce a rectified signal, the amplitude of which is a direct function of inverter output current magnitude. This dc signal is applied to the Zener diode reference bridge to regulate inverter current output; it is also paralleled to an overcurrent trip circuit.

The Zener diode reference bridge receives a rectified dc signal, representing voltage output, from the circuitry in the ac output filter. A variance in voltage output unbalances the bridge, providing an error signal of proper polarity and magnitude to the buck-boost amplifier via the magnetic amplifier. The buck-boost amplifier, through its bias voltage output, compensates for voltage variations. When inverter current output reaches 200 to 250 percent of rated current, the rectified signal applied to the bridge from the current sensing transformers is of sufficient magnitude to provide an error signal, causing the buck-boost amplifier to operate in the same manner as during an overvoltage condition. The bias output of the buck-boost amplifier, controlled by the error signal, will be varied to correct for any variation in inverter voltage or a beyond-tolerance increase in current output. When inverter current output reaches 250 percent of rated current, the overcurrent trip circuit is activated.

The overcurrent trip circuit monitors a rectified dc signal representing current output. When total inverter current output exceeds 250 percent of rated current, this circuit will disconnect an inverter in 15±5 seconds. If current output of any single phase exceeds 300 percent of rated current, this circuit will disconnect an inverter in 5±1 seconds.

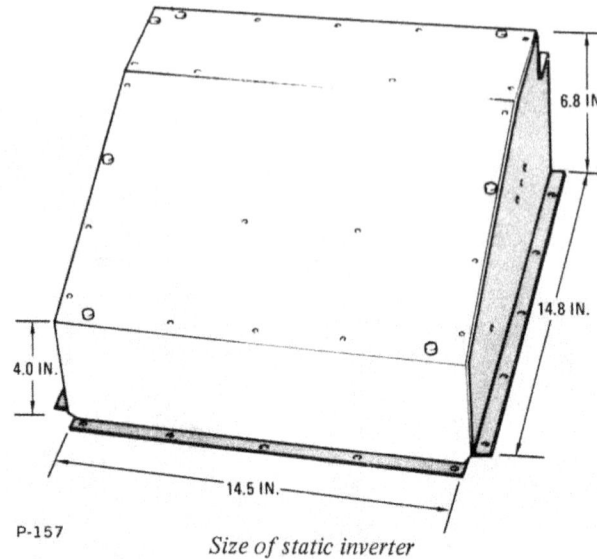

P-157 *Size of static inverter*

The disconnect is provided through relays located in the motor switch circuits that connect the inverters to the ac buses.

Dc power to the inverter is supplied from the main dc buses through the dc line filter. The filter reduces the high-frequency ripple in the input, and the 25 to 30 volts dc is applied to the silicon-controlled rectifiers.

The silicon-controlled rectifiers are alternately set by the 1600 pulses-per-second signal from the magnetic amplifier to produce a dc square wave with an on-time of greater than 90 degrees from each rectifier. This is filtered and supplied to the buck-boost amplifier where it is transformer-coupled with the amplified 1600 pulses-per-second output of the magnetic amplifier to develop a filtered 35 volts dc which is used for amplification in the power inversion stages.

The buck-boost amplifier also provides a variable bias voltage to the eight-stage power inversion section. The amplitude of this bias voltage is controlled by the amplitude and polarity of the feedback signal from the Zener diode reference bridge which is referenced to output voltage and current. This bias signal is varied by the error signal to regulate inverter voltage and maintain current output within tolerance.

The demodulator circuit compensates for any low-frequency ripple in the dc input to the inverter. The high-frequency ripple is attenuated by the input filters. The demodulator senses the 35-volt dc output of the buck-boost amplifier and the current input to the buck-boost amplifier. An input dc voltage drop or increase will be reflected in a drop or increase in the 35-volt dc output of the buck-boost amplifier, as well as a drop or increase in current input to the buck-boost amplifier. A sensed decrease in the buck-boost amplifier voltage output is compensated for by a demodulator output, coupled through the magnetic amplifier to the silicon-controlled rectifiers. The demodulator output causes the silicon-controlled rectifiers to conduct for a longer time, thus increasing their filtered dc output. An increase in buck-boost amplifier voltage output caused by an increase in dc input to the inverter is compensated for by a demodulator output coupled through the magnetic amplifier to the silicon-controlled rectifiers causing them to conduct for shorter periods, thus producing a lower filtered dc output to the buck-boost amplifier. In this manner, the 35-volt dc input to the power inversion section is maintained at a relatively constant level irrespective of the fluctuations in dc input voltage.

The low-voltage control circuit samples the input voltage to the inverter and can terminate inverter operation. Since the buck-boost amplifier provides a boost action during a decrease in input voltage to the inverter, in an attempt to maintain a constant 35 volts dc to the power inversion section and a regulated 115-volt inverter output, the high boost required during a low-voltage input would tend to overheat the solid-state buck-boost amplifier. As a precautionary measure, the low-voltage control will terminate inverter operation by disconnecting operating voltage to the magnetic amplifier and the first power inversion stage when input voltage decreases to between 16 and 19 volts dc.

A temperature sensor with a range of 32° to $248^O F$ is installed in each inverter and will illuminate a light in the caution and warning system at an inverter overtemperature of $190^O F$. Inverter temperature is telemetered to the ground.

BATTERY CHARGER

A constant-voltage, solid-state battery charger is located in the CM lower equipment bay. It is provided 25 to 30 volts from both main dc buses and 115 volts 400-cps 3-phase from either of the ac buses. All three phases of ac are used to boost the 25 to 30-volt dc input and produce 40 volts dc for charging. In addition, Phase A of the ac is used to supply power for the charger circuitry. The logic network in the charger, which consists of a two-stage differential amplifier (comparator), Schmitt trigger, current-sensing resistor, and a voltage amplifier, sets up the initial condition for operation. The first stage of the comparator is on, with the second stage off, thus setting the Schmitt trigger first stage to on with the second stage off. Maximum base drive is provided to the current amplifier which turns on the switching transistor. With the switching transistor on, current flows from the transformer rectifier through the switching transistor, current sensing resistor, and switch choke to the battery being charged. Current lags voltage due to switching choke action. As current flow increases, the voltage drop across the sensing resistor increases, and at a specific level sets the first stage of the comparator off and the second stage on. The voltage amplifier is set off to reverse the Schmitt trigger to first stage off and second stage on. This sets the current amplifier off, which in turn sets the switching transistor off. This terminates power from the source, causing the field in the choke to continue collapsing, discharging into the battery, then through the switching diode and the current sensing resistor to the opposite side of the choke. As the electromagnetic field in the choke decreases, current through the sensing resistor decreases, reducing the voltage drop across the resistor. At some point, the decrease in voltage drop across the sensing resistor reverses the comparator circuit, setting up the initial condition and completing one cycle of operation. The output load current, due to the choke action, remains relatively

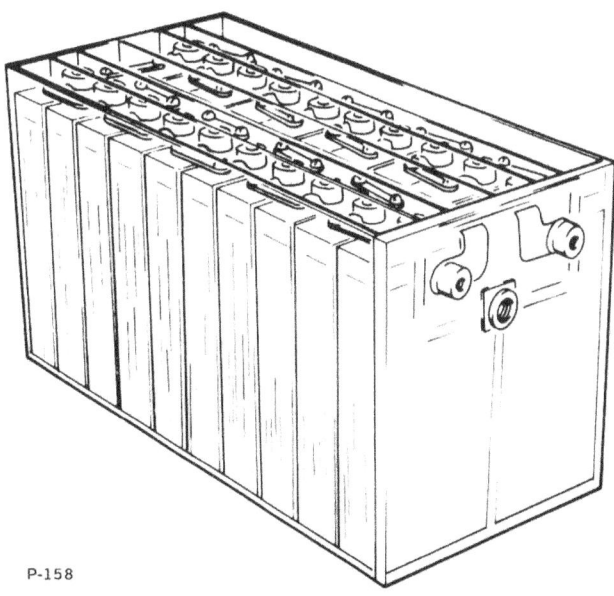

Entry and post-landing battery

constant except for the small variation through the sensing resistor. This variation is required to set and reset the switching transistor and Schmitt trigger through the action of the comparator.

Battery charger output is regulated by the sensing resistor until battery voltage reaches approximately 36 volts. At this point, the biased voltage control network is unbiased, and in conjunction with the sensing resistor provides a signal for cycling the battery charger. As battery voltage increases, the internal impedance of the battery increases, decreasing current flow from the charger. At 39 volts minimum, the battery is considered fully charged and current flow becomes negligible.

POWER DISTRIBUTION

Dc and ac power distribution is provided by two redundant buses in each system. A single-point ground on the spacecraft structure is used to eliminate ground loop effects. Sensing and control circuits are provided for monitoring and protection of each system.

Dc power is distributed with a two-wire system and a series of interconnected buses, switches, circuit breakers, and isolation diodes. The dc negative buses are connected to the single-point ground. The buses consist of the following:

1. Two main dc buses (A and B), powered by the three fuel cell powerplants and/or entry and post-landing Batteries A, B, and C.

2. Two battery buses (A and B), each powered by its respective entry and post-landing battery. Battery C can power either or both buses if the Batteries A or B fail.

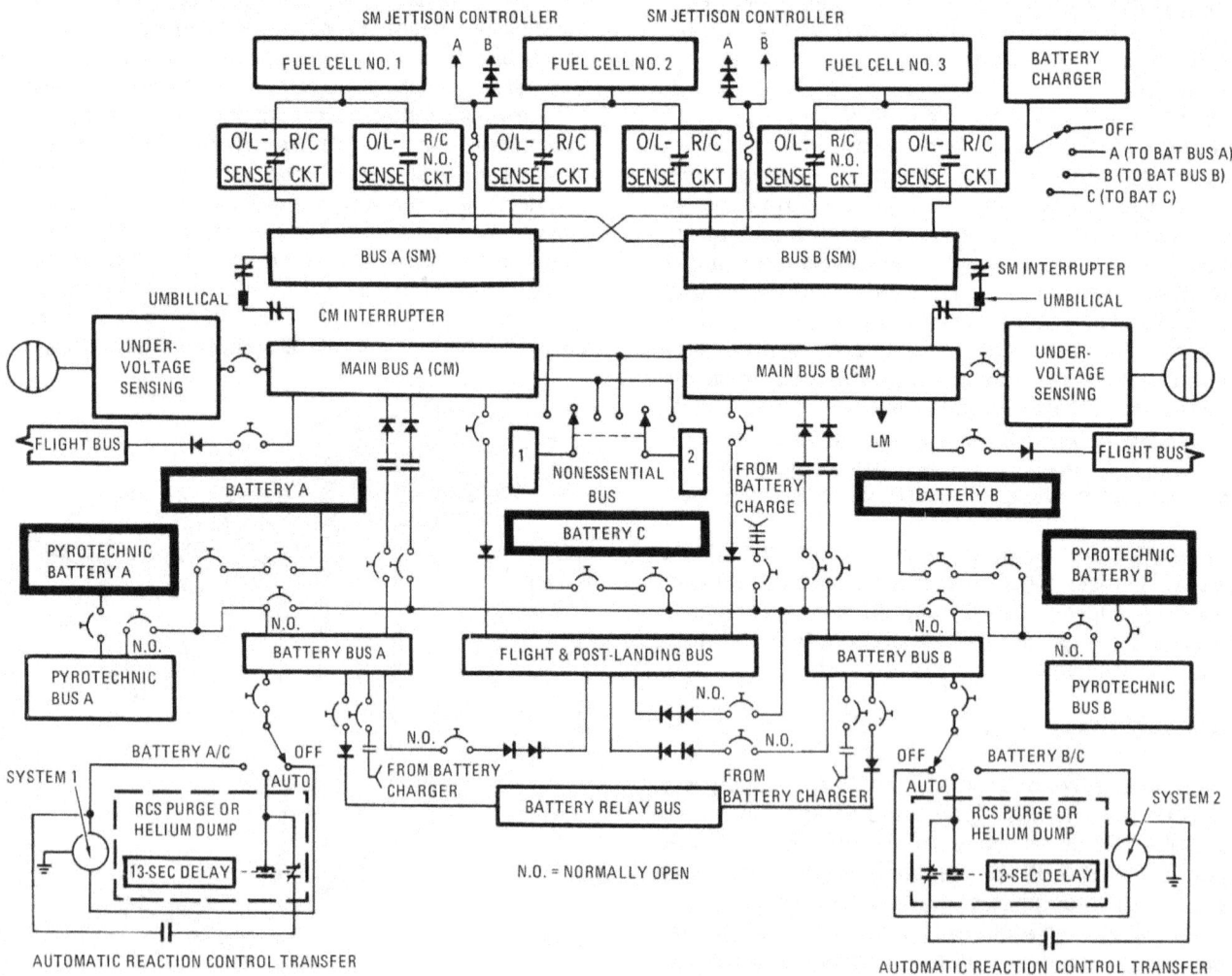

DC power distribution

3. Flight and post-landing bus, powered through both main dc buses and diodes, or directly by the three entry and post-landing batteries through dual diodes.

4. Flight bus, powered through both main dc buses and isolation diodes.

5. Nonessential bus, powered through either dc Main Bus A or B.

6. Battery relay bus, powered by two entry and post-landing batteries (A and B) through the individual battery buses and isolation diodes.

7. Pyro buses, isolated from the main electrical power subsystem when powered by the pyro batteries. Entry batteries can be connected to the A or B pyro system in case of loss of a pyro battery.

8. SM jettison controllers, powered by the fuel cell powerplants and completely isolated from the main electrical power subsystem until activated during CSM separation.

Power from the fuel cell powerplants can be connected to the main dc buses through six motor switches (part of overload/reverse current circuits in the SM) which are controlled by switches in the CM. Fuel cell power can be connected to either or both of the main dc buses. When an overload occurs, the overload-reverse current circuits in the SM automatically disconnect the fuel cell powerplants from the overloaded bus and provide visual displays for isolation of the trouble. A reverse current condition will disconnect the malfunctioning powerplant from the dc system. Dc undervoltage sensing circuits are provided to indicate bus low-voltage conditions. If voltage drops below 26.25 volts dc, the applicable dc undervoltage light on the caution and warning panel will illuminate. Since each bus is capable of handling all loads, an undervoltage condition should not occur except in an isolated instance (if too many electrical units are

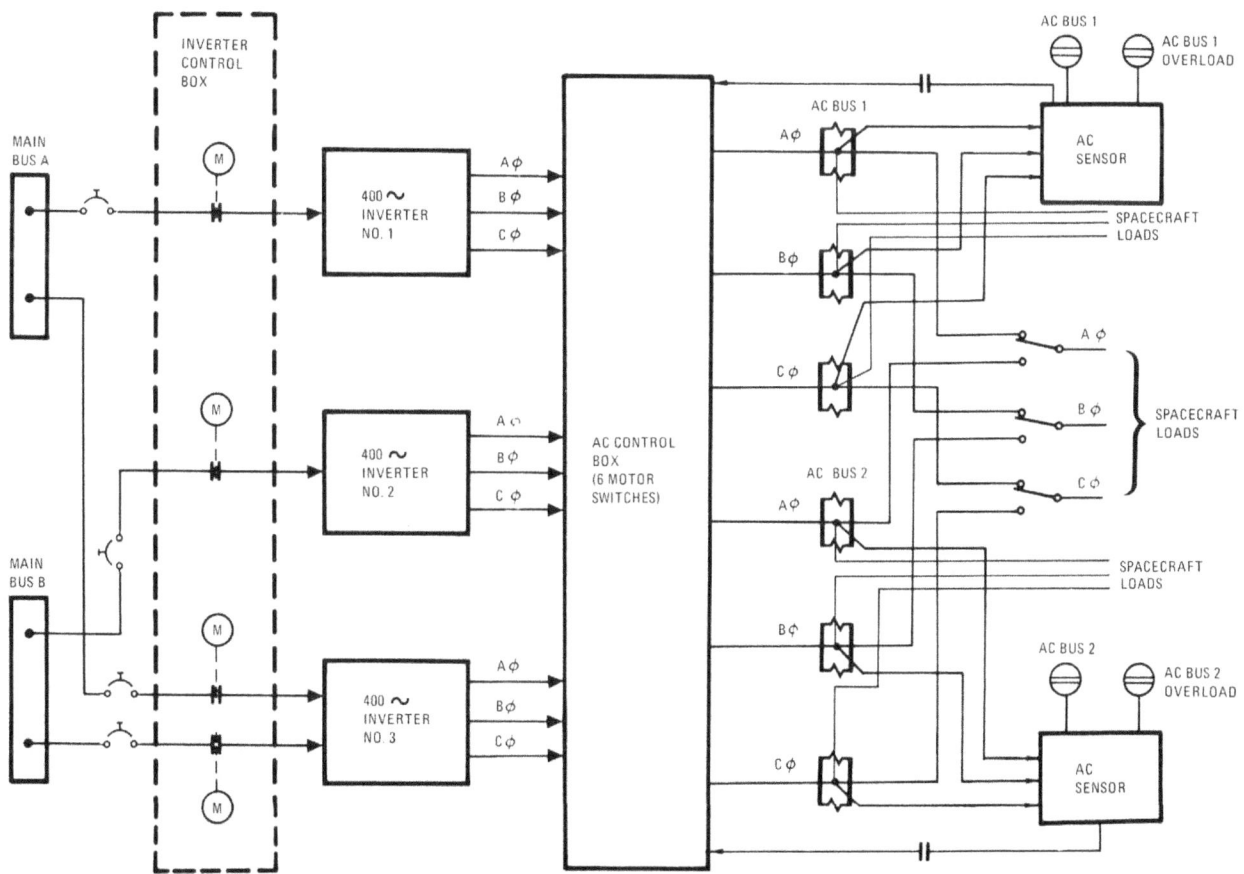

AC power distribution

placed on the bus simultaneously or if a malfunction exists in the subsystem). A voltmeter is provided to monitor voltage of each main dc bus, the battery charger, and each of the five batteries. An ammeter monitors current output of the powerplants, batteries, and battery charger.

During high power demand or emergencies, supplemental power to the main dc buses can be supplied from batteries A and B via the battery buses and directly from battery C. During entry, spacecraft power is provided by the three entry and post-landing batteries which are connected to the main dc buses before CM-SM separation.

The nonessential bus permits nonessential equipment to be shut off during a shortage of power (two fuel cell powerplants out). The flight bus distributes power to in-flight telecommunications equipment. The flight and post-landing bus distributes power to some of the in-flight telecommunications equipment and float bag No. 3 controls. The post-landing bus receives power from the fuel cells or entry and post-landing batteries through the main dc buses. After completion of reaction control subsystem purge during main chute descent, the entry batteries supply power to the post-landing bus directly through individual circuit breakers.

The battery relay bus provides dc power to the dc and ac sensing unit, the fuel cell, inverter control circuits, and some of the indicators on the main display console. The pyrotechnic batteries supply power to ordnance devices used during the course of the mission. The three fuel cell powerplants supply power to the SM jettison controllers for the SM separation maneuver.

Ac power is distributed with a four-wire system via two redundant buses, 1 and 2. The ac neutral bus is connected to the single-point ground. Ac power is provided by one or two of the solid-state 115/200-volt 400-cps 3-phase inverters. Dc power is routed to the inverters through the main dc buses. Inverter No. 1 is powered through dc Main Bus A, inverter No. 2 through dc Main Bus B, and inverter No. 3 through either dc Main Bus A or B by switch selection. Each of these circuits has a separate circuit breaker and a power control motor switch. The three inverters are identical and are provided with overtemperature circuitry. A light indicator in the caution and warning group illuminates at 190°F to indicate overtemperature. Inverter output is routed through a series of control motor switches to the ac buses. Six switches control motor switches which operate contacts to connect or disconnect the inverters from the ac buses. The motor switch circuits are designed to prevent connecting two inverters to the same ac bus at the same time. Ac loads receive power from either ac bus through bus selector switches. In some instances, a single phase is used for operation of equipment and in others all three. Over- or undervoltage and overload sensing circuits are provided for each bus. Inverters are automatically disconnected during overvoltage or overload. Ac bus voltage fail and overload lights in the caution and warning group indicate voltage or overload malfunctions. Phase A voltage of each bus is telemetered to ground stations.

ENVIRONMENTAL CONTROL

The environmental control subsystem provides a controlled environment for three astronauts for up to 14 days. For normal conditions, this environment includes a pressurized cabin (5 pounds per square inch), a 100-percent oxygen atmosphere, and a cabin temperature of 70 to 75 degrees. For use during critical mission phases and for emergencies, the subsystem provides a pressurized suit circuit. The environmental control unit, a major part of the environmental control subsystem, is produced by Garrett Corp.'s AiResearch Division, Los Angeles.

The subsystem provides oxygen and hot and cold water, removes carbon dioxide and odors from the CM cabin, provides for venting of waste, and dissipates excessive heat from the cabin and from operating electronic equipment. It is designed so that a minimum amount of crew time is needed for its normal operation.

The environmental control unit is the heart of the environmental control subsystem. It is a compact grouping of equipment about 29 inches long, 16 inches deep, and 33 inches at its widest point. It is mounted in the left-hand equipment bay. The unit contains the coolant control panel, water chiller, two water-glycol evaporators, carbon dioxide-odor absorber canisters, and suit heat exchanger, water separator, and compressors. The oxygen surge tank, water glycol pump package and reservoir, and control panels for oxygen and water are adjacent to the unit.

The subsystem is concerned with three major elements: oxygen, water, and coolant (water-glycol). All three are interrelated and intermingled with other subsystems. These three elements provide the major functions of spacecraft atmosphere and thermal control and water management through four major subsystems: oxygen, pressure suit circuit, water, and water-glycol. A fifth subsystem, post-landing ventilation, also is part of the environmental control subsystem; it provides outside air for breathing and cooling after the command module has splashed down in the ocean.

The oxygen subsystem controls the flow of oxygen within the CM, stores a reserve supply for use during entry and emergencies, regulates the pressure of oxygen supplied to subsystem and

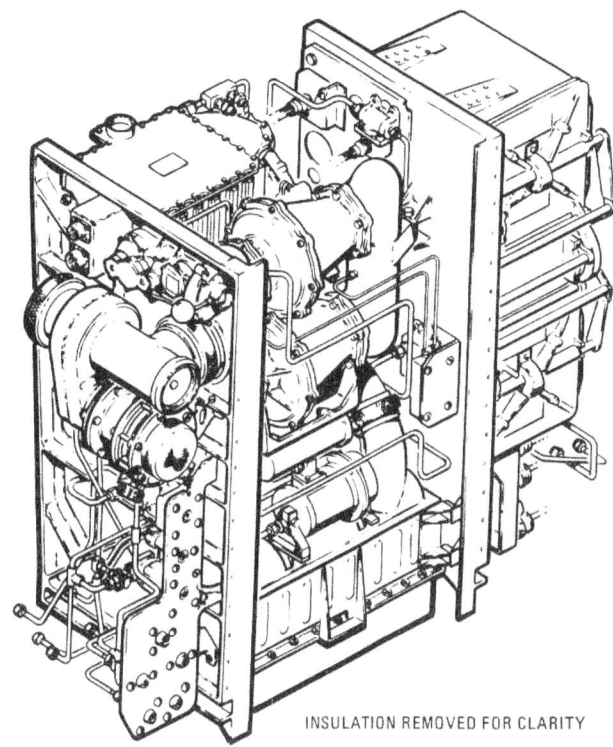

INSULATION REMOVED FOR CLARITY

P-161

Environmental control unit

pressure suit circuit components, controls cabin pressure, controls pressure in water tanks and the glycol reservoir, and provides for purging the pressure suit circuit.

The pressure suit circuit provides a continuously conditioned atmosphere. It automatically controls suit gas circulation, pressure, and temperature, and removes debris, excess moisture, and carbon dioxide from both suit and cabin gases.

The water subsystem collects and stores potable water, delivers hot and cold water to the crew, and augments the waste water supply for evaporative cooling. The waste water section of the subsystem collects and stores water extracted from the suit heat exchanger and distributes it to the evaporators for cooling.

The water-glycol subsystem provides cooling for the pressure suit circuit, the potable water chiller, and the spacecraft equipment, as well as heating or cooling for the cabin atmosphere.

ATMOSPHERE CONTROL

The three astronauts in the command module are in their space suits and connect to the pressure suit circuit when they first enter the CM. They remain in the suits at least until after they have attained earth orbit confirmation. They also are in their suits whenever critical maneuvers are performed or thrusting is being applied to the spacecraft.

Cabin atmosphere is 60-percent oxygen and 40-percent nitrogen on the launch pad to reduce fire hazard (fire propagates more rapidly in pure oxygen that in a mixed-gas atmosphere). The mixed atmosphere, supplied by ground equipment, will gradually become changed to pure oxygen after launch as the environmental control subsystem supplies oxygen to the cabin to maintain pressure and replenish the atmosphere.

During pre-launch and initial orbital operation, the suit circuit supplies oxygen at a flow rate slightly more than needed for breathing and suit leakage. This results in the suit being pressurized slightly above cabin pressure, which prevents cabin gases from entering and contaminating the suit circuit. The excess oxygen in the suit circuit is vented into the cabin.

Cabin pressure drops from sea level (14.7 psi) as the spacecraft rises during ascent and at a pressure of 6 psi automatic equipment begins operating to keep cabin pressure at that level. As the cabin pressure decreases, oxygen from the suit circuit is dumped into the cabin; in other words, the suit circuit is maintained at a constant differential pressure relative to cabin pressure.

During normal space operations, cabin pressure is maintained at a nominal 5 psia by a cabin pressure regulator at flow rates up to 1.3 pounds of oxygen per hour. In case a high leak rate develops, an emergency cabin pressure regulator supplies oxygen at high flow rates to maintain cabin pressure above 3.5 psia for more than 15 minutes, long enough for the crew to don their suits.

Before entry, the suit circuit is isolated from the cabin; cabin pressure is maintained by the cabin pressure regulator until the ambient (outside) pressure rises to a maximum of 0.9 psi above the cabin pressure. At that point a relief valve will open to allow the outside air to flow into the cabin. As the cabin pressure increases, the oxygen demand regu-

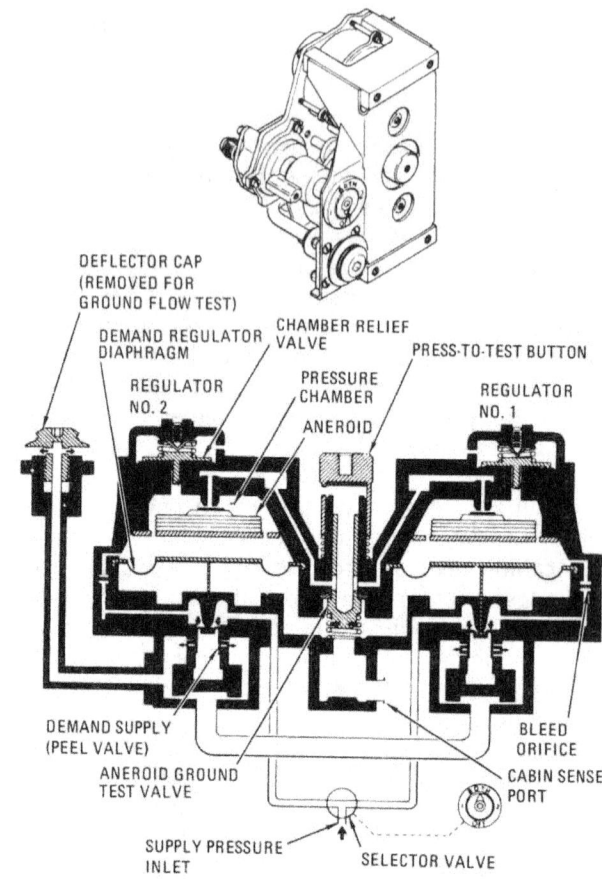

Emergency cabin pressure regulator

lator admits oxygen into the suit circuit to maintain suit pressure slightly above cabin pressure.

Oxygen normally is supplied from the cryogenic tanks in the service module. After CM-SM separation, it is supplied from the oxygen surge tank in the command module. In addition an oxygen repressurization package is provided to allow rapid repressurization of the cabin during space flight and to serve as an added entry oxygen supply. After landing, the cabin is ventilated with outside air through the post-landing fan and valves.

The oxygen in the suit circuit becomes heated and contaminated with carbon dioxide, odors, and moisture. This oxygen is circulated through the absorber assembly where the carbon dioxide and odors are removed, then through the heat exchanger where it is cooled and the excess moisture removed. In addition, any debris that might get into the circuit is trapped by the debris trap or in the absorber assembly.

WATER MANAGEMENT

The potable and waste water tanks are partially filled before launch to assure an adequate supply during early stages of the mission. Through the rest of the mission until CM-SM separation, the fuel cell powerplants supply potable water. A portion of the water is chilled for drinking and food preparation, and the remainder is heated and delivered through a separate valve in the food preparation unit. Provision is made to sterilize the potable water.

THERMAL CONTROL

Spacecraft heating and cooling is performed through two water-glycol coolant loops. The water-glycol, initially cooled through ground equipment, is pumped through the primary loop to cool operating electric and electronic equipment and the suit and cabin heat exchangers. The water-glycol also is circulated through a reservoir in the CM to provide a heat sink (heat-absorbing area) during ascent.

The water-glycol is a heat-absorbing medium; it picks up excess heat from operating equipment and the heat exchangers and is routed to the service module, where it passes through radiator tubes on the outside skin. The glycol mixture radiates its heat to space in its passage through these tubes, which are exposed to the cold of space. Then the mixture, now cold again, returns to the CM and repeats the cycle.

During ascent the radiators are heated by aerodynamic friction so a bypass valve is used to shut off the SM portion of the water-glycol loop. From liftoff until 110,000 feet excess heat is absorbed by the coolant and by pre-chilling of the structure; above 110,000 feet the excess heat is rejected by evaporating water in the primary glycol evaporator.

Temperature in the cabin is controlled by the way the water-glycol is routed. Normally it passes through the space radiators and returns to pick up and dissipate heat, thus cooling the cabin. If heating is needed, the coolant can be routed so that it returns to the cabin heat exchanger after absorbing heat from the operating equipment; this heat would be absorbed into the cabin gas circulated through the cabin heat exchanger by dual fans.

The secondary water-glycol loop is used when additional cooling is needed and before entry. Dual-loop operation may be used to "cold-soak" the CM interior for the plunge into the atmosphere.

EQUIPMENT

Environmental Control Unit (Garrett Corporation's AiResearch Division, Los Angeles) — Located in the command module left-hand equipment bay. Unit weighs 158 pounds and is 29 inches long, 16 inches high, and 33 inches wide. It consists of a water chiller, water-glycol evaporators, lithium hydroxide canisters, and suit heat exchanger and compressors.

Water-Glycol Reservoir — This aluminum tank contains a bladder under oxygen pressure of 20 pounds per square inch (psi) from the 20-psi oxygen supply system. The bladder stores one gallon of water-glycol and has a volume of 210 cubic inches. The tank is 7.13 by 13.38 by 4.67 inches and weighs 4-1/2 pounds. The reservoir is used to replenish the system and as a spare accumulator.

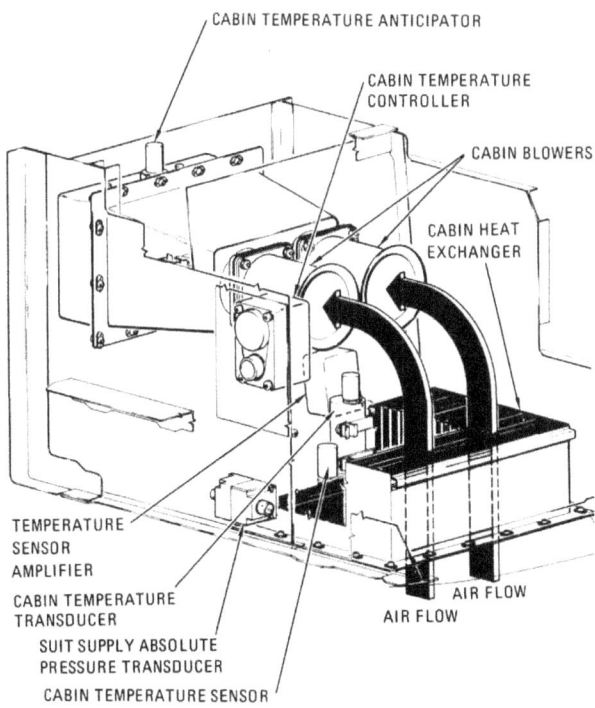

Cabin temperature unit

Water Chiller — It consists of stainless steel coil tubing with a 1/4-inch water inlet and outlet and 5/8-inch water-glycol inlet and outlet. The tubing holds about a tenth of a gallon of water. The water-glycol flows around the tubing, which contains the water, at 20 gallons an hour at about 45 degrees F to cool the water. The cooled water is used for drinking.

Evaporators — Two evaporators, one for the primary and the other for the secondary coolant system, are made of special corrosion-resistant stainless steel plate and fin passages for the water glycol arranged in a series of stacks alternated with sintered Feltmetal wicks. Each wick pad is fed water through a plate which has tiny holes (5/1000 of an inch in diameter). Each evaporator is 8 by 4.7 by 6.62 inches and weighs 18 pounds.

The wicks are vented to the very low space pressure and water boils at 35 to 40 degrees F. Its evaporation cools the plates, through which the water-glycol passes, thus cooling the water-glycol to between 37 and 45 degrees F. The water-glycol flow is about 24 gallons an hour. About 8000 Btu per hour can be removed.

Lithium Hydroxide Canisters — There are two canisters in aluminum housings of 8-1/2 by 20 by 7-1/2 inches. The canisters, a diverter valve, and inlet and outlet ducts weigh 19.7 pounds. The canisters have removable lithium hydroxide elements. The elements are alternately changed, one every 12 hours. The elements absorb carbon dioxide and also contain activated charcoal, which absorbs odors.

Suit Heat Exchanger — The suit heat exchanger is made of two separate stacks of stainless steel fins and plates. One set is connected to the primary coolant system and the other is connected to the secondary coolant system. The unit is 15 by 11 by 5.2 inches. It cools suit gas to 50 to 55 degrees F and controls humidity by removing excess water. The water is collected by metal wicks and transported to the waste water storage tank.

Accumulators — Two reciprocating water pumps on the suit heat exchangers collect condensate from the suit circuit and pump it into the waste water tank. One accumulator is operated at a time; the other is standby. On automatic mode, a pump goes through a cycle every 10 minutes.

Suit Compressors (AiResearch) — Two centrifugal blowers made of aluminum are conical with a diameter of 6-1/2 inches and a length of 7/8 inches. One is used at a time. It circulates gasses through the suit circuit at a rate of 30 cubic feet per minute during normal operation. Each weighs 10.8 pounds. They operate on 3-phase, 110-volt, 400-Hertz power. Power consumption is 85 watts during normal operation.

Cabin Heat Exchanger — The plate fin, stainless steel, sandwich construction unit is 5.7 by 2.23 by 16.2 inches. It uses water-glycol as heat-transfer medium. It controls cabin temperature by cooling gas that flows through it. It is in the left-hand forward equipment bay.

Oxygen Surge Tank — The Inconel (nickel-steel alloy) tank has a diameter of 13 inches and is 14

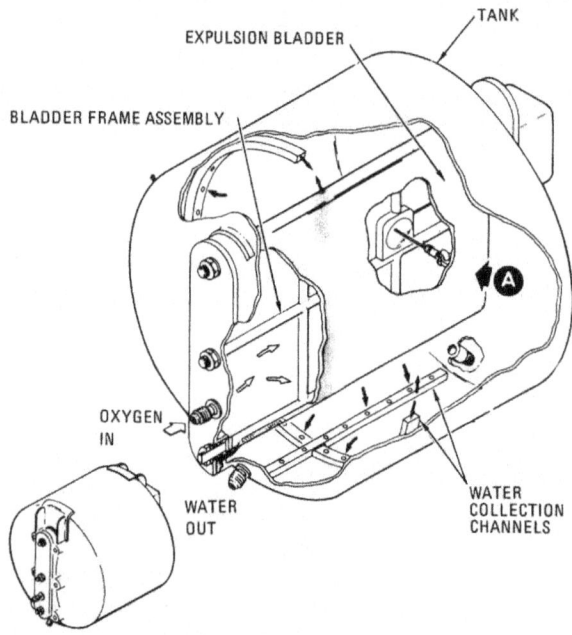
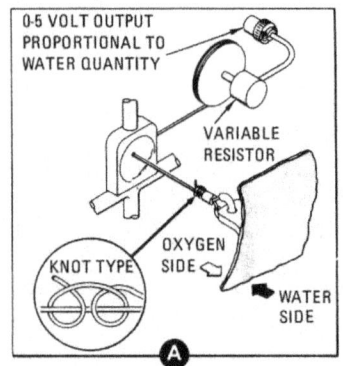

Potable water tank

inches high. It weighs 8.86 pounds. It holds 3.7 pounds of oxygen at a pressure of about 900 pounds per square inch. The volume is 0.742 cubic foot. It provides oxygen during entry. In emergencies, it can supply oxygen at a high flow rate. It is in the left-hand equipment bay of the command module.

Repressurization Unit — There are three bottles, each containing one pound of oxygen, in an aluminum case with a repressurization valve connected to them. The oxygen is stored at 900 pounds per square inch. Used in conjunction with the oxygen surge tank, it can repressurize the cabin from 0 to 3 pounds per square inch in about 2 minutes. It can also be used with three face masks stored just below the bottles. With the masks, the pressure is reduced to 100 pounds per square inch to the face mask regulator. There is also a direct reading pressure gauge to show the pressure. The unit is below the hatch in the command module.

Potable Water Tank — Aluminum tank with a bladder kept at a pressure of about 20 pounds per square inch by the 20 psi oxygen system. It has a diameter of 12-1/2 inches, and is about 12-1/2 inches deep. It weighs 7.9 pounds. It holds 17 quarts of drinking water and is used for storage of water from the fuel cells. It is in the aft compartment of the command module.

Waste Water Tank — The aluminum tank with a bladder has a diameter of about 12-1/2 inches and is 25 inches deep. It holds 28 quarts. It stores waste water from the suit heat exchanger to be used for cooling purposes through evaporation. It is in the aft compartment of the command module.

Coldplates — Two aluminum sheets about one-eighth of an inch apart are bonded together and have thousands of tiny posts. Water-glycol flows through the assembly absorbing heat from electronic equipment attached to the plates. The plates' sizes depend on the equipment they cool. Largest coldplate is about 2 by 3 feet; the smallest is about 2 by 10 inches.

Space Radiators — Two aluminum panels about 49 square feet each are around the outside surface of the service module in a 130-degree arc. Each panel has five tubes through which water-glycol flows. There is also a secondary tube for the secondary

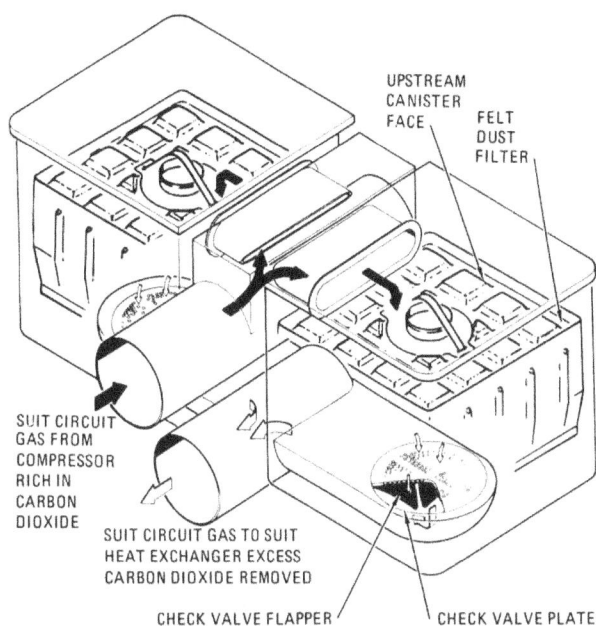

P-165

Operational schematic of carbon dioxide canister (cover removed)

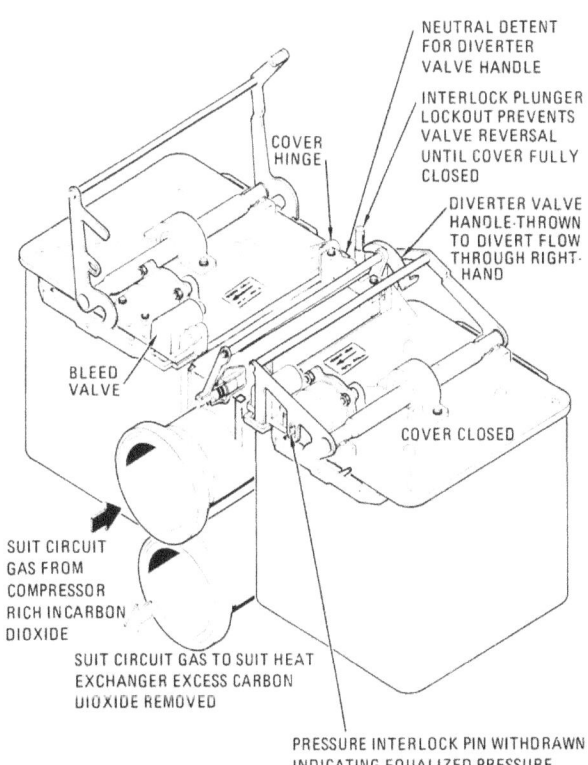

P-166

Operational schematic of carbon dioxide canister

coolant systems. As the water-glycol flows through the tubes, its heat is rejected through radiation to space. About 4415 Btu per hour can be removed through each panel.

Water Glycol Pumps (AiResearch) — Aluminum housing of 12.9 by 8.4 by 9.89 inches contains three centrifugal-type pumps, two for the primary system and one for the secondary coolant system, and two bellow-type stainless steel accumulators, one for the primary and one for the secondary. The primary accumulator has a volume of 60 cubic inches; the secondary has 35 cubic inches. Only one pump is used at a time. They operate off 3-phase, 110-volt, 400-Hertz power. They pump water glycol through the system.

Glycol — Ethylene glycol, one of a large class of dihydroxy alcohols, is mixed with water (62.5 percent glycol to 37.5 percent water) to carry heat to the space radiator from cabin, space suits, electronic equipment, and the potable water chiller. Fluid can also provide heat or cooling for the cabin.

DETAILED DESCRIPTION

OXYGEN SUBSYSTEM

The oxygen subsystem shares the oxygen supply with the electrical power subsystem. Approximately 640 pounds of oxygen is stored in two cryogenic tanks located in the SM. Heaters in the tanks pressurize the oxygen to 900 psig for distribution to the using equipment.

Oxygen is delivered to the CM through two separate supply lines, each of which enters at an oxygen inlet restrictor assembly. Each assembly contains a filter, a capillary line, and a check valve. The filters provide final filtration of gas entering the CM. The capillaries, which are wound around the hot glycol line, serve two purposes: they restrict the total oxygen flow to 9 pounds per hour to prevent starvation of the fuel cells, and they heat the oxygen to prevent it from entering the CM as a liquid. The check valves serve to isolate the two supply lines.

After passing the inlet check valves, the two lines merge and a single line is routed to the oxygen-SM supply valve. This valve is used in flight as a shutoff valve to back up the inlet check valves during entry. It is closed before CM-SM separation.

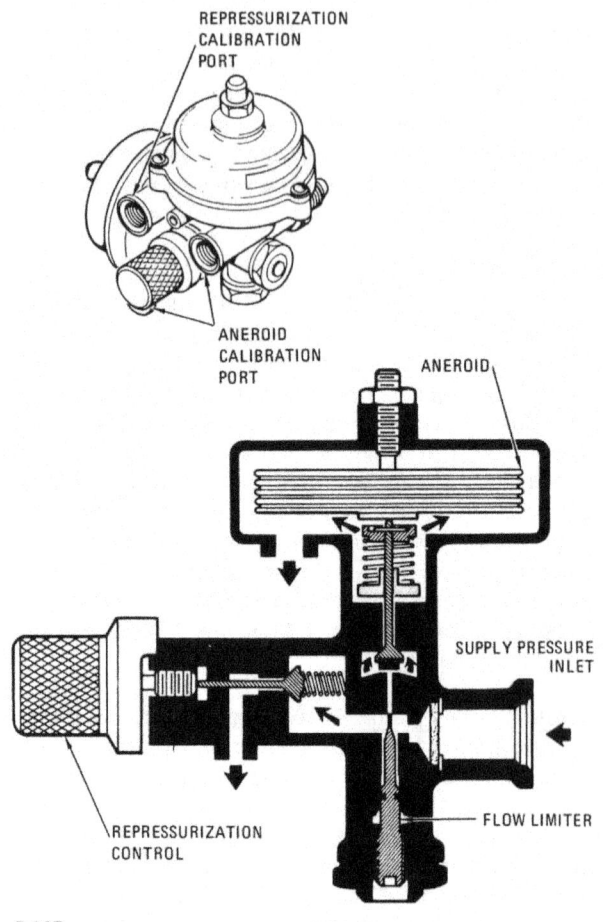

Cabin pressure regulator

The outlet of the supply valve is connected in parallel to the oxygen-surge tank valve and to a check valve on the oxygen control panel. The surge tank valve is closed only when it is necessary to isolate the surge tank from the system. The surge tank stores approximately 3.7 pounds of oxygen at 900 psig for use during entry, and for augmenting the SM supply when the operational demand exceeds the flow capacity of the inlet-restrictors. A surge tank pressure relief and shutoff valve prevents overpressurization of the surge tank, and provides a means for shutting off the flow in case the relief valve fails. A pressure transducer puts out a signal proportional to surge tank pressure for telemetry and for display to the crew.

An oxygen entry valve is used to control the flow of oxygen to and from the oxygen repressurization package. The package consists of three one-pound capacity oxygen tanks connected in parallel; a toggle-type fast-acting repressurization valve for dumping oxygen into the cabin at very high flow rates, and a toggle valve and regulator for supplying oxygen to the emergency oxygen face masks. Opening the repressurization valve, with the entry valve in the "fill" position, will dump both the package tanks and the surge tank at a rate that will pressurize the command module from 0 to 3 psia in one minute. When the entry valve is in the "on" position, the package tanks augment the surge tank supply for entry and emergencies.

The main regulator reduces the supply pressure to 100 ± 10 psig for use by subsystem components. The regulator assembly is a dual unit which is normally operated in parallel. Selector valves at the inlet to the assembly provide a means of isolating either of the units in case of failure, or for shutting them both off. Integral relief valves limit the downstream pressure to 140 psig maximum. The output of the main regulator passes through a flowmeter, then is delivered directly to the water and glycol tank pressure regulator and through the oxygen supply valve in parallel to the cabin pressure regulator, emergency cabin pressure regulator, the oxygen demand regulator, the direct oxygen valve, and the water accumulator valves.

The output of the flowmeter is displayed on an oxygen flow indicator which has a range of 0.2 to 1.0 pound per hour. Nominal flow for metabolic consumption and cabin leakage is approximately 0.43 pound per hour. Flow rates of 1 pound per hour or more with a duration in excess of 16.5 seconds will illuminate a light on the caution and warning panel to alert the crew to the fact that the oxygen flow rate is greater than is normally required. It does not necessarily mean that a malfunction has occurred, since there are a number of flight operations in which a high oxygen flow rate is normal.

The water and glycol tank pressure regulator assembly also is a dual unit, normally operating in parallel, which reduces the 100-psi oxygen to 20±2 psig for pressurizing the positive expulsion bladders in the waste and potable water tanks and in the glycol reservoir. Integral relief valves limit the

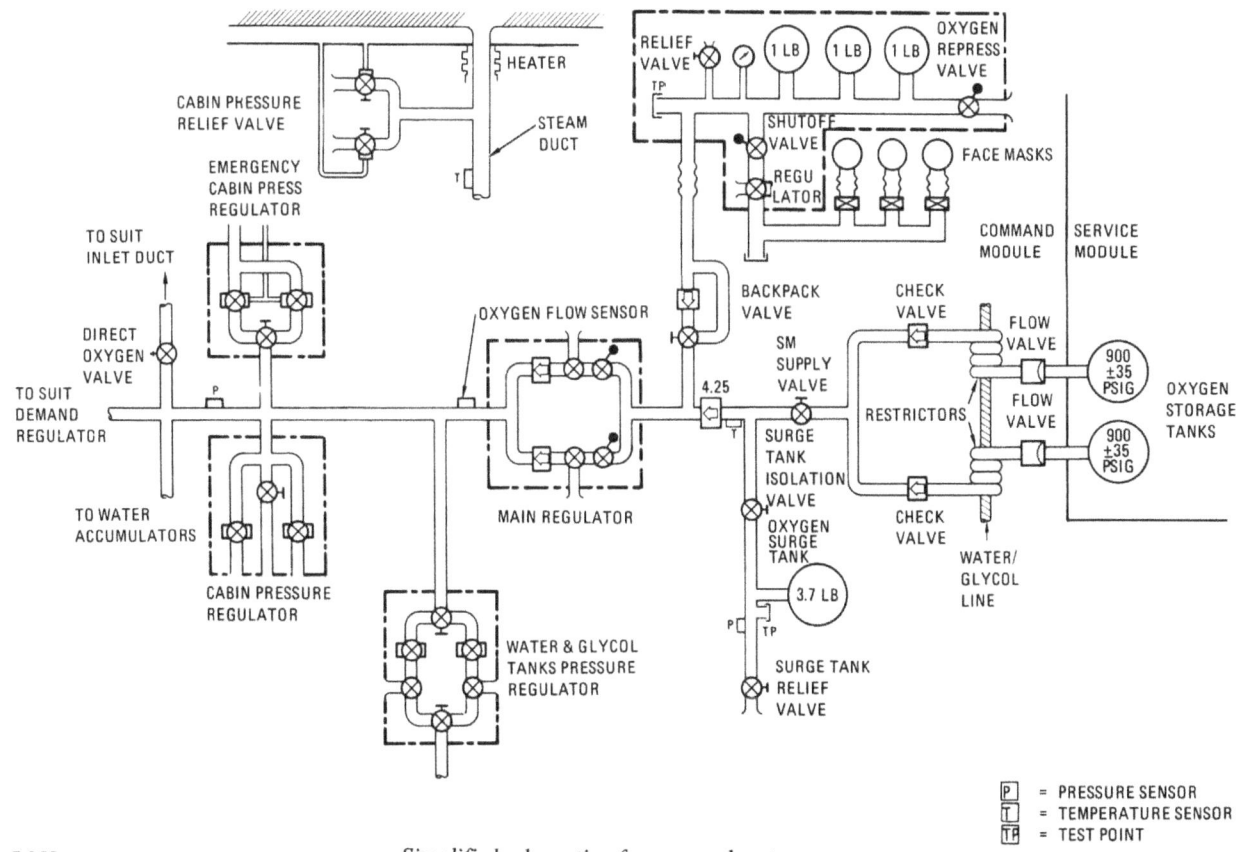

Simplified schematic of oxygen subsystem

downstream pressure to 25±2 psi above cabin pressure. Inlet and outlet selector valves are provided for selecting either or both regulators and relief valves, or for shutting the unit off.

The cabin pressure regulator controls the flow of oxygen into the cabin to make up for depletion of the gas due to metabolic consumption, normal leakage, or repressurization. The assembly consists of two absolute pressure regulators operating in parallel, and a manually operated cabin repressurization valve. The regulator is designed to maintain cabin pressure at 5±0.2 psia with losses up to 1.3 pounds per hour. Losses in excess of this value will result in a continual decrease in cabin pressure. When cabin pressure falls to 3.5 psia minimum, the regulator will automatically shut off to prevent wasting the oxygen supply. Following depressurization, the cabin can be repressurized by manually opening the cabin repressurization valve. This will result in a minimum flow of 6 pounds per hour.

An emergency cabin pressure regulator provides emergency protection for the crew in the event of a severe leak in the cabin. The regulator valve starts to open when cabin pressure decreases to 4.6 psia; and at 4.2 psia, the valve is fully open, flooding the cabin with oxygen. The regulator supplies oxygen to the cabin at flow rates up to 0.66 pound per minute to prevent rapid decompression in case of cabin puncture. The valve can provide flow rates that will maintain cabin pressure above 3.5 psia for a period of 15 minutes, against a leakage rate equivalent to 1/4-inch-diameter cabin puncture. The valve is normally used during shirtsleeve operations, and is intended to provide time for donning pressure suits before cabin pressure drops below 3.5 psia. During pressure suit operations, the valve is shut off to prevent unnecessary loss of oxygen.

An oxygen demand regulator supplies oxygen to the suit circuit whenever the suit circuit is isolated from the cabin and during depressurized operations. It also relieves excess gas to prevent overpressurizing the suits. The assembly contains redundant regulators, a single relief valve for venting excess suit pressure, an inlet selector valve for selecting either or both regulators, and a suit test valve for performing suit integrity tests.

Each regulator section consists of an aneroid control and a differential diaphragm housed in a reference chamber. The diaphragm pushes against a rod connected to the demand valve; the demand valve will be opened whenever a pressure differential is sensed across the diaphragm. In operation, there is a constant bleed flow of oxygen from the supply into the reference chamber, around the aneroid, and out through the control port into the cabin. As long as the cabin pressure is greater than 3.75 psia (nominal), the flow of oxygen through the control port is virtually unrestricted, so that the pressure within the reference chamber is essentially that of the cabin. This pressure acts on the upper side of the diaphragm, while suit pressure is applied to the underside of the diaphragm through the suit sense port. The diaphragm can be made to open the demand valve by either increasing the reference chamber pressure or by decreasing the sensed suit pressure.

Increased pressure occurs during depressurized operations. As the cabin pressure decreases, the aneroid expands. At 3.75 psia the aneroid will have expanded sufficiently to restrict the outflow of oxygen through the control port, thus increasing the reference chamber pressure. When the pressure rises approximately 3 inches of water pressure above the sensed suit pressure, the demand valve will be opened.

Decreased pressure occurs whenever the suit circuit is isolated from the cabin, and cabin pressure is above 5 psia. In the process of respiration, the crew

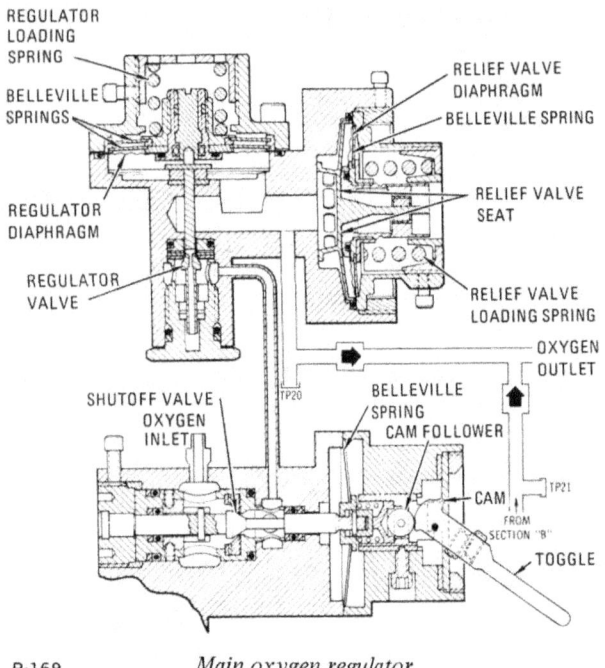

Main oxygen regulator

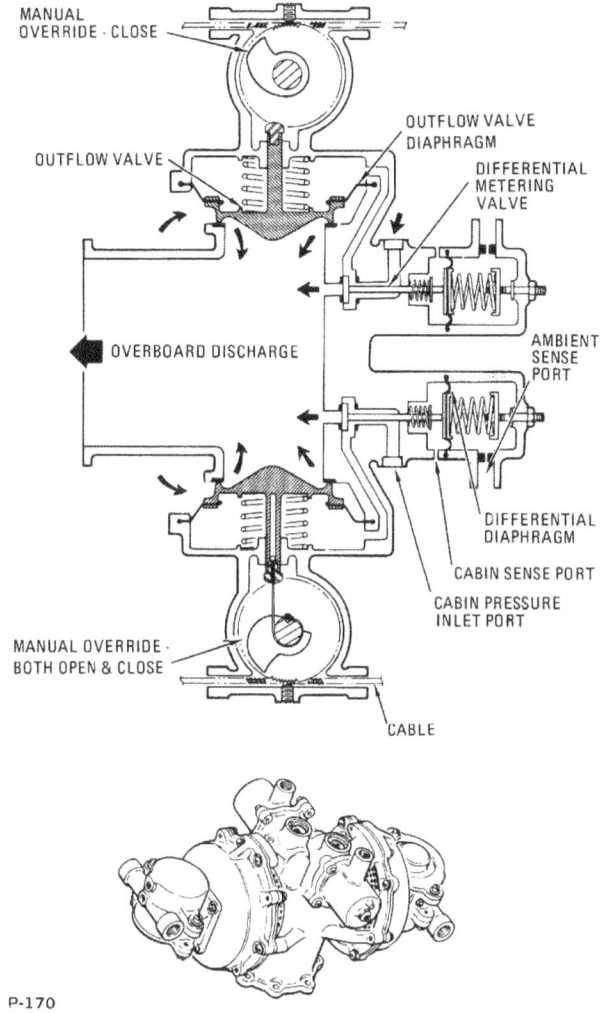

Cabin pressure relief valve

cabin pressure decreases to 3.75 psia, the reference chamber pressure is increased by the throttling effect of the expanding aneroid. The reference chamber pressure is applied, through ducts, to two relief valve loading chambers which are arranged in tandem above the relief valve poppet. The pressure in the loading chambers acts on tandem diaphragms which are forced against the relief valve poppet. The relief value of the valve is thus increased to 3.75 psia plus 2 to 9 inches of water pressure.

The suit test valve provides a means for pressurizing and depressurizing the suit circuit, at controlled rates, for performing suit integrity tests. In the "Press" position the valve supplies oxygen through a restrictor to pressurize the suit circuit to a nominal 4 psi above the cabin in not less than 75 seconds. The maximum time required for pressurizing or depressurizing the suits depends on the density of the suit and cabin gases. It will take longer to pressurize or depressurize during prelaunch than in orbit because of the higher density of the gas at sea-level pressure. In the "Depress" position the valve will depressurize the suits in not less than 75 seconds. Moving the valve from "Press" to "Off" will dump the suit pressure immediately. Also, if any one of the three suits is vented to the cabin while the valve is in the "Press" position, all three suits will collapse immediately. This is due to the restrictor in the pressurizing port which prevents the demand regulator from supplying the high oxygen flow rate required for maintaining the pressure in the other two suits.

The direct oxygen valve is a manual metering valve with a flow capability of zero to 0.67 pound per minute. The primary purpose is for purging the pressure suit circuit.

PRESSURE SUIT CIRCUIT

The pressure suit circuit is a circulating gas loop which provides the crew with a continuously conditioned atmosphere throughout the mission. The gas is circulated through the circuit by two centrifugal compressors which are controlled by individual switches. Normally only one of the compressors is operated at a time; however, the individual switches provide a means for connecting either or both of the compressors to either ac bus.

A differential pressure transducer connected across the compressors provides a signal to an indicator on the main display console, to telemetry,

will exhale carbon dioxide and water vapor. In circulating the suit gases through the carbon dioxide and odor absorber and the suit heat exchanger, the carbon dioxide and water are removed. The removal reduces the pressure in the suit circuit, which is sensed by the regulator on the underside of the diaphragm. When the pressure drops approximately 3 inches of water pressure below the cabin pressure, the diaphragm will open the demand valve.

The regulator assembly contains a poppet-type relief valve which is integral with the suit pressure sense port. During operations where the cabin pressure is above 3.75 psia, the relief valve is loaded by a coil spring which allows excess suit gas to be vented whenever suit pressure rises to 2 to 9 inches of water pressure above cabin pressure. When the and to the caution and warning system, which will illuminate a light at a differential pressure of 0.22 psi or less. Another differential pressure transducer is connected between the suit compressor inlet manifold and the cabin; the output is displayed on the indicator. A switch on the main display console selects the output of either transducer for display on the indicator. A pressure transducer connected to the compressor inlet manifold provides a signal to another indicator and to telemetry.

The gas leaving the compressor flows through the carbon dioxide and odor absorber assembly. The assembly is a dual unit containing two absorber elements in separate compartments with inlet and outlet manifolds common to both. A diverter valve in the inlet manifold provides a means of isolating one compartment (without interrupting the gas flow through the suit circuit) to replace a spent absorber. An interlock mechanism between the diverter valve handle and the cover handles is intended to prevent opening both compartments at the same time. The absorber elements contain lithium hydroxide and activated charcoal for removing carbon dioxide and odors from the suit gases. Orlon pads on the inlet and outlet sides trap small particles and prevent absorbent materials from entering the gas stream.

From the filter the gas flows through the suit heat exchanger where the gases are cooled and the excess moisture is removed. The heat exchanger assembly is made up of two sets of broad flat tubes through which the coolant from the primary and secondary loops can be circulated. The coolant flow or bypass is controlled by two valves located on the coolant control panel. The space between the tubes forms passages through which the suit gases flow. The coolant flowing through the tubes absorbs some of the heat from the suit gases. As the gases are cooled to about 55°F, the excess moisture condenses and is removed from the heat exchanger by one or both of a pair of water accumulator pumps.

The water accumulators are piston-type pumps actuated by oxygen pressure (100 psi) on the discharge stroke and by a return spring for the suction stroke. The oxygen flow is controlled by two water accumulator selector valve assemblies on the coolant control panel. Each valve assembly contains a selector valve, a solenoid valve, and an integral bypass. Oxygen flow can be controlled

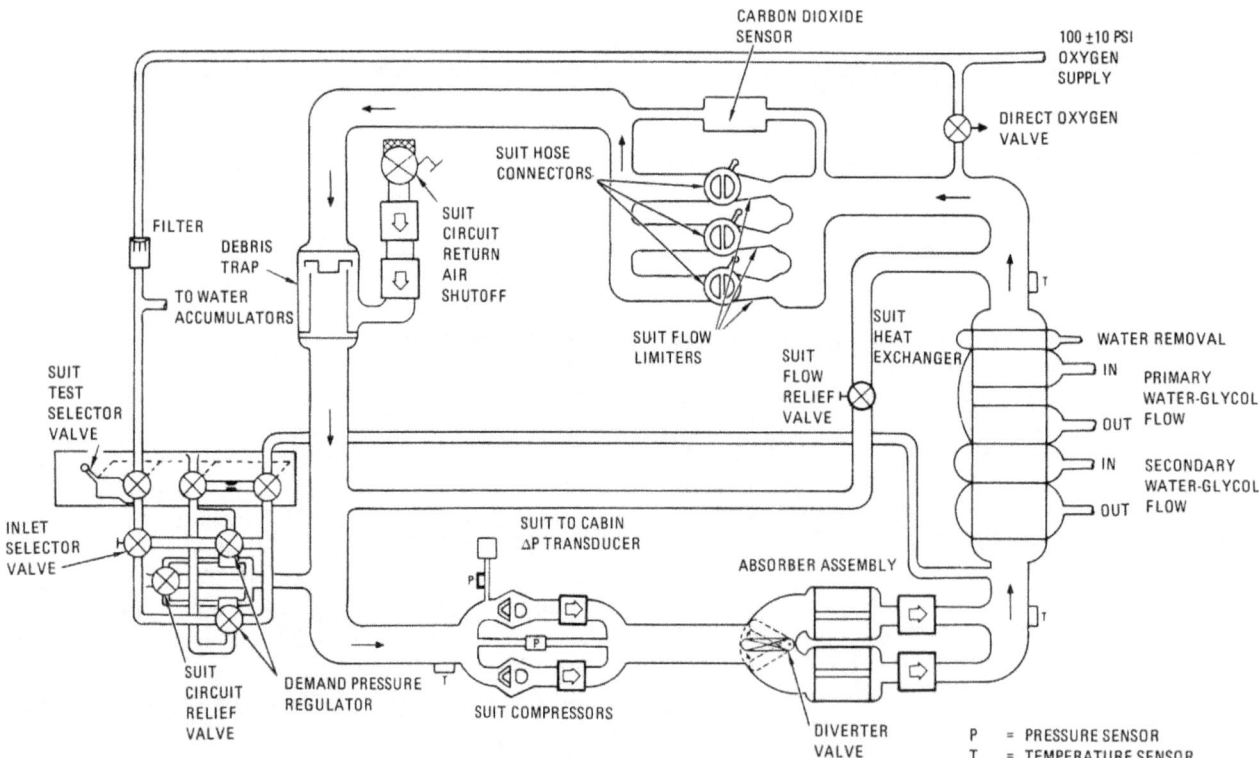

Simplified schematic of suit circuit

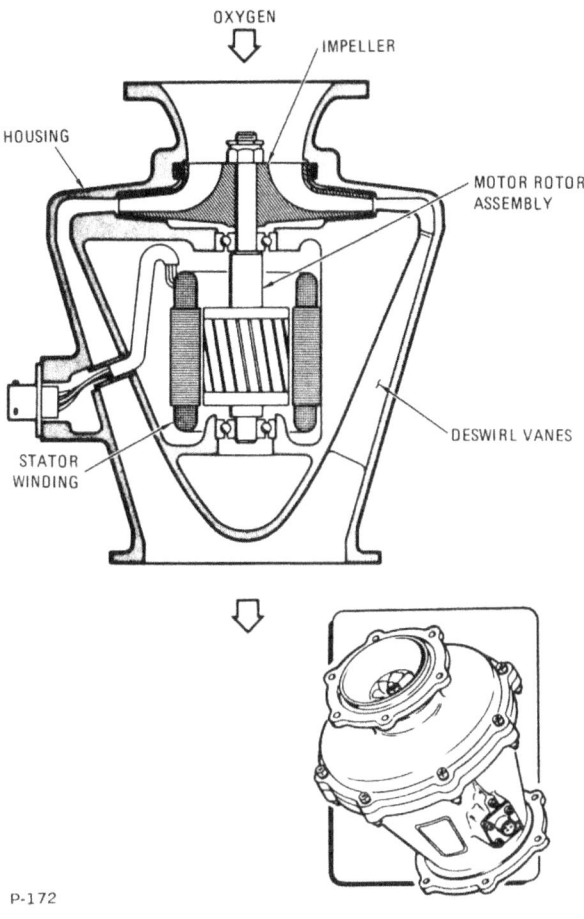

Suit compressor

(for at least 5 minutes) to maintain pressure in the circuit while the torn suit is being repaired.

Flow control valves are part of the suit hose connector assembly. These valves provide a means for adjusting the gas flow through each suit individually. When operating in a shirtsleeve environment with the inlet hose disconnected from the suit, approximately 12 cubic feet of suit gas per minute flows into the cabin.

A suit flow relief valve is installed between the suit heat exchanger outlet and the compressor inlet, and is intended to maintain a relatively constant pressure at the inlets to the three suits by relieving transient pressure surges. A control is provided for manually closing the valve; the valve is normally off throughout the mission.

automatically by the solenoid valve through signals from the central timing equipment. These signals will cause one of the accumulators to complete a cycle every ten minutes. If it becomes necessary to cycle the accumulators at more frequent intervals the solenoid valve can be controlled manually.

The cool gas (55°F nominal) flows from the heat exchanger through the suit flow limiters and the flow control valves into the suits. The suit temperature is measured at the heat exchanger outlet, and is displayed on the main display console and telemetered.

A suit flow limiter is installed in each suit supply duct to restrict the gas flow rate through any one suit. The flow limiter is a tube with a Venturi section sized to limit flow to 0.7 pound per minute. The limiter offers maximum resistance to gas flow through a torn suit, when cabin pressure is near zero psia. The oxygen demand regulator will supply oxygen at flow rates up to 0.67 pound per minute

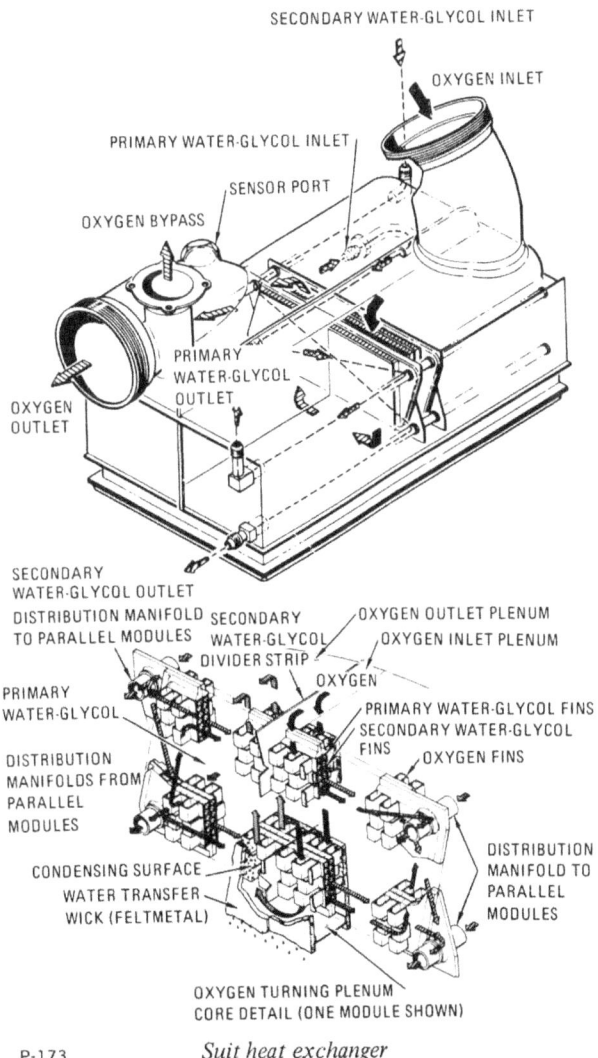

Suit heat exchanger

127

Gas leaving the suits flows through the debris trap assembly into the suit compressor. The debris trap is a mechanical filter for screening out solid matter that might otherwise clog or damage the system. The trap consists of a stainless steel screen designed to block particles larger than 0.040 inch, and a bypass valve which will open at differential pressure of 0.5 inch of water pressure in the event the screen becomes clogged.

A suit circuit return valve is installed on the debris trap upstream of the screen. It permits cabin gases to enter the suit circuit for scrubbing. The valve consists of two flapper-type check valves and a manual shutoff valve, in series. The shutoff valve provides a means of isolating the suit circuit from the cabin manually by means of a remote control. This is done to prevent inducting cabin gases into the suit circuit in the event the cabin gases become contaminated. The valve is located at the suit compressor inlet manifold, which is normally 1 to 2 inches of water pressure below cabin pressure. The differential pressure causes cabin gases to flow into the suit circuit. The reconditioned cabin gases are recirculated through the suits or cabin. During emergency operation, the valves prevent gases from flowing into the depressurized cabin from the suit circuit.

A carbon dioxide sensor is connected between the suit inlet and return manifold. It is connected to an indicator on the main display console, to telemetry, and to the caution and warning system and will activate a warning if the carbon dioxide partial pressure reaches 7.6 millimeters of mercury.

WATER SUBSYSTEM

The water subsystem consists of two individual fluid management networks which control the collection, storage, and distribution of potable and waste water. The potable water is used primarily for metabolic and hygienic purposes. The waste water is used solely as the evaporant in the primary and secondary glycol evaporators. Although the two networks operate and are controlled independently, they are interconnected in a manner which allows potable water to flow into the waste system under certain conditions.

Potable water produced in the fuel cells is pumped into the CM at a flow rate of approximately 1.5 pounds per hour. The water flows through a check valve to the inlet ports of the potable tank inlet and waste tank inlet valves. The check valve at the inlet prevents loss of potable water after CM-SM separation.

The potable tank inlet is a manual shutoff valve used to prevent the flow of fuel cell water into the potable system in the event the fuel cell water becomes contaminated.

The waste tank inlet is an in-line relief valve with an integral shutoff valve. The relief valve allows potable water to flow into the waste water tank whenever the potable water pressure is 6 psi above waste water pressure. This pressure differential will occur when the fuel cells are pumping water, and

Suit distribution duct and hose connectors

either the potable water tank is full, or the potable tank inlet valve is closed; or when the waste water tank is completely empty and the glycol evaporators are demanding water for cooling. In the latter case, the water flow is only that quantity which is demanded. The shutoff valve is used to block flow in case the relief valve fails. If such a failure occurs, potable water can flow through the valve (provided the potable water pressure is higher than the waste), until the two pressures are equal. Reverse flow is prevented by a check valve.

In the event that both water tanks are full at the time the fuel cells are pumping, the excess potable water will be dumped overboard through a pressure relief valve. This is a dual unit with a selector valve for placing either or both relief valves on-stream or shutting the unit off.

Water flows from the control panel to the potable water tank, the food preparation water unit, and the water chiller. Chilled water is delivered to the food preparation water unit and to the drinking water dispenser through the drinking water supply valve.

The water chiller cools and stores 0.5 pound of potable water for crew consumption. The water chiller is designed to supply 6 ounces of 50°F water every 24 minutes. The unit consists of an internally baffled reservoir containing a coiled tube assembly which is used as the coolant conduit. The baffles are used to prevent the incoming hot water from mixing with and raising the temperature of the previously chilled water.

The food preparation water unit heats potable water for use by the crew and allows manual selection of hot or cold potable water; the cold potable water is supplied by the water chiller. The unit consists of an electrically heated water reservoir and two manually operated valves which meter

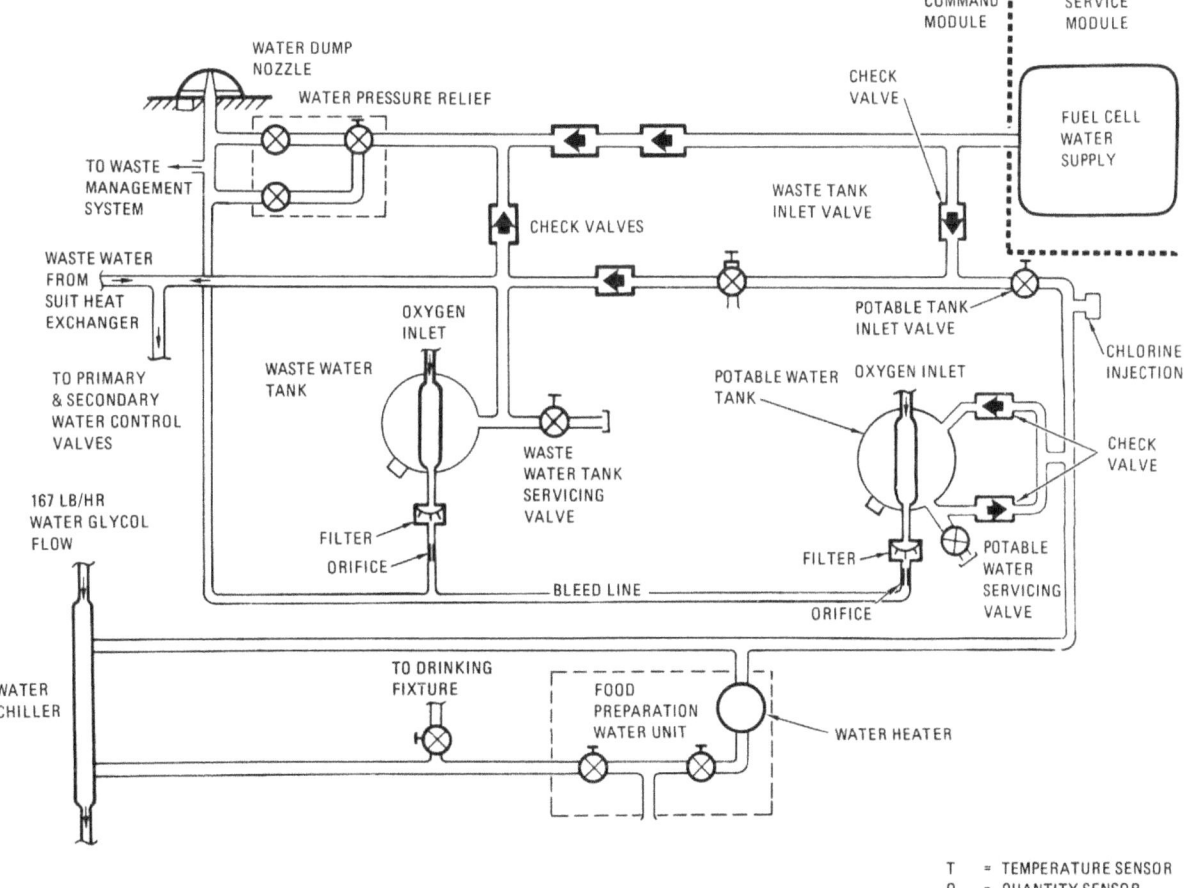

Schematic of water management subsystem

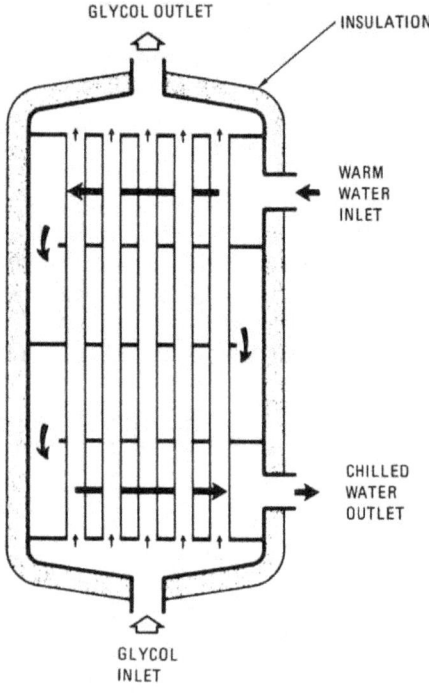

Water chiller

being trapped within the tank by the expanding bladder. Quantity transducers provide signals to an indicator on the main display console.

Bacteria from the waste water system can migrate through the isolating valves into the potable water system. A syringe injection system provides for periodic injection of bactericide to kill bacteria in the potable water system.

Waste water extracted from the suit heat exchanger is pumped into the waste water tank, and is delivered to the evaporator control valves. When the tank is full, excess waste water is dumped overboard through the water pressure relief valve. The evaporator control valves consist of a manually operated inlet valve and a solenoid valve. The primary solenoid valve can be controlled automatically or manually. The secondary solenoid valve is controlled automatically.

water in 1-ounce increments. The insulated reservoir has a capacity of 2.5 pounds of water. Thermostatically controlled heating elements in the reservoir heat the water and maintain it at 154°F nominal. Two metering valves dispense either hot or cold water, in 1-ounce increments, through a common nozzle. The hot water delivery rate is approximately 10 ounces every 30 minutes.

The drinking water supply valve is used to shut off the flow of water to the drinking water dispenser (water pistol), in case of a leak in the flex hose.

The waste water and potable water are stored in positive expulsion tanks, which with the exception of capacity are identical in function, operation, and design. The positive expulsion feature is obtained by an integrally supported bladder, installed longitudinally in the tank. Water collector channels, integral with the tank walls, prevent water from

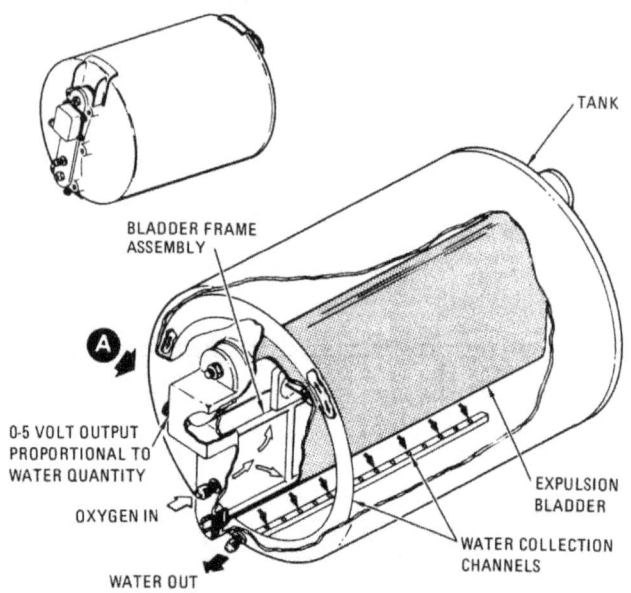

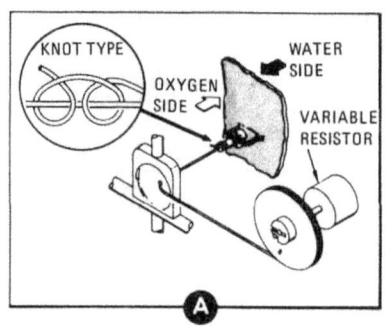

Waste water tank

WATER-GLYCOL COOLANT SUBSYSTEM

The water-glycol coolant subsystem consists of two independently operated closed coolant loops. The primary loop is operated continuously throughout the mission unless damage to the equipment necessitates shutdown. The secondary loop is operated at the discretion of the crew, and provides a backup for the primary loop. Both loops provide cooling for the suit and cabin atmospheres, the electronic equipment, and a portion of the potable water supply. The primary loop also serves as a source of heat for the cabin atmosphere when required.

The coolant is circulated through the loops by a pumping unit consisting of two pumps, a full-flow filter, and an accumulator for the primary loop, and a single pump, filter, and accumulator for the secondary loop. The purpose of the accumulators is to maintain a positive pressure at the pump inlets by accepting volumetric changes due to changes in coolant temperature. If the primary accumulator leaks, it can be isolated from the loop. Then the reservoir must be placed in the loop to act as an accumulator. Accumulator quantity is displayed on the main display console. A switch on the console permits either of the pumps to be connected to either ac bus. The secondary permits either of the pumps to be connected to either ac bus. The secondary pump also has a switch which allows it to be connected to either ac bus.

The output of the primary pump flows through a passage in the evaporator steam pressure control valve to de-ice the valve throat. The coolant next flows through a diverter valve, through the radiators, and returns to the CM. The diverter valve is placed in the "Bypass" position before launch to isolate the radiators from the loop, and before CM-SM separation to prevent loss of coolant when the CSM umbilical is cut. Otherwise it is in the normal operating position.

Coolant returning to the CM flows to the glycol reservoir valves. From pre-launch until after orbit insertion, the reservoir inlet and outlet valves are open and the bypass valve is closed, allowing coolant to circulate through the reservoir. This provides a quantity of cold coolant to be used as a heat sink during the early stage of launch. After orbit insertion, the reservoir is isolated from the primary loop to provide a reserve supply of coolant for refilling the loop in the event a leak occurs.

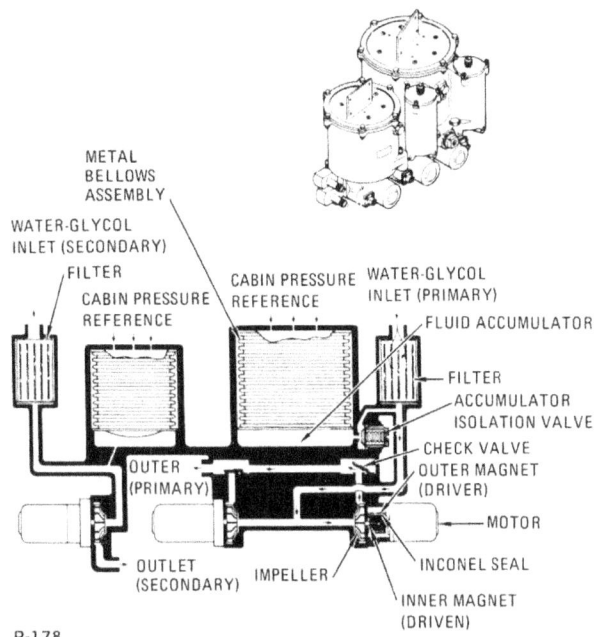

Water-glycol pump assembly

The coolant flow from the evaporator divides into two branches. One carries a flow of 33 pounds per hour to the inertial measurement unit and into the coldplate network. The other branch carries a flow of 167 pounds per hour to the water chiller through the suit heat exchanger primary glycol valve and the suit heat exchanger to the primary cabin temperature control valve.

The primary cabin temperature control valve routes the coolant to either the cabin heat exchanger or to the coldplate network. The valve is positioned automatically by the cabin temperature control, or manually by means of an override control on the face of the valve. The valve is so constructed that in the cabin full cooling mode, the flow of coolant from the suit heat exchanger (167 pounds per hour) is routed first through the cabin heat exchanger and then through the thermal coldplates where it joins with the flow (33 pounds per hour) from the inertial measurement unit. In the cabin full heating mode, the total flow (200 pounds per hour) is routed through the thermal coldplates first, where the water-glycol absorbs heat; from there it flows through the cabin heat exchanger. In the intermediate valve position, the quantity of cool or warm water-glycol flowing through the heat exchanger is reduced in proportion to the demand for cooling or heating. Although the amount of water-glycol flowing through the cabin heat exchanger will vary, the total flow through the thermal coldplates will always be total system flow. An orifice restrictor is installed between the cabin temperature control valve and the inlet to the coldplates. Its purpose is to maintain a constant flow rate through the coldplates by reducing the heating mode flow rate to that of the cooling mode flow rate. Another orifice restrictor, located in the coolant line from the inertial measurement unit, maintains a constant flow rate through this component regardless of system flow fluctuations. The total flow leaving the primary cabin temperature valve enters the primary pump and is recirculated.

The output of the secondary pump flows through a passage in the secondary evaporator steam pressure control valve for de-icing the valve throat. The coolant next flows through a diverter valve, through the radiators, and returns to the CM. This valve also is placed in the bypass position before CM-SM separation to prevent loss of coolant when the CSM umbilical is severed. After returning to the CM the coolant flows through the secondary evaporator, the suit heat exchanger secondary glycol valve, and the suit heat exchanger to the secondary cabin temperature control valve. The secondary cabin temperature control valve regulates the quantity of coolant flowing through the cabin heat exchanger in the cooling mode (there is not heating capability in the secondary loop). The coolant from the secondary cabin temperature control valve and/or the cabin heat exchanger then flows through redundant passages in the coldplates and returns to the secondary pump inlet.

The heat absorbed by the coolant in the primary loop is transported to the radiators where a portion is rejected to space. If the quantity of heat rejected by the radiators is excessive, the temperature of the coolant returning to the CM will be lower than desired (45°F nominal). If the temperature of the coolant entering the evaporator drops below a nominal 43°F, the mixing mode of temperature control is initiated. The automatic control opens the glycol evaporator temperature valve, which allows a sufficient quantity of hot coolant from the pump to mix with the coolant returning from the

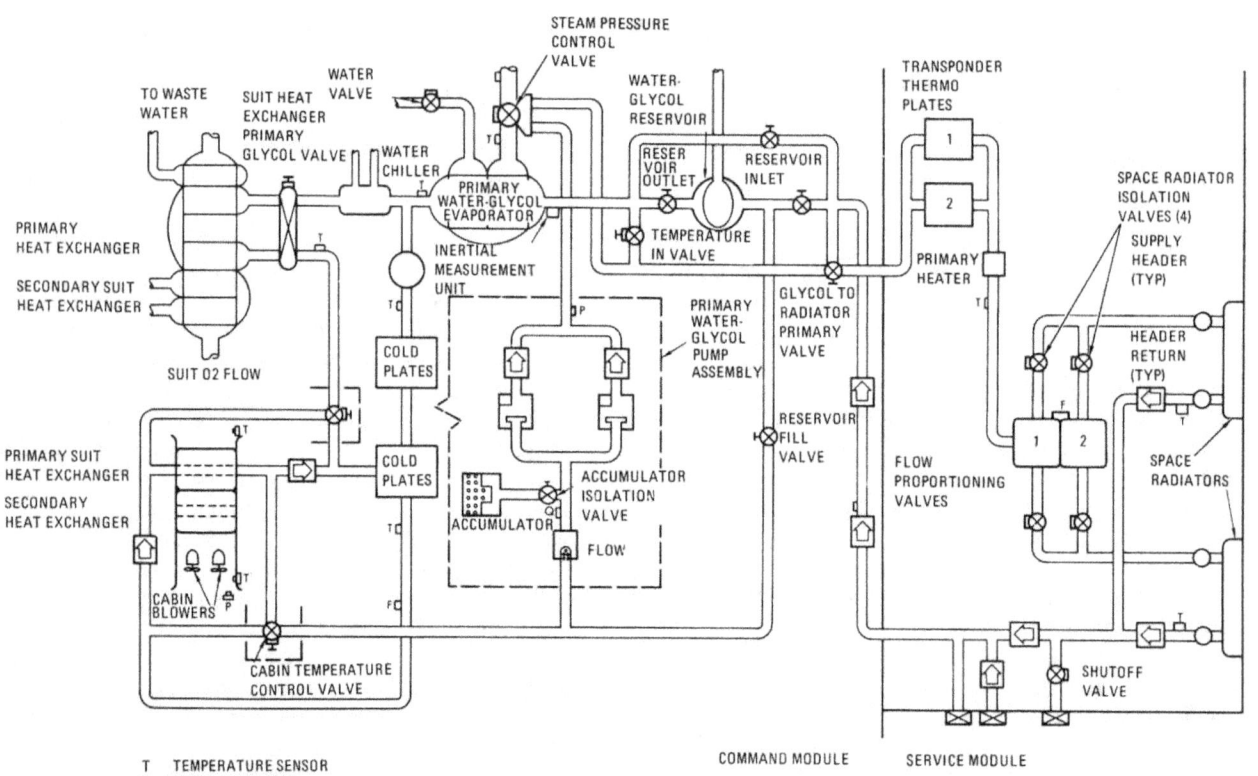

Schematic of primary water-glycol subsystem

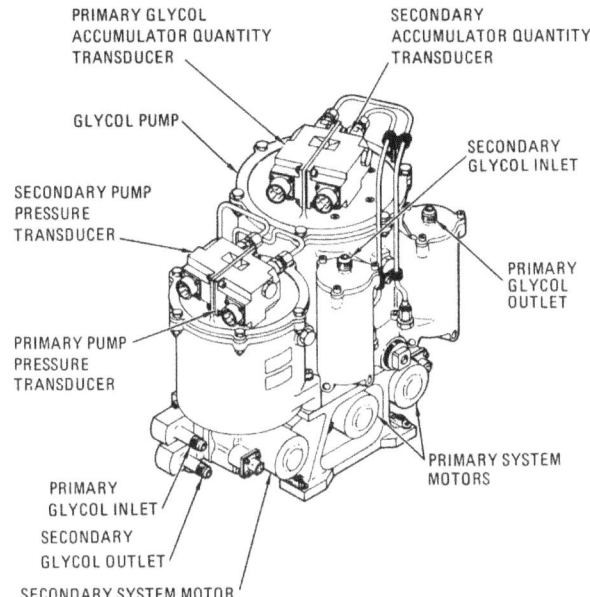

Glycol pump assembly

controlled remotely, using evaporator outlet temperature as an indicator. The secondary evaporator is controlled automatically.

Each coolant loop includes a radiator circuit. The primary radiator circuit consists basically of two radiator panels in parallel with a flow-proportioning control for dividing the flow between them, and a heater control for adding heat to the loop. The secondary circuit consists of a series loop utilizing some of the area of both panels, and a heater control for adding heat to the loop.

The radiator panels are an integral part of the SM skin and are located on opposite sides of the SM in Sectors 2 and 3 and in Sectors 5 and 6. With the radiators being diametrically opposite, it is possible that one primary panel may face deep space while the other faces the sun, earth, or moon. These

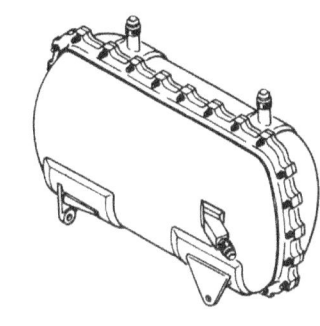

radiators to produce a mixed temperature at the inlet to the evaporator between 43° and 48°F. There is no mixing mode in the secondary loop. If the temperature of the coolant returning from the secondary radiator is lower than 45°F nominal, the secondary radiator inlet heater will be turned on to maintain the outlet temperature between 42° and 48°F.

If the radiators fail to radiate a sufficient quantity of heat, the coolant returning to the CM will be above the desired temperature. When the temperature of the coolant entering the evaporator rises to 48° to 50.5°F, the evaporator mode of cooling is initiated. The glycol temperature control opens the steam pressure valve allowing the water in the evaporator wicks to evaporate, using some of the heat contained in the coolant for the heat of vaporization. A temperature sensor at the outlet of the evaporator controls the position of the steam pressure valve to establish a rate of evaporation that will result in a coolant outlet temperature between 40° to 43°F. The evaporator wicks are maintained in a wet condition by wetness control which uses the wick temperature as an indication of water content. As the wicks become dryer, the wick temperature increases and the water control valve is opened. As the wicks become wetter, the wick temperature decreases and the water valve closes. The evaporative mode of cooling is the same for both loops. The steam pressure valve can be

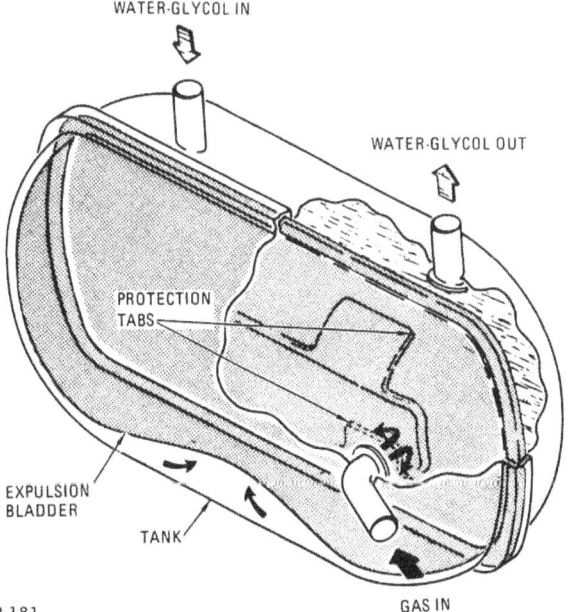

Water-glycol reservoir

133

extremes in environments mean large differences in panel efficiencies and outlet temperatures. The panel facing deep space can reject more heat than the panel receiving external radiation; therefore, the overall efficiency of the subsystem can be improved by increasing the flow to the cold panel. The higher flow rate reduces the transit time of the coolant through the radiator, which decreases the quantity of heat radiated.

The flow through the radiators is controlled by a flow-proportioning valve. When the differential temperature between the outlets of the two panels exceeds 10°F, the flow-proportioning valve is positioned to increase the flow to the colder panel.

The flow-proportioning valve assembly contains two individually controlled valves, only one of which can be in operation. When the switches are

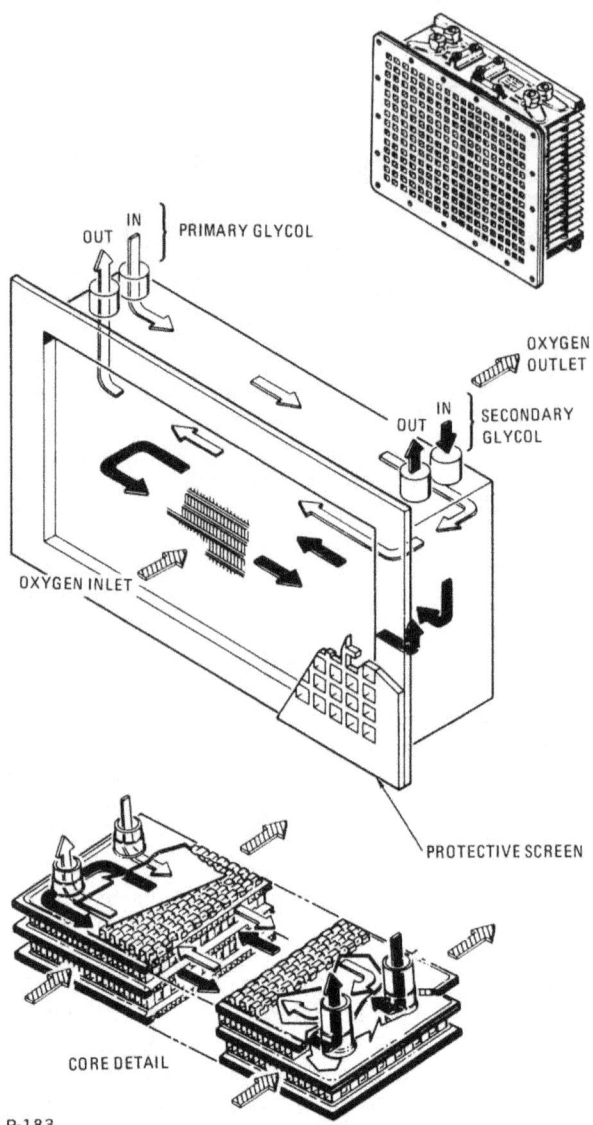

Cabin heat exchanger

on automatic the flow controller selects the No. 1 valve and positions the appropriate radiator isolation valves. Manual selection and transfer also is possible. Automatic transfer will occur when the temperature differential exceeds 15°F, providing a failure has occurred. In the absence of a failure, the transfer signal will be inhibited. In situations where the radiator inlet temperature is low and the panels have a favorable environment for heat rejection, the radiator outlet temperature starts to decrease and thus the bypass ratio starts to increase. As more flow is bypassed, the radiator outlet temperature decreases until the -20°F minimum desired temperature would be exceeded. To prevent this from occurring, a heater is automatically turned on when

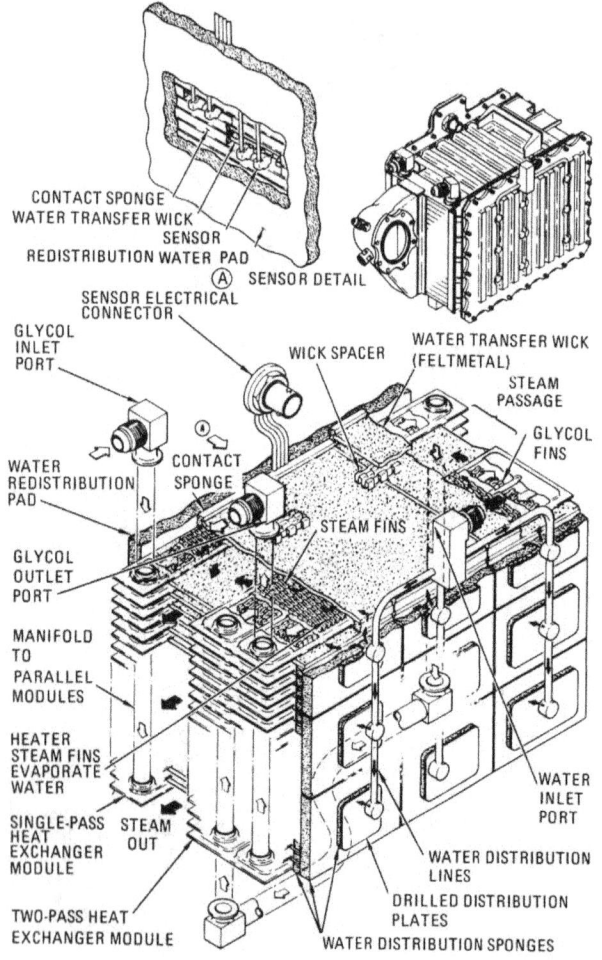

Glycol evaporator

radiator mixed outlet temperature drops to -15°F and remains on until -10°F is reached. The controller provides only on-off heater control which results in a nominal 450 watts being added to the coolant each time the heater is energized. The crew can switch to a redundant heater system if the temperature decreases to -20°F.

If the radiator outlet temperature falls below the desired minimum, the effective radiator surface temperature will be controlled passively by the selective stagnation method. The two primary circuits are identical, consisting of five tubes in parallel and one downstream series tube. The two panels, as explained in the flow proportioning control system, are in parallel with respect to each other. The five parallel tubes of each panel have manifolds sized to provide specific flow rate ratios in the tubes, numbered 1 through 5. Tube 5 has a lower flow rate than Tube 4, and so on, through Tube 1 which has the highest flow. For equal fin areas, therefore, the tube with the lower flow rate will have a lower coolant temperature. During minimum CM heat loads, stagnation begins to occur in Tube 5 as its temperature decreases; for as its temperature decreases, the fluid resistance increases, and the flow rate decreases. As the fin area around Tube 5 gets colder, it draws heat from Tube 4 and the same process occurs with Tube 4. In a fully stagnated condition, there is essentially no flow in Tubes 3, 4, and 5, and some flow in Tubes 1 and 2, with most of it in Tube 1.

When the CM heat load increases and the radiator inlet starts to increase, the temperature in Tube 1 increases and more heat is transferred through the fin toward Tube 2. At the same time, the glycol evaporator temperature valve starts to close and force more coolant to the radiators, thus helping to thaw the stagnant portion of the panels. As Tube 2 starts to get warmer and receives more flow it in turn starts to thaw Tube 3, and so on. This combination of higher inlet temperatures and higher flow rates quickly thaws out the panel. The panels automatically provide a high effectiveness (completely thawed panels operating at a high average fin temperature) at high heat loads, and a low effectiveness (stagnated panels operating at a low average fin temperature) at low heat loads.

The secondary radiator consists of four tubes which are an integral part of the radiator panel structure. Each tube is purposely placed close to the hottest primary radiator tubes (i.e., Tube 1 and

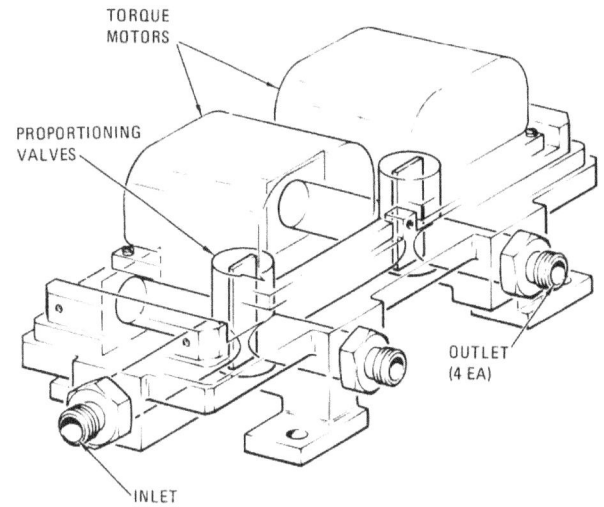

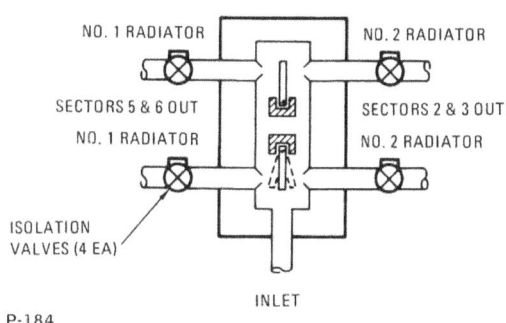

Space radiator flow proportioning valves

the downstream series tube on each panel) to keep the water-glycol in the secondary tubes from freezing while the secondary circuit is inoperative. The selective stagnation principle is not utilized in the secondary radiator because of the narrower heat load range requirements. This is also the reason the secondary radiator is a series loop. Because of the lack of this passive control mechanism, the secondary circuit depends on the heater control system at low heat loads and the evaporator at high heat loads for control of the water-glycol temperature.

The secondary heaters differ from the primary in that they can be operated simultaneously. When the secondary outlet temperature reaches 43°F the No. 1 heater comes on, and at 42°F the No. 2 heater comes on; at 44°F No. 2 goes off, and at 45°F No. 1 goes off.

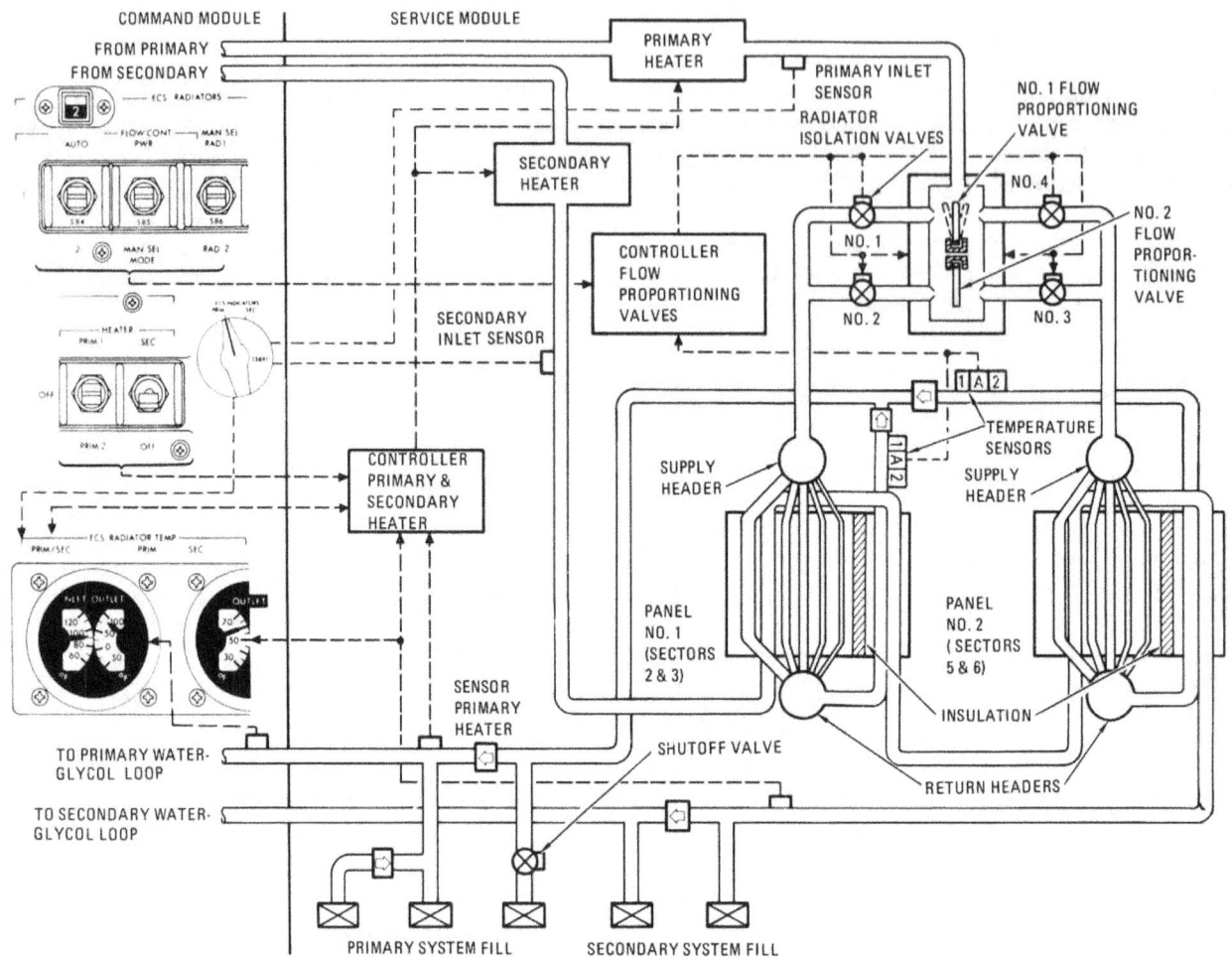

Schematic of radiator subsystem

LAUNCH ESCAPE

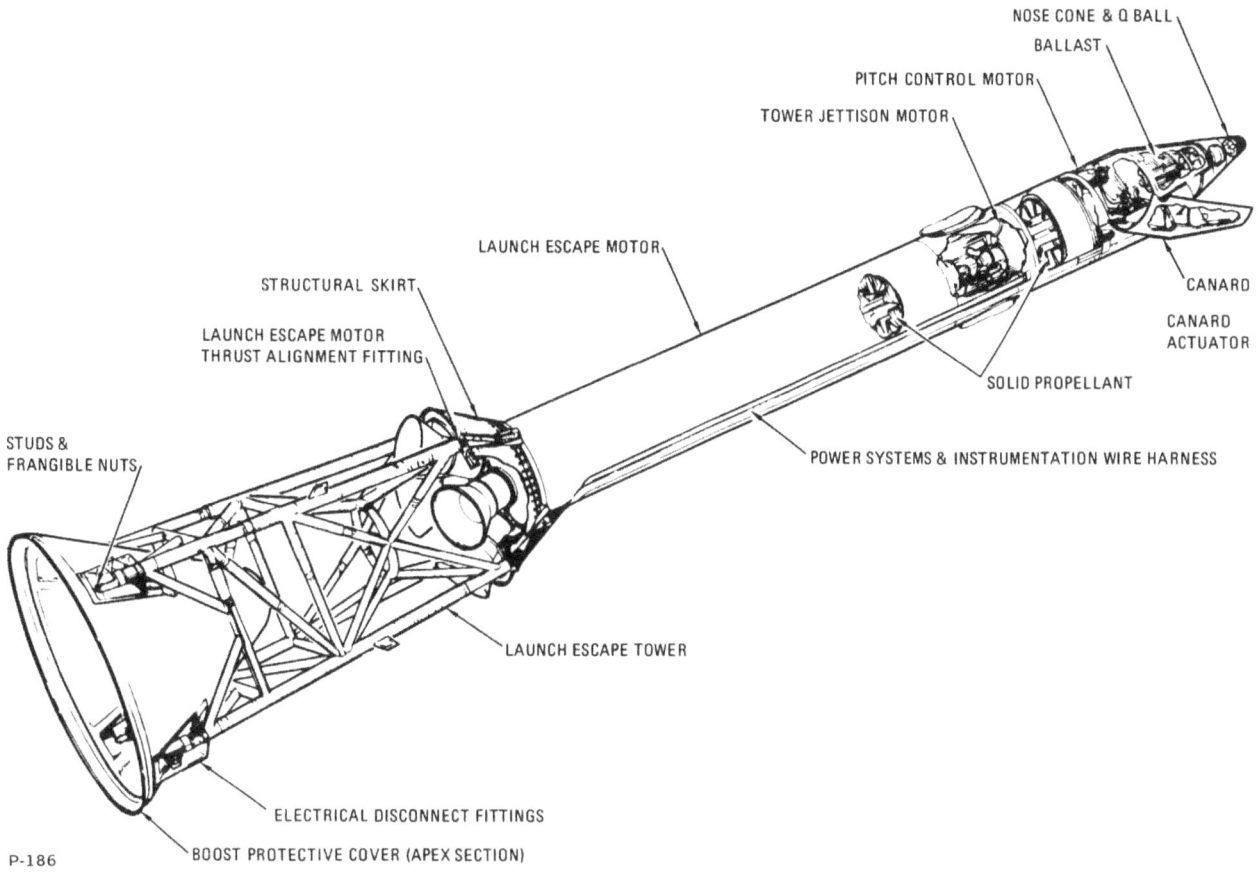

Launch escape subsystem

The launch escape subsystem will take the command module containing the astronauts away from the launch vehicle in case of an emergency on the pad or shortly after launch. The subsystem carries the CM to a sufficient height and to the side, away from the launch vehicle, so that the earth landing subsystem can operate.

The subsystem looks like a large rocket connected to the command module by a lattice-work tower. It is 33 feet long and weighs about 8,000 pounds. The maximum diameter of the launch escape assembly is four feet.

The forward or rocket section of the subsystem is cylindrical and houses three solid-propellant rocket motors and a ballast compartment topped by a nose cone containing instruments. The tower is made of titanium tubes attached at the top to a structural skirt that covers the rocket exhaust nozzles and at the bottom to the command module by means of an explosive connection.

A boost protective cover is attached to the tower and completely covers the command module. This cover protects the command module from the rocket exhaust and also from the heating generated by launch vehicle boost through the atmosphere. It remains attached to the tower and is carried away when the launch escape assembly is jettisoned.

The subsystem is activated automatically by the emergency detection system in the first 100 seconds or manually by the astronauts at any time from the pad to jettison altitude. With the Saturn V, the subsystem is jettisoned at about 295,000 feet, or about 30 seconds after ignition of the second stage; with the Saturn IB, the subsystem is jettisoned at about 275,000 feet, about 20 seconds after second stage ignition.

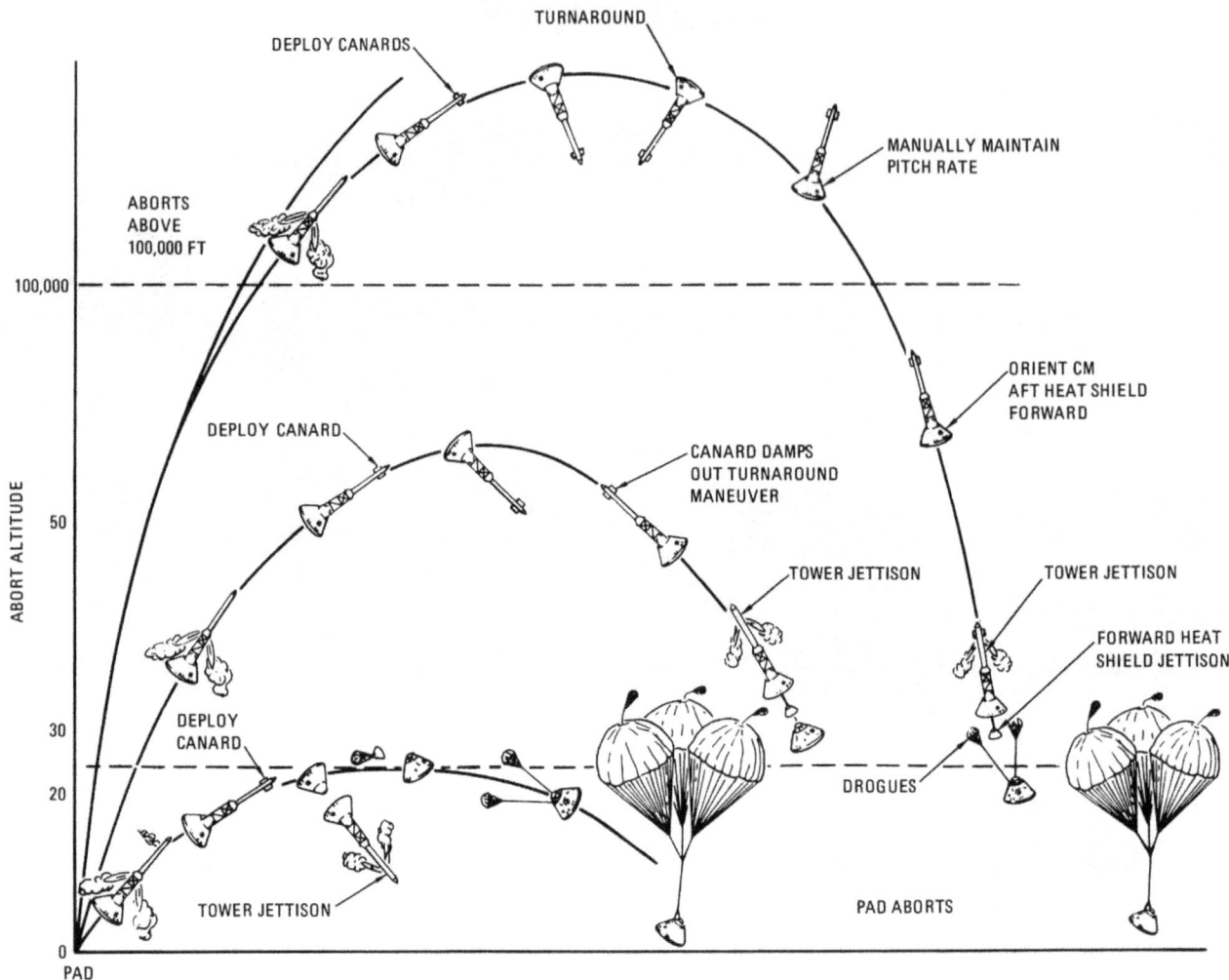

How launch escape subsystem operates at different altitudes

After receiving an abort signal, the booster is cut off (after 40 seconds of flight), the CM-SM separation charges fired, and the launch escape motor ignited. The launch escape motor lifts the CM and the pitch control motor (used only at low altitudes) directs the flight path off to the side.

Two canards (wing-like surfaces at the top of the structure) are deployed 11 seconds after the abort is started. The aerodynamic forces acting on the canard surfaces turn the CM so that its blunt end is forward. Three seconds later on extreme low-altitude aborts, or at approximately 24,000 feet on high-altitude aborts, the tower separation devices are fired and the jettison motor is started. These actions carry the launch escape subsystem assembly away from the CM's landing trajectory. Four-tenths of a second after tower jettisoning the CM's earth landing subsystem is activated and begins its sequence of operations to bring the module down safely.

During a successful launch the launch escape structure is jettisoned by the astronauts at the prescribed altitude. If an abort is necessary after the launch escape assembly has been jettisoned, it is performed with the service propulsion engine in the service module. In that case, the initiation of abort automatically separates the launch vehicle from the spacecraft, the service module reaction control engines fire in an ullage maneuver (which settles fuel and oxidizer so that they will flow properly to the engine), and the service propulsion engine ignites to thrust the spacecraft away from the launch vehicle.

The emergency detection system operates from the time of umbilical separation until 100 seconds after liftoff. It is designed to detect emergency conditions of the launch vehicle, display the information to the astronauts, and, if the system is on automatic, start an abort. Under certain conditions (excessive vehicle rates or two booster engines out), the system initiates an abort signal. This signal resets the event timer, activates the launch escape subsystem, and (after 30 to 40 seconds of flight) cuts off the launch vehicle engines. A "lockout" system prevents the emergency detection system from operating before liftoff.

A manual abort can be initiated before or during launch by the commander's translation control located on the arm of his couch. In an abort from the pad, the launch escape subsystem will carry the command module to a height of about 4,000 feet before it is jettisoned.

The nose cone of the launch escape assembly contains an instrument package called the Q-ball. The Q-ball has eight static ports (openings) through which pressure changes are measured. The instruments use this information to determine aerodynamic incidence angle and dynamic pressure data. The instruments send information on the angle of attack to an indicator of the CM's main display console and to the launch vehicle guidance.

The canards are two deployable surfaces and operating mechanisms which are faired (attached in a smooth line) into the outer skin of the launch escape assembly just below the nose cone. The operating mechanism is inside the structure. Each canard is mounted on two hinges and is deployed by a gas-operated actuator. Eleven seconds after the abort signal is received by the master events sequence controller, an electric current fires cartridges to open the canards. Gas from the cartridges causes a piston to retract, operating the opening mechanism. The canard surfaces mechanically lock in place when fully opened.

The launch escape and pitch control motors are produced by Lockheed Propulsion Co., Redlands, Calif. The tower jettison motor is produced by Thiokol Chemical Corp., Elkton, Md.

EQUIPMENT

<u>Launch Escape Motor</u> (Lockheed Propulsion Co.) — Solid rocket motor about 15-1/2 feet long and 26 inches in diameter in steel case;

Subsystem is lowered onto command module for test at White Sands

weighs 4700 pounds. Thrust is about 147,000 pounds on pad, increasing with altitude. Propellant is a composite of polysulfides. It provides thrust required to rescue crew from a dangerous situation in the early launch phase.

Pitch Control Motor (Lockheed Propulsion Co.) — Solid rocket motor 2 feet long and 9 inches in diameter in steel case; weighs 50 pounds. Motor generates 2,400 pounds of thrust for half a second. Propellant is a composite of polysulfides. Provides an initial pitch maneuver toward the Atlantic Ocean in case of an abort on the pad or at very low altitude.

Tower Jettison Motor (Thiokol Chemical Corp.) — Solid rocket motor 4-1/2 feet long and 26 inches in diameter in a high carbon chrome-molybdenum steel-forged case. Motor generates 31,500 pounds of thrust for 1 second. Propellant

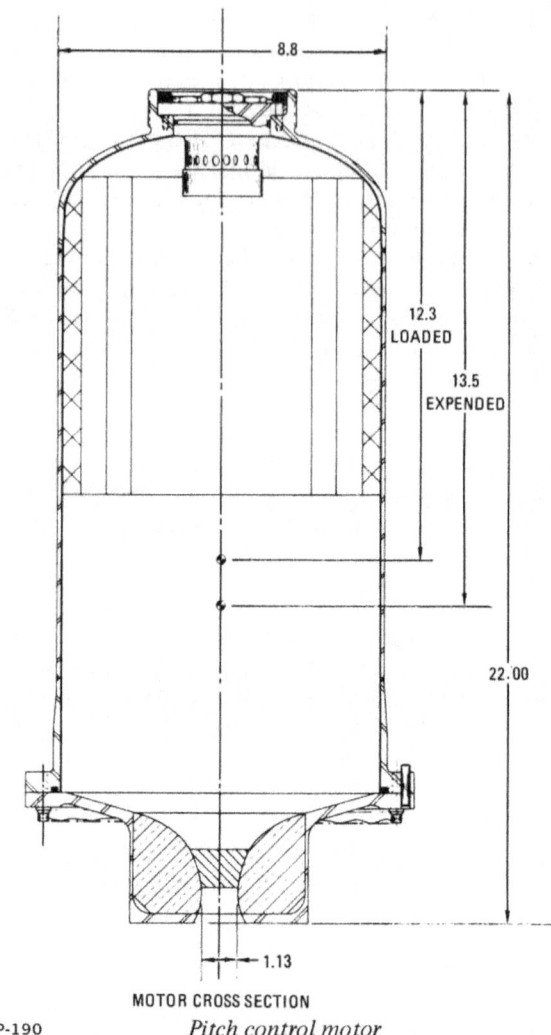

P-190 *Pitch control motor*

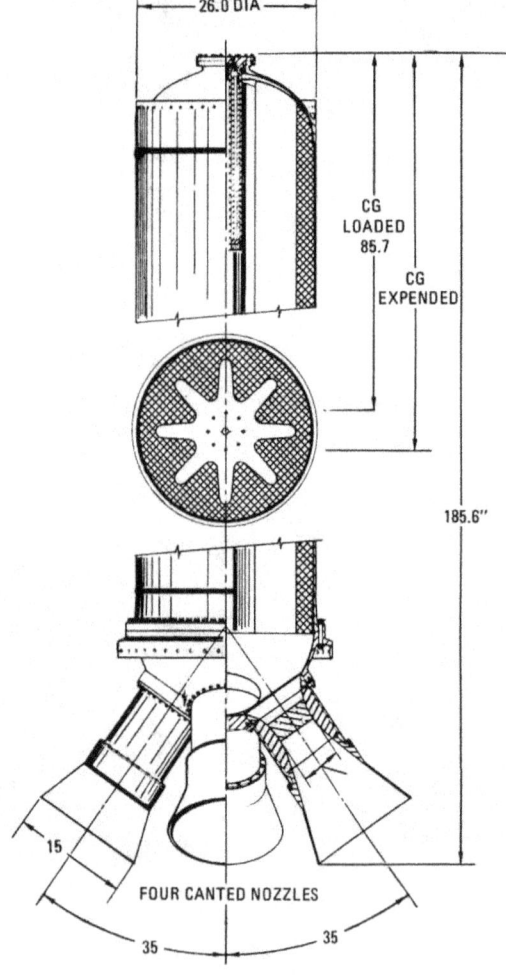

P-189 *Launch escape motor*

is a composite of polysulfides. It jettisons the launch escape tower either after the first stage when the launch escape tower is no longer required or after a launch abort and before parachutes are deployed for recovery.

Launch Escape Tower — Welded titanium tubing truss structure, 10 feet long, about 3 feet square at the top and 4 feet square at the base where it is connected to the command module. It weighs about 500 pounds including all attachments, wiring, and insulation.

Boost Protective Cover — It is made of layers of impregnated fiberglass, honeycomb cored-laminated fiberglass, and cork. It has 12 "blow-out" ports for reaction control motors, vents, and an 8-inch diameter window in front of the

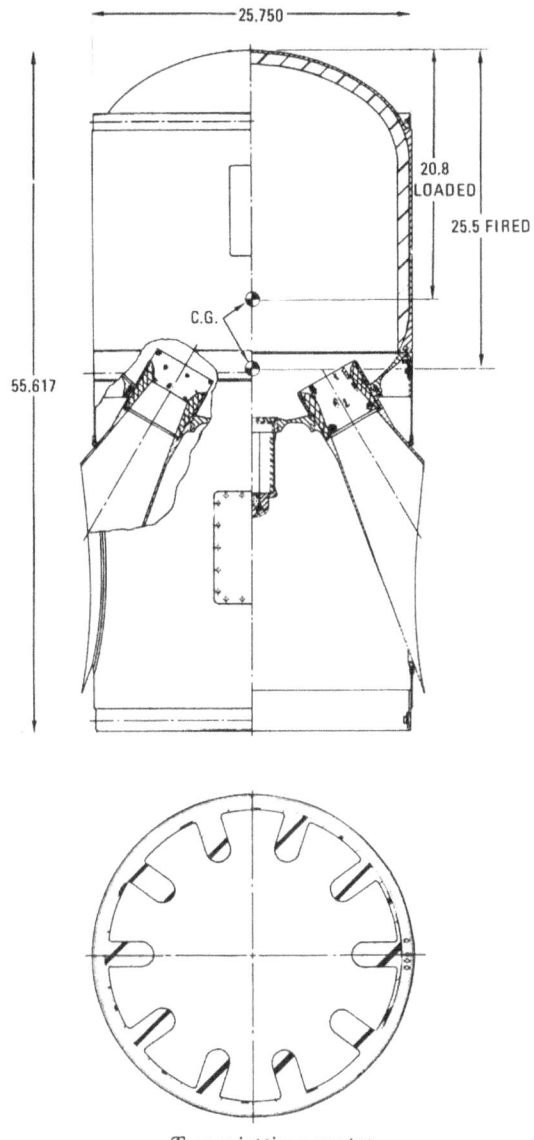

Tower jettison motor

at the base. It is made up of differential pressure transducers and electronic modules. It measures the differential of dynamic pressures about the pitch and yaw axes in order to monitor the angle of attack of the space vehicle.

Master Events Sequence Controllers — Two boxes about 14 by 10 by 8 inches, each containing time delays, relays, fuses, and fusistors, located in the forward right-hand equipment bay of the command module. It controls numerous launch abort functions as well as numerous normal mission functions.

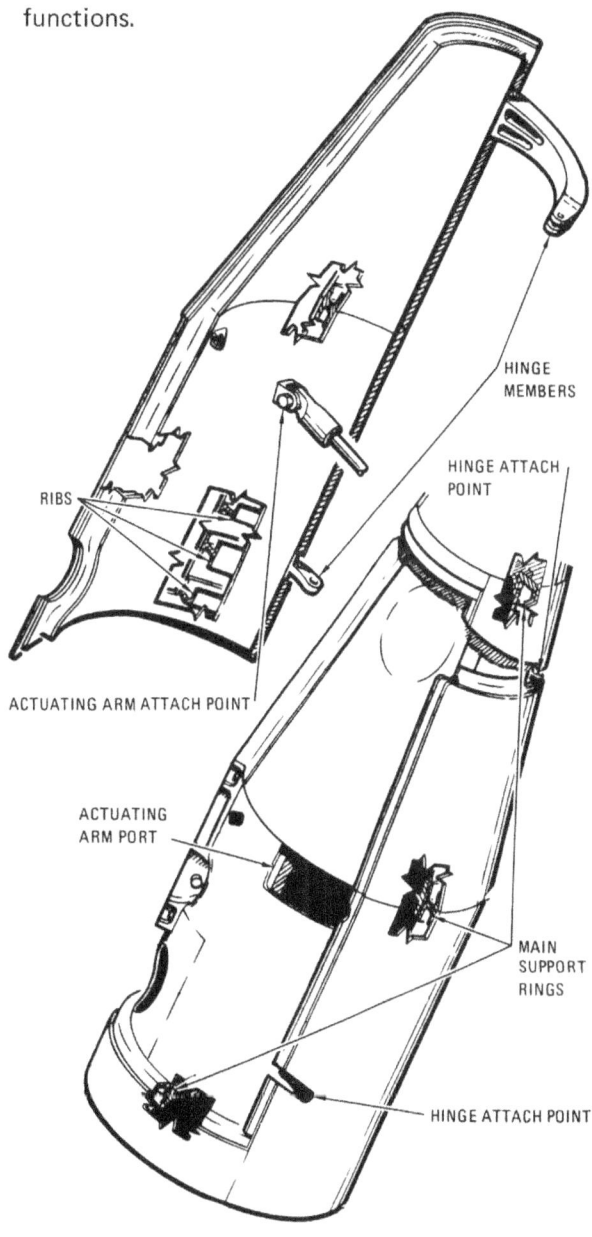

Canard structure

commander's forward viewing window. It completely covers the command module to prevent charring of external surfaces during boost out of the earth's atmosphere. It is jettisoned with the launch escape tower assembly.

Canards — Two metallic clamshell aerodynamic control surfaces about 4 feet long, which are deployed only during launch escape abort modes. They orient the command module so that the heat shield is forward and the parachutes aft. They are deployed to a fixed position 11 seconds after abort initiation.

Q-Ball — Aluminum nose cone of the space vehicle. It is 13.37 inches long with a 13.3-inch diameter

DETAILED DESCRIPTION

STRUCTURE

The launch escape subsystem consists of a boost protective cover, tower, launch escape motor, tower jettison motor, pitch control motor, ballast compartment, canards, nose cone, and Q-ball.

The boost protective cover protects the command module from the rocket exhaust of the launch escape motor and from the aerodynamic heating of boost. It is made of layers of resin-impregnated fiberglass covered with cork and fits over the entire command module like a glove. It is 11 feet tall, 13 feet in diameter, and weighs about 700 pounds.

The apex section of the boost cover is attached to the launch escape tower legs. A passive tension tie connects the apex of the boost cover and the docking probe of the command module. During a normal launch, the thrust of the tower jettison motor pulls the cover away and snaps this tension tie, leaving the probe with the command module. During an abort, however, the master events sequence controller and the lunar docking events sequence controller send signals which fire the ordnance devices that separate the docking ring from the CM. When the launch escape assembly is jettisoned the tension tie then pulls the probe and docking ring away with the boost protective cover.

The launch escape tower is made of titanium tubes of 2-½- and 3-½-inch diameter covered with Buna N rubber insulation to protect it against the heat of the rocket motor exhaust. The four legs of the tower fit in wells in the forward structure of the command module. They are fastened with studs and frangible (brittle) nuts. These nuts contain a small charge which breaks them to separate the module from the tower when the launch escape assembly is jettisoned.

The launch escape motor is the largest of the three rockets in the subsystem. It is approximately 15-1/2 feet long (including nozzles), 26 inches in diameter, and weighs 4,700 pounds. Two-thirds of this weight is in its polysulfide solid propellant. The motor produces about 147,000 pounds of thrust in its 3.2 seconds of burning, enough to lift the 13,000-pound command module and carry it a mile away from the launch vehicle. In order to provide a nominal thrust vector angle of approximately 2.75 degrees from the center of gravity excursion line, the throat area of one of the two

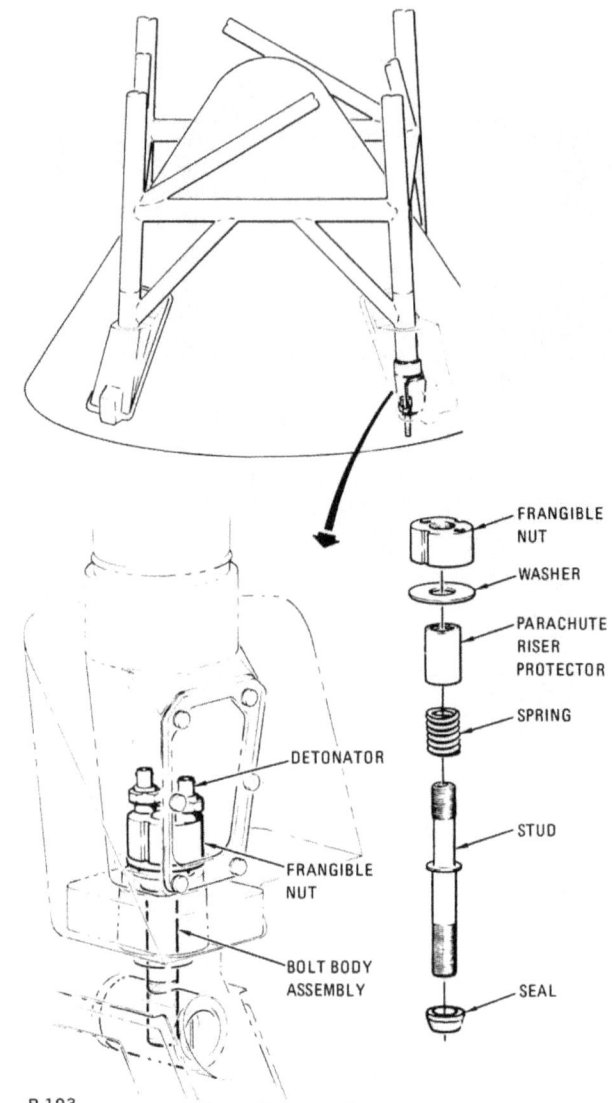

Tower leg attachment

nozzles in the pitch plane is approximately 5 percent larger than the two nozzles in the yaw plane; and the throat area of the second nozzle in the pitch plane is approximately 5 percent smaller than the two nozzles in the yaw plane. The motor skirt, to which the tower is attached, is 31 inches in diameter and is made of titanium.

Mounted atop the launch escape motor is the tower jettison motor, which is used in all missions to jettison the subsystem. The jettison motor is a little over 4-½ feet long, about 26 inches in diameter, and weighs about 525 pounds. Its solid propellant also is polysulfide. It burns for a little over 1 second and produces about 31,500 pounds of thrust. The motor has two fixed nozzles, located 180 degrees apart, which project from its steel case at a downward and outward angle. These skewed nozzles produced a resultant thrust vector angle of approximately 4 degrees which tips the tower and pulls it free of the CM.

The pitch control motor is used during an abort to push the launch escape assembly carrying the CM to one side, away from the launch vehicle. It is 2 feet long, 9 inches in diameter, and weighs about 50 pounds. Its polysulfide solid propellant burns only for a little more than a half second and produces about 2,400 pounds of thrust.

The tapered ballast compartment and nose cone top the assembly. The ballast compartment contains lead and depleted uranium weights. The nose cone contains the Q-ball instrumentation. Both the ballast compartment and nose cone are made of Inconel (a heat-resistant nickel alloy) and stainless steel.

The Q-ball provides an electrical signal to a display on the main display console and to the ground. The Q-ball has eight static ports (openings) for measuring pressure changes which are a function of angle of attack. The pitch and yaw pressure-change signals are electronically summed in the Q-ball and displayed on the indicator. The Q-ball information provides a basis for crew abort decision in the event of slow launch vehicle divergence.

EMERGENCY DETECTION SYSTEM

The emergency detection system monitors critical conditions of launch vehicle powered flight. Emergency conditions are displayed to the crew on the main display console to indicate necessity for abort. The system includes provisions for a crew-initiated abort with the use of the launch escape subsystem or with the service propulsion subsystem after tower jettison. The crew can start an abort during countdown until normal spacecraft separation from the launch vehicle. Also included in the system are provisions for an automatic abort in case of the following time-critical conditions:

1. Loss of thrust on two or more engines on the first stage of the launch vehicle.

2. Excessive vehicle angular rates in any of the pitch, yaw, or roll planes.

When an abort is initiated (either manual or automatic), the system cuts off booster engines except for the first 40 seconds of flight in the case of the uprated Saturn I, or the first 30 seconds of flight in the case of the Saturn V. Range safety requirements impose the time restrictions.

The emergency detection system automatic abort circuits in the spacecraft are activated automatically at liftoff and deactivated 100 seconds after liftoff. Switches on main display console deactivate the entire automatic abort capability or the "two engines out" and "excessive rates" portions of the system independently. These three switches are placed in the automatic position before liftoff and are switched off before first stage separation. The two automatic abort circuits also are deactivated automatically in the instrument unit just before the first stage inboard engine cutoff as a backup to manual deactivation.

The electrical circuits that control the launch vehicle status lights are in the instrument unit. The "LV Rate" light will illuminate when launch vehicle roll, pitch, or yaw rates are in excess of predetermined limits. The red "LV Guid" light illuminates to indicate loss of attitude reference in the guidance unit. The yellow "LV Engines" lights illuminate when an engine is developing less than the required thrust output. The engine lights provide four cues: ignition, cutoff, engine below thrust, and physical stage separation. A yellow "S-II Sep" light will illuminate at second stage first-plane separation and is extinguished at second-plane separation on vehicles launched by the Saturn V.

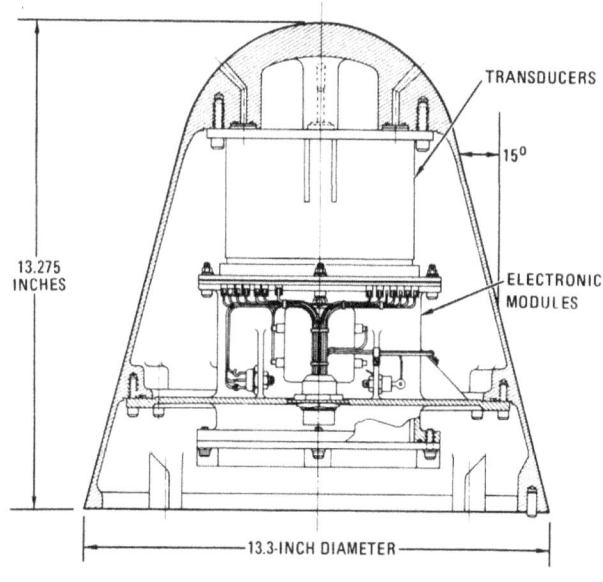

Nose cone and Q-ball

The "Abort" light is a red lamp assembly containing four bulbs that provide high-intensity illumination. This light is illuminated if an abort is requested by the launch control center for a pad abort or an abort during liftoff via up-data link. The "Abort" light also can be illuminated after liftoff by the range safety officer, or via the up-data link from the manned spacecraft flight network.

The emergency detection system will automatically initiate an abort signal when two or more first stage engines are out or when launch vehicle excessive rates are sensed by gyros in the instrument unit. The abort signals are sent to the master events sequence controller, which initiates the abort procedure.

LAUNCH ESCAPE TOWER JETTISON

After second stage ignition, the launch escape tower is jettisoned. Normally both of the tower jettison switches will be used to initiate this function; however, either one will initiate the tower jettison circuits. The frangible nut assemblies which attach the tower legs to the CM each include two detonators which are fired at activation of the jettison switches. The tower jettison circuits also ignite the tower jettison motor. The cues which the flight crew will use when initiating tower jettison are the number one engine status light for the Saturn IB and the "S-II Sep" light for the Saturn V. The crew will use the digital events timer in conjunction with the visual light cues to jettison the tower at the correct time. If the tower jettison motor should fail to ignite, the launch escape motor can be used to jettison the tower.

When the launch escape tower is jettisoned, the emergency detection system automatic abort circuits are disabled.

ABORTS

Abort procedures fall into several categories termed modes. Mode 1 aborts are those using the launch escape subsystem; Modes 2, 3, and 4 aborts are those using the service propulsion subsystem. There are subdivisions within each category, principally due to the altitude of the spacecraft at the time of abort.

Launch escape subsystem abort procedures are controlled automatically by the master events sequence controller. The controller also commands portions of the aborts using the service propulsion subsystem. The sequence commanded by the controller for a Mode 1a abort (pad or low altitude) is:

1. Relay booster engine cutoff signal to the launch vehicle's instrument unit.

2. Reset and start the commander's digital events timer.

3. Deadface (cut off the flow of current) the CM-SM umbilical.

4. Pressurize the CM reaction control subsystem.

Launch escape tower jettison

5. Transfer electrical control from the SM reaction control engines to the CM reaction control engines.

6. Transfer entry and post-landing battery power to the main dc bus tie.

7. Fire ordnance devices to cut CM-SM tension and ties and to activate the CM-SM umbilical guillotine.

8. Ignite the launch escape and pitch control motors.

9. Start rapid CM reaction control subsystem propellant dumping and purging.

10. Deploy launch escape subsystem canards.

11. Activate the earth landing subsystem controller.

12. Jettison the launch escape tower.

13. Fire charges to separate the docking ring.

14. Jettison the forward heat shield.

15. Fire the mortar to deploy the heat shield drag parachute.

16. Deploy the drogue parachutes.

17. Purge CM reaction control subsystem.

18. Release the drogue parachutes.

19. Deploy the pilot parachutes (which pull out the main parachutes).

20. Deploy the two VHF recovery antennas and the flashing beacon light.

21. After splashdown, release the main parachutes.

The sequence differs slightly for other aborts using the launch escape subsystem. On high-altitude aborts, for example, the pitch control motor is not fired, dumping of the CM reaction control propellant follows the normal entry procedure (during descent on the main parachutes), and the reaction jet engine control is not cut off (enabling the stabilization and control subsystem to control the CM's attitude automatically).

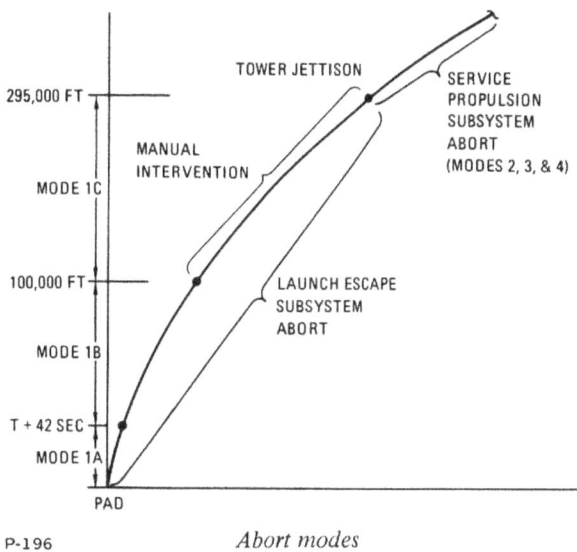

Abort modes

Aborts using the service propulsion subsystem, since they are at a much higher altitude or in space, follow procedures more like a normal entry.

Test at White Sands of launch escape subsystem's ability to carry CM to safety during pad abort

REACTION CONTROL

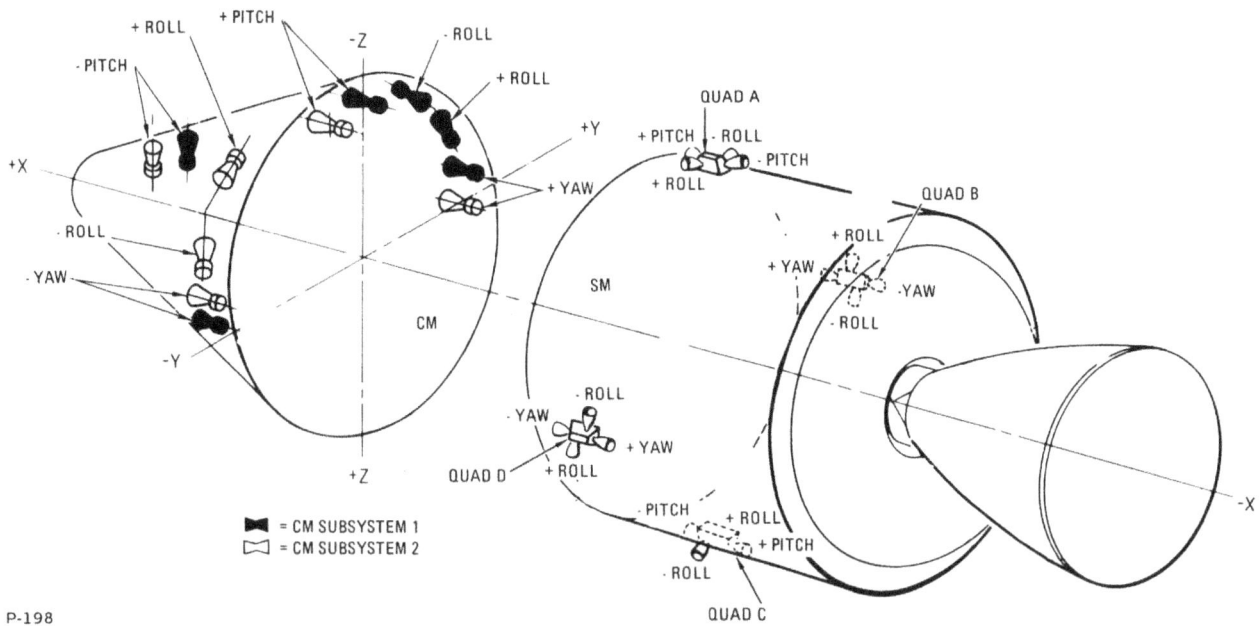

Location of reaction control subsystem engines

The reaction control subsystem provides the thrust for normal and emergency attitude maneuvers of the Apollo spacecraft. Operation of the subsystem is in response to automatic control signals from the stabilization and control subsystem in conjunction with the guidance and navigation subsystem. It can be controlled manually by the crew. The reaction control subsystem consists of CM and SM reaction control systems.

SM REACTION CONTROL SYSTEM

The SM reaction control system consists of four similar, independent systems (quads) located 90 degrees apart around the service module. It provides thrust required for three-axis stabilization and control of the spacecraft during earth orbit, translunar trajectory abort, transposition and docking, and translunar, lunar orbital, and transearth flight. It also may be used for minor course corrections both on the translunar and transearth flights.

The system provides the small velocity changes required for service propulsion subsystem propellant-settling maneuvers (ullage). Only roll axis control is provided during service propulsion engine thrusting. In addition, it provides velocity changes for spacecraft separation from the third stage during high-altitude or translunar injection abort, for separation from the boost vehicle after injection of the spacecraft into translunar trajectory, LM rendezvous in lunar orbit, and for CM-SM separation.

The four quads can be operated simultaneously or in pairs during spacecraft maneuvers. Each quad is mounted on a honeycomb structural panel about 8 feet long and 3 feet wide. It becomes part of the integrated service module structure when hinged and bolted in place. Center lines of engine mounts are offset about 7 degrees from the Y and Z axes. The cluster of four engines for each quad is rigidly mounted in a housing on the outside of the honeycomb panel. Laterally mounted (roll) engines are used for rotating the vehicle about the X axis. Longitudinally mounted engines are used for rotating the vehicle about the Y and Z axes and translational maneuvers along the X axis. Roll engines are offset to minimize engine housing frontal area to reduce boost heating effects. All engines in each cluster are canted 10 degrees outward to reduce the effects of exhaust plume impingement on the service module structure.

Each engine provides approximately 100 pounds of thrust and uses hypergolic propellant. The fuel is monomethyl hydrazine (MMH) and the oxidizer is nitrogen tetroxide (N_2O_4). The engines are produced by The Marquardt Corp., Van Nuys, Calif.

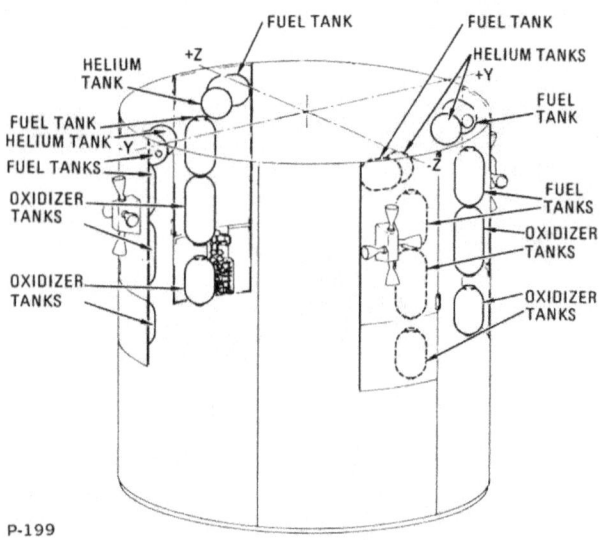

SM reaction control subsystem quads

The reaction control engines may be pulse-fired (in bursts) to produce short-thrust impulses or fired continuously to produce a steady thrust. The short-pulse firing is used for attitude-hold and navigation alignment maneuvers. Attitude control can be maintained with two adjacent quads operating.

Each quad contains a pressure-fed, positive-expulsion propellant feed system. The propellant tanks (two fuel and two oxidizer) are located on the inside of the structural panel; feed lines are routed through the panel to the engines. The propellant tanks are produced by Bell Aerosystems Co., Buffalo, N.Y., a division of Textron, Inc.

Helium is used to pressurize the propellant tanks; a single helium tank is located on the inside of the panel. Helium entering the propellant tanks around the positive-expulsion bladders forces the propellant in the tanks into the feed lines. Oxidizer and fuel are thus delivered to the engines. The fuel valve on each engine opens approximately 1/500th of a second before the oxidizer valve to provide proper ignition characteristics. Each valve contains orifices which meter the propellant flow to obtain the proper (2 to 1) mixture ratio. The propellants are hypergolic; that is, they ignite when they come in contact in the engine combustion chamber without an ignition system.

CM REACTION CONTROL SYSTEM

The CM reaction control system is used after CM-SM separation and for certain abort modes. It provides three-axis rotational and attitude control to orient and maintain the CM in the proper entry attitude before encountering aerodynamic forces. During entry, it provides the torque (turning or twisting force) required to control roll attitude.

The system consists of two independent, redundant systems, each containing six engines, helium and propellant tanks, and a dump and purge system. The two systems can operate in tandem; however, one can provide all the impulse needed for the entry maneuvers and normally only one is used.

The 12 engines of the system (produced by North American Rockwell's Rocketdyne Division, Canoga Park, Calif.) are located outside the crew compartment of the command module, 10 in the aft compartment and 2 in the forward compartment. The nozzle of each engine is ported through the heat shield of the CM and matches the mold line. Each engine produces approximately 93 pounds of thrust.

Operation of the CM reaction control engines is similar to the SM. Propellant is the same (monomethyl hydrazine and nitrogen tetroxide) and helium is used for pressurization. Each of the redundant CM systems contains one fuel and one oxidizer tank similar to the fuel and oxidizer tanks of the SM system. Each CM system has one helium tank.

The helium is isolated from the system by squib valves before entry; these are valves which contain small explosive charges (squibs). These valves are activated before CM-SM separation.

High-pressure helium flows through regulators (to reduce the pressure) and check valves to the propellant tanks, where it maintains pressure around the positive-expulsion bladders in each tank. The propellants are forced into the feed lines, through a burst diaphragm, and to the engines. The diaphragm must rupture for propellant to reach the engines; it is additional assurance that the engines cannot be fired inadvertently.

Oxidizer and fuel is fed to the 12 engines by a parallel feed system. The injector valve on each engine contains orifices which meter the fuel and oxidizer so that a flow ratio of 2 oxidizer to 1 fuel is obtained.

The engines may need heating before use so that the oxidizer doesn't freeze when it comes in contact with the injector valve. Astronauts monitor

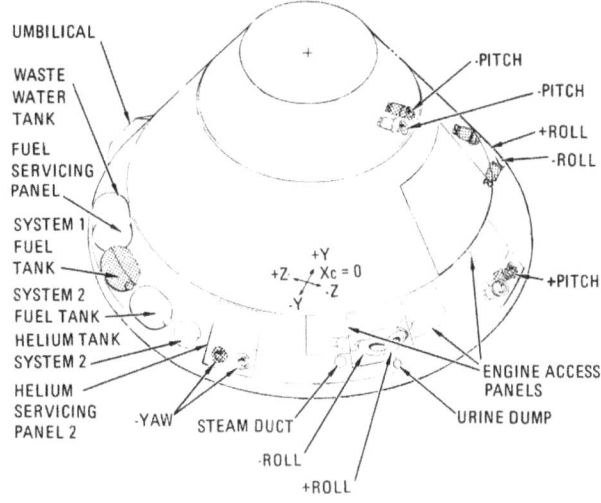

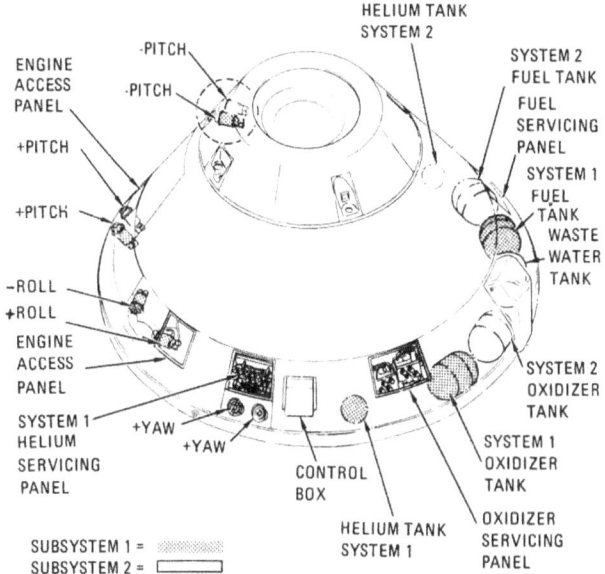

Location of CM reaction control subsystem components
P-200

the temperature of the engines on a cabin display and turn on the engine injector valve direct coils which act as heaters if necessary.

Because the presence of hypergolic propellant can be hazardous at CM splashdown, the propellant remaining in the fuel and oxidizer tanks is disposed of by burning during the final descent on the main parachute. After all propellant is disposed of the feed lines are purged with helium. The burn and purge operations are controlled manually by the crew except during an abort in the early part of boost (up to 42 seconds after liftoff), when dumping and purging is automatic.

EQUIPMENT

SERVICE MODULE

Helium Tanks (Airite Div., Sargent Industries, El Segundo, Calif.) — The four spherical tanks are made of titanium and weigh 11.5 pounds each. Each has an internal volume of 910 cubic inches. The helium is pressurized to 4150 psig. The outside diameter is 12.37 inches, a wall thickness of 0.135 inch, and a capacity of 1.35 pounds. The tanks store helium used to pressurize the propellant tanks.

Primary Fuel Tanks (Bell Aerosystems Co., Buffalo, N.Y.) — There are four cylindrical titanium tanks with domed ends, one tank for each quad of engines. The tanks have Teflon bladders. Each tank is 23.717 inches long with an outside diameter of 12.62 inches. Wall thickness is 0.017 to 0.022 inch. Combined propellant ullage volume is 69.1 pounds, resulting in tank pressure no greater than 215 psia at 85 degrees. The tanks store fuel (monomethyl hydrazine) and supply it on demand to the engines.

Primary Oxidizer Tanks (Bell) — There are four cylindrical titanium tanks with domed ends, one tank for each quad of engines. The tanks have Teflon bladders. Each tank is 28.558 inches long with an outside diameter of 12.62 inches. Wall thickness is 0.017 to 0.022 inch. Combined propellant and ullage volume is 137 pounds resulting in tank pressure no greater than 215 psia at 85 degrees. The tanks store oxidizer (nitrogen tetroxide) and supply it on demand to the engines.

Secondary Fuel Tanks (Bell) — There are four cylindrical titanium tanks with domed ends, one tank for each quad of engines. The tanks have Teflon bladders. Each tank is 17.329 inches long with an outside diameter of 12.65 inches. Wall thickness is 0.022 to 0.027 inch. Combined propellant and ullage volume is 45.2 pounds, resulting in tank pressure no greater than 205 psia at 105 degrees. The tanks store fuel and supply it upon demand to the engines.

Secondary Oxidizer Tanks (Bell) — There are four cylindrical titanium tanks with domed ends, one for each quad of engines. The tanks have Teflon bladders. Each tank is 19.907 inches long and has an outside diameter of 12.65 inches. Wall thickness is 0.022 to 0.027 inch. Combined propellant and ullage volume is 89.2 pounds, resulting in tank pressure no greater than 205 psia at 105 degrees. The tanks store oxidizer and supply it on demand to the engines.

Engines (Marquardt) — There are 16 radiation-cooled engines grouped in clusters of four 90 degrees apart on the outside of the service module. They are the only nonablative engines on the command and service module. The thrust chambers are pure molybdenum, and nozzle extensions are a cobalt-base alloy. Each engine is 13.400 inches long and weighs 5 pounds. Nozzle exit diameter is 5.6 inches. Each engine has a nominal thrust of 100 pounds. Service life of each engine is 1000 seconds: any combination of pulsed (intermittent) and continuous operation up to a maximum of 500 seconds of steady-state firing. Minimum firing time is 12 milliseconds. Each engine is capable of 10,000 operation cycles. The engines are used for translation and rotational maneuvers and for obtaining star sightings.

COMMAND MODULE

Helium Tanks (Menasco Manufacturing Co., Burbank, Calif.) — The four spherical tanks are

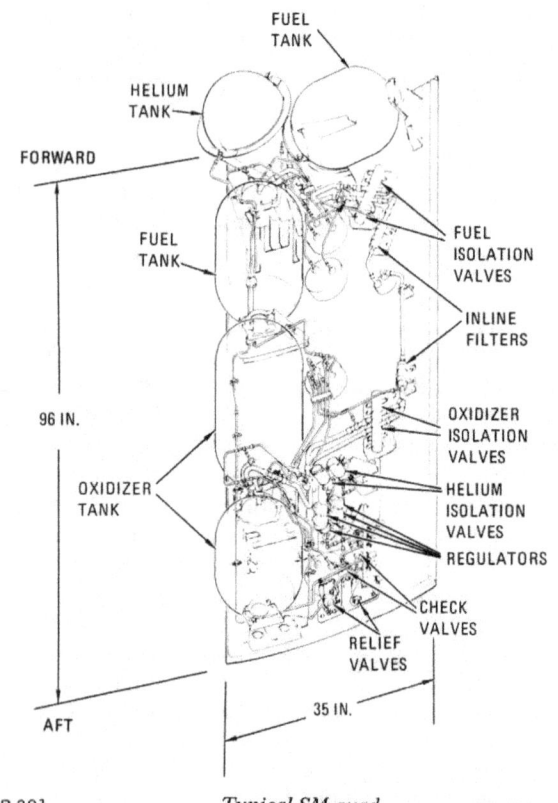

P-201   *Typical SM quad*

made of titanium. Each has a volume of 365 cubic inches. The helium is pressurized to 4150 psig. Outside diameter of each is 9.2 inches, and wall thickness is 0.102 inch. Capacity of each is 0.57 pound. The tanks store helium to pressurize the propellant tanks.

Fuel Tanks (Bell) — There are two titanium tanks identical to the secondary fuel tanks on the service module system.

Oxidizer Tanks (Bell) — There are two titanium tanks identical to the secondary oxidizer tanks on the service module system.

Engines (Rocketdyne) — There are 12 engines — 10 in the aft equipment compartment and two in the apex cover area. They are ablative engines and are installed with scarfed (smoothed into the surface) ablative nozzle extensions. Each engine is 12.65 inches long and weighs 8.3 pounds. Nozzle exit diameter is 2.13 inches. Thrust is 93 pounds. Service life for each engine is 200 seconds. Its minimum firing time is approximately 12 milliseconds. Each engine is capable of 3000 operational cycles. The primary use of the engines is for rotation maneuvers, rate damping, and attitude control during entry.

DETAILED DESCRIPTION

SM REACTION CONTROL SYSTEM

The SM system is composed of four separate, individual quads, each containing pressurization, propellant, rocket engine, and temperature control systems.

The pressurization system regulates and distributes helium to the propellant tanks. It consists of a helium storage tank, isolation valves, pressure regulators, check valves, relief valves, and lines necessary for filling, draining, and distribution of the helium.

The helium supply is contained in a spherical storage tank, which holds 1.35 pounds of helium at a pressure of about 4150 psia. Isolation valves between the helium tank and pressure regulators contain two solenoids: one is energized momentarily to latch the valve open magnetically; the other is energized momentarily to unlatch the valve, and spring pressure and helium pressure forces the valve closed. The isolation valves in each quad are individually controlled by switches on the main display console. The valves are normally open to pressurize the system. They are held open by a magnetic latch rather than by the application of power which conserves power and prevents overheating of the valve coil. Indicators above each valve switch show gray when the valve is open (the normal position) and diagonal lines when the valve is closed. The valve is closed in the event of a pressure regulator unit problem and during ground servicing.

Helium pressure is regulated by two assemblies connected in parallel, with one assembly downstream of each isolation valve. Each assembly incorporates two (primary and secondary) regulators connected in series. The secondary regulator remains open if the primary regulator functions properly. If the primary regulator fails open, the secondary regulator will maintain slightly higher but acceptable pressures.

Two check valve assemblies, one for oxidizer and one for fuel, permit helium flow to the tanks and prevent propellant or propellant vapor flow into the pressurization system if seepage or failure occurs in the propellant tank bladders. Filters are incorporated in the inlet to each check valve assembly and each test port.

The helium relief valve contains a diaphragm, filter, a bleed device, and the relief valve. The diaphragm is installed to provide a more positive seal against helium than that of the actual relief valve. The diaphragm ruptures at 228 psia. The filter retains any fragments from the diaphragm and prevents particles from flowing onto the relief valve seat. The relief valve will open at 236 psia and dump excessive pressure overboard. The relief valve will reseat at 220 psia.

A pressure bleed device vents the cavity between the diaphragm and relief valve in the event of any leakage across the diaphragm, or upon completion of checkout of the relief valve. The bleed device is normally open and will be fully closed when the pressure increases to 150 psia; it will be fully opened when the pressure decreases to 20 psia.

The propellant system consists of two oxidizer tanks, two fuel tanks, two oxidizer and two fuel isolation valves, a fuel and oxidizer inline filter, and associated distribution plumbing.

The oxidizer supply is contained in two titanium alloy, hemispherically domed cylindrical tanks. The tanks are mounted to the SM structural panel. The

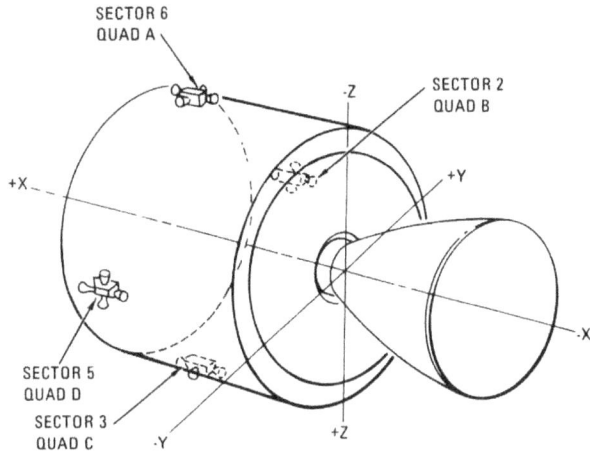

P-202 *Location of SM quads*

primary tank is about 28-½ inches long, 12-½ inches in diameter, and holds 137 pounds of oxidizer. The secondary tank is about 20 inches long, 12-½ inches in diameter, and holds 89 pounds of oxidizer.

Each tank contains a diffuser tube assembly and a Teflon bladder for positive expulsion of the oxidizer. The bladder is attached to the diffuser tube at each end of each tank. The diffuser tube acts as the propellant outlet.

When the tanks are pressurized, the helium surrounds the entire bladder, exerting a force which causes the bladder to collapse about the propellant, forcing the oxidizer into the diffuser tube assembly and out of the tank outlet into the manifold, providing expulsion during zero gravity.

The fuel supply is contained in two tanks that are similar in material, construction, operation, and diameter to oxidizer tanks. The primary tank is about 23-½ inches long and holds 69 pounds of fuel; the secondary tank is about 17 inches long and holds 45 pounds of fuel.

Isolation valves in the fuel and oxidizer tank lines in each quad are controlled by switches on the main display console. Each isolation valve contains solenoids and indicators that operate in the same manner as the helium isolation valves. The primary tank valves are normally open and the secondary valves closed. When a propellant quantity indicator displays 43 percent propellant remaining, the secondary valves are opened and the primary valves are closed. The valves may be closed to prevent fluid flow in the event of a failure such as line rupture or a runaway thruster.

Quad panels installed on service module in Downey clean room

Propellant distribution plumbing is identical in each quad. Each quad contains separate similar oxidizer and fuel plumbing networks. Propellant in each network is directed from the supply tanks through manifolds for distribution to the four engines in the cluster.

Filters are installed in the fuel and oxidizer lines between the propellant isolation valves and the engine manifold to prevent any particles from flowing into the engine injector valves and engine injector.

The SM reaction control engines are radiation cooled, pressure fed, bipropellant thrust generators which can be operated in either the pulse or steady-state mode.

Each engine consists of a fuel and oxidizer injector control valve which controls the flow of propellant by responding to automatic or manual electrical commands and an injector head assembly which directs the flow of the propellant from each control valve to the combustion chamber. A filter is at the inlet of each fuel and oxidizer solenoid injector valve. An orifice in the inlet of each fuel and oxidizer solenoid injector valve meters the propellant flow to obtain a nominal 2:1 oxidizer-fuel ratio.

The propellant solenoid injector valves use two coaxially wound coils, one for automatic and one for direct manual operation. The automatic coil is used when the thrust command originates from the controller reaction jet assembly, which is the electronic circuitry that selects the required automatic coils to be energized for a given maneuver. The direct manual coils are used when the thrust command originates at the rotation control, direct ullage pushbutton, service propulsion subsystem abort, or the SM jettison controller.

The main chamber portion of the injector will allow 8 fuel streams to impinge upon 8 oxidizer streams for main chamber ignition. There are 8 fuel holes around the outer periphery of the injector which provide film cooling to the combustion chamber walls.

The injector contains a precombustion chamber in which a single fuel and a single oxidizer stream impinge upon each other. The precombustion chamber provides a smoother start transient. There are 8 fuel holes around the outside of the precombustion chamber providing cooling to its walls.

The combustion chamber is constructed of unalloyed molybdenum which is coated with molybdenum disilicide to prevent oxidation of the base metal. Cooling of the chamber is by radiation and fuel film cooling.

The nozzle extension with integral stiffener rings is machined from a cobalt base alloy.

Each of the engine mounts contain two electrical strip heaters. Each heater contains two electrical elements. One element in each heater is controlled by a secondary temperature therm-o-switch that is set to open at 118°F and close at 70°F. When a switch on the main display console for that quad is set for the secondary system, dc power is supplied to the therm-o-switch in each heater of that quad and will automatically open and close according to the temperature.

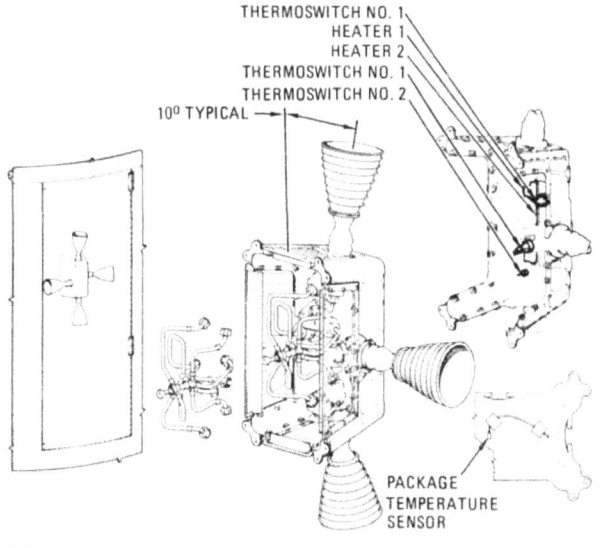

SM reaction control engine housing

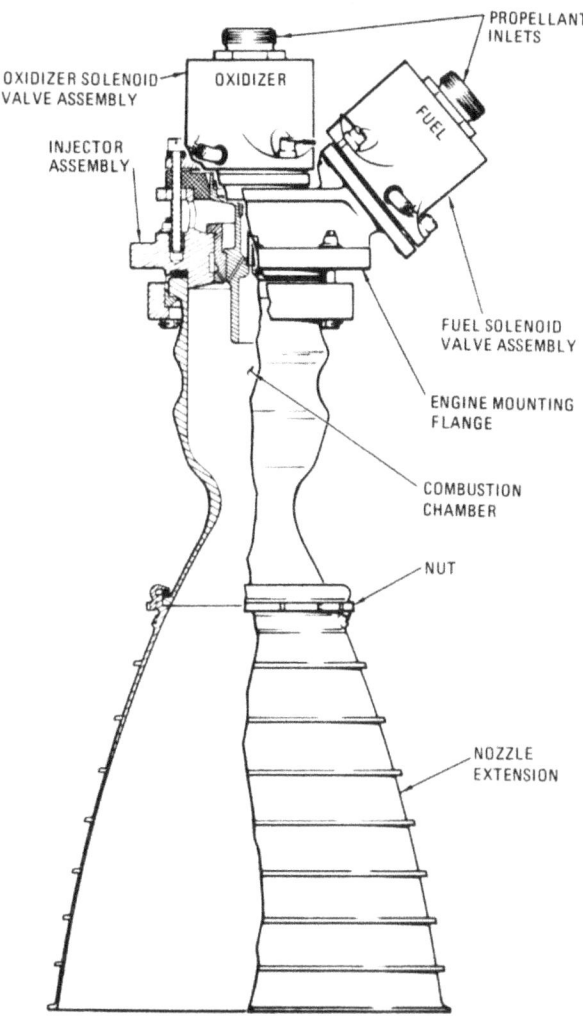

SM reaction control engine

When the switch is set for the primary heater, power is supplied to the redundant element in each heater for that quad. This therm-o-switch is a higher temperature switch and will automatically open at 134°F and close at 115°F. The heaters provide propellant temperature control by conductance to the engine housing and engine injector valves.

A gauge on the main display console is used to monitor the package temperature of any one of the four SM quads.

The helium tank supply pressure and temperature for each quad is monitored by a pressure/temperature ratio transducer. This provides a signal to a switch on the main display console. When the switch is positioned to a given SM quad, the pressure/temperature ratio signal is transmitted to a propellant quantity gauge, and the propellant quantity remaining for that quad is indicated in percent.

The helium tank temperature for each quad is monitored by a helium tank temperature transducer. A switch allows the crew to monitor either the helium tank temperature/pressure ratio as a percentage of quantity remaining, or helium tank temperature which can be compared against the helium supply pressure readout. Helium tank temperature is not displayed in the first Block II spacecraft, although it is telemetered to the ground.

In the SM reaction control system, the main buses cannot supply electrical power to one leg of the channel enable switches and controller reaction jet assembly until the contacts of the subsystem latching relay are closed. These are closed after separation of the spacecraft from the third stage, or to prepare for a service propulsion subsystem abort.

CM REACTION CONTROL SYSTEM

The CM reaction control system is composed of two separate, normally independent systems, called System 1 and System 2. They are identical in operation, each containing pressurization, propellant, rocket engine and temperature control systems.

The pressurization system consists of a helium supply tank, two dual pressure regulator assemblies, two check valve assemblies, two pressure relief valve assemblies, and associated distribution plumbing.

The total high-pressure helium for each system is contained in a spherical storage tank about 9 inches in diameter and containing 0.57 pound of helium at a pressure of 4150 psia.

Two squib-operated helium isolation valves are installed in the plumbing from each helium tank to confine the helium into as small an area as possible to reduce helium leakage until the system is used. Two squib valves are employed in each system to assure pressurization.

The pressure regulators used in the CM systems are similar in type, operation, and function to those used in the SM system. The difference is that the regulators in the CM system are set at a higher pressure than those of the SM system: 291 psia against 181 psia for the primary regulators and 291 against 187 for the secondary regulators.

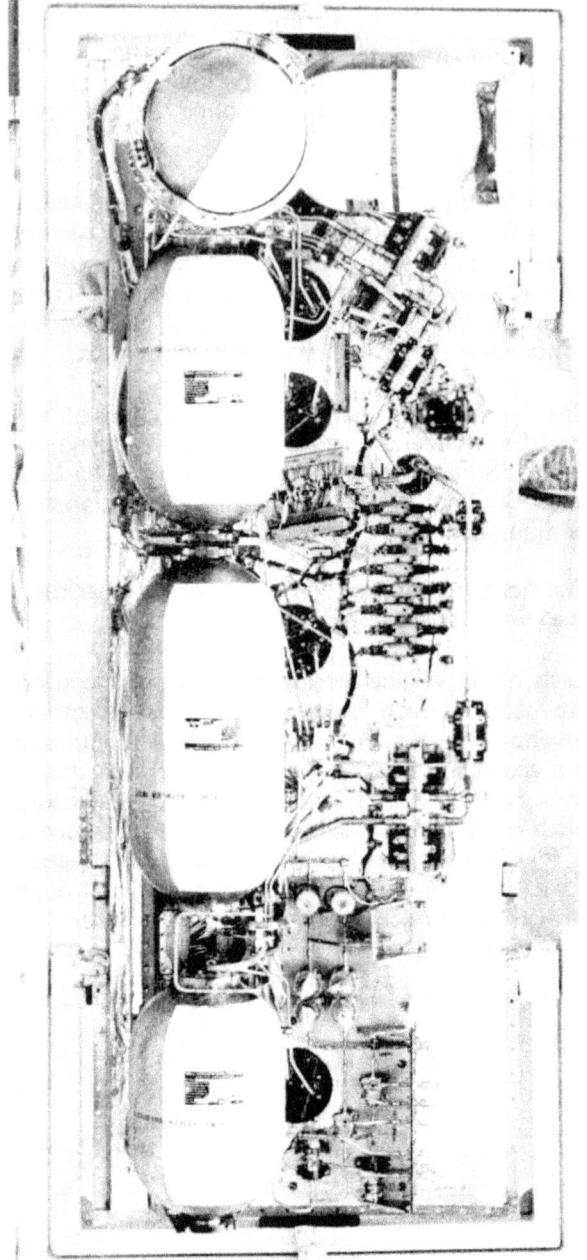

P-206 *SM quad prepared for installation*

The check valve assemblies used in CM system are identical in type, operation, and function to those used in the SM system. The helium relief valves also are similar to those in the SM system except that the rupture pressure of the diaphragm in the CM system is higher (340 psia instead of 228) and the relief valve relieves at a higher pressure (346 psia instead of 236 psia).

Each propellant system consists of one oxidizer tank, one fuel tank, oxidizer and fuel isolation valves, oxidizer and fuel diaphragm isolation valves, and associated distribution plumbing.

The oxidizer supply is contained in a single titanium alloy, hemispherical-domed cylindrical tank in each subsystem. These tanks are identical to the secondary oxidizer tanks in the SM system.

Each tank contains a diffuser tube assembly and a Teflon bladder for positive expulsion of the oxidizer similar to that of the SM secondary tank assemblies. The bladder is attached to the diffuser tube at each end of the tank. The diffuser tube acts as the propellant outlet.

When the tank is pressurized, the helium gas surrounds the entire bladder, exerting a force which causes the bladder to collapse about the propellant, forcing the oxidizer into the diffuser tube assembly and out of the tank outlet into the manifold.

The fuel supply is contained in a single titanium alloy, hemispherical-domed cylindrical tank in each subsystem that is identical to the SM secondary fuel tanks.

The diaphragms are installed in the lines from each tank to confine the propellants to as small an area as possible throughout the mission.

When the helium isolation squib valves are opened, regulated helium pressure pressurizes the propellant tanks creating the positive expulsion of propellants into the respective manifolds to the diaphragms, which rupture and allow the propellants to flow on through the propellant isolation valves to the injector valves on each engine. A filter will prevent any diaphragm fragments from entering the engine injector valves.

When the diaphragms are ruptured, the propellant flows to the propellant isolation valves. These are controlled by a single switch on the main display console. Each propellant isolation valve contains two solenoids, one that is energized momentarily to latch the valve open magnetically, and one that is energized momentarily to unlatch the magnetic latch. Spring force and propellant pressure close the valve. An indicator on the main display console shows gray indicating that the valves are open (the normal position) and diagonal lines when either valve is closed. The valves are closed in the event of a line rupture or runaway thruster.

The distribution lines contain 16 explosive-operated (squib) valves which permit the helium and propellant distribution configuration to be changed for various functions. Each squib valve is actuated by an explosive charge and detonated by an ignitor. After ignition of the explosive device, the valve remains open permanently. Two squib valves are used in each subsystem to isolate the high-pressure helium supply. Two squib valves are used to interconnect System 1 and 2 regulated helium supply which assess pressurization of both systems during dump-burn and helium purge operation. Two squib valves in each subsystem permit helium gas to bypass the propellant tanks and allow helium purging of the propellant subsystem. One squib valve in the oxidizer system permits both oxidizer systems to become common. One squib valve in the fuel system permits both fuel systems

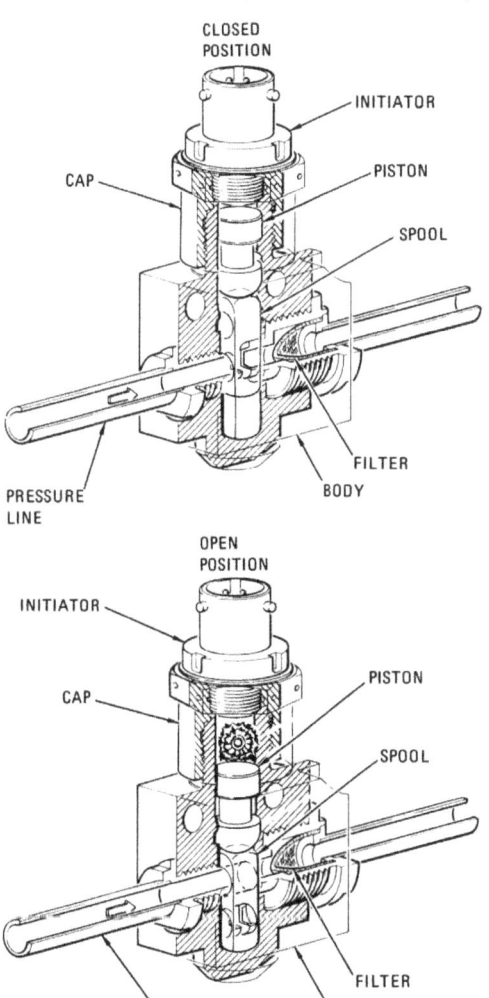

Typical CM squib valve

to become common. Two squib valves in the oxidizer system and two in the fuel system are used to dump the respective propellant in the event of an abort from the pad up to 42 seconds after liftoff.

The CM reaction control subsystem engines are ablative-cooled, bipropellant thrust generators that can be operated in either pulse or steady-state mode.

Each engine consists of fuel and oxidizer injector valves which control the flow of propellants, an injector and a combustion chamber in which the propellants are burned to produce thrust.

The injector valves use two coaxially wound coils, one for automatic and one for direct manual control. The automatic coil is used when the thrust command originates in guidance and control electronics. The direct manual coil is used when the thrust command originates at the astronaut hand rotation control. The engine injector valves are spring-loaded closed and energized open.

The automatic coils in the fuel and oxidizer injector valves are connected in parallel from guidance and control electronics. The direct manual coils in the fuel and oxidizer injector valves provide a direct backup to the automatic system. They are connected in parallel.

The injector contains 16 fuel and 16 oxidizer passages that impinge on a splash plate within the combustion chamber. This pattern is referred to as an unlike impingement splash-plate injector.

The thrust chamber assembly consists of the combustion chamber ablative sleeve, throat insert, ablative body, asbestos, and a fiberglass wrap. The engine is ablative-cooled.

The CM reaction control engines are mounted within the structure of the CM. The nozzle extensions extend through the CM heat shield and are made of ablative material. They match the mold line of the CM.

Temperature of the CM engines before activation is controlled by energizing injector valve direct coils on each engine. Temperature sensors are mounted on 6 of the 12 engine injectors. The temperature transducers have a range from $-50°$ to $+50°F$. The temperature transducers from the System 1 and 2 engine injectors provide inputs to two rotary switches located in the lower equipment bay of the CM. The specific engine injector temperature is monitored as dc voltage on the voltmeter in the bay. If any one of the engines registers less than $48°F$, the direct manual heating coils of all 12 engines are switched on. If $48°F$ (approximately 5 volts on the dc voltmeter) is reached from the coldest instrumented engine before 20 minutes, the valves are turned off. If 20 minutes pass before $+48°F$ is reached, the valves are turned off then. The heaters prevent the oxidizer from freezing at the engine injector valves and the 20-minute time limit assures that the warmest engines will not be overheated.

All automatic thrust commands for CM attitude are generated from the controller reaction jet assembly. These commands may originate at the rotation controls, the stabilization and control subsystem, or the CM computer. If the controller reaction jet assembly is unable to provide commands to the automatic coil of the CM engines, switches on the main display console will provide power to the rotation controls for direct coil control. The CM-SM separation switches automatically energize relays in the reaction control system control box that transfer the controller reaction jet assembly and direct manual inputs from the SM

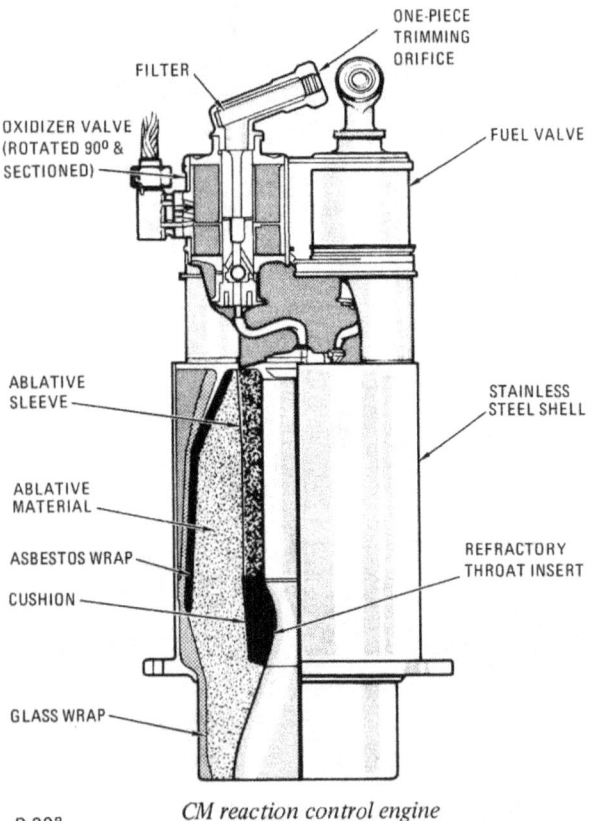

CM reaction control engine

P-208

engines to the CM engines. These functions also occur automatically on any launch escape subsystem abort.

The transfer motors in the control box are redundant to assure that the direct manual inputs are transferred from the SM engines to the CM engines, in addition to providing a positive deadface.

The RCS transfer motors may also be activated by a transfer switch placed to "CM" position; this is a manual backup to the automatic transfer.

CM Systems 1 and 2 also may be checked out before CM-SM separation by use of the transfer switch.

There are two sequences of propellant jettison. One sequence is used in the event of an abort while the vehicle is on the launch pad and through the first 42 seconds of flight. The second is used for all other conditions.

The sequence of events before and during a normal entry is as follows:

1. The CM system is pressurized by manual switching which fires the helium isolation squib valves in both System 1 and 2.

2. The CM reaction control engines provide attitude control during entry; and at approximately 24,000 feet, a barometric switch is activated unlatching the subsystem latching relay, inhibiting any further commands from the controller reaction jet assembly.

3. When the main parachute is fully deployed, a crewman will turn on the CM reaction control propellant dump switch, simultaneously initiating the two helium interconnect squib valves, the fuel interconnect squib valve, and the oxidizer interconnect squib valve, and energizing the fuel and oxidizer injector valve

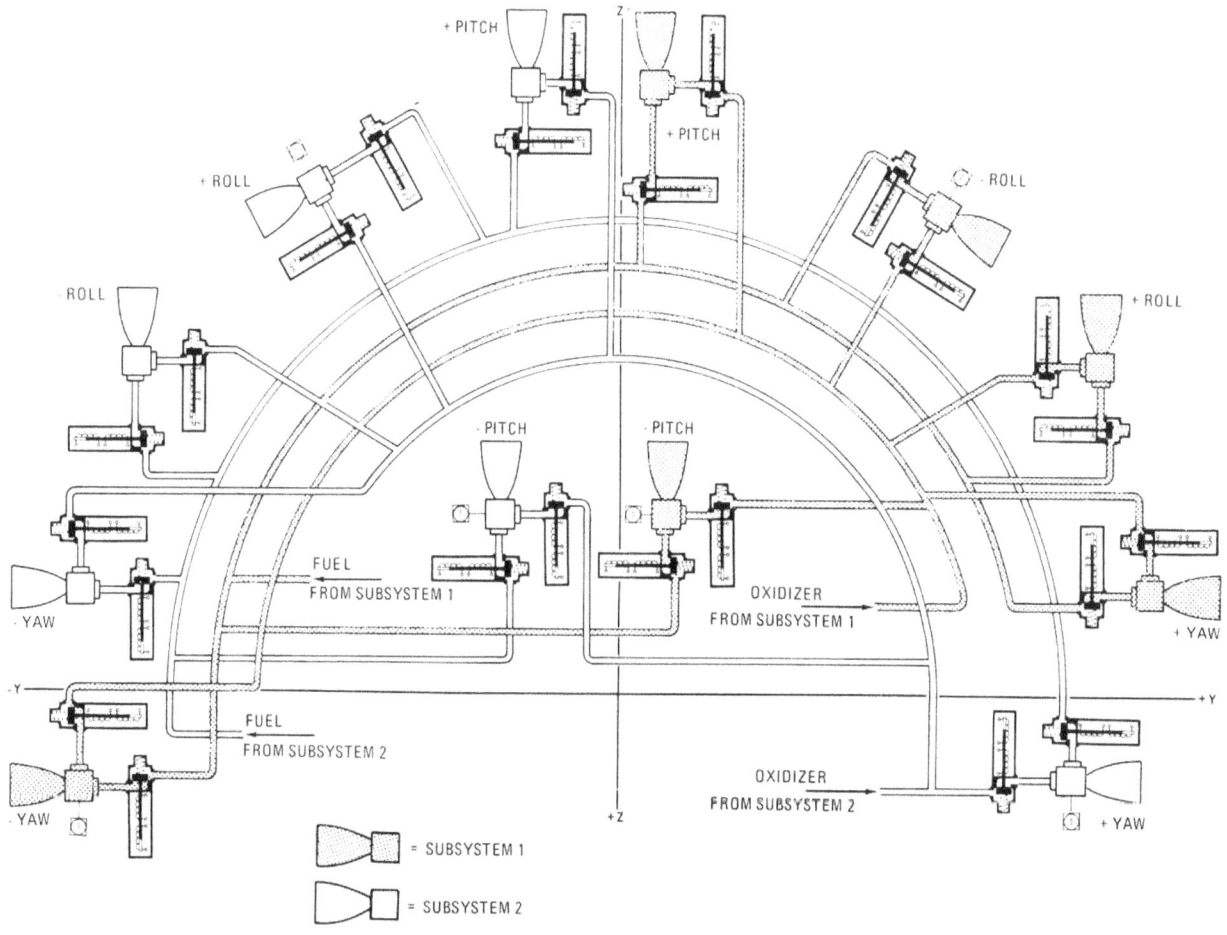

Schematic of reaction control engines

direct manual coils on 10 of the 12 CM engines. (The two forward or pitch engines are not energized because their plume might impinge on the parachutes.) The remaining propellant is burned through the 10 engines. The length of burn time will vary depending on the amount of propellant remaining. If an entire propellant load remained, a nominal burn time would be 88 seconds through 10 engines. In the worst case (only 5 of the 12 engines burning), a nominal burn time would be 155 seconds.

4. Upon completion of propellant burn, the CM propellant purge switch is turned on initiating the four helium bypass squib valves to allow the regulated helium pressure to bypass around each fuel and oxidizer tank bladder and purge the lines and manifolds out through the 10 engines. Purging requires approximately 15 seconds (until helium is depleted).

5. In case of a switch failure, the remaining propellants may be burned by manipulating the two rotation controllers so that 10 of the 12 CM engines will fire.

6. If the purge switch fails, the CM "helium dump" pushbutton would be pressed to initiate the four helium bypass squib valves, purge the lines and manifolds out through 10 of the 12 engines, and deplete the helium source pressure.

7. After purging, the direct coils of the CM engine injector valves are switched off manually.

The sequence of events during an abort from the pad up to 42 seconds after liftoff is controlled automatically by the master event sequence controller by manually rotating the translation control counterclockwise. The following events occur simultaneously:

1. The CM-SM transfer motor-driven switches are automatically driven upon receipt of the abort signal, transferring the logic circuitry from SM reaction control engines to CM engines.

2. When the abort signal is received, the two squib-operated helium isolation valves in each system are initiated, pressurizing Systems 1 and 2.

3. The squib-operated helium interconnect valve for the oxidizer and fuel tanks are opened even if only one of the two squib helium isolation valves opens. Both subsystems are pressurized as a result of the helium interconnect squib valve.

4. The solenoid-operated fuel and oxidizer isolation shutoff valves are closed to prevent fuel and oxidizer from flowing to the thrust chamber assemblies.

5. The squib-operated fuel and oxidizer interconnect valves are opened. Even if only one of the two oxidizer or fuel overboard dump squib valves opens, the oxidizer and fuel manifolds of each system are common as a result of the oxidizer and fuel interconnect squib valves.

6. The squib-operated oxidizer overboard dump valves are opened and route the oxidizer to blowout plug in the aft heat shield of the CM. The oxidizer shears a pin due to the pressure buildup and blows the plug out, dumping the oxidizer overboard. The entire oxidizer supply is dumped in approximately 13 seconds.

7. Five seconds after abort initiation, the squib-operated fuel overboard dump valves are initiated open and route the fuel to a fuel blow out plug in the aft heat shield of the CM. The fuel shears a pin due to the pressure buildup and blows the plug out, dumping the fuel overboard. The entire fuel supply is dumped in approximately 13 seconds.

8. Thirteen seconds after the fuel dump sequence was started, the fuel and oxidizer bypass squib valves in Systems 1 and 2 are opened to purge the fuel and oxidizer systems through the fuel and oxidizer overboard dumps.

SERVICE PROPULSION

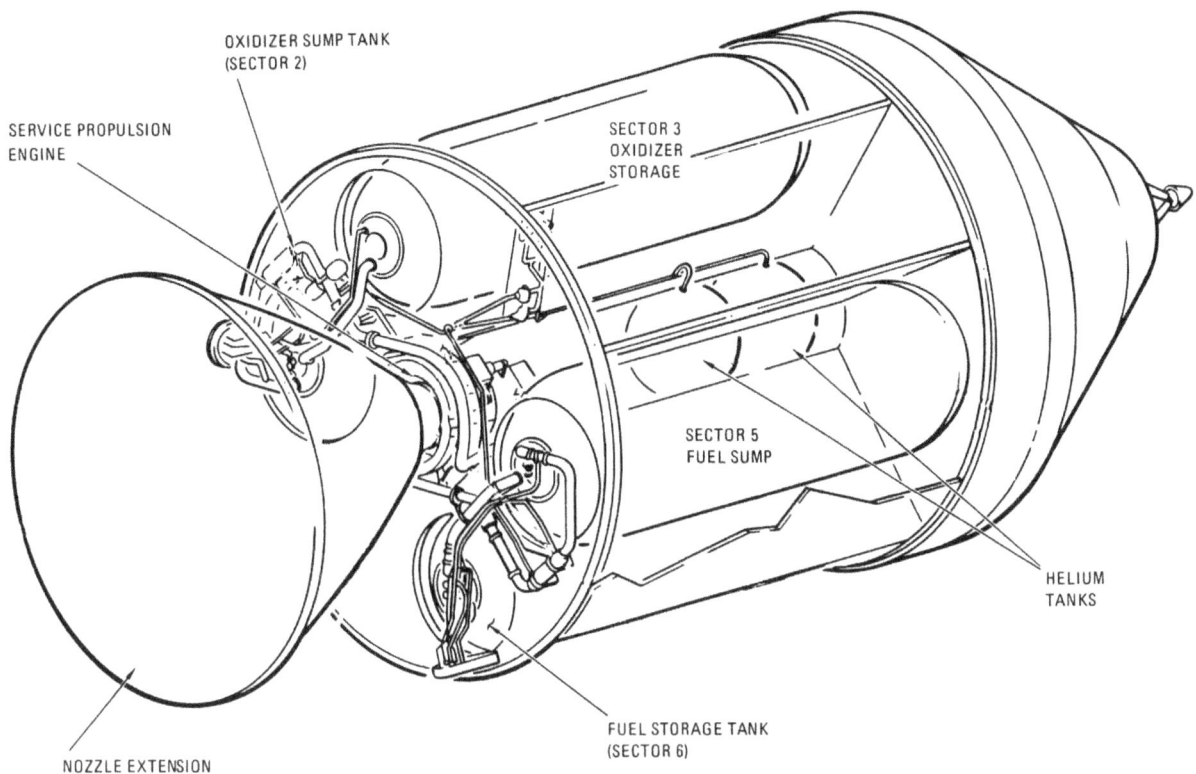

Location of main components of service propulsion subsystem

The service propulsion subsystem provides the thrust for all major velocity changes throughout a mission. These include the retrobraking maneuver for insertion into an orbit around the moon, the thrust for injection into the transearth trajectory, major course corrections, and the power to return the CM to earth's atmosphere during an abort after the launch escape subsystem has been jettisoned.

The subsystem includes a single large rocket engine, its pressurization and propellant subsystems, a bipropellant valve assembly, a thrust mount assembly with a gimbal actuator assembly, and the propellant utilization and gauging subsystem. Displays and sensing devices enable the crew and ground stations to monitor subsystem performance.

All of the components of the service propulsion subsystem except the controls are located in the service module. Control of engine firing normally is automatic, but there are provisions for manual override. Subsystem components occupy about three-quarters of the space in the SM and make up more than 41,500 pounds of its 55,000-pound weight.

The service propulsion engine is 3 feet, 5 inches long with a radiation-cooled extension nozzle of 9 feet 4 inches. The engine is gimbaled (can be turned) and provides 20,500 pounds of thrust in vacuum. Its propellant is composed of fuel of 50-percent hydrazine and 50-percent unsymmetrical dimethylhydrazine (UDMH) and an oxidizer of nitrogen tetroxide. The propellant is hypergolic; that is, the fuel and oxidizer ignite and burn on contact.

Its major components are a bipropellant valve assembly, injector, propellant lines, electrical wire harness, ablative thrust chamber, nozzle extension, thrust mount, gimbal ring, and gimbal actuator assembly. The engine is produced by Aerojet-General Corp., Sacramento, Calif.

The service propulsion engine responds to automatic firing commands from the guidance and navigation subsystem or to commands from manual controls. The engine assembly is gimbal-mounted to allow engine thrust-vector alignment with the spacecraft's center of mass to preclude tumbling. Thrust-vector alignment control is maintained automatically by the stabilization and control subsystem or manually by the crew. The engine has no throttle, producing a single-value thrust.

The engine's propellant supply is contained in four tanks of similar size and construction, each almost filling one of the sectors of the SM. There are two tanks each for fuel and for oxidizer; one is a storage tank and one a sump tank (which feeds the engine). Total propellant is 15,723 pounds of fuel and 25,140 pounds of oxidizer. (Although the volume in the fuel and oxidizer tanks is identical, the oxidizer weighs a great deal more than the fuel. The tanks, made of titanium, are built by General Motors Corp.'s Allison Division, Indianapolis, Ind.

The storage and sump tanks for fuel and oxidizer are connected in series by a single transfer line. Regulated helium from the pressurization subsystem enters the fuel and oxidizer storage tanks and forces the fluids into a transfer line to a sump tank standpipe. The pressure forces the fluids in the sump tanks into a propellant retention reservoir. This reservoir retains enough propellant to permit starting of the engine in zero gravity when the sump tanks are full without an ullage maneuver. The ullage maneuver is one in which reaction control engines are fired to give the spacecraft positive thrust and settle the propellant in the bottom of the tanks, thus assuring liquid flow through the feed lines.

The helium which pressurizes the propellant tanks is contained in two spherical tanks located in the center section of the SM just above the engine. Valves isolate the helium during non-thrusting periods and allow the gas into the tanks during thrusting periods. Helium pressure in the tanks is reduced by regulators in the pressure lines and then directed to the tanks through check valves. These valves permit the helium to flow into the tanks and prevent a reverse flow of propellant. Heat exchangers transfer heat from the propellant to the helium so that the gas and the propellant will be the same temperature in the tanks. Relief valves open to vent the gas if pressure in the tanks becomes too high.

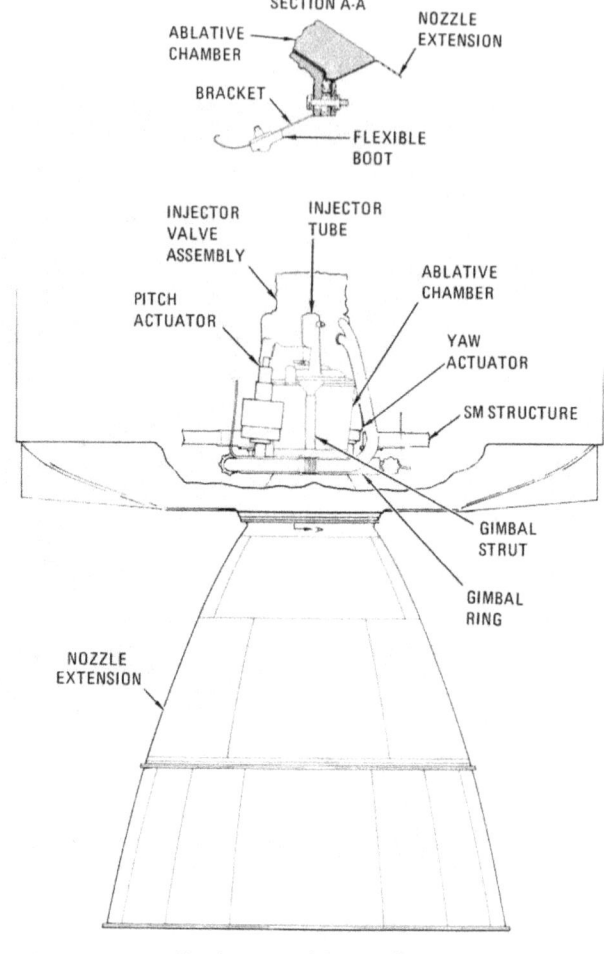

P-211 *Service propulsion engine*

Propellant quantity is measured by primary and secondary sensing systems that are completely independent. Accurate measurement is possible only during periods of thrusting.

EQUIPMENT

Rocket Engine (Aerojet-General Corp.) — The ablative combustion-chamber engine is mounted in the center section of the service module. It is conical shaped and gimballed. The engine and nozzle extension has an overall length of about 12 feet 10 inches and weighs about 650 pounds. The nozzle extension, made of columbium and titanium, is radiant-cooled, is more than 9 feet long, and has an exit diameter of about 7 feet. The engine has a nominal 20,500-pound thrust. Its service life is 750 seconds and can be fired for a minimum of 0.4 second. It can be restarted 50 times. Nitrogen tetroxide is the oxidizer. Fuel is a

Titanium fuel tank is installed in service module at Space Division's clean room, Downey, Calif.

blended hydrazine. The engine provides for velocity changes along the X axis of the spacecraft.

Helium Tanks — Two spherical pressure vessels are mounted in the center section of the service module. Each has an internal volume of 19.4 cubic feet. Maximum operating pressure is 4400 psia; normal pressure is 3600 psia. Aluminum alloy walls are 0.46 inch thick; tank diameter is 40 inches. The tanks store helium for pressurization of service propulsion system propellant tanks.

Oxidizer Tanks (General Motors Corp.'s Allison Division, Indianapolis, Ind.) — Two hemispherically domed cylindrical tanks located in sectors 2 and 5 of the service module. Tanks (one storage and one sump) are made of titanium. The storage tank is about 13 feet long with an inside diameter of 45 inches. The sump tank is slightly shorter, with an inside diameter of 51 inches. The wall thickness of both tanks is 0.054 inch. Internal volume of each tank is 175 cubic feet, and the nominal working pressure is 175 psia. Storage tank capacity is 11,285 pounds; sump tank is 13,924 pounds. The tanks store propellants during non-thrusting periods and supply propellants during thrusting.

Quantity Sensing System (Simmonds Precision Products, Tarrytown, N.Y.) — Primary system consists of cylindrical capacitance probes 13 feet long located in the tanks. When the liquid level changes, the probe signals register on the command module displays. The auxiliary system consists of point sensors that send signals when a point sensor is uncovered. Oxidizer tank probes are 2-plate capacitance probes. Fuel tank probe is a Pyrex glass rod. The system operates only during thrusting by measuring liquid level by height.

DETAILED DESCRIPTION

Principal components of the service propulsion subsystem include the engine, the pressurization subsystem, the propellant subsystem, the propellant utilization and gauging subsystem, and the flight combustion stability monitor.

ENGINE

The engine is a non-throttleable rocket engine which burns hypergolic propellant to produce a thrust of 20,500 pounds in vacuum. It is 3 feet 5 inches long and is mounted in the center section of

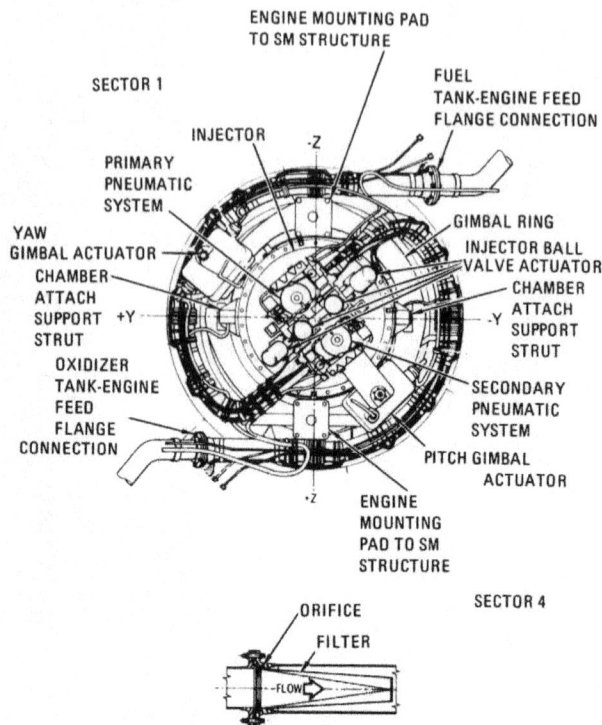

P-213   *Top view of service propulsion engine*

the service module. It has a nozzle extension 9 feet 4 inches long which extends out the aft (bottom) of the SM. The bell-shaped nozzle extension, which is made of columbium and titanium, has an exit diameter of 7 feet 10-1/2 inches and is cooled by radiation (dissipating its heat to space). Total weight of the engine is about 650 pounds.

The engine's combustion chamber is lined with an ablative (heat resistant) material which extends from the injector attachment pad to the nozzle extension. The ablative material consists of a liner, a layer of insulation, and metal attachment flanges for mounting the injector.

The engine injector is bolted to the ablative thrust chamber. Propellant is distributed through concentric annuli machined orifices in the injector assembly which are covered by concentric closeout rings. Alternate radial manifolds welded to the back of the injector body distribute propellant to the annuli. The injector is baffled to provide combustion stability.

The engine has no ignitor, since the propellant is hypergolic. Fuel and oxidizer are injected into the combustion chamber when they impinge, atomize, and ignite. The engine can fire for nearly 8-1/2 minutes and can be restarted as many as 36 times.

The bipropellant valve assembly consists of two nitrogen pressure vessels, two injector prevalves, two nitrogen regulators, two nitrogen relief valves, four solenoid control valves, four actuators, and eight bipropellant ball valves.

The nitrogen tanks are mounted on the bipropellant valve assembly to supply pressure to the injector prevalves. One tank is in the primary pneumatic control system (A) and the other tank is in the secondary pneumatic control system (B). The tanks each contain 5.8 cubic inches on nitrogen — enough to operate the valves 43 times with an initial nominal pressure of 2500 psi.

The injector prevalves are two-position, solenoid-operated valves, one for each pneumatic control system, and identified as A and B. The valve is energized open and spring-loaded closed. The prevalves, when energized open, allow nitrogen supply tank pressure to be directed through the regulator into a relief valve and to a pair of solenoid control valves.

The single-stage regulator is installed in each pneumatic control system between the injector prevalves and the solenoid control valves. The regulator reduces the nitrogen pressure to 190-230 psi. The pressure relief valve is located downstream from the regulator to limit the pressure applied to the solenoid control valves in case a regulator malfunctions. The orifice between the injector prevalve and regulator restricts the flow of nitrogen and allows the relief valve to relieve the pressure overboard in the event the regulator malfunctions, preventing damage to the solenoid control valves and actuators.

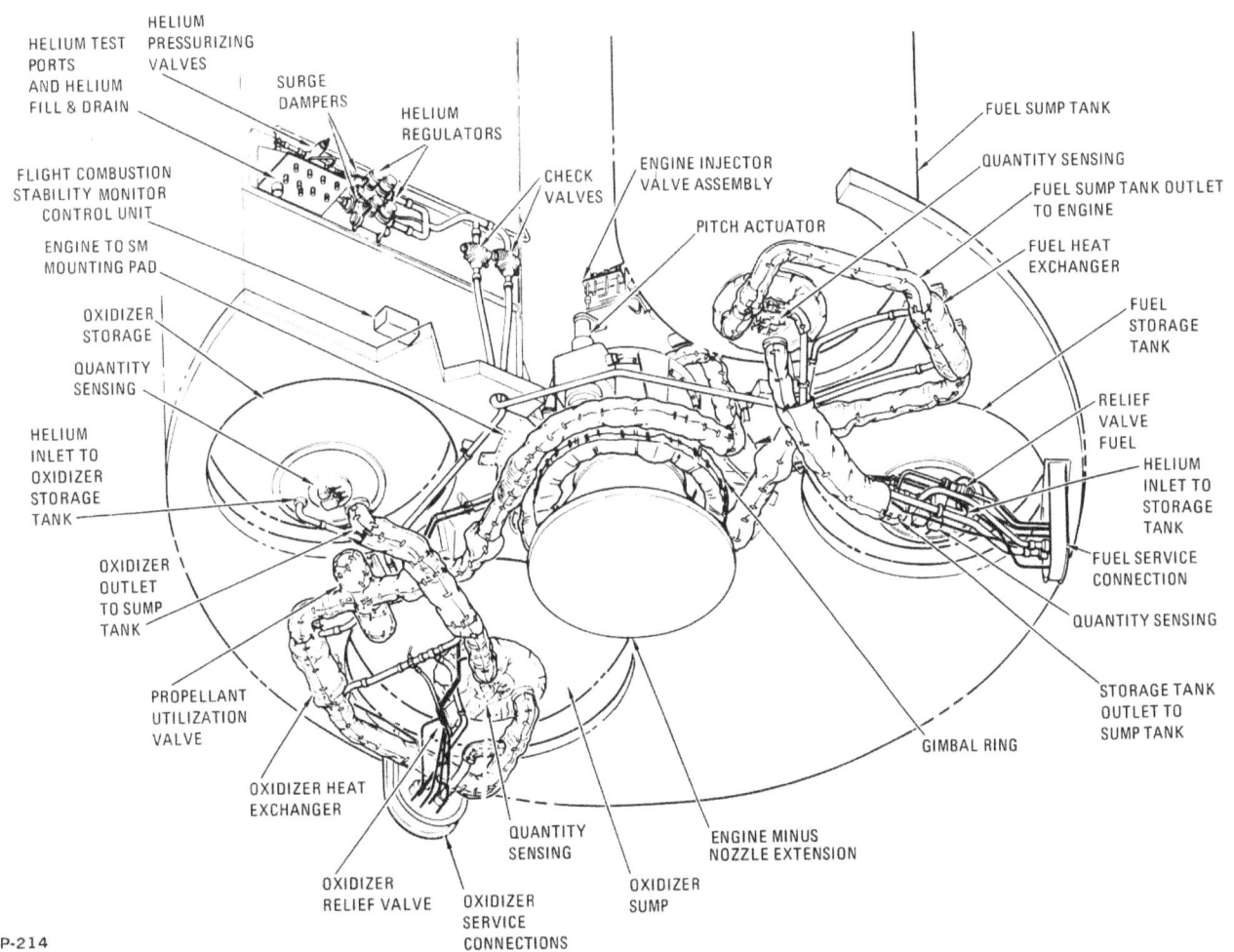

Aft view of service propulsion engine components and line insulation

Four solenoid-operated, three-way, two-position control valves are used for actuator control. Two solenoid control valves are located in each pneumatic control system. The solenoid control valves in the primary system are identified as 1 and 2 and the two in the secondary system are identified as 3 and 4. The solenoid control valves in the primary system control actuator and ball valves 1 and 2. The two solenoid control valves in the secondary system control actuator and ball valves 3 and 4.

Four piston-type, pneumatically operated actuators control the eight propellant ball valves. Each actuator piston is mechanically connected to a pair of propellant ball valves, one fuel and one oxidizer. When the solenoid control valves are opened, pneumatic pressure is applied to the opening side of the actuators. The spring pressure on the closing side is overcome and the actuator piston moves. Utilizing a rack and pinion gear, linear motion of the actuator connecting arm is converted into rotary motion, which opens the propellant ball valves. When the engine firing signal is removed from the solenoid control valves, the solenoid control valves close, removing the pneumatic pressure source from the opening side of the actuators. The actuator spring pressure then forces the actuator piston to move in the opposite direction, causing the propellant ball valves to close. The piston movement forces the remaining nitrogen on the opening side of the actuator back through the solenoid control valves where it is vented overboard.

Each actuator contains a pair of linear position transducers. One supplies information on the position of the ball valve to the main display console and the other information to telemetry.

The eight propellant ball valves are used to distribute fuel and oxidizer to the engine injector assembly. Four linked pairs, each pair consisting of one fuel and one oxidizer ball valve, are controlled by single actuator, and arranged in a series-parallel configuration. The parallel redundancy assures the engine ignition; the series redundancy assures thrust termination. When nitrogen pressure is applied to the actuators, each propellant ball valve is rotated, aligning the ball to a position that allows propellant to flow to the engine injector assembly. The mechanical arrangement is such that the oxidizer ball valves maintain an 8-degree lead over the fuel ball valves upon opening, which results in smoother engine starting.

Check valves are installed in the vent port outlet of each of the four solenoid control valves, spring pressure vent port of the four actuators, and the ambient vent port of the two nitrogen pressure regulator assemblies to protect the seals of these components from the hard vacuum of space.

The thrust mount assembly consists of a gimbal ring, engine-to-vehicle mounting pads, and gimbal ring-to-combustion chamber assembly support struts. The thrust structure is capable of providing ±10 degrees inclination about the Z axis (yaw) and ±6 degrees about the Y axis (pitch).

Thrust vector (direction of thrust) control of the service propulsion engine is achieved by dual, servo, electromechanical actuators. The gimbal actuators can provide control around the Z axis (yaw) of ±4.5 degrees in either direction from a +1-degree offset, and around the Y axis (pitch) of ±4.5 degrees in either direction from a -2-degree null offset. The reason for the offset is the offset center of mass of the SM.

Each actuator assembly consists of four electromagnetic particle clutches, two dc motors, a bull gear, jack-screw and ram, ball nut, two linear position transducers, and two velocity generators. The actuator assembly is a sealed unit and encloses those portions protruding from the main housing. One motor and a pair of clutches (extend and retract) in each actuator are identified as System 1 and the other as System 2.

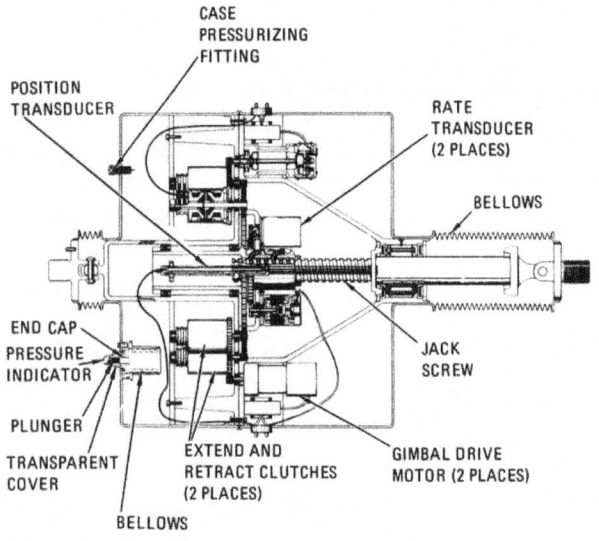

P-215 *Engine gimbal actuator*

An overcurrent relay in each primary and secondary gimbal motor is controlled by a switch on the main display console. Power from the battery bus is applied to the motor-driven switch within the overcurrent relay of the primary or secondary system. One of the motor switch contacts then supplies power from the main bus to the gimbal motor. When the switch is released, it spring-loads to the "on" position which activates the overcurrent sensing circuitry of the primary or secondary relay that monitors the current to the gimbal motor.

The overcurrent relay of the primary or secondary system is used to monitor current to the gimbal motor for variable current flow during the initial gimbal motor start and normal operation for main dc bus and gimbal motor protection.

Using No. 1 yaw system as an example, the operation of the current monitoring system follows this sequence: If the relay senses an overcurrent to gimbal motor No. 1, the monitor circuitry within

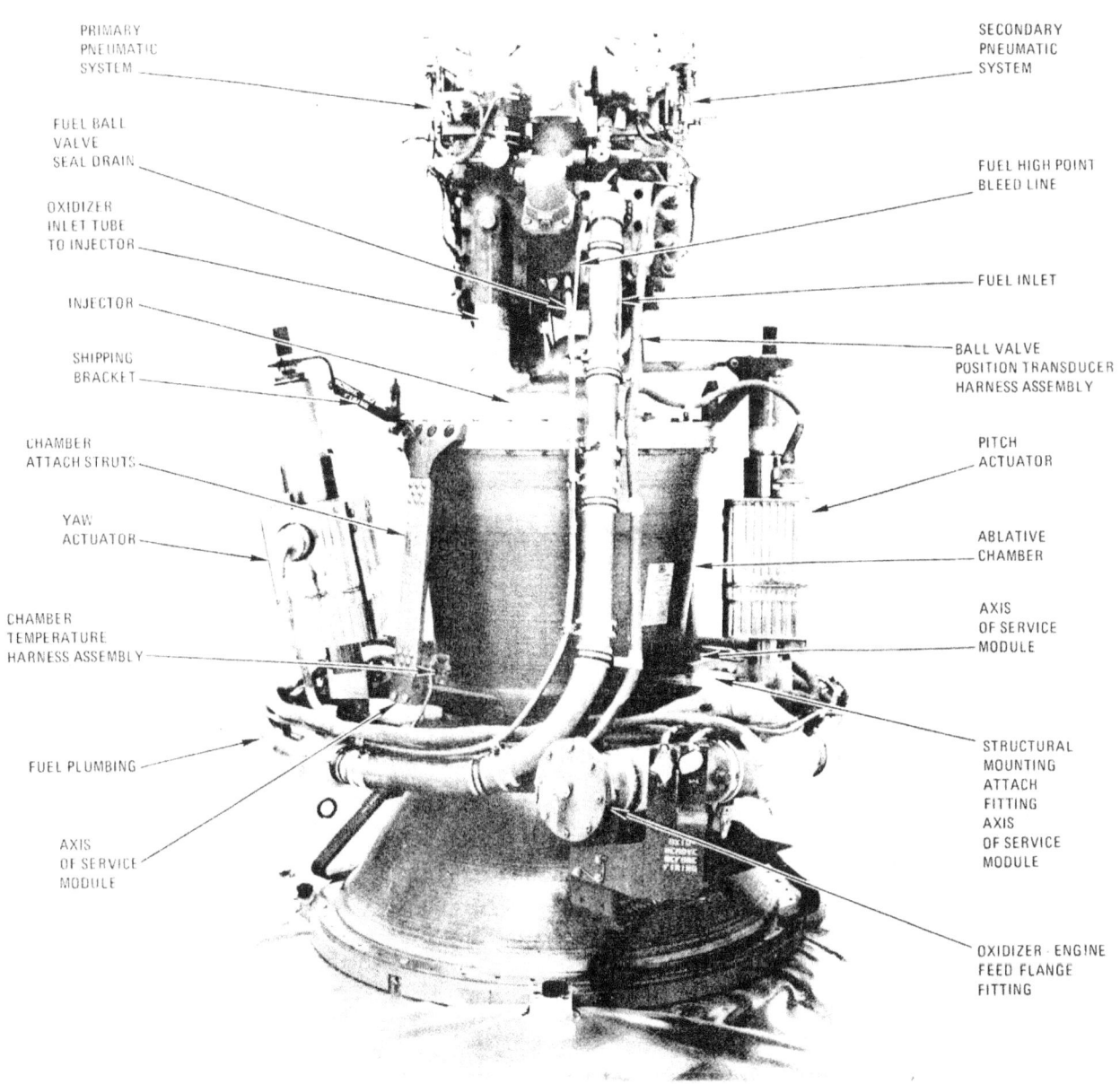

Service propulsion engine

the relay will drive the motor-driven switch, removing power from gimbal motor No. 1. Simultaneously a fail signal is sent from the relay to the stabilization and control subsystem which opens upper relay contacts to remove inputs from the No. 1 clutches and closes lower relay contacts to apply inputs to the No. 2 clutches within the same actuator. Simultaneously a signal is sent to illuminate a caution and warning light to indicate the primary gimbal motor has failed.

The crew would then switch on the No. 2 yaw system. This would apply power to the motor-driven switch from battery Bus B within the over-current relay of the secondaries. The motor switch then supplies power from Main Bus B through the motor switch contact to the secondary gimbal motor. When the secondary switch is released, it spring-loads to the "on" position which activates the overcurrent sensing circuitry for the secondary. If the relay senses an overcurrent to gimbal motor No. 2, the monitor circuitry within the relay will drive the motor-driven switch, removing power from the motor. There is no fail signal in this case; however, the yaw No. 2 caution and warning light will illuminate to inform of secondary gimbal motor failure. If the No. 2 system has failed due to an overcurrent, that specific actuator is inoperative if the No. 1 system has previously failed.

This switching procedure is controlled automatically when the gimbal drive switches are turned to automatic operation.

The clutches are of a magnetic-particle type. The gimbal motor drive gear meshes with the gear on the clutch housing. The gears on each clutch housing mesh and as a result, the clutch housings counter-rotate. The current input is applied to the electromagnet mounted to the rotating clutch housing from the stabilization and control subsystem, the CM computer, or the manual control. A quiescent current may be applied to the electromagnet of the extend and retract clutches, preventing any movement of the engine during the boost phase of the mission with the gimbal motors off. The gimbal motors will be turned on before jettisoning the launch escape tower to support the service propulsion subsystem abort after the launch escape tower has been jettisoned. It will be turned off again as soon as possible to reduce the heat increase that occurs due to the gimbal motor driving the clutch housing with quiescent current applied to the clutch.

Before any stabilization and control subsystem or manual thrusting period the thumbwheels are set to position the engine. The thumbwheels are for backup; they do not position the engine. In any thrusting mode, the current input required for a gimbal angle change (to maintain the engine thrust vector through the center of mass) to the clutches will increase above the quiescent current. This increases the current into the electromagnets that are rotating with the clutch housings. The dry powder magnetic particles can become magnetized or demagnetized readily. The magnetic particles increase the friction force between the rotating housing and the flywheel, causing the flywheel to rotate. The flywheel arrangement is attached to the clutch output shaft allowing the clutch output shaft to drive the bull gear. The bull gear drives a ball nut which drives the actuator jackshaft to an extend or retract position, depending on which clutch housing electromagnet the current input is supplied to. The larger the excitation current, the higher the clutch shaft rotation rate.

Meshed with the ball nut pinion gear are two tachometer-type rate transducers. When the ball nut is rotated, the rate transducer supplies a feedback to the summing network of the thrust vector

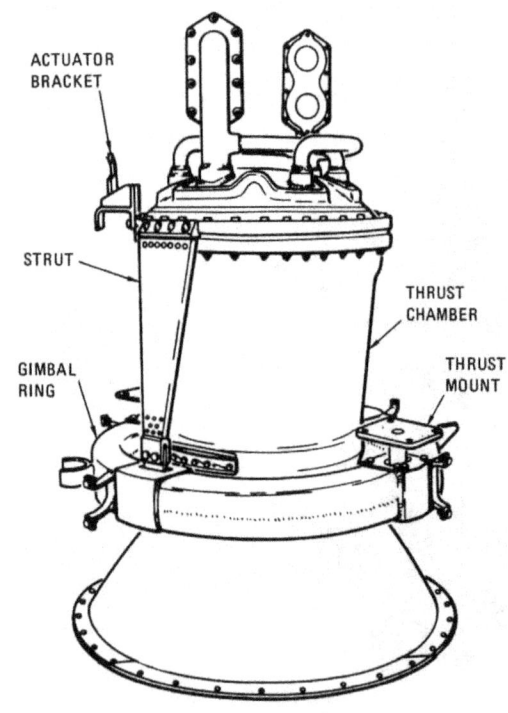

P-217 *Engine thrust chamber*

control logic to control the driving rates of the jackscrew (acting as a dynamic brake to prevent over- and undercorrecting). There is one rate transducer for each system.

The jackscrew contains two position transducers arranged for linear motion and connected to a single yoke. The position transducers are used to provide a feedback to the summing network and the CM displays. The operating system provides feedback to the summing network that reduces the output current to the clutch, resulting in proportional rate change to the desired gimbal angle position and a return to a quiescent current, in addition to providing a signal to the display. The remaining position transducer provides a feedback to the redundant summing network of the thrust vector logic for the redundant clutches, in addition to the display if the secondary system is the operating system.

A snubbing device provides a hard stop for an additional 1-degree travel beyond the normal gimbal limits.

Twelve electrical heaters are used in the engine propellant feed line brackets and bipropellant valve assemblies. The heaters are controlled by the crew and are used to keep the feedlines from getting too cold and freezing the propellant. Displays in the

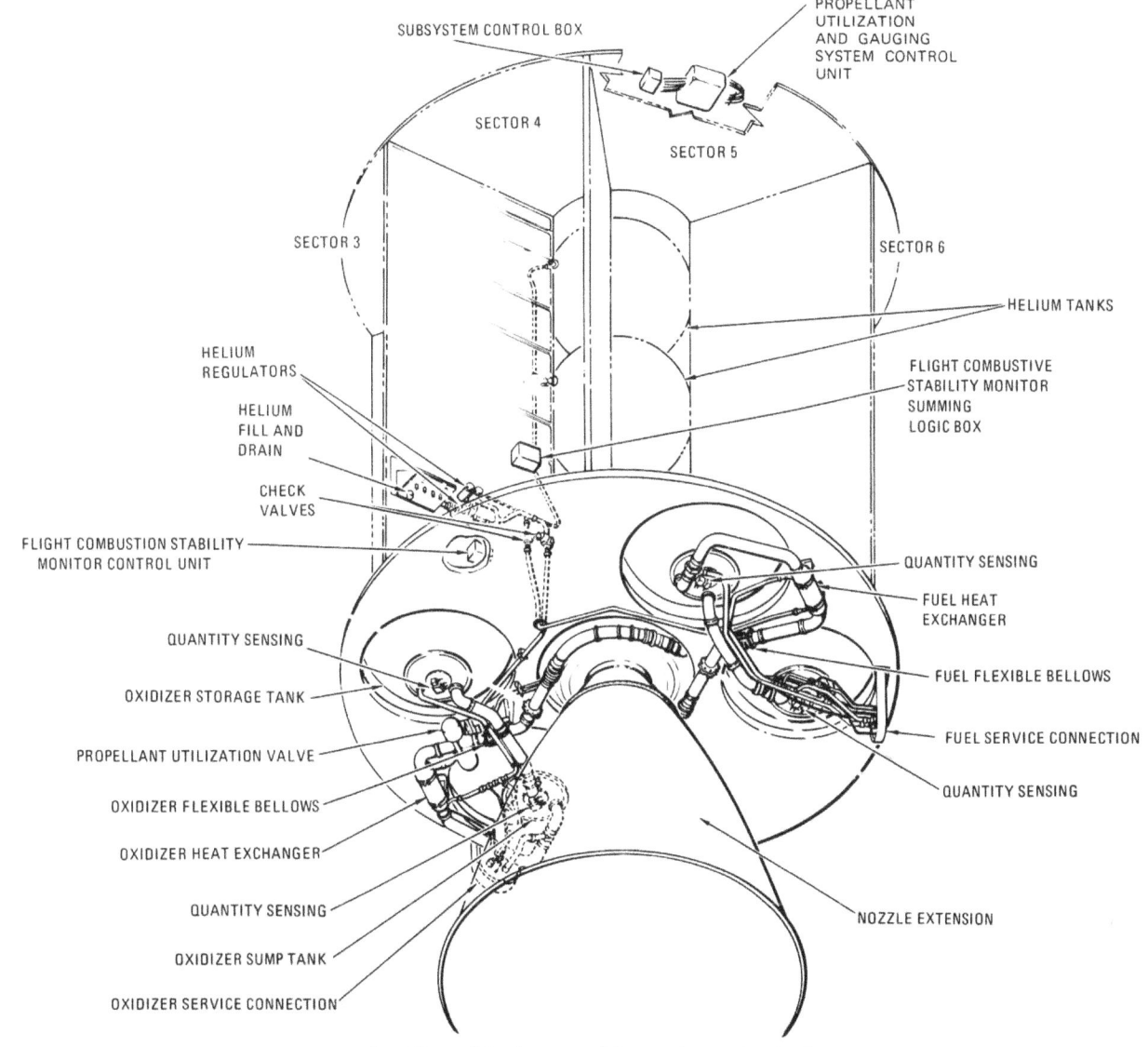

Location of service propulsion engine components

crew compartment show the feedline temperature and swiches on the main display console control the heaters.

PRESSURIZATION SUBSYSTEM

The pressurization subsystem consists of two helium tanks, two helium pressurization valves, two dual pressure regulator assemblies, two dual check valve assemblies, two pressure relief valves, and two heat exchangers. The critical components are redundant to increase reliability.

The two helium supply spherical pressure vessels contain 19.6 cubic feet of gas each, pressurized to 3600 psia. The tanks are located in the center section of the SM above the engine.

The helium valves are continuous-duty, solenoid-operated type. The valves are energized open and spring-loaded closed and can be controlled automatically or manually. An indicator above each valve control switch on the main display console shows the position of the valve. When the valves are closed, the indicator shows diagonal lines (the indication during non-thrusting periods). When the valves are open, the indicator shows gray (the indication during thrusting periods).

Pressure is regulated by an assembly downstream from each helium pressurizing valve. Each assembly contains a primary and secondary regulator in series, and a pressure surge damper and filter installed on the inlet to each regulating unit. These regulators reduce the pressure of the helium gas from 3600 psia to 186 psia nominal.

The primary regulator is normally the controlling regulator. The secondary regulator is normally open during a dynamic flow condition. The secondary

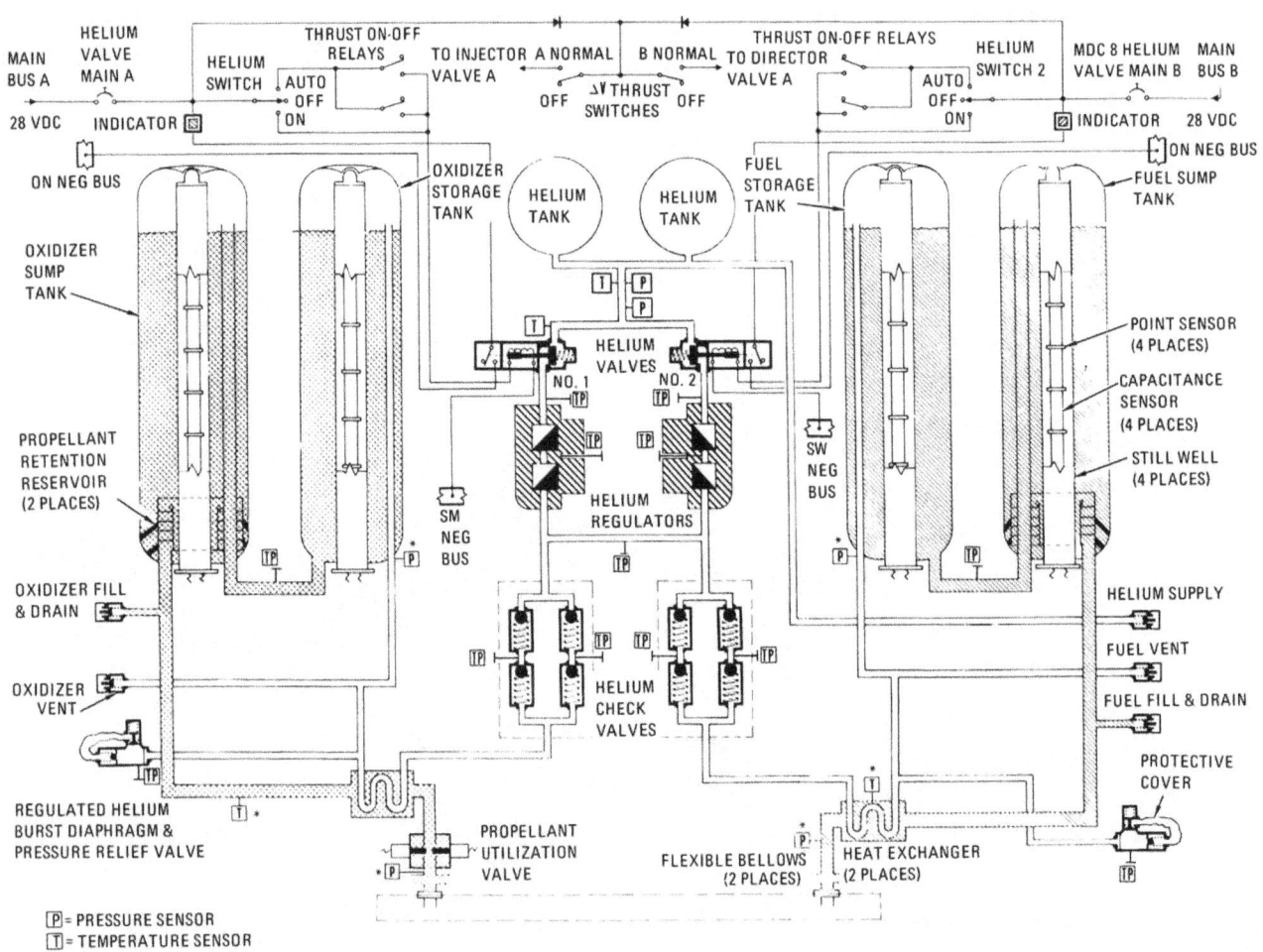

Schematic of service propulsion subsystem's propellant tank system

regulator will not become the controlling regulator until the primary regulator allows a higher pressure than normal. All regulator pressures are in reference to a diaphragm assembly that is vented to ambient.

Only one of the parallel regulator assemblies regulates helium pressure under dynamic conditions. The downstream pressure causes the second assembly to lock up (close). When the regulated pressure decreases below the lockup pressure of the non-operating assembly, that assembly becomes operational.

Each assembly contains four independent check valves connected in a series-parallel configuration for added redundancy. The check valves prevent the reverse flow of propellant liquid or vapor and permit helium pressure to be directed to the propellant tanks.

The pressure relief valves consist of a relief valve, a diaphragm, and a filter. In the event that excessive helium or propellant vapor ruptures the diaphragm, the relief valve opens and vents the system. The relief valve will close and reseal after the excessive pressure has returned to the operating level. The diaphragm provides a more positive seal of helium than a relief valve. The filter prevents any fragments from the diaphragm from entering onto the relief valve seat. The relief valve opens at a pressure of 212 psi after the diaphragm ruptures at about 213 psi. The valve will close when pressure drops to 208 psi.

A pressure bleed device is incorporated between the diaphragm and relief valve. The bleed valve vents the cavity between the diaphragm and relief valve in the event of any leakage from the diaphragm. The bleed device is normally open and will close when the pressure increases to 150 psi.

The line-mounted, counterflow heat exchangers consist of the helium pressurization line coiled

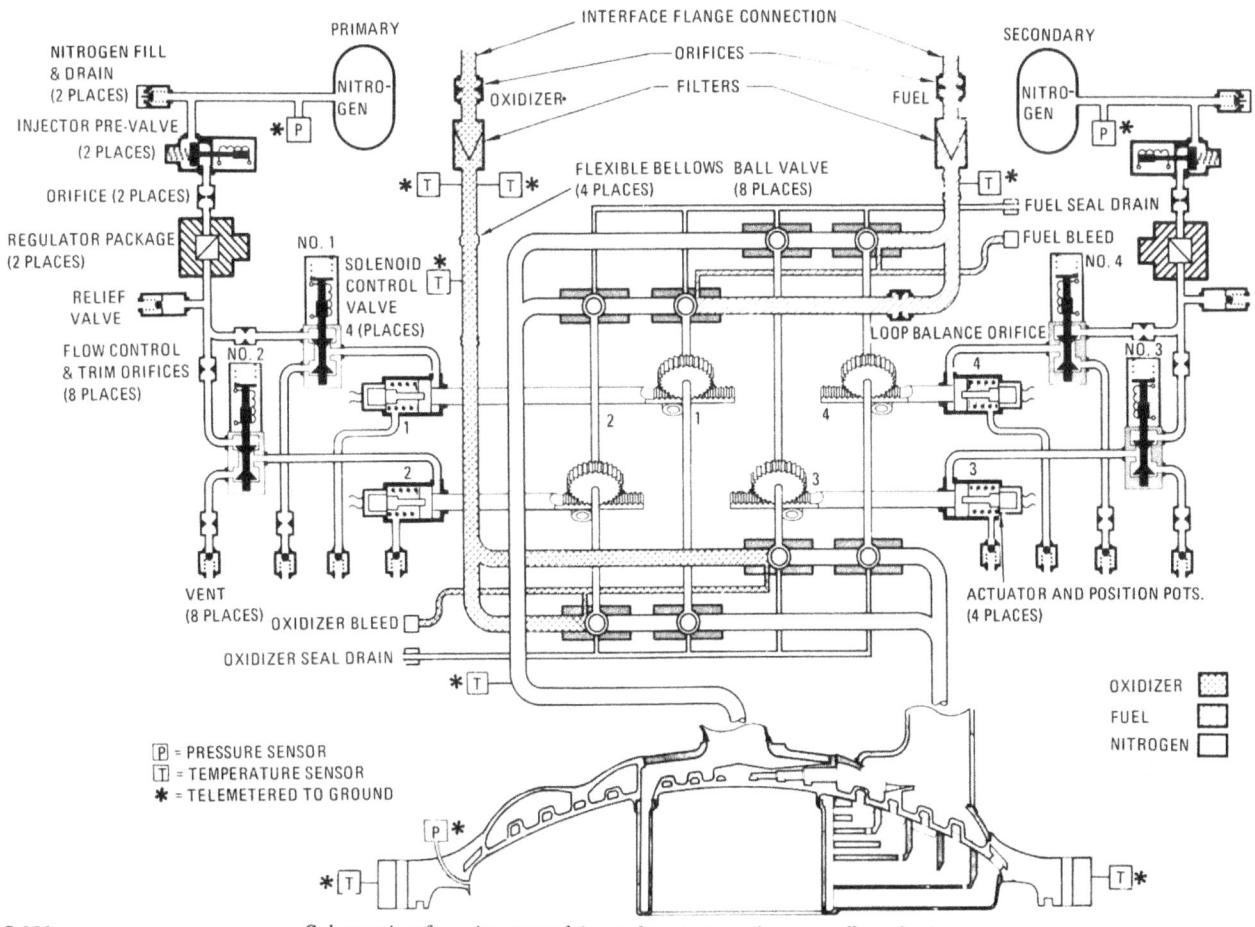

Schematic of service propulsion subsystem engine propellant feed

within an enlarged section of the propellant supply line. The helium gas, flowing through the coiled line, absorbs heat from the propellant and approaches the same temperature when it gets to the tanks.

PROPELLANT SUBSYSTEM

This subsystem consists of two fuel tanks (storage and sump), two oxidizer tanks (storage and sump), and propellant feed lines.

The propellant supply is contained in four hemispherical-domed cylindrical tanks constructed of titanium. The tanks each occupy a different sector of the service module. The storage tanks are pressurized by helium. An outlet transfers the propellant and helium from the storage tanks through transfer lines to the sump tanks. Standpipes in the sump tanks allow the propellant and helium from the storage tanks to pressurize the sump tanks. The propellant in the sump tanks flows through an umbrella screen assembly; into a retention reservoir, to the outlet, and to the engine.

The umbrella-shaped screen assembly and retention reservoir are installed in the exit end of the sump tanks. The reservoir retains a quantity of propellant at the tank outlet and in the engine plumbing during zero gravity. Normal engine ignition when the sump tanks are full is accomplished without an ullage maneuver. For all other conditions, an ullage maneuver is performed before engine ignition to assure that gas is not trapped below the screens.

The propellant feed lines contain flexible bellows assemblies to align the tank feed plumbing to the engine plumbing.

PROPELLANT UTILIZATION AND GAUGING SUBSYSTEM

The subsystem consists of a primary and auxiliary sensing system, a propellant utilization valve, a control unit, and a display unit.

Propellant quantity is measured by two separate sensing systems: primary and auxiliary. The primary quantity sensors are cylindrical capacitance probes mounted the length of each tank. In the oxidizer tanks, the probes consist of a pair of concentric electrodes with oxidizer used as the dielectric. In the fuel tanks, a Pyrex glass probe, coated with silver on the inside, is used as one conductor of the capacitor. Fuel on the outside of the probe is the other conductor. The Pyrex glass forms the dielectric. The auxiliary system utilizes point sensors mounted at intervals along the primary probes to provide a step function impedance change when the liquid level passes their location.

Propellant is measured by the primary system through the probe's capacitance, a function of propellant height.

The auxiliary propellant measurement system uses seven point sensors in the storage tanks and eight in the sump tanks. The point sensors consist of concentric metal rings which present a variable impedance depending on whether they are covered or uncovered by the propellants remaining and are integrated by a rate flow generator which integrates the servos at a rate proportional to the nominal flow rate of the fuel and oxidizer. A mode selector senses when the propellant crosses a sensor and changes the auxiliary servos from the flow rate generator mode to the position mode. The system moves to the location specified by the digital-to-analog converter for 0.9 second to correct for any difference. The system then returns to the flow rate

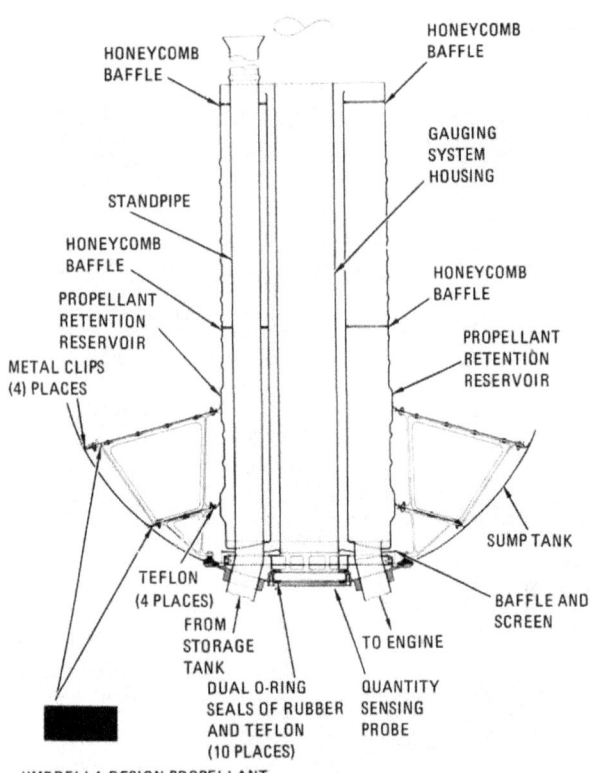

P-221 *Propellant retention reservoir*

generator mode until the next point sensor is reached. A non-sequential pattern detector detects false or faulty sensor signals. If a sensor has failed, the information from that sensor is blocked from the system, preventing disruption of system computation.

Propellant flow rate is converted in a quantity measurement and transmitted to displays on the main display console and to telemetry. These displays are updated during thrusting as point sensors are uncovered. Any deviation from the nominal oxidizer to fuel ratio (1.6:1 by mass) is displayed by the "unbalance" indicator in pounds. The indicator is marked to identify the required change in oxidizer flow rate to correct any unbalance condition.

When the sensor switches are set normally, the output of both sensor systems is continually compared in the comparator network. If a differential of 3 ±1 percent occurs between total primary and total auxiliary fuel or oxidizer, a caution and warning indicator is illuminated. The output of the oxidizer sump tank servoamplifier and the primary

Service propulsion engine is prepared for installation in service module

potentiometer of the unbalance indicator are compared in the comparator network and if a nominal 600 pounds or a critical unbalance is reached, a caution and warning light is illuminated

When the primary or auxiliary sensor system is selected on the switch, the output of the oxidizer sump tank servoamplifier and the output the primary or auxiliary potentiometer (whichever system is selected) in the unbalance meter are compared in the comparator network and if a nominal 600 pounds or a critical unbalance is reached, the caution and warning light is illuminated.

Once the warning light is illuminated, the crew can determine whether there is a malfunction within the quantity and indicating systems or if there is a true unbalance condition existing by use of a self-test portion of the system.

If an unbalance condition exists, the crew will use the propellant utilization valve to return the propellants to a balanced condition. The propellant utilization valve housing contains two sliding gate valves within the housing. One of the sliding gate valves is primary, and the other is secondary. Stops are provided in the valve housing for the full increase or decrease portions of the primary and secondary sliding gate valves.

The secondary propellant utilization valve has twice the travel of the primary propellant utilization valve to compensate for primary propellant utilization valve failure in any position. The propellant utilization valve controls are on the main display console.

FLIGHT COMBUSTION STABILITY MONITOR

The flight combustion stability monitor accelerometers are mounted to the SPS engine injector to monitor the engine for vibration buildup characteristic of combustion instability.

Three accelerometers in the monitor package provide signals to a box assembly which amplifies them. When the vibration level exceeds 180 ± 18 g's peak-to-peak for 70 ± 20 milliseconds, a level detector triggers a power switch which sends power to the summing logic.

The summing logic will trip if there are two or more rough combustion signals received; the normally closed contacts will open removing power from the inverter in the thrust logic and shutting down the engine. At the same time, a caution and warning light will illuminate. The engine would not shut down in the manual thrust mode.

Trigger circuits in the flight combustion stability monitor provide power to the voting logic relay coils continuously once unstable combustion is sensed. During CM computer or stabilization and control subsystem thrust, power is still applied to the voting logic relays even though the engine is shut down and combustion instability no longer can be sensed.

The flight combustion stability monitor can be reset after triggering engine shutdown, and it also can be bypassed by the crew so that it will not automatically shut down the engine.

TELECOMMUNICATIONS

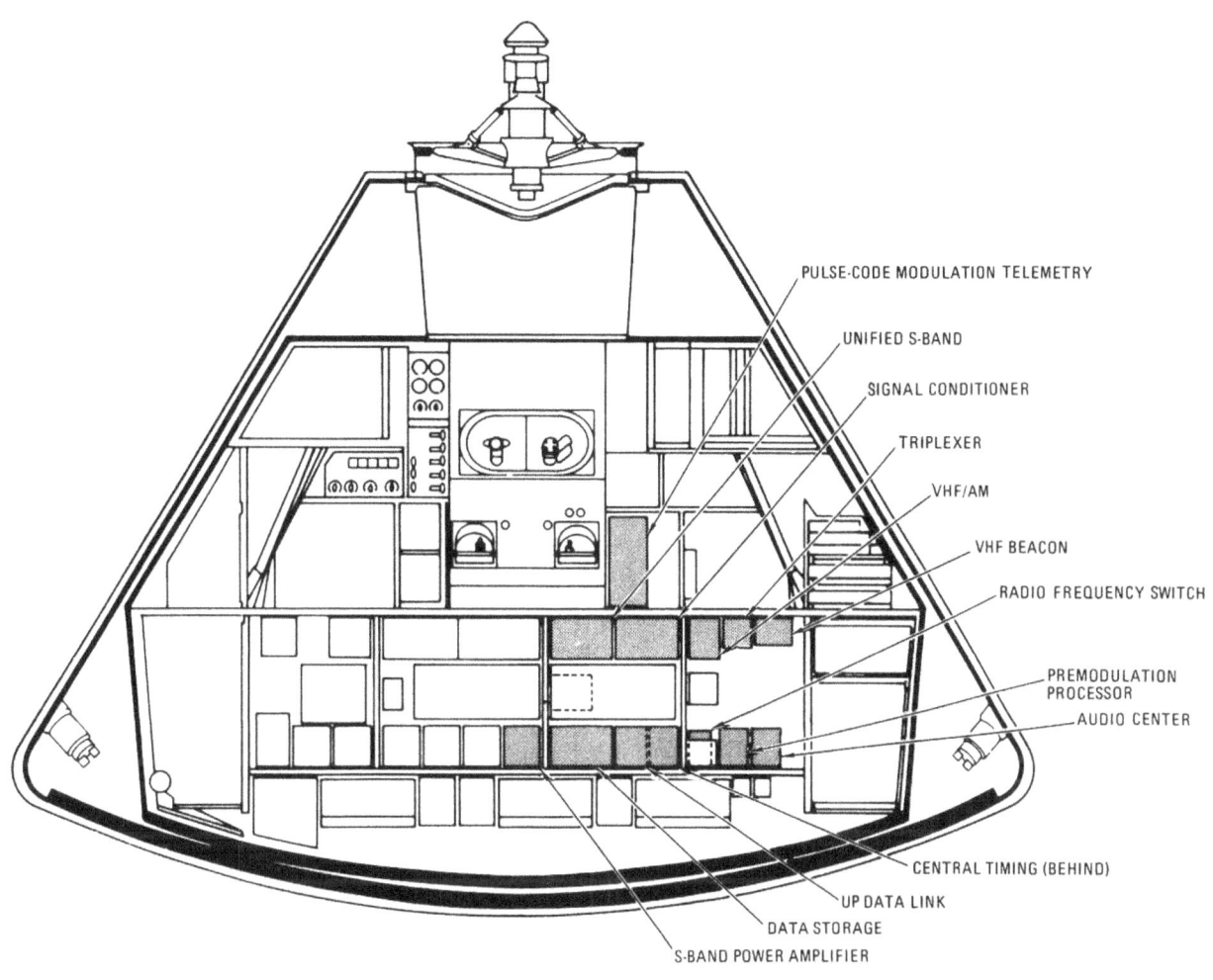

Location of main telecommunications equipment

The telecommunications subsystem provides voice, television, telemetry, and tracking and ranging communications between the spacecraft and earth, between the CM and LM, and between the spacecraft and astronauts wearing the portable life support system. It also provides communications among the astronauts in the spacecraft and includes the central timing equipment for synchronization of other equipment and correlation of telemetry equipment.

For convenience, the telecommunications subsystem can be divided into four areas: intercommunications (voice), data, radio frequency equipment, and antennas. Most of the components of the telecommunications subsystem are produced by the Collins Radio Co., Cedar Rapids, Iowa.

INTERCOMMUNICATIONS

The astronauts headsets are used for all voice communications. Each headset has two independently operated earphones and two microphones with self-contained pre-amplifiers. Each astronaut has an audio control panel on the main display console which enables him to control what comes into his headset and where he will send his voice. The headsets are connected to the audio panels by separate umbilical cables. These cables also contain wiring for the biomedical sensors in the constant-wear garment.

The three headsets and audio control panels are connected to three identical audio center modules.

The audio center is the assimilation and distribution point for all spacecraft voice signals. The audio signals can be routed from the center to the appropriate transmitter or receiver, the launch control center (for pre-launch checkout), the recovery forces intercom, or voice tape recorders.

Two methods of voice transmission and reception are possible: the VHF/AM transmitter-receiver and the S-band transmitter and receiver. Transmission is controlled by either the push-to-talk switch located in the astronaut's umbilical cable or the voice-operated relay circuitry during recovery operations. The push-to-talk switch also can be used like a telegraph key for emergency transmission.

The VHF/AM equipment is used for voice communications with the manned space flight network during launch, ascent, and near-earth phases of a mission. The S-band equipment is used during both near-earth and deep-space phases of a mission. When communications with earth are not possible, a limited amount of audio signals can be stored on tape. During recovery, the VHF/AM and recovery intercom equipment are used to maintain contact with ground stations and with frogmen.

DATA

The spacecraft structure and subsystems contain sensors which gather data on their status and performance. Biomedical, TV, and timing data also is gathered. These various forms of data are assimilated into the data system, processed, and then transmitted to the ground. Some data from the operational systems, and some voice communications, may be stored for later transmission or for recovery after landing. Stored data can be transmitted to the ground simultaneously with voice or real-time data.

Signals from some of the instrumentation sensors are fed into signal conditioning (converting) equipment. These signals, and others which are already conditioned or don't need to be, are then sent to a

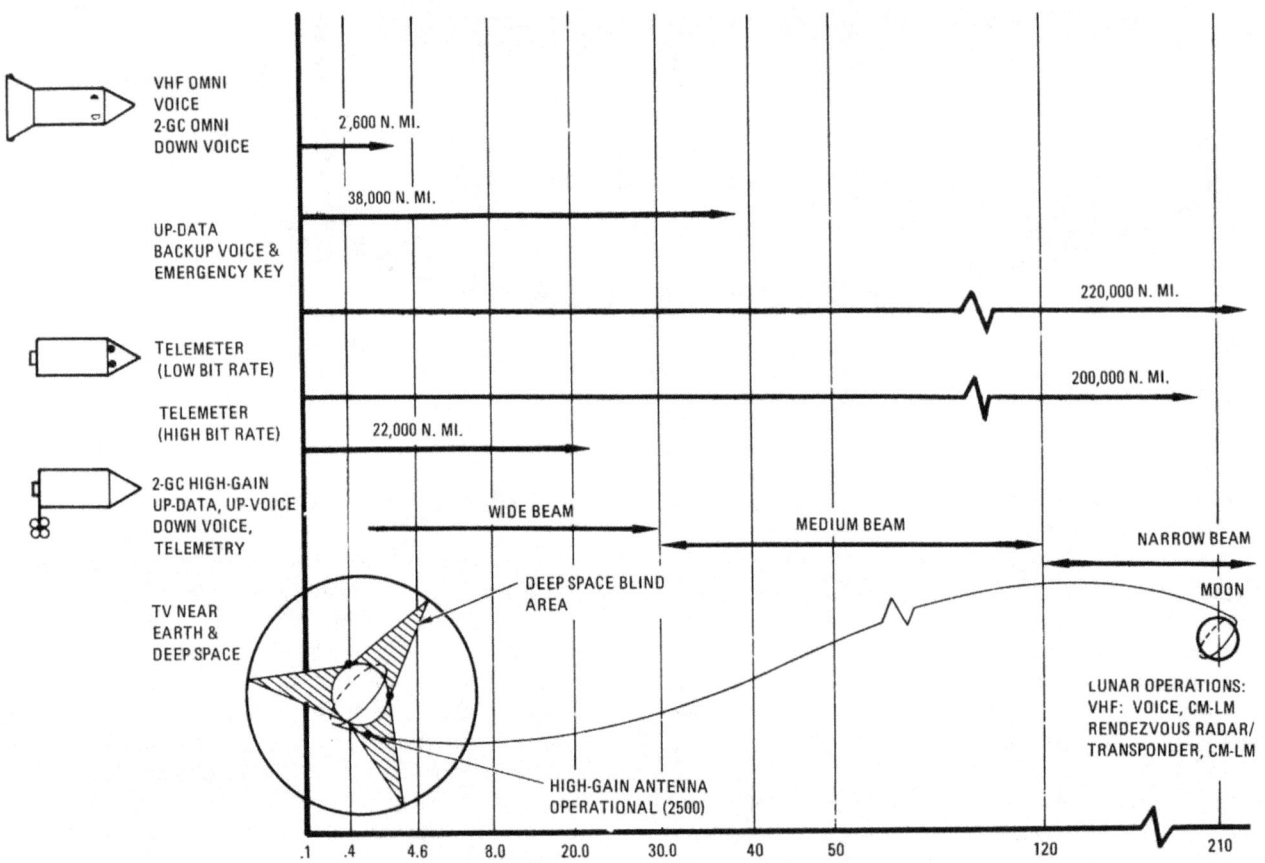

CSM communication ranges

data distribution panel, which routes them to CM displays and to pulse-code modulation telemetry equipment. The latter combines them into a single signal and sends it to the premodulation processor.

The premodulation processor is the assimilation, integration, and distribution center for nearly all forms of spacecraft data. It accepts signals from telemetry, data storage, TV, central timing, and audio center equipment. It modulates, mixes, and switches these signals to the appropriate transmitter or to data storage.

Voice and data command signals from the ground received over the S-band receiver also are supplied to the processor, which routes them to the audio center equipment or the up-data link (ground command) system. Up-data is of three types: guidance and navigation data for updating the CM computer, timing data for updating the central timing equipment, and real-time commands. The commands give the ground limited control over certain spacecraft telecommunications functions.

RADIO FREQUENCY EQUIPMENT

The radio frequency equipment is the means by which voice information, telemetry data, and ranging and tracking information are transmitted and received. The equipment consists of two VHF/AM transceivers (transmitter-receiver) in one unit, the unified S-band equipment (primary and secondary transponders and an FM transmitter), primary and secondary S-band power amplifiers (in one unit), a VHF beacon, an X-band transponder (for rendezvous radar), and the premodulation processor.

The equipment provides for voice transfer between the CM and the ground, between the CM and LM, between the CM and extravehicular astronauts, and between the CM and recovery forces. Telemetry can be transferred between the CM and the ground, from the LM to the CM and then to the ground, and from extravehicular astronauts to the CM and then to the ground. Ranging information consists of pseudo-random noise and double-doppler ranging signals from the ground to the CM and back to the ground and X band radar signals from the LM to the CM and back to the LM. The VHF beacon equipment emits a 2-second signal every 5 seconds for line-of-sight direction-finding to aid recovery forces in locating the CM after landing.

ANTENNAS

There are nine antennas on the command and service modules, not counting the rendezvous radar antenna which is an integral part of the rendezvous radar transponder.

These antennas can be divided into four groups: VHF, S-band, recovery, and beacon. The two VHF antennas (called scimitars because of their shape) are omni-directional and are mounted 180 degrees apart on the service module. There are five S-band antennas, one mounted at the bottom of the service module and four located 90 degrees apart around the command module. The high-gain antenna is stowed within the SLA until the transposition and docking maneuver, when it is deployed so that it is at right angles to the module. It can be steered through a gimbal system and is the principal antenna for deep-space communications. The four S-band antennas on the command module are mounted flush with the surface of the module and are used for S-band communications during near-earth phases of mission, as well as for a backup in deep space. The two VHF recovery antennas are located in the forward compartment of the command module, and are deployed automatically shortly after the main parachutes. One of these antennas also is connected to the VHF recovery beacon.

EQUIPMENT

All communication and data system units are in the command module lower equipment bay, mounted to coldplates for cooling.

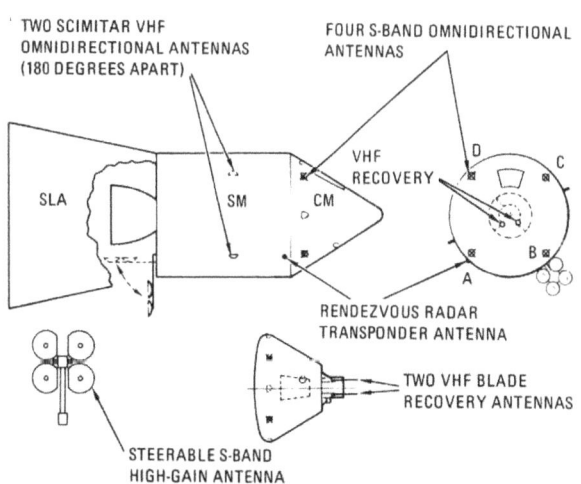

P-225 *Location of antennas*

Audio Center (Collins Radio Co., Cedar Rapids, Iowa) — Center weighs 7.9 pounds and is 4.7 by 4.56 by 8.65 inches. It is a 28-volt, 20-watt gasket-sealed box with three identical headset amplifiers, one for each crewman. It provides communication among astronauts, between astronauts and launch pad personnel, and post-landing recovery frogmen, recording of audio signals in conjunction with tape recording equipment, and relaying of audio signals.

Central Timing Equipment (General Time Corp.) — This 10-pound unit provides time correlation of all spacecraft time-sensitive functions. It also generates and stores the real-time day, hour, minute, and second mission elapsed time in binary-coded decimal format for onboard recording and transmission to the Manned Space Flight Network. It is normally synchronized to a continuously generated signal from the guidance and navigation computer. If the signal is lost, backup is provided by synchronizing to a self-contained source. The unit contains two power supplies for redundancy. Each is supplied from a different power source and through separate circuit breakers. The two power supplies provide parallel 6-volt dc outputs, either one of which is sufficient to power the entire unit.

Data Storage Equipment (Leach Corp., Azusa, Calif.) — This 40-pound unit is 22 by 9.5 by 6 inches and operates from 115-volt, 3-phase, 400-Hertz and 28-volt dc power. Tape is one-inch, 14-track Mylar. It operates at three speeds: 3.75, 15, and 120 inches per second. While being played back, the tape speed is selected automatically to Provide an apparent 51.2 KBPS PCM data. It plays in a single direction, but a rewind mode is provided. It stores data and voice information during powered phases of mission and during periods of lost communications and plays them back later.

Digital Ranging Generator (RCA) — This generator, used with the VHF/AM transmitter-receiver, supplements the lunar module rendezvous radar system by providing distance-to LM information in the CM. It has solid-state circuitry and is housed in a machined-aluminum case 4 by 6 by 8.5 inches. It weighs 6-1/2 pounds.

Premodulation Processor (Collins) — It is of solid state design and modular construction and has redundant circuitry. It weighs 14.5 pounds and is 4.7 by 6 by 10.5 inches. It requires 12.5 watts at 28 volts dc. It provides the interface connection between the spacecraft data-gathering equipment and the S-band RF electronics. It accomplishes signal modulation and demodulation, signal mixing, and the proper switching of signals so that the correct intelligence corresponding to a given mode of operation is transmitted.

Pulse-Code Modulation Telemetry Equipment (Radiation, Inc., Melbourne, Fla.) — This unit weighs 42.1 pounds and is 13 by 7 by 14 inches. The 115/200 volt, 3-phase, 400-Hertz unit has two modes of operation: high-bit rate of 51.2 kilobits per second (normal mode) and low-bit rate of 1.6 kilobits per second. It receives and samples analog, parallel digital, and serial digital information, which consists of biomedical, operation, and scientific data, and converts it to a single serial output for transmission to earth.

S-Band Flush-Mounted Antennas (Amecom Division of Litton Systems, Inc.) — Four antennas, each 3.5 inches long and 2.1 inches in diameter with 3.12-inch diameter mounting ring, are mounted on the command module. Each weighs 2.5 pounds. They are omnidirectional right-hand polarized Helix antennas in a loaded cavity. The

P-226 *Technican at Radiation, Inc., checks out pulse-code modulation telemetry unit*

astronaut selects the antenna he will use. They transmit and receive all S-band signals during near-earth operation and are backup for the high-gain antenna in deep space.

2-Gigahertz High-Gain Antenna (Textron's Dalmo Victor Company, Belmont, Calif.) — An array of four 31-inch diameter parabolas clustered around an 11-inch-square wide-beam horn. It weighs 94 pounds and is 65 by 64 by 33.78 inches. It is mounted to the aft bulkhead of the service module and is used for deep-space communications with the unified S-band equipment. Actual deployment takes place during transposition and docking phase of the mission when the spacecraft lunar adapter panels are opened. After deployment, the antenna is capable of automatically or manually tracking the RF signal within the travel limits of its gimbaling system. It has three modes of operation: wide, medium, and narrow beam.

S-Band Power Amplifier (Collins) — This traveling-wave-tube power amplifier is housed in a sealed and pressurized case 5.75 by 5.56 by 22.26 inches and weighs 32 pounds. It has two independent amplifiers, either of which can amplify outputs of either the phase-modulated or frequency-modulated unified S-band transmitters. It operates on 3-phase, 115/200 volt, 400 Hertz power. Its two power outputs are 2.8 and 11.2 watts. The amplifier increases the low-power output of the unified S-band equipment to high power.

Signal Conditioner (North American Rockwell's Autonetics Division) — This 45-pound, 1191-cubic-inch electronics package is in the lower equipment bay. It requires 28 volts dc and consumes about 35 watts. It transforms signals from sensors and transducers to basic instrumentation analog (coded measurements) voltage level. Signals are then distributed to telemetry and command module displays.

TV Camera (RCA Astro Electronics for CM) — The 85-cubic-inch camera weighs 4.5 pounds and requires 6.75 watts at 28 volts dc. It has a 1-inch vidicon tube. Its frame rate is 10 frames per second with 320 lines per frame. Its wide angle lens is 160 degrees. The telephoto lens is 90 degrees. It can be operated at three locations within the command module. It transmits to receiving stations on earth via the unified S-band equipment, and the signals are processed there to be compatible with commercial television. (The

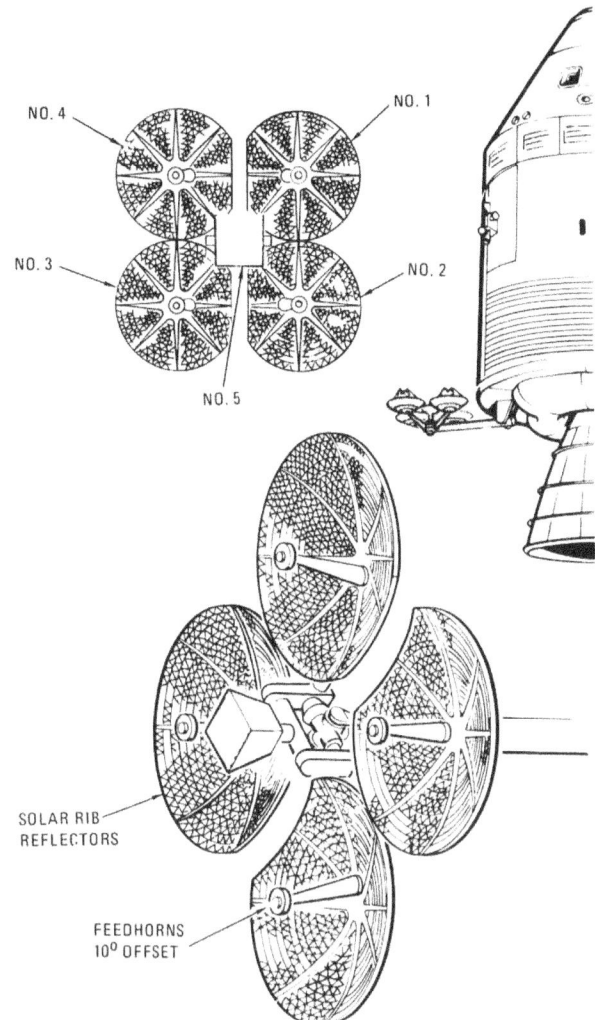

High-gain antenna construction

TV camera to be used on the lunar surface, supplied by Westinghouse Electric's Aerospace Division, Baltimore, is stowed in the LM and is part of its equipment.)

Unified S-Band Equipment (Motorola, Inc., Military Electronics Division, Scottsdale, Ariz.) — Two phase-locked transponders and one frequency-modulated transmitter are housed in single, gasket-sealed, machined aluminum case, 9.5 by 6 by 21 inches. The unit weighs 32 pounds, operates from 400 Hertz power, with RF output of 300 milliwatts. It is used for voice communications, tracking and ranging, transmission of pulse-code-modulated data and television, and reception of up data. It is the normal communications link to the spacecraft from the earth.

Up-Data Link Equipment (Motorola) — This 21-pound device is 6 by 18.3 by 9.6 inches. It receives, verifies, and distributes digital updating information sent to the spacecraft from the Manned Space Flight Network at various times throughout the mission to update or change the modes of the telecommunications systems. Data is received by the S-band receiver and routed to the up-data link.

VHF/AM Transmitter-Receiver (RCA Defense Electronic Products Communications System Division, Camden, N.J.) — It is one case housing dual transmitters and receivers for simplex or duplex operation. The enclosure contains 11 subassemblies, two coaxial relays, and two band-pass filters mounted within a three-piece hermetically sealed case. Powered from 28 volts dc, the 5-watt (RF output) unit weighs 13-1/2 pounds and is 6 by 4.7 by 12 inches. It is used for voice communication between the command module and earth during near-earth and recovery phases of mission, for voice and data communication between the command module and lunar module, and between the command module and astronauts engaged in extravehicular activity.

VHF Omnidirectional Antennas (North American Rockwell's Columbus Division) — Two scimitar antennas are located on the service module 180 degrees apart. Each is 19 by 9 inches and 1-inch thick and weighs 12.5 pounds. They are stainless steel with a fiberglass skin and a covering of Teflon. They provide for two-way voice communication between the command module and earth, between the command module and the lunar module, between the command module and extravehicular activity personnel, and for voice and biomedical telemetry from the lunar module to the command module.

VHF Recovery Beacon (Collins) — The beacon is a solid-state, 1000-Hertz tone-modulated AM transmitter. It transmits for two seconds followed by three seconds of no transmission. It weighs 2.7 pounds and is 4 by 4 by 6.75 inches. The 3-watt unit is powered by 28 volts dc. It emits signals to provide line-of-sight direction finding for recovery forces.

VHF Recovery Antennas (North American Rockwell's Los Angeles Division) — Two antennas are mounted on the forward compartment of the

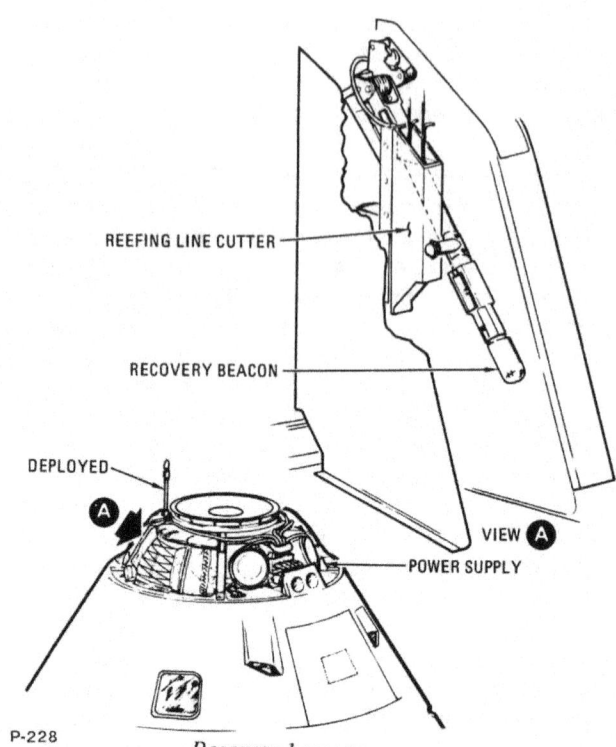

Recovery beacon

command module. No. 1 is connected to VHF recovery beacon. No. 2 is used during recovery with VHF/AM transmitter-receiver for voice communication or transmission of line-of-sight beacon signal. Each antenna is a quarter-wave ground plane antenna with a 10-inch radiating element, and weighs 3 pounds. They are not extended until 8 seconds after the main chutes are deployed.

VHF Triplexer (Rantec, Inc.) — It weighs 1.7 pounds and is 3.93 by 3.33 by 4.6 inches. It allows transmission and reception of two RF frequencies via one antenna.

DETAILED DESCRIPTION

INTERCOMMUNICATION EQUIPMENT

The audio center equipment provides the necessary audio signal amplification and switching for communication among the three astronauts, communication between one or more astronauts and extravehicular personnel, recording of audio signals in conjunction with tape recording equipment, and relaying of audio signals. The audio center equipment consists of three electrically identical sets of circuitry which provide parallel selection, isolation, gain control, and amplification of all voice

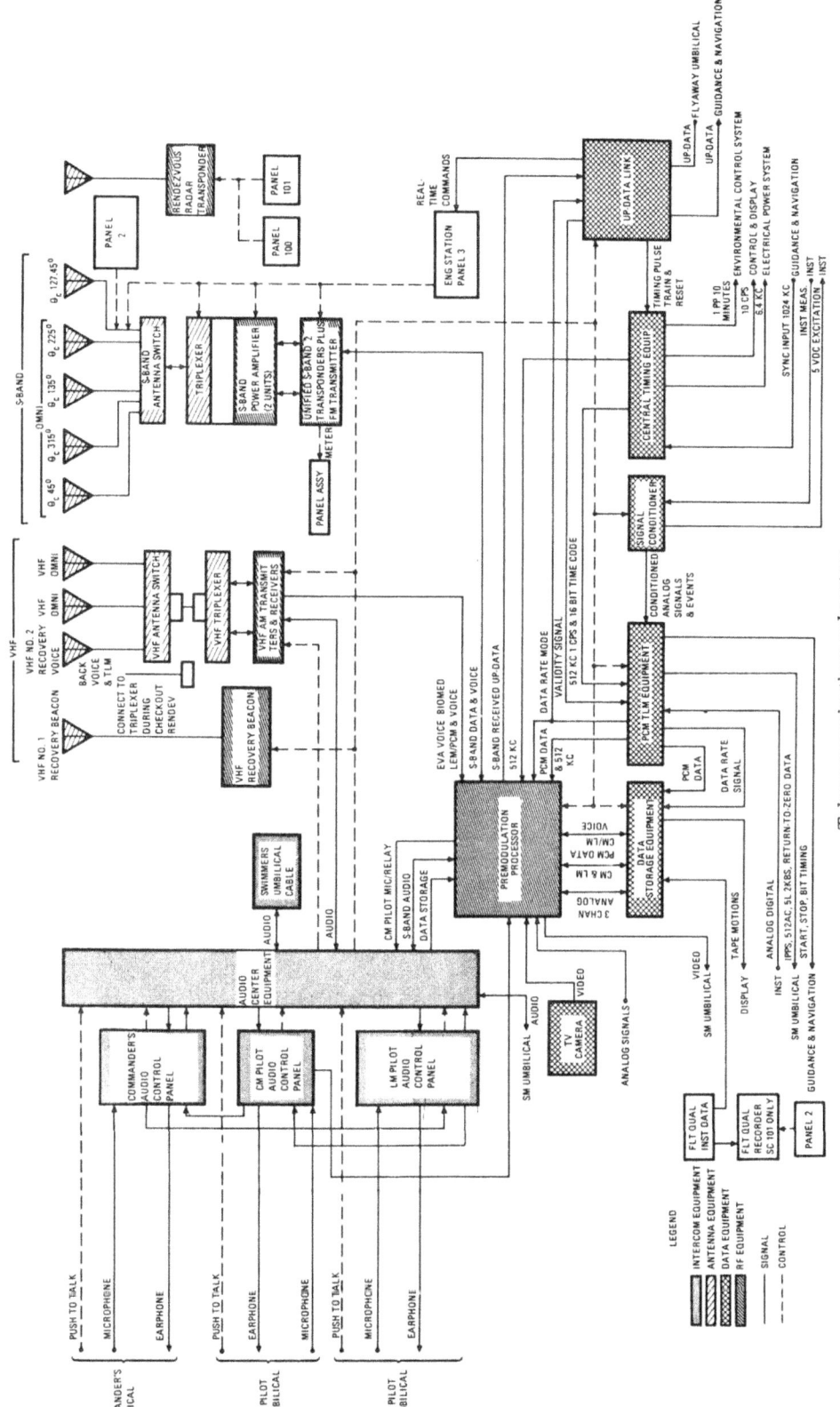

Telecommunications subsystem

communications. Each set of circuitry contains an isolation pad, diode switch, and gain control for each receiver input and an intercom channel; an isolation pad and diode switch for each transmitter modulation output and an intercom channel; an earphone amplifier and a microphone amplifier; and voice-operated relay circuitry with externally controlled sensitivity.

The equipment operates with three remote control panels to form three audio stations, each providing an astronaut with independent control. Each station can accommodate a second astronaut for emergency operation. Any or all of the transmitters can be turned on at each station. A "hot mike" enables continuous intercrew communication. When a transmitter is turned on, the corresponding receiver also is turned on. Sidetone is provided in all transmit modes.

Audio signals are provided to and from the VHF/AM transmitter-receiver equipment, unified S-band, and the intercom bus. The intercom bus is common to all three stations and provides for the hardline cable communications among crewmen and with the launch control center and recovery force swimmers.

Voice communication is controlled by the VHF/AM, S-band, and intercom switches on the audio control panels. Each of these switches has three positions: T/R (transmission and reception of voice signals), RCV (reception only), and Off. The Power switch of each station energizes the earphone amplifier to permit monitoring. The operation of the microphone amplifier in each station is controlled by the voice-operated relay keying circuit or the push-to-talk button on the communications cable or rotation controller. The voice operated relay circuit is energized by the "Vox" position of the Mode switch on each audio control panel. When energized, this circuit will enable both the intercom and accessed transmitter keying circuits. The "Intercom/PTT" position permits activation of the VHF/AM, and S-band voice transmission circuits by the push-to-talk key while the intercom is on continuously. The "PTT" position permits manual activation of the intercom or intercom and transmitter keying circuits by depression of the "I'COM" or "XMIT" side of the communication cable switch, respectively.

Five potentiometer controls are provided on each audio control panel: The "Vox Sens" control is

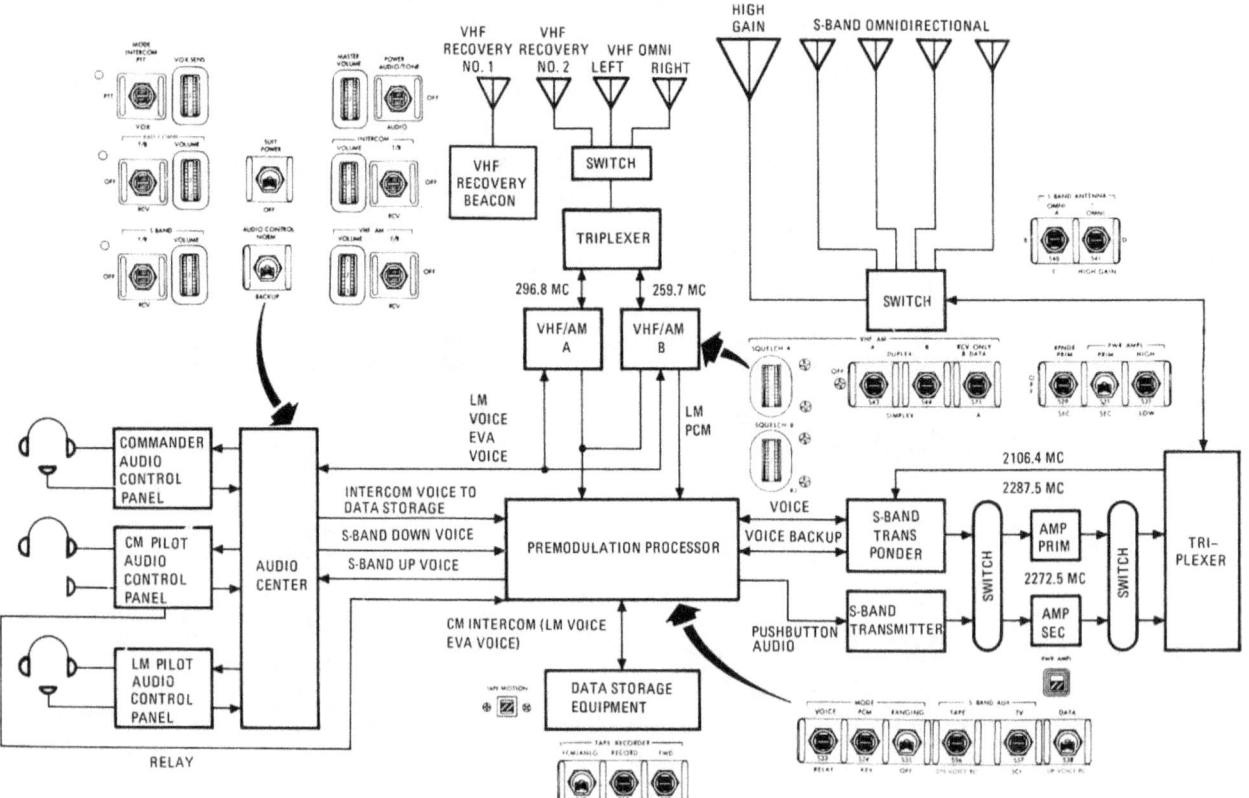

Audio system relationship to other systems

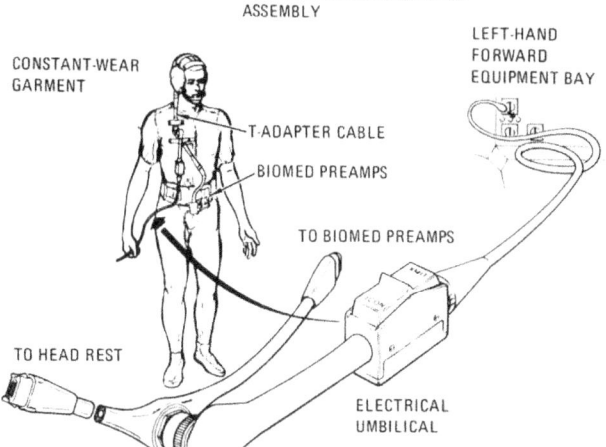

P-231   *Communications cable*

used to adjust the sensitivity of the voice-operated relay circuitry, determining the amplitude of the voice signal necessary to trigger the keying circuit. The "S-band, VHF/AM, and Intercom" volume controls are used to control the signal levels from the respective units to the earphone amplifier. The "Master Volume" controls the level of the amplified signal going to the earphones.

The intercom bus connects to ground support equipment, recovery interphone (swimmer umbilical), and the premodulation processor which in turn routes the signal to the data storage equipment for recording.

DATA EQUIPMENT

Instrumentation equipment consists of various types of sensors and transducers which monitor environmental and operational systems, and experimental equipment. The output of these sensors and transducers are conditioned into signals suitable for the spacecraft displays and for telemetry to the ground.

Various digital signals, including event information, guidance and navigation data, and a time signal from the central timing equipment, also are telemetered to the ground.

Many of the signals emanating from the instrumentation sensors are in forms of levels which are unsuitable for use by the displays or telemetry equipment. Signal conditioners are used to convert these to forms and levels that can be used. Some signals are conditioned at or near the sensor by individual conditioners located throughout the spacecraft. Other signals are fed to the signal conditioning equipment, a single electronics package located in the lower equipment bay. In addition, the signal conditioning equipment also supplies 5-volt dc excitation power to some sensors. The equipment can be turned on or off by the crew, but that is the only control they have over instrumentation equipment for operational and flight qualification measurements.

Operational measurements are those normally required for a routing mission and include three categories: in-flight management of the spacecraft, mission evaluation and systems performance, and pre-flight checkout. The operational instrumentation sensors and transducers measure pressure, temperature, flow, rate, quantity, angular position, current, voltage, frequency, RF power, and "on-off" type events.

Flight qualification measurements depend on mission objectives and the state of hardware development. Most of these measurements will be pulse-code modulated along with the operational measurements and transmitted to the ground. Other flight qualification measurements will be stored for post-flight analysis.

The central timing equipment provides precision square wave timing pulses of several frequencies to correlate all time-sensitive functions. It also generates and stores the day, hour, minute, and second of mission elapse and time in binary-coded decimal format for transmission to the ground.

In primary or normal operation, the command module computer provides a 1,024-kiloHertz sync pulse to the central timing equipment. This automatically synchronizes the central timing equipment with the computer. If this pulse fails, the central timing equipment automatically switches to the secondary mode of operation with no time lapse and operates using its own crystal oscillator. The central timing equipment contains two power supplies for redundancy. Each is supplied from a different power source and through separate circuit breakers.

The timing signals generated by the central timing equipment go to the following equipment:

Pulse code modulation equipment	For synchronization of an internal clock, frame synchronization, and time-correlation of pulse-code modulated data
Premodulation processor	For S-band emergency key transmission
Electrical power subsystem inverters	For synchronization of 400-cycle ac power
Digital event timer	Pulse for digital clock
Environmental control subsystem	To discharge water from astronauts suits (1 pulse every 10 minutes)
Scientific data equipment	For time-correlation of data

The signal conditioning equipment is a hermetically sealed unit contained in a single electronics package in the lower equipment bay. Its function is to accept and process a variety of inputs from various systems within the spacecraft and produce analog signals compatible with pulse-code modulation or displays and to provide excitation voltages to some of the instrumentation sensors and transducers. The package contains dc differential amplifier assemblies, dc differential bridge amplifier assemblies, an ac to dc converter assembly, dc active attenuator assemblies, and redundant power supplies. The signal conditioning equipment contains an error detection circuit which automatically switches to the redundant power supply if the primary power supply voltages go out of tolerance. A switch on the main display console allows the crew

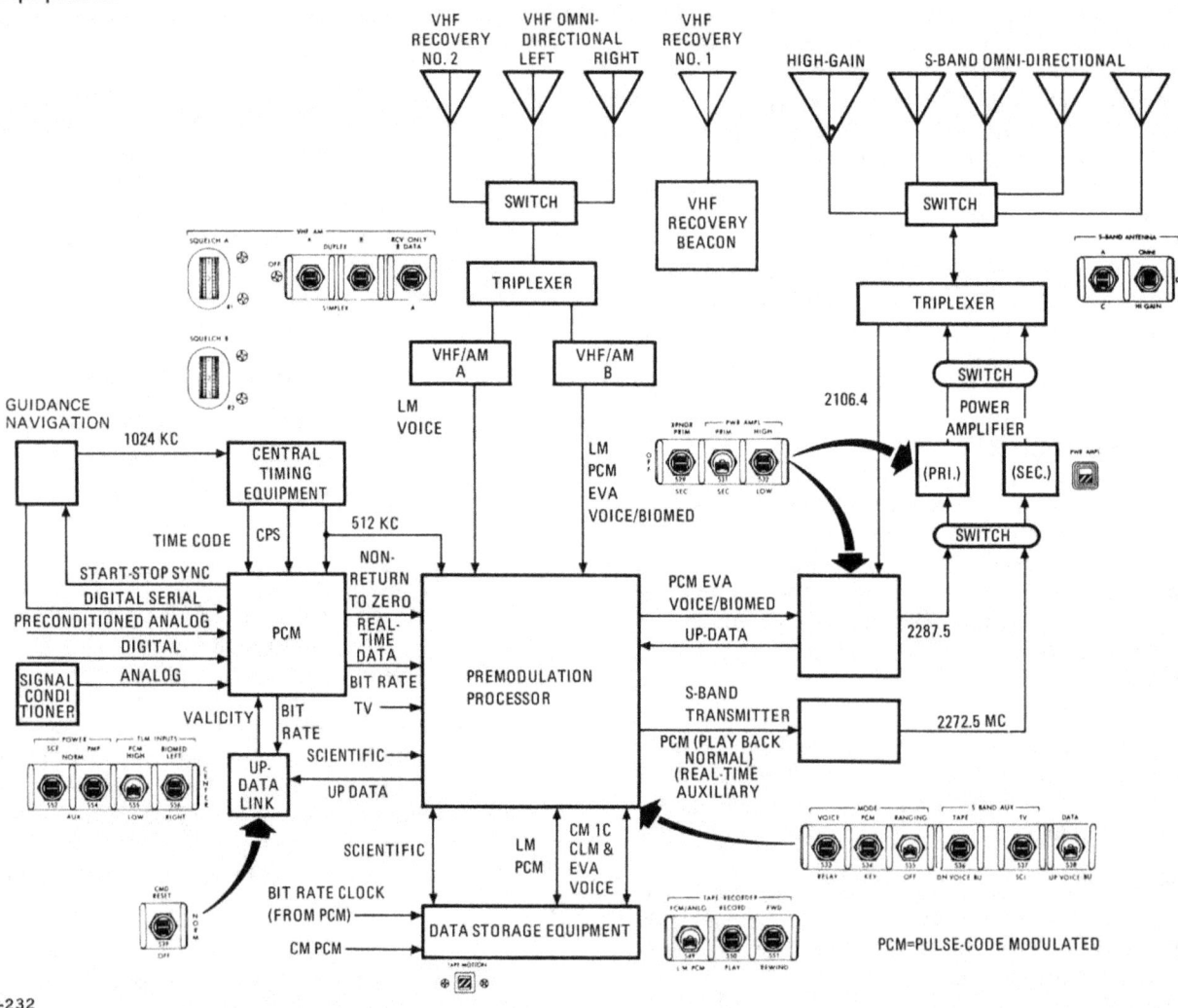

Data system relationship to other systems

to switch between either power supply. The equipment requires 28-volt dc power and consumes 35 watts maximum with a full complement of modules.

The pulse-code modulation telemetry equipment converts data inputs from various sources throughout the spacecraft into one serial digital signal. This single-output signal is routed to the premodulation processor for transmission to the ground or to data storage equipment. The pulse-code modulated telemetry equipment is located in the lower equipment bay. Incoming signals are of three general types: high-level analog, parallel digital, and serial digital. Two modes of operation are possible: the high- (normal) bit-rate mode of 51.2 kilobits per second and the low- (reduced) bit-rate mode of 1.6 kilobits per second.

The analog multiplexer can accommodate 365 high-level analog inputs in the high-bit rate mode. These analog signals are gated through the multiplexer, the high-speed gates, and are then fed into the coder. In the coder, the 0- to 5-volt analog signal is converted to an 8-bit binary digital representation of the sample value. This 8-bit word is parallel-transferred into the digital multiplexer where it is combined with 38 external 8-bit digital parallel inputs to form the output in non-return-to zero format.

This digital parallel information is transferred into the output register where it is combined with the digital serial input, and then transferred serially into the data transfer buffer. From here the information is passed on to the premodulation processor for preparation for transmission.

The pulse-code modulation telemetry equipment receives 512-kiloHertz and 1-Hertz timing signals from the central timing equipment. If this source fails, its programmer uses an internal timing reference. The timing source being used is telemetered. Two calibration voltages are also telemetered as a confidence check of the operation of the telemetry equipment.

Television equipment consists of a small, portable TV camera that can be hand-held or mounted in three locations in the command module. The camera is connected to a 12-foot cable for use throughout the CM. The camera is controlled by a switch on the camera handle and an automatic light control switch on the back. Power required by the camera is 6.75 watts at 28 volts dc. The composite video signal is sent from the camera to the premodulation processor where it is then sent to the SM umbilical for hard-line communications before liftoff.

The TV video signal from the premodulation processor also may be routed to the S-band FM transmitter and its associated power amplifier for transmission to the ground for reception of TV during flight.

The data storage equipment stores data for delayed playback or recovery with the spacecraft. Information is recorded during powered flight phases and when out of communications and is played back (dumped) when over selected S-band stations. The equipment, located in the lower equipment bay, has tape speeds of 3.75, 15, and 120 inches per second. The tape speed is selected automatically based on the data rate. The tape has fourteen parallel tracks: four CM pulse code modulated digital data, one digital clock, one LM pcm data, one CM-LM voice, three scientific data, and four spare.

The flight qualification recorder is a 14-track magnetic tape recorder used to record certain flight qualification measurements during critical phases of the mission. This data will be used for post-flight analysis only; it cannot be played back or transmitted during flight. It will be activated during the ascent and entry phases of the mission and during service propulsion engine firings. Flight qualification data is recorded in analog form. Two recording tracks (one record head in each of two record-head stacks)

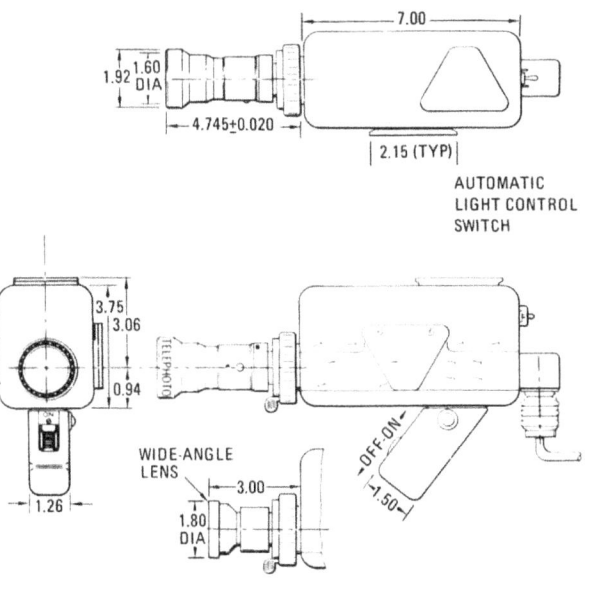

Apollo television camera

are used for reference and time code recording. To accomplish this, an elapsed time code generator is used to modulate a narrow-band voltage-controlled oscillation. The output of the voltage-controlled oscillator is then mixed with the output of a 501-kiloHertz reference oscillator. This composite signal is presented to each of the two record heads through two direct record amplifiers.

The recorder operates at a record speed of 15 inches per second and a rewind speed of 120 inches per second. The 15 inches per second record speed allows a total of 30 minutes recording time per reel of tape. A sensor will automatically halt the tape motion and remove power from the electronic circuits when the end of the tape is reached in either direction. The crew controls operation of the recorder through a switch on the main display console.

The up-data link equipment receives, verifies and distributes digital information sent to the spacecraft by the ground to update or change the status of operational systems. The up-data link consists of detecting and decoding circuitry, a buffer storage unit, output relay drivers, and a power supply. It provides the means for the ground to update the computer and the central timing equipment, and to select certain vehicle function. Up-data information is transmitted to the spacecraft as part of the 2-gigaHertz S-band signal. When this signal is received by the unified S-band receiver, the 70-kiloHertz subcarrier containing the up-data information is extracted and sent to the up-data discriminator in the premodulation processor. The resulting composite-audio frequency signal is routed to the sub-bit detector in the up-data link which converts it to a serial digital signal. The digital output from the sub-bit detector is fed to the remaining up-data link

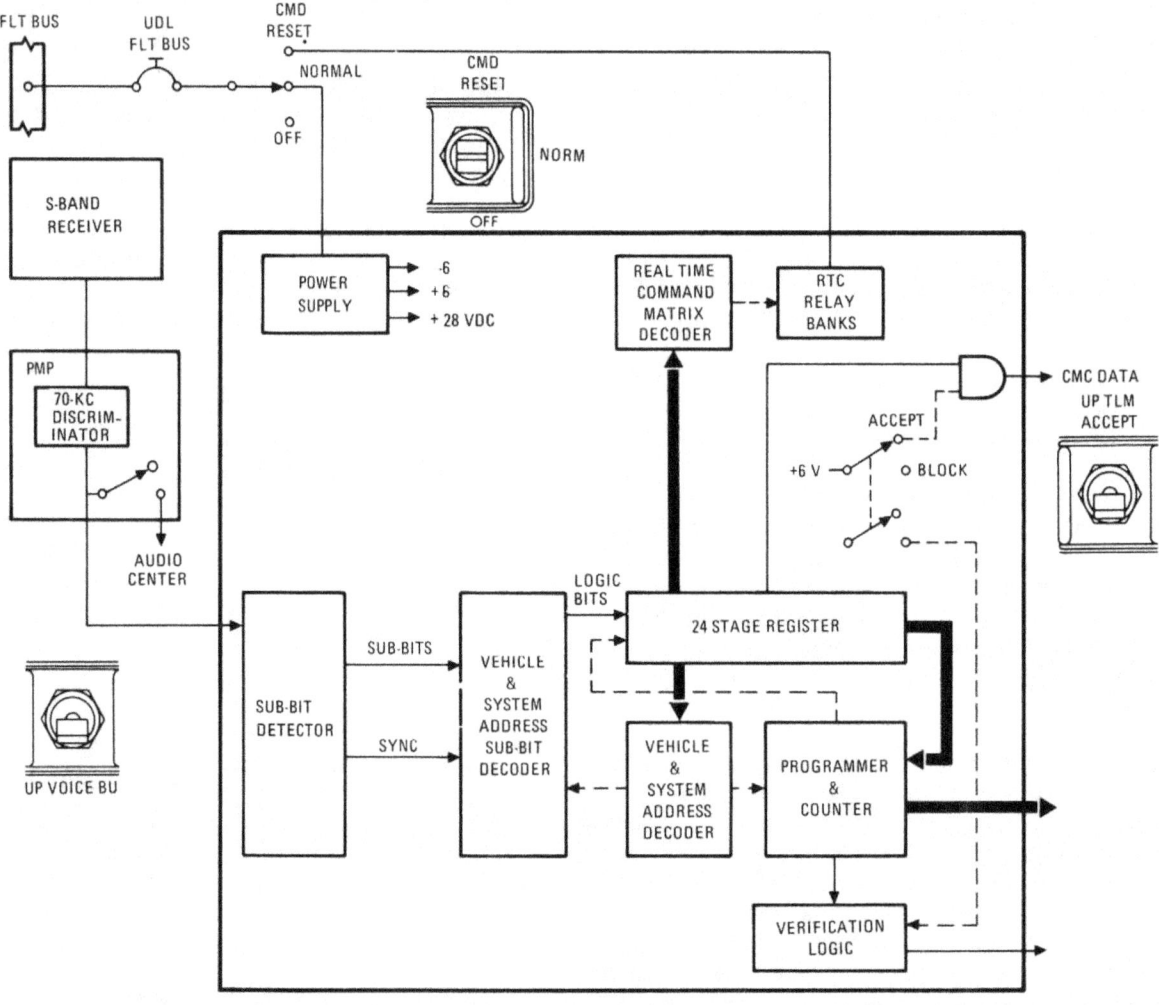

Block diagram of up-data link

circuitry, which checks and stores the digital data, determines its proper destination, and transfers it to the appropriate system or equipment.

RADIO FREQUENCY ELECTRONICS EQUIPMENT

The radio frequency electronics equipment group includes all telecommunications which functions as radio frequency transmitters or receivers. The group includes VHF/AM transmitter-receiver equipment, unified S-band equipment, the S-band power amplifier, the premodulation processor, VHF recovery beacon equipment, and the rendezvous radar transponder.

The VHF/AM transmitter-receiver equipment provides two-way voice communications among the CM, the ground, the LM, astronauts outside the CM, and recovery forces, relay of two-way voice from either the LM or extravehicular astronauts to the ground (via the S-band); reception of pulse-code modulated data from the LM; and reception of biomedical data from extravehicular astronauts. It is contained in a single enclosure consisting of 11 subassemblies, 2 coaxial relays, and 2 bandpass filters mounted within a 3-piece hermetically sealed case in the lower equipment bay.

The equipment includes two independent VHF/AM transmitters and two independent VHF/AM receivers. The transmitters and receivers operate on different frequencies and one receiver accepts data as well as voice. The receiver circuits are isolated up to the final common output.

The VHF/AM transmitter-receiver equipment is controlled by switches on the main display console and push-to-talk buttons. A squelch control varies the level of squelch sensitivity. The transmitters and receivers connect with the main display console, the audio center, and the triplexer. The equipment is connected through the triplexer and antenna control switch to either of the VHF omni-directional antennas in the service module or the VHF recovery antenna No. 2 in the command module.

The unified S-band equipment consists of two transponders, an FM transmitter, and their power supplies contained in a single electronics package in the lower equipment bay. It is used for voice communications, tracking and ranging, transmission of pulse-code modulated data, and reception of up-data. It also provides the sole means for transmission of TV.

S-band tracking is by the two-way or double-doppler method. In this technique, a stable carrier of known frequency is transmitted to the spacecraft where it is received by the phase-locked receiver, multiplied by a known ratio, and then re-transmitted to the ground for comparison. Because of this, S-band equipment is also referred to as the S-band transponder.

To determine spacecraft range, the ground station phase-modulates the transmitted carrier with a pseudo-random noise binary ranging code. This code is detected by the spacecraft's S-band receiver and used to phase-modulate the carrier transmitted to the ground. The ground station receives the carrier and measures the amount of time delay between transmission of the code and reception of the same code, thereby obtaining an accurate measurement of range. Once established, this range can be continually updated by the double-doppler measurements. The ground stations also can transmit up-data commands and voice signals to the spacecraft by means of two subcarriers: 70 kiloHertz for up-data and 30 kiloHertz for up-voice.

The S-band transponder is a double-superheterodyne phase-lock loop receiver that accepts a phase-modulated radio frequency signal containing the up-data and up-voice subcarriers, and a pseudo-random noise code when ranging is desired. This signal is supplied to the receiver via the triplexer in the S-band power amplifier equipment and presented to three separate detectors: the narrow-band loop phase detector, the narrow-band coherent amplitude detector, and the wide-band phase detector. In the wide-band phase detector, the intermediate frequency is detected, and the 70-kiloHertz up-data and 30-kiloHertz up-voice subcarriers are extracted, amplified, and routed to the up-data and up-voice discriminators in the premodulation processor. When operating in a ranging mode, the pseudo-random noise ranging signal is detected, filtered, and routed to the S-band transmitter as a signal input to the phase modulator. In the loop-phase detector, the intermediate frequency signal is filtered and detected by comparing it with the loop reference frequency. The resulting dc output is used to control the frequency of the voltage-controlled oscillator. The output of the voltage controlled oscillator is used as the reference frequency for receiver circuits as well as for the transmitter. The coherent amplitude detector provides the automatic gain control for receiver sensitivity control. In addition, it detects the amplitude modulation of the carrier introduced by the high-gain antenna system. This detected output is

	PRE-LAUNCH	LAUNCH, EARTH ORBIT, TRANSLUNAR INJECTION	TRANSLUNAR FLIGHT 4000 MI	LUNAR ORBIT	LM DESCENT, LUNAR SURFACE, ASCENT	TRANSEARTH INJECTION, TRANSEARTH FLIGHT	DESCENT, LANDING, RECOVERY
VOICE	HARDLINE S BAND	S BAND VHF/AM+	S BAND	S BAND	S-BAND VHF/AM*	S BAND	S BAND VHF/AM
TV	HARDLINE	S BAND	S BAND	S BAND	S BAND	S BAND	
BIOMEDICAL	HARDLINE S BAND*	S BAND	S BAND VHF/AM+	S BAND	S BAND VHF/AM+	S BAND	
UP-DATA	HARDLINE S BAND*	S BAND	S BAND	S BAND	S BAND	S BAND	
DOWN	HARDLINE S BAND*	S BAND	S BAND	S BAND	S BAND	S BAND	
RANGING		S BAND	S BAND	S BAND	S BAND	S BAND	
BEACON							VHF-BCN

*BACKUP
+LM EXTRAVEHICULAR COMMUNICATIONS

Communication equipment use by mission phase

returned to the antenna control system to point the high-gain antenna to the ground station. When the antenna points at the ground station, the amplitude modulation is minimized. An additional function of the detector is to select the auxiliary oscillator to provide a stable carrier for the transmitter, whenever the receiver loses lock.

The S-band transponders can transmit a phase-modulated signal. The initial transmitter frequency is obtained from one of two sources: the voltage-controlled oscillator in the phase-locked S-band receiver or the auxiliary oscillator in the transmitter. Selection of the excitation is controlled by the coherent amplitude detector.

The S-band equipment also contains a separate FM transmitter which permits scientific, television, or playback data to be sent simultaneously to the ground while voice, real-time data, and ranging are being sent via the transponder.

The S-band power amplifier equipment is used to amplify the radio frequency output from the S-band transmitters when additional signal strength is required for adequate reception by the ground. The amplifier equipment consists of a triplexer, 2 traveling-wave tubes for amplification, power supplies, and the necessary switching relays and control circuitry. The S-band power amplifier is contained in a single electronics package located in the lower equipment bay.

All received and transmitted S-band signals pass through the triplexer. The S-band carrier received by the spacecraft enters the triplexer from the S-band antenna equipment. The triplexer passes the signal straight through to the S-band receiver. The output signal from the S-band transponder enters the S-band power amplifier where it is either bypassed directly to the triplexer and out to the S-band antenna equipment, or amplified first and then fed to the triplexer. There are two power amplifier modes of operation: low power and high power. The high-power mode is automatically chosen for the power amplifier connected to the FM transmitter.

The premodulation processor equipment provides the connection between the airborne data-gathering equipment and the radio frequency electronics. The processor accomplishes signal modulation and demodulation, signal mixing, and the proper switching of signals so that the correct intelligence corresponding to a given mode of operation is transmitted. It requires a maximum power of 12.5 watts at 28-volt dc power.

The VHF recovery beacon provides a line-of-sight direction-finding signal to aid in locating the spacecraft after landing. It is located in the lower equipment bay. The beacon signal is an interrupted carrier, modulated by a 1000-Hertz square wave. The signal is transmitted for 2 seconds, then interrupted for 3 seconds. The signal from the VHF recovery beacon is fed to VHF recovery antenna No. 1, which is deployed automatically when the main chutes are deployed.

The rendezvous radar transponder is located in the service module. Its function is to receive the X-band tone-modulated continuous wave signal from the LM rendezvous radar, and transmit back to it a phase-coherent return signal. The return signal is offset in fundamental carrier frequency from the received signals and contains the same modulation components phase-related with respect to the received signal.

The transponder is a part of the LM radar subsystem which consists of a rendezvous radar in the LM, the transponder in the CM, and a landing radar mounted in the descent stage of the LM. During the descent to the lunar surface, the LM and CSM maintain continuous radar contact through the rendezvous radar-transponder link. At the end of the lunar stay, the rendezvous radar in the LM is used to track the transponder in the orbiting CSM to obtain CSM orbital conditions, which are used to calculate the launching of the LM into a rendezvous trajectory. In the rendezvous phase, the LM and CSM again maintain radar contact to obtain information needed by the LM for course correction, and rendezvous operations. With the aid of the CSM transponder, the LM can rendezvous with a less powerful rendezvous radar transmitter, using the CSM phase-coherent transponder to achieve the required rendezvous range capability and to minimize tracking errors.

ANTENNAS

The antenna equipment group contains all of the spacecraft antennas and ancillary equipment used in the telecommunication subsystem.

The VHF omni-directional antennas and ancillary equipment consists of two VHF scimitar antennas, a VHF triplexer, a VHF antenna switch, and the necessary signal and control circuits. This equipment radiates and picks up radio frequency signals in the VHF spectrum. The portable life-support communication equipment also can be checked through this equipment.

The VHF triplexer is a passive, three-channel filtering device which enables three items of VHF transmitting and receiving equipment to utilize one VHF antenna simultaneously. The three-channel filters are composed of two tuned cavities each, which

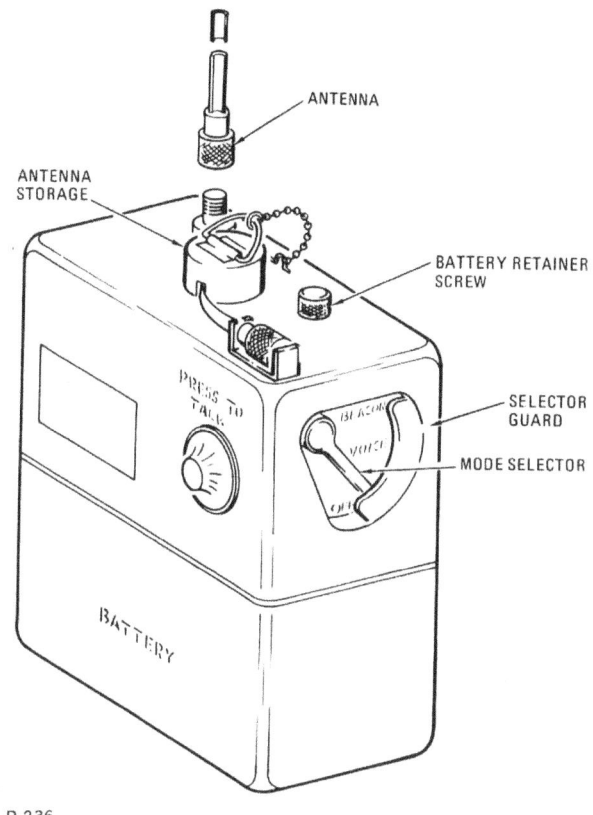

P-236

Survival beacon/transceiver radio

function as bandpass filters. No power is required by the device and there are no external controls.

The VHF scimitar antennas are omni-directional with approximately hemispherical radiation patterns. Because of its characteristic shape, this type of VHF antenna is called a scimitar. These two antennas are located on opposite sides of the service module, one near the +Y axis (called the right VHF antenna) and the other near the -Y axis (called the left VHF antenna). Because of their approximate hemispherical radiation patterns, full omni-directional capabilities can be obtained only by switching from one antenna to the other.

The S-band high-gain antenna is provided for use with the unified S-band equipment to provide sufficient gain for two-way communications at lunar distances. To accomplish this, the antenna can be oriented manually or automatically toward ground stations for maximum operational efficiency. The antenna has three modes of operation for transmission and two for reception. The gain and beamwidths of these modes are:

Mode	Gain	Beamwidth
Wide - transmit	9.2 db	40°
Wide - Receive	3.8 db	40°
Medium - Transmit	20.7 db	11.3°
Medium - Receive	22.8 db	4.5°
Narrow - Transmit	26.7 db	3.9°
Narrow - Receive	23.3 db	4.5°

The antenna is deployed during transposition and docking when the SLA panels are opened. After deployment, the positioning circuitry is enabled. Manual controls, position readouts, and a signal strength meter on the main display console allow normal positioning of the antenna for initial signal acquisition. After acquisition, the antenna automatically tracks the radio frequency signal within the travel limits of its gimbaling system. The operational modes can be selected by the crew.

The antenna consists of a four-parabolic dish array whose attendant feed horns are offset 10 degrees for the desired propagation pattern and a cluster of four feed horns enclosed in the center enclosure. In the wide mode, the center feed horns are used for transmission and reception of signals. In the medium mode, one of the parabolic dish-reflector antennas is used for transmission and all four of the dish antennas are used for reception of S-band signals. The narrow mode employs the four parabolic dish antennas for transmission and reception of S-band signals.

The four S-band omni-directional antennas transmit and receive all S-band signals during the near-earth operational phase and back up the high-gain S-band antenna in lunar operations. The antennas are flush-mounted, right-hand polarized helical, and in a loaded cavity. They are rated at 15 watts continuous wave.

There are two VHF recovery antennas (No. 1 and No. 2) stowed in the forward compartment of the CM. Each antenna consists of a quarter-wave stub, 11 inches long, and a ground plane. They are automatically deployed 8 seconds after main parachute deployment during the descent phase of the mission. VHF recovery antenna No. 1 is connected to the VHF recovery beacon equipment through a coaxial connector on the coaxial bracket. VHF recovery antenna No. 2 is used with the VHF/AM transmitter-receiver equipment; therefore, it is also connected to the VHF antenna switch through a

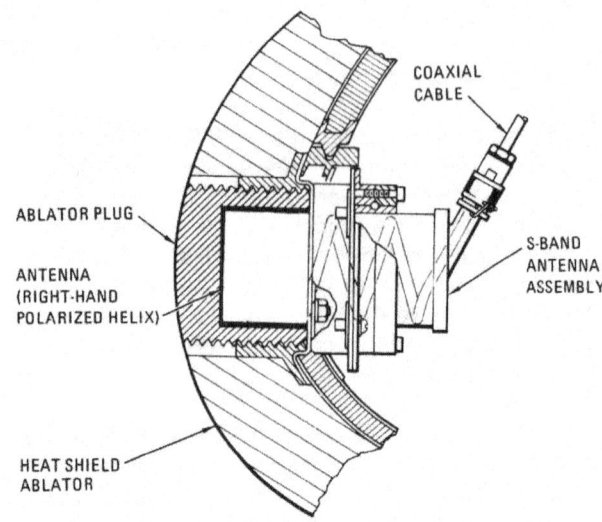

P-237 *S-band flush-mounted antenna*

connector on the coaxial bracket. The purpose of the coaxial bracket is to allow either of the VHF recovery antennas to be used with the survival transceiver. This required that the coaxial cable from one of the antennas be manually disconnected at the coaxial bracket and reconnected to the survival transceiver.

GUIDANCE AND CONTROL

The Apollo spacecraft is guided and controlled by two interrelated subsystems. One is the guidance and navigation subsystem. The other is the stabilization and control subsystem.

The two subsystems provide rotational, line-of-flight, and rate-of-speed information. They integrate and interpret this information and convert it into commands for the spacecraft's propulsion subsystems.

The guidance and navigation subsystem contains three major elements. They are the inertial, optical, and computer subsystems.

The inertial subsystem senses any changes in the velocity and angle of the spacecraft and relays this information to the computer. The computer digests the information and transmits any necessary signals to the spacecraft engines.

The optical subsystem is used to obtain navigation sightings of celestial bodies and landmarks on the earth and moon. It passes this information along to the computer for guidance and control purposes.

The computer subsystem uses information from a number of sources to determine the spacecraft's position and speed and, in automatic operation, to give commands for guidance and control. Data fed into the computer include: telemetry information from the ground regarding velocity, attitude, and position in space; a fixed memory which permanently stores navigation tables, trajectory parameters, programs, and constants; an erasable memory which stores intermediate results of computation, auxiliary program information, and variable data supplied by the guidance and control, and other subsystems of the spacecraft.

The other of the two interrelated subsystems is the stabilization and control subsystem. In general, it operates in these three ways: it determines the spacecraft's attitude (its angular position); it maintains the spacecraft's attitude; it controls the

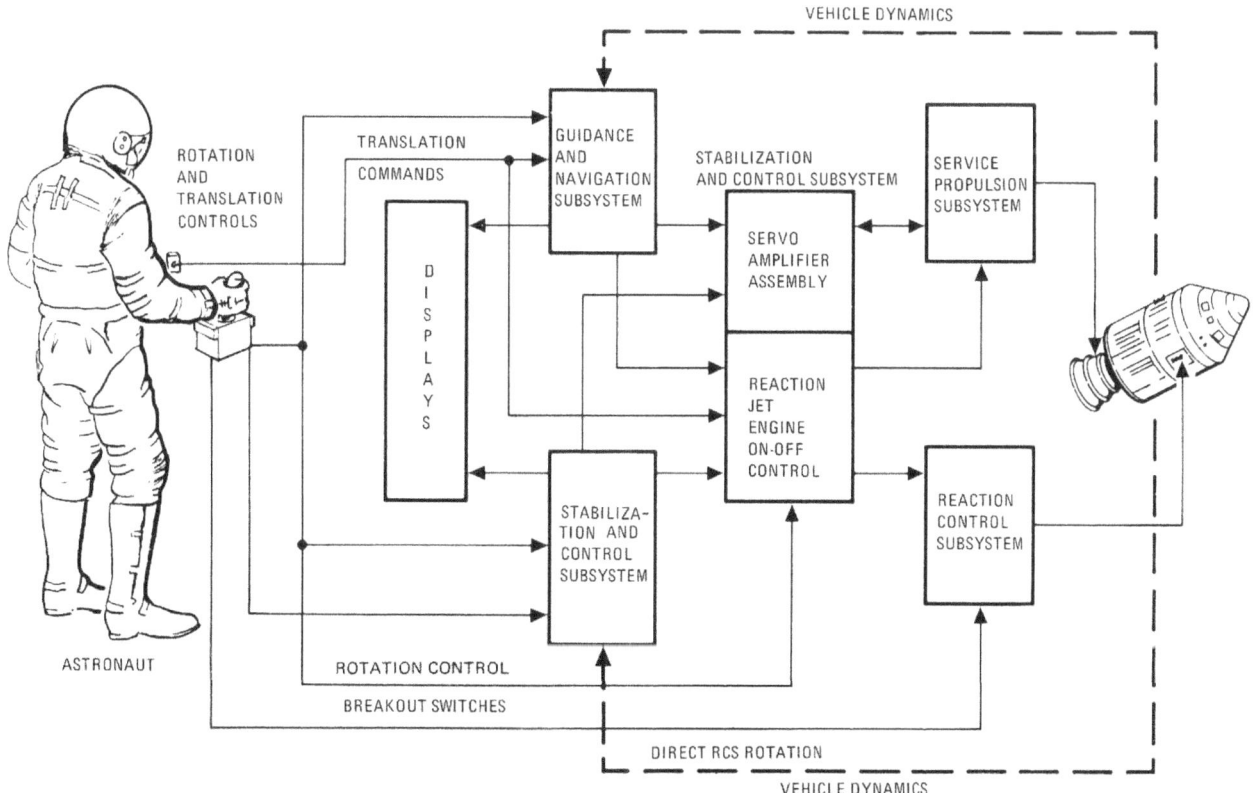

Guidance and control functional flow

direction of thrust of the service propulsion engine.

Both of the subsystems are used by the computer in the command module to provide automatic control of the spacecraft. This is done electronically. Manual control of the spacecraft's attitude and thrust is provided mainly through the stabilization and control subsystem equipment.

The space trajectory of Apollo is established essentially by the firing of engines, either its own or those of a launch vehicle. "Flying" Apollo, in the sense that an airplane is flown, is not possible because there is no atmosphere in space. The speed, altitude, and flight path angle of Apollo at the instant the last booster engine cuts off determines the characteristics of the spacecraft's orbit: its apogee (maximum altitude), perigee (minimum altitude), velocity and position at any point, and the time to complete an orbit.

Any additional engine firing automatically changes the flight path in space. To maneuver Apollo, the engines must be fired. The direction, amount, and duration of thrust of the service propulsion engine, for example, causes a change in the shape of the orbit about the earth, tilts the plane in which Apollo orbits the earth, or slows it down to permit entry and return to earth. During the flight to the moon, the firing changes the path to the moon (course correction), or slows the spacecraft at the proper time to permit the moon's gravity to pull Apollo into an orbit about the moon.

SPACECRAFT ATTITUDE

Two indicators on the main display console show the spacecraft's attitude. The indicators show the angle, or angular position, in reference to star sightings. These two indicators are called flight director attitude indicators. They tell the spacecraft's total angle (or position), the attitude errors, and rates of change.

The term total attitude refers to a combination of information from two sources.

One of the sources is the stable platform of the inertial measurement unit.

This provides total attitude information by holding the gimbaled, gyro-stabilized stable platform to a fixed inertial reference (maintained by periodic star sightings). In other words, no matter how the spacecraft moves, the stable platform retains its fixed, star-sighted position.

The second source is a gyro display coupler. This is a device which gives a reading of the spacecraft's actual attitude as compared with an attitude which the crew desires to maintain. A reading of the desired attitude is given by manually dialing roll, pitch, and yaw dials of the attitude set display in the console. The gyro display coupler then shows attitude errors by comparing total attitude information against the manually dialed attitude.

Information about attitude error also is obtained by comparison of the inertial measurement unit's gimbal angles with computer-reference angles. Another source of this information is gyro assembly No. 1, which senses any spacecraft rotation about any of the three axes.

Total attitude information goes to the command module computer as well as to the attitude indicators on the console.

Attitude control of the spacecraft is provided for the purpose of maintaining a certain angle or for changing it during maneuver. If a specific attitude or orientation is desired, attitude error signals are sent to the reaction jet engine control assembly. Then the proper reaction jet fires in the direction necessary to return the spacecraft to the desired position.

THRUST CONTROL

The computer in the command module provides primary control of thrust. The flight crew pre-sets thrusting and spacecraft data into the computer by means of the display keyboard. The forthcoming commands include time and duration of thrust. Accelerometers sense the amount of change in velocity obtained by the thrust.

Thrust direction control is required because of center of gravity shifts caused by depletion of propellants in service propulsion tanks. This control is accomplished through electro-mechanical actuators which position the service propulsion engine, which rides on gimbals. Automatic control commands may originate in either the guidance and navigation subsystem or the stabilization and control subsystem. There is also provision for manual controls.

STABILIZATION AND CONTROL

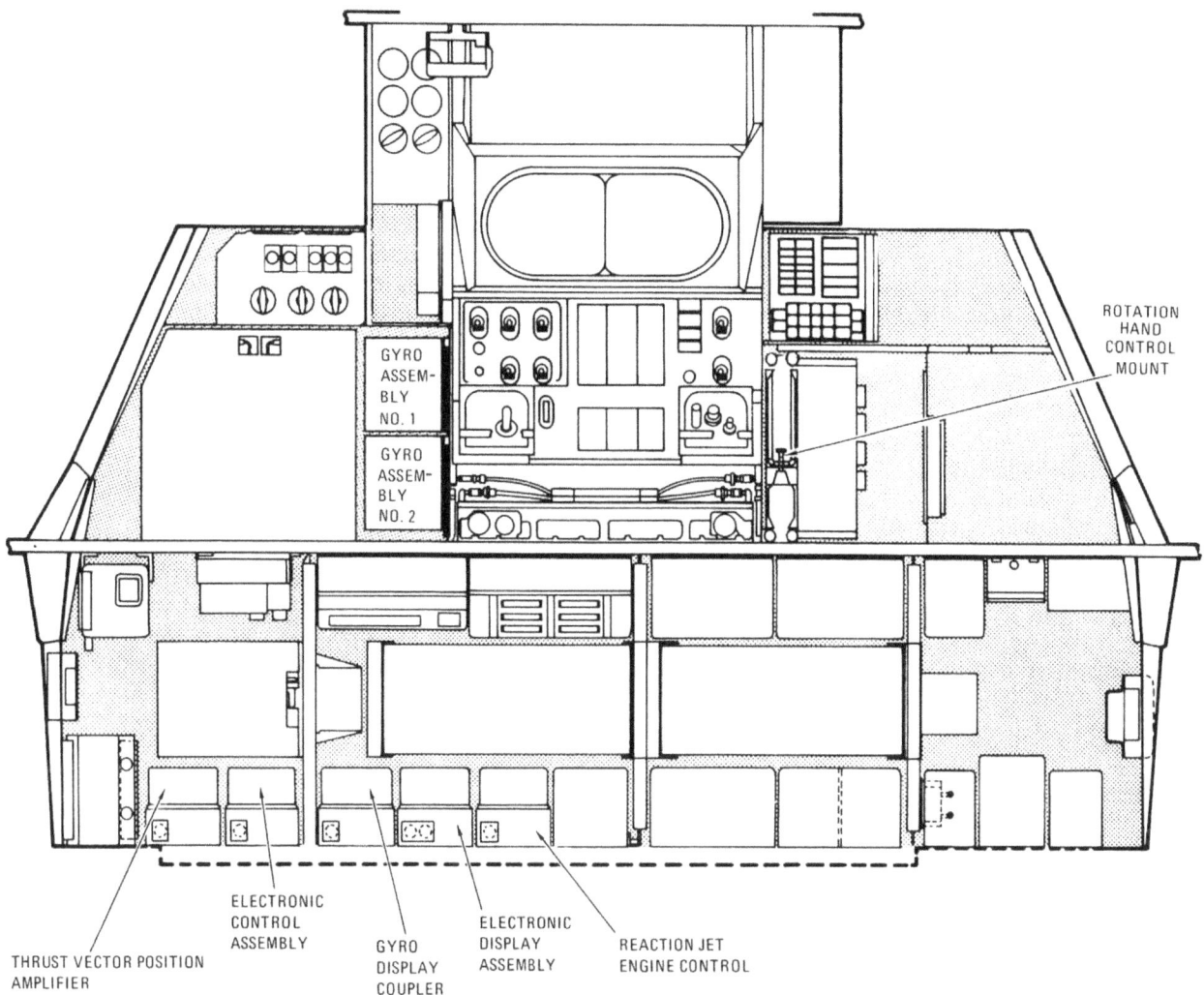

Location of main components of stabilization and control subsystem

The stabilization and control subsystem provides control and monitoring of the spacecraft's attitude, and control of the firing direction or thrust vector of the service propulsion engine. It also serves as a backup system for inertial reference. The subsystem may be operated either automatically or manually. It is produced by Honeywell, Inc., Minneapolis, Minn.

All the control functions of this subsystem are backups to the guidance and navigation subsystem.

This subsystem is divided into three basic elements. It provides attitude reference, attitude control, and thrust direction control.

An electronic control assembly is used for attitude control and thrust direction control. This assembly has the circuits required for utilizing the rate and attitude error sensors. It also has the manual commands necessary to maintain backup stabilization and control in all axes (pitch, yaw, and roll).

An electronic display assembly provides the logic for establishing signal sources to be displayed and displays to be used. This assembly also provides for monitoring, isolation, and signal conditioning for telemetry of display signals.

There are three hand controls. One is for line-of-flight thrusting. Two are rotation controls for angular thrusting. The control for line-of-flight (or rectilinear motion) of the spacecraft also enables the crew to initiate abort during launch. It can also be used to transfer spacecraft control from the guidance and navigation subsystem to the stabilization and control subsystem.

The rotation control gives the crew manual control of spacecraft rotation in either direction about its three control axes. Also, it may be used for manual thrust direction control in pitch and yaw conditions when the service propulsion subsystem is thrusting.

The flight data display associated with the stabilization and control subsystem includes the flight director attitude indicators. They show the spacecraft attitude, attitude error, and angular rate information. Also located on the main display console is a gimbal position and fuel pressure indicator which shows either the service propulsion engine pitch and yaw gimbal angles or the booster propellant pressures. This provides a means for manually trimming the gimbals of the service propulsion subsystem engine.

EQUIPMENT

Gyro Assemblies (Honeywell, Inc., Minneapolis) — Two units, each 6 by 7 by 14-1/4 inches weighing 22 pounds 7 ounces, are on the left-hand side of the navigation station in the lower equipment bay. Each contains three body-mounted attitude gyros mounted along body axes to sense attitude displacement. Components of each assembly are contained in a welded aluminum alloy enclosure designed to provide vacuum sealing and electrical continuity. In the attitude-hold mode, any displacement causes one gyro assembly to signal the reaction control system jets to restore the original attitude. The signal is displayed for the crew on the flight director attitude indicator. The other gyro assembly provides signals for display of body rates on either or both flight director attitude indicators, for rate damping, and for computation of inertial attitude changes.

Rotation Controls (Honeywell) — Two identical three-axes rotation controllers with control sticks provide proportional rate command signals for attitude or manual thrust vector maneuvers. Motion of the controls is analogous to rotation of the spacecraft about its axes. Direct angular acceleration command signals are provided when the direct reaction control system switch is actuated and the control is displaced to full stop position. Controls can be mounted on left and right couches and at the navigation position in the lower equipment bay. Each control is in a 3 by 5.5 by 7.26-inch metal box with a 5.25-inch aluminum stick handle. Each weighs 7 pounds, including cable and connector. A trigger-type push-to-talk switch is in the stick handle. There are redundant locking devices on each control. The control unit includes six breakout switches to provide on-off command and signals to the command module computer, stabilization control system minimum impulses, acceleration commands, caging of attitude gyros, and to enable proportional rate commands in the electronics; three transducers to command spacecraft rotation rates during proportional rate control and to command service propulsion engine gimbal position in pitch and yaw during manual thrust vector control; and direct switches in each axis for each direction of rotation. Direct switches produce acceleration commands through the direct solenoids on the reaction control system engines, bypassing stabilization control system electronics.

Translation Control (Honeywell) — This control with a T-handle stick provides a means of accelerating along one or more of the spacecraft axes. Motion of the control is analogous to translation motion of the spacecraft. The metal box measures 3.78 by 6.3 by 3.8 inches with a 3-inch rubber boot and aluminum T-handle. It weighs 5 pounds 11 ounces. The control is mounted with its axis approximately parallel to those of the spacecraft. Redundant switches close for each direction of control displacement. Clockwise switches transfer spacecraft control from the command module computer to the stabilization control system. One of the switches supplies 28 volts dc to a portion of the primary guidance navigation and control system. Redundant counterclockwise switches provide for initiation of manual abort during the launch phase. A discrete signal from switch closure is fed to the master events sequence controller, which initiates other abort functions.

Flight Director Attitude Indicator (Honeywell) — There are two indicators, each in a metal box 6.9 inches in diameter and 9.33 inches long weighing 9 pounds. They have glass cover plates. One is on Panel 1 in front of the commander

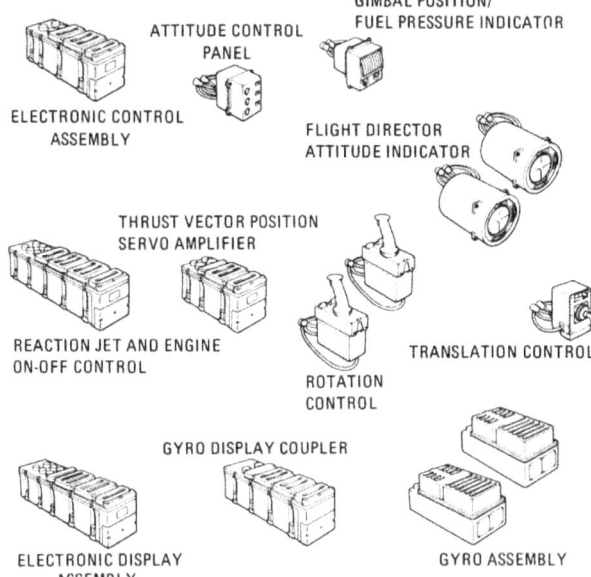

P-240 *Major stabilization and control components*

and the other on a panel to the right and directly above it, just over the display and keyboard. The display is a ball with circles showing pitch attitude, yaw attitude, and roll attitude, and three scales — one above and below the ball and a third to the right of it — showing pitch, yaw, and roll rate with needles showing pitch, yaw, and roll error. Attitude and attitude error signals to the displays are supplied by the stabilization and control system or the guidance and navigation system. Rate signals to the displays are supplied by the stabilization and control system only.

Attitude Set Control Panel (Honeywell) — This 3.5 by 4.8 by 4.4-inch panel weighs 3 pounds 6 ounces and is in the lower left-hand corner of the main display panel facing the commander. It has three windows to display roll, pitch, and yaw in degrees and three thumbwheels to the left of the windows. The panel provides the means for manually inserting desired attitude information into the stabilization control system in the form of three angles. It receives signals that represent the actual attitude of the spacecraft relative to an arbitrary inertial (fixed) reference frame. Output signals are provided, which represent the attitude error or the difference between the actual and desired total attitude of the spacecraft. These signals can be used to drive the attitude error needles on a flight director attitude indicator, thereby providing the astronaut with a visual indication of the spacecraft attitude error and to align the gyro display coupler to a fixed reference frame.

Gimbal Position and Fuel Pressure Indicator (Honeywell) — The indicator is in a 4.25 by 4.5 by 4.82-inch metal box with a single window showing four meter movements. The unit weighs 2 pounds 14 ounces and is directly below the flight director attitude indicator in front of the commander. It contains redundant indicators for both the pitch and yaw channels. During boost phases it displays second-stage fuel pressure on the redundant pitch indicators and third-stage fuel pressure on the two yaw indicators. The gimbal position indicator consists of two dual servometric meter movements mounted within a common hermetically sealed case. For a stabilization control system velocity change mode, manual service propulsion system engine gimbal trim capability is provided. Desired gimbal trim angles are set in with the pitch and yaw trim thumbwheels.

DETAILED DESCRIPTION

CONTROLS AND DISPLAYS

Stabilization and control subsystem controls and displays consist of two rotation controls, a translation control, an attitude set control panel, a gimbal position and fuel pressure indicator, two flight director attitude indicators, and two gyro assemblies.

Two identical rotation controls give the astronauts control of the spacecraft's rotation in either direction around all three axes. The controls are connected in parallel so that they operate in a redundant fashion without switching. Each axis of control performs three functions:

1. Breakout Switches — A switch closure occurs whenever the control is moved 1.5 degrees from its rest position. Separate switches are provided in each axis and for each direction of rotation. These six switches are used to provide command signals to the CM computer, stabilization and control minimum impulses, acceleration commands, attitude gyro caging, and proportional rate commands in the electronics.

2. Transducers — These produce alternating current signals proportional to the rotation control displacement from the null position which are used to command spacecraft rotation rates

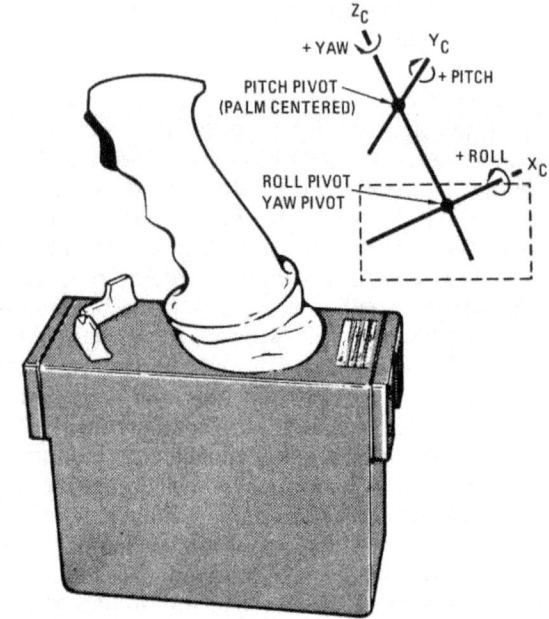

Rotation control

and to command service propulsion engine gimbal position in pitch and yaw during manual thrust vector control. All three transducers can be used simultaneously.

3. Direct Switches — A switch closure occurs whenever the control is moved a nominal 11 degrees from its null position (hardstops limit control movement to ±11.5 degrees from null in all axes). Separate switches are provided in each axis and for each direction of rotation. These switches are enabled by placing the DIRECT RCS switch to ON. Direct switch closure will produce acceleration commands through the direct solenoids on the RCS engines. All SCS electronics are bypassed.

The rotation control has a tapered female dovetail on each end of the housing which mates with mounting brackets on the couch armrests and at the navigation station in the lower equipment bay. When attached to the armrests, the input axes are approximately parallel with spacecraft body axes. The control is spring-loaded to null in all axes. A trigger-type push-to-talk switch also is located in the control grip. Redundant locking devices are provided on each control.

The translation control gives the astronauts a means of accelerating the spacecraft in either direction along any of the three axes. The control is mounted with its axes approximately parallel to those of the spacecraft. Redundant switches close for each direction of control displacement. These switches command the CM computer and the reaction jet and engine control. A mechanical lock inhibits the commands.

The controls T-handle may be rotated in either direction about the centerline of the shaft on which it is mounted. Hardstops for these rotations are ±17 degrees from null with detent positions encountered at a nominal ±12 degrees. In the detent position the hand can be removed and the T-handle will not return to null.

The clockwise switches will transfer spacecraft control from the CM computer to the stabilization

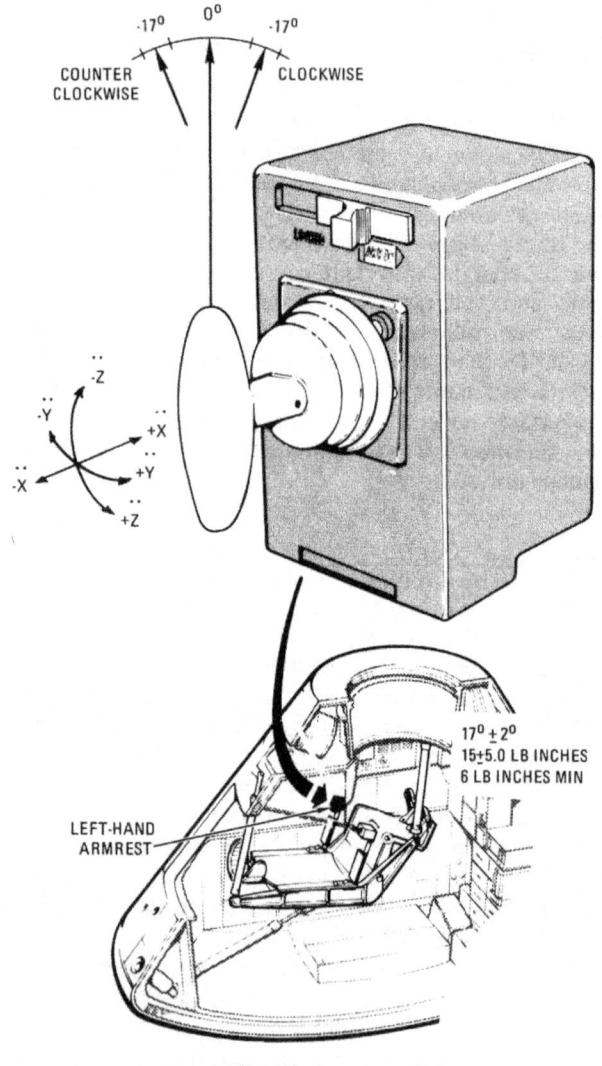

Translation control

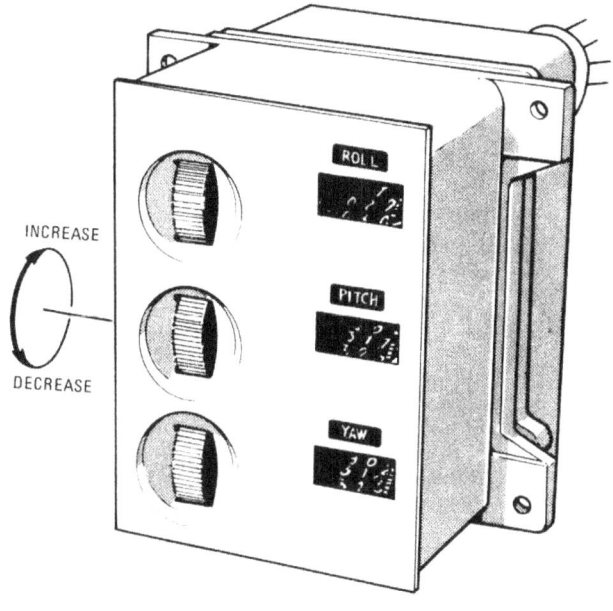

Attitude set display

The gimbal position and fuel pressure indicator contains redundant indicators for both the pitch and yaw channels. During the boost phases the indicators display fuel and oxidizer pressures for the Saturn second and third stages. Second stage fuel pressure (or third stage oxidizer pressure depending on the launch vehicle configuration) is on the redundant pitch indicators while third stage fuel pressure is on the two yaw indicators. The gimbal position indicator consists of two dual servometric meter movements mounted within a common hermetically sealed case. Thumbwheels enable crewmen to set service propulsion engine gimbal angler for stabilization and control subsystem velocity change maneuvers. Desired gimbal trim angles are set in with the pitch and yaw trim thumbwheels. The indicator displays service propulsion engine position relative to actuator null and not body axes. The range of the engine pitch and yaw gimbal displays is ±4.5 degrees. This range is graduated with marks at each 0.5 degree and reference numerals at each 2-degree division. The range of the fuel pressure scale is 0 to 50 psi with graduations at each 5-psi division and reference numerals at each 10-psi division.

and control subsystem. It may also transfer control between certain submodes within the stabilization and control subsystem. The redundant counterclockwise switches initiate abort during the launch phase. A discrete signal from switch closure is fed to the master events sequence controller which initiates other abort functions.

The attitude set control panel provides thumbwheels to position resolvers for each of the three axes. The resolvers are mechanically linked with indicators to provide a readout of the dialed angles. The signals to these attitude set resolvers are from the inertial measurement unit or the gyro display coupler.

The panel counters indicate resolver angles in degrees, and allow continuous rotation from 000 through 359 to 000 without reversing direction. There are graduation marks every 0.2 degree. Pitch and roll are marked continuously between 0 and 359.8 degrees. Yaw is marked continuously from 0 to 90 degrees and from 270 to 359.8 degrees; it is also marked with 0.2-degree graduation marks from 270 to 0 to 90 degrees and is numbered at 180 degrees. Readings increase for an upward rotation of the thumbwheels. One revolution of the thumbwheel produces a 20-degree change in the resolver angle and a corresponding 20-degree change in the counter reading.

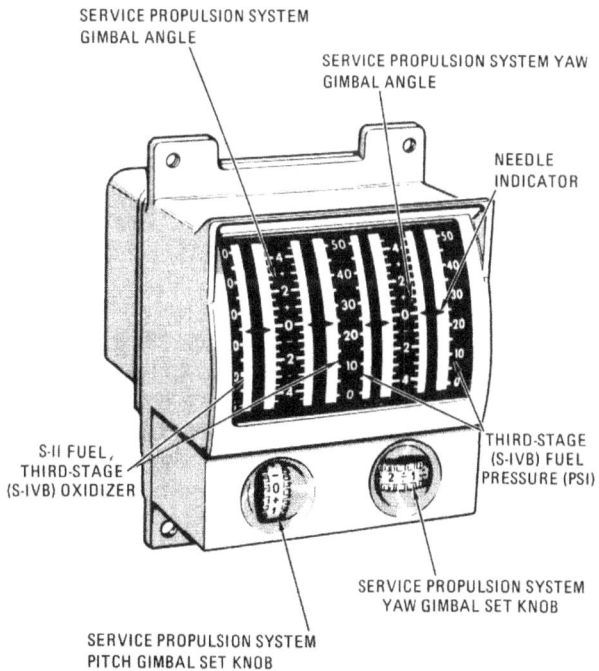

Fuel pressure/gimbal position indicator

The body rate (roll, yaw, or pitch) displayed on either or both flight director attitude indicators is derived from the body-mounted attitude gyros. Positive angular rates are indicated by a downward displacement of the pitch rate needle and by leftward displacement of the yaw and roll rate needles. The angular rate displacements are related to the direction of motion required by the rotation control to reduce the indicated rates to zero. The angular rate scales are marked with graduations at null and full range, and at 1/5, 2/5, 3/5, and 4/5 of full range. Full-scale deflection ranges are obtained with the FDAI SCALE switch and are 1, 5, and 10 degrees per second in pitch and yaw, and 1, 5, and 50 degrees per second in roll. Servometric meter movements are used for the three rate indicator needles.

The indicator's attitude error needles show the difference between the actual and desired spacecraft attitude. Positive attitude error is indicated by a downward displacement of the pitch error needle, and by a leftward displacement of the yaw and roll error needles. The attitude error needle displacements also are related to the direction of motion required by the rotation control to reduce the error to zero. The ranges of the error needles are 5 degrees or 50 degrees for full-scale roll error, and 5 degrees or 15 degrees for pitch and yaw error. The error scale factors are selected by a switch that also establishes the rate scales. The pitch and yaw attitude error scales contain graduation marks at null and full scale, and at 1/3 and 2/3 of full scale. The roll attitude scale contains marks at null, 1/2, and full scale. The attitude error indicators also use servometric meter movements.

Spacecraft orientation with respect to a selected inertial reference frame is also displayed on the attitude indicator ball. This display contains three servo control loops that are used to rotate the ball about three independent axes. These axes correspond to inertial pitch, yaw, and roll. The control loops can accept inputs from either the inertial measurement unit gimbal resolvers or gyro display coupler resolvers.

Pitch attitude is represented on the ball by great semicircles. The semicircle displayed under the inverted wing symbol is the inertial pitch at the time of readout. The two semicircles that make up a great circle correspond to pitch attitudes of θ and $\theta+180$ degrees.

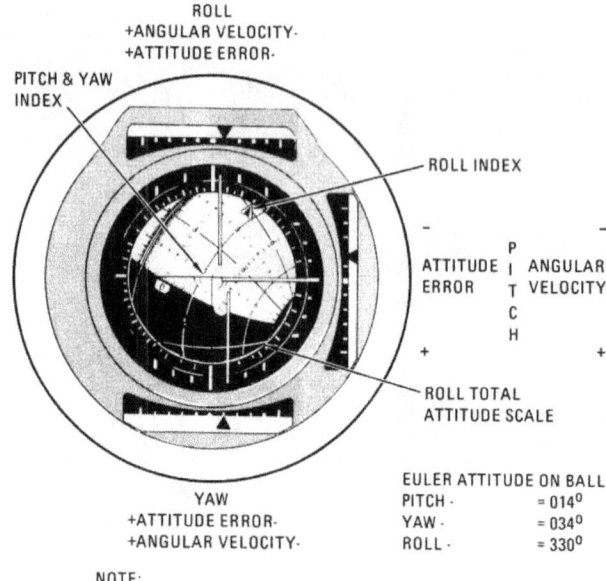

P-245 *Flight director attitude indicator*

Yaw attitude is represented by minor circles. The display readout is similar to the pitch readout. Yaw attitude circles are restricted to the intervals of 270 to 360 degrees (0) and 0 (360) to 90 degrees.

Roll attitude is the angle between the wing symbol and the pitch attitude circle. The roll attitude is more accurately displayed on a scale attached to the indicator mounting under a pointer attached to the roll (ball) axis.

The last digits of the circle markings are omitted. Thus, for example, 3 corresponds to 30, and 33 corresponds to 330. The ball is symmetrically marked about the 0-degree yaw and 0/180-degree pitch circles. Marks at 1-degree increments are provided along the entire yaw 0-degree circle. The pitch 180-degree semicircles have the same marking increments as the 0-degree semicircle. Numerals along the 300- and 60-degree yaw circles are spaced 60-pitch degrees apart. Numerals along the 30-degree yaw circle are spaced 30-pitch degrees apart. Red areas of the ball indicate gimbal lock.

Each gyro assembly contains three body-mounted attitude gyros mounted so that the input axis of one gyro is parallel to one of the spacecraft body axes. Gyro assembly No. 2 provides signals for display of body rates on either or both attitude indicators; rate damping for stabilization and control subsystem control configurations (excluding acceleration command and minimum impulse), and computation of inertial attitude changes.

The gyro assembly gyros provide signals equivalent to body attitude errors (deviations). These signals can be used for attitude control or display on the attitude indicator. The gyros also can provide backup rate signals for the functions of gyro assembly No. 2.

The Block II stabilization and control subsystem uses a switching concept as opposed to the "mode select" switching used in Block I Apollo spacecraft. Functional switching means the manual operation of a number of independent switches to configure the subsystem for various mission functions (e.g., course correction, velocity changes, entry, etc.). Mode switching would, for example, use one switch labeled "entry" to accomplish automatically all the necessary system gain changes, etc., for that mission phase. Mode selection simplifies crew tasks but limits system flexibility. Functional switching offers flexibility in selection of subsystem elements and allows part of a failed signal path to be turned off without affecting the total signal source.

There are two types of controls selectable from the main display console: stabilization and control subsystem or CM computer. The computer is the primary method of control and the stabilization and control subsystem is the backup. Attitude control is obtained from the reaction control engines and thrust vector control from the service propulsion engine.

ATTITUDE REFERENCE

The gyro display coupler provides signals to either of the attitude indicators for display of spacecraft total attitude and attitude errors. Angular velocity for display will always be supplied from either of the two gyro assemblies. The spacecraft total attitude display requires a connection between either gyro assembly and the attitude ball. This combination provides a backup attitude reference system for accurate display of spacecraft position relative to a given set of reference axes. Spacecraft attitude errors can be developed using the attitude set control panel connection between the gyro display coupler and the attitude indicator error needles. This combination provides a means of aligning the attitude reference system to a fixed reference while monitoring the alignment process on the error needles; it can also be used in conjunction with manual maneuvering of the spacecraft.

The gyro display coupler can be operated in the following configurations:

Alignment — Provides a means of aligning the coupler to a given reference.

Euler — Computes total inertial attitude and body attitude error from body rate signal inputs.

Non-Euler — Computes digital body rate signals from dc body rate signal inputs.

Entry ≤ 0.05 G — Computes roll axis attitude about the entry roll stability axis from body rate signal inputs.

The alignment function is used to align the coupler Euler angles (shafts) to the desired inertial reference selected by the attitude-set thumbwheels (resolvers). This is done by interfacing the coupler resolvers with the ASCP resolvers in each axis to generate error signals that are proportional to the difference between the resolver angles. These error signals are fed back to the gyro display coupler to drive the resolver angular difference to zero. During this operation all other functions for the coupler are inhibited.

In the Euler configuration, the coupler accepts pitch, yaw, and roll dc rate signals from either gyro assembly and transforms them to Euler angles to be displayed on either attitude ball. The coupler Euler angles also are sent to the attitude set control panel to provide an Euler angle error, which is transformed to body angle errors for display on either attitude error indicator.

Non-Euler pitch, yaw, and roll dc body rate signals from either gyro assembly are converted to digital body rate signals and sent to the CM computer. Power is removed from both attitude ball-drive circuits when this configuration is selected.

In the entry ≤ 0.05 G operation, the coupler accepts yaw and roll dc body rate signals from either gyro assembly and computes roll attitude with respect to the stability axis to drive the roll stability indicator on the entry monitor system or from gyro assembly No. 1 and computes roll attitude with respect to the stability axis to drive either attitude ball in roll only.

The purpose of displaying total attitude, attitude error, and rate is to monitor the spacecraft and system functions. Since a single source can supply more than one type of information, the mission function required at a particular time will normally dictate what type of information is required from the source.

The flight director attitude indicator may be modified by an orbital rate display-earth and lunar unit. This unit is inserted electrically in the pitch channel between the electronic display assembly and attitude indicator to provide a local vertical display in the pitch axis. Local vertical is attitude with respect to the body (earth or moon) the spacecraft is orbiting. Controls on the unit permit selection of earth or lunar orbits and orbital attitude adjustment.

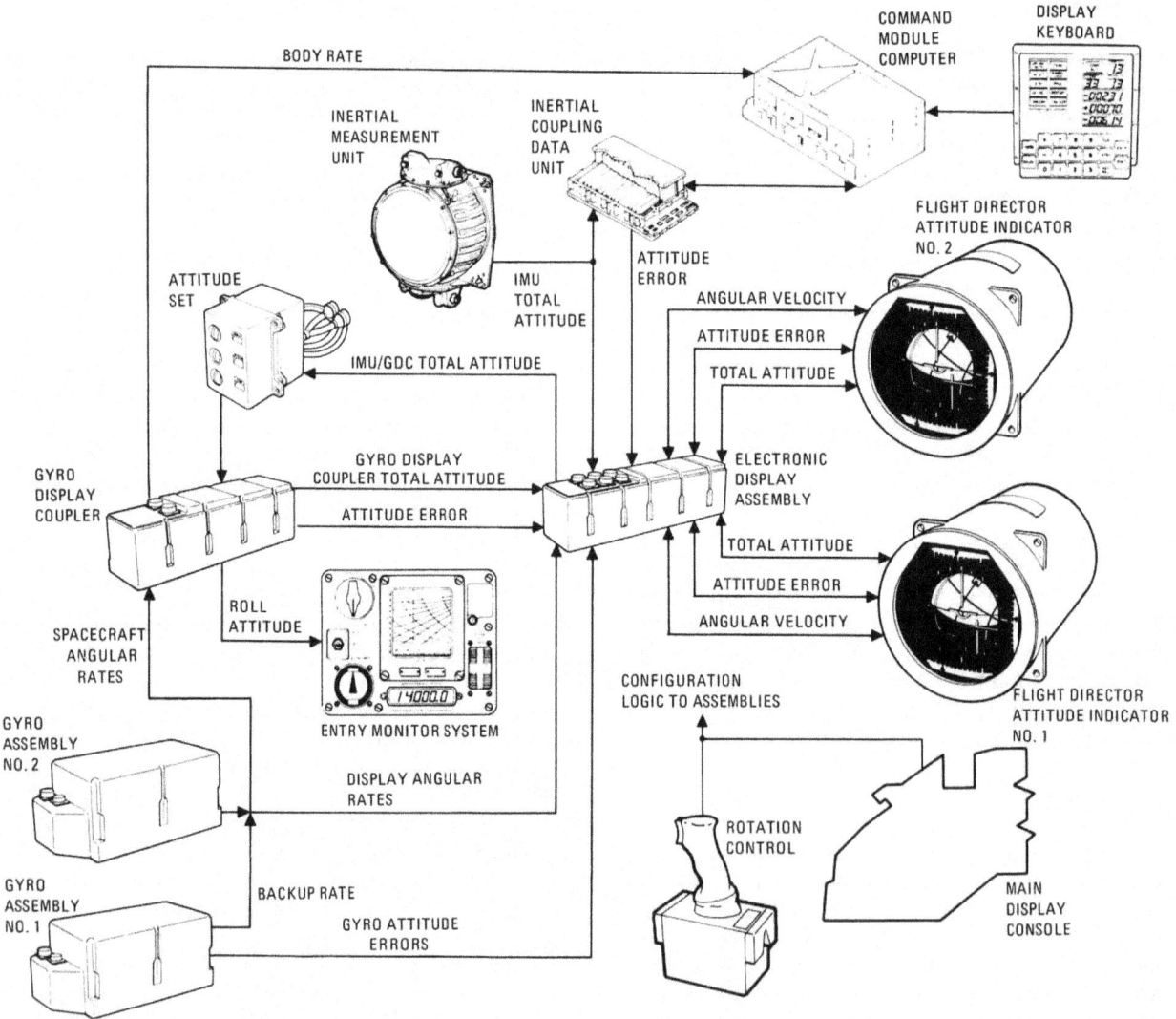

Attitude reference equipment and flow

ATTITUDE CONTROL

The attitude and translation control portion of the stabilization and control subsystem uses the reaction control subsystem. The reaction control subsystem and stabilization and control equipment are described only in relation to attitude control. Stabilization and control equipment used for attitude control includes:

1. Gyro assembly No. 1, where three body-mounted attitude gyros provide pitch, yaw, and roll attitude error signals for use when automatic attitude hold is desired.

2. Gyro assembly No. 2, whose three body-mounted attitude gyros provide pitch, yaw, and roll rate damping for automatic control and proportional rate maneuvering.

3. The rotational controllers which enable crewmen to control the spacecraft attitude simultaneously in three axes.

4. The translation controller which enables the crew to command simultaneous accelerations along all three spacecraft axes and is also used to initiate several transfer commands.

5. The electronics control assembly which contains the electronics used for automatic, proportional rate, and minimum impulse capabilities.

6. The reaction jet engine control which contains the automatic reaction control subsystem logic and the solenoid drivers that provide commands to the automatic coils of the reaction control engines.

The reaction control subsystem provides the rotation control torques and translational thrusts for all attitude control functions. The reaction control engine is operated by applying excitation to a pair (fuel and oxidizer) of solenoid coils. Each engine has two pairs of solenoid coils, one automatic and the other direct.

Commands to the reaction control engines are initiated by switching a ground through the solenoid driver to the low side of the automatic coils. The solenoid drivers receive commands from the automatic reaction control subsystem logic circuitry contained in the reaction jet engine control. The automatic reaction control subsystem logic activates the command source selected and commands the solenoid drivers necessary to perform the desired attitude control function. The logic receives reaction control commands from the CM computer (for rotation and translation), the electronics control assembly (rotation for either automatic, proportional rate, or minimum impulse control), the rotation controls (for continuous rotational acceleration), and the translation control (for translational acceleration).

Commands to the direct coils have priority over those to the automatic coils. The direct coils receive commands from the rotation control when the "Direct RCS" switch is actuated and the control is deflected 11 degrees about one or more of its axes. The direct coils are used to command an ullage maneuver before a service propulsion engine firing of normal ullage methods are not available; this is controlled through a pushbutton on the main display console. The master events sequence controller also can command an ullage maneuver to enable separation of the CSM from the third stage. Direct coils on the SM reaction control engines are activated by the SM jettison controller for SM-CM separation. CM direct coils are activated by a switch on the main display console for reaction control propellant dumping during the final descent on the main parachutes.

The stabilization and control subsystem can be placed in various configurations for attitude control, depending on crew selection of control panel switches. The configuration desired is selected independently for each axis. Both automatic and manual control can be selected.

Automatic control involves rate damping and attitude hold. In rate damping large spacecraft rates are reduced to a small range (rate deadband) and held within the range. In attitude hold, angular deviations about the body axes are kept within certain limits (attitude deadband). If attitude hold is selected in pitch, yaw, and roll, the control can be defined as maintaining a fixed inertial reference. Rate damping is used in the mechanization of the attitude hold configuration.

Attitude hold uses the control signals provided by the body-mounted attitude gyros. These signals are summed in the electronics control assembly switching amplifier. The switching amplifier has two output terminals that provide commands to the automatic reaction control subsystem logic: one terminal provides positive rotation commands and the other negative commands. If the magnitude of the signal input is smaller than a specific value, neither output is obtained. The input level required to obtain an output is referred to as the switching amplifier deadband. This can be interpreted as a rate deadband or an attitude (minimum) deadband. The deadband limits are a function of the control loop gains which depend on the position of the "Rate" switch: low is ±0.2 degree per second and high is ±2 degrees per second. An additional deadband can be selected for attitude control: ±0.2 to 4.2 degrees in low and ±4 to 8 degrees in high.

When the summation of rate and attitude signals exceeds the switching amplifier deadband, a rotation command is sent to the reaction control subsystem and the engines are automatically fired for the duration needed to correct the deviation.

Manual attitude control involves proportional rate, minimum-impulse, acceleration command, and direct control. These commands are initiated by operation of either rotation control. With the exception of direct, the rotation control commands go through the reaction control subsystem automatic coils.

Proportional rate provides the ability to command a spacecraft rate that is directly related to

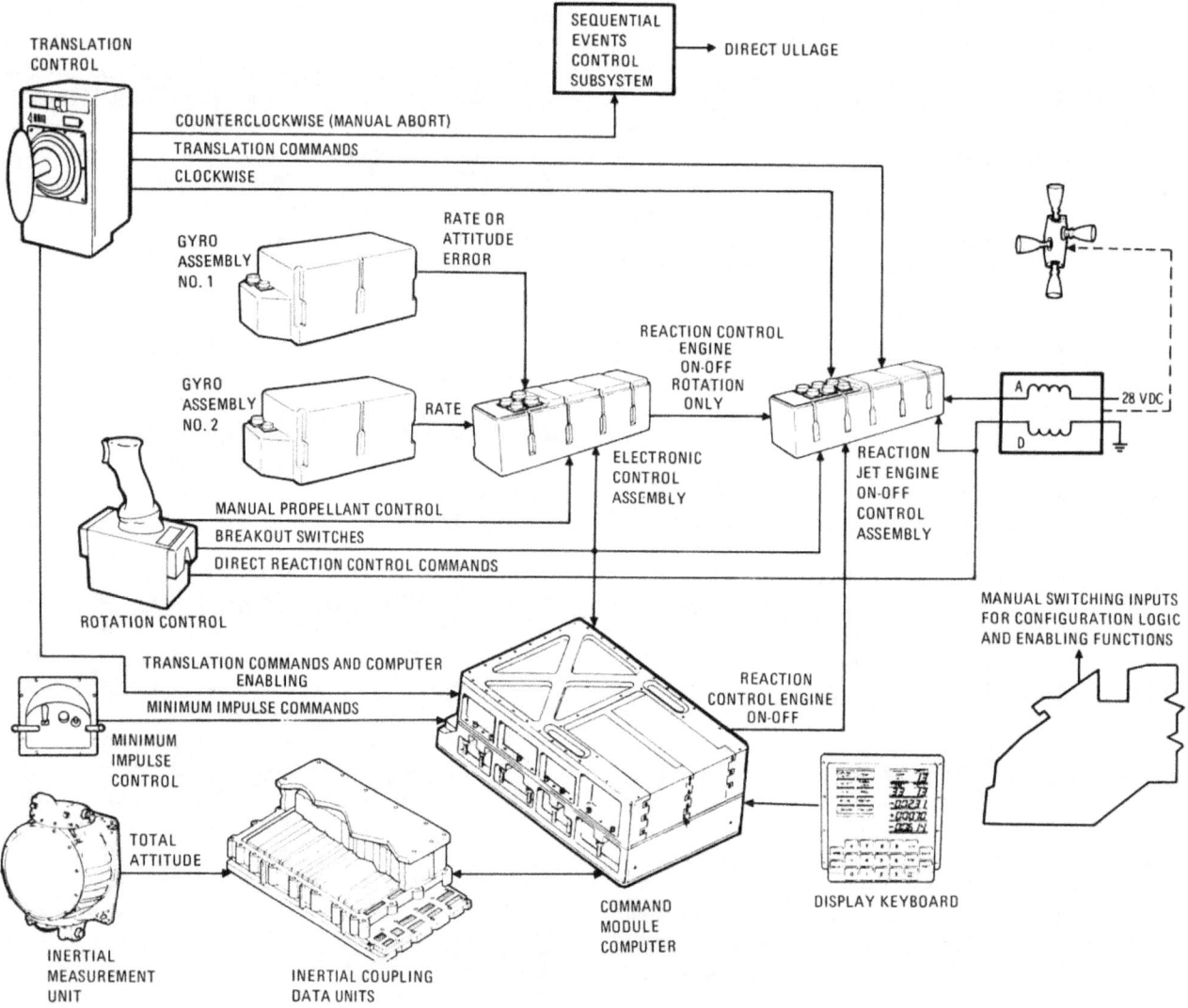

Attitude control equipment and flow

the amount of rotation control stick deflection. This capability is obtained by summing the control's transducer output with the body-mounted attitude gyro signal in the electronics control assembly; when the stick is deflected, an error is developed at the switching amplifier input that results in an acceleration command. The command is present until the gyro signal is large enough to reduce the error to less than the deadband. The spacecraft will then coast at a constant rate until the rotation control input is removed.

Minimum impulse provides the ability to make small changes in the spacecraft rate. When minimum impulse is enabled in an axis, the output of the switching amplifier in that axis is inhibited. Thus, the spacecraft (attitude) is in free drift in the axis where minimum impulse is enabled if direct control is not being used. Minimum impulse is commanded by the rotation control breakout switch. When minimum impulse is selected in the roll axis, one-half of the roll solenoid drivers are inhibited for minimum impulse commands. A roll minimum impulse command is executed by two reaction control engines unless one of the "Channel Roll" switches is turned off, which reduces the command to a single engine.

When acceleration command is activated and a breakout switch is closed, continuous commands are sent to the appropriate reaction control subsystem automatic coils. The selection of acceleration command in an axis inhibits all other inputs to the automatic reaction control subsystem logic for that axis. This differs from minimum impulse selection in that translation control commands are available during minimum impulse control.

Direct rotation control is available as backup to any other control, including CM computer when the "Direct RCS" switch is on. When the rotation control stick is deflected to hard stop, a direct switch is closed and the voltage is routed to the direct coils on the appropriate reaction control engines. The control's direct switch also routes a signal to the automatic reaction control subsystem logic that inhibits all automatic coil commands in the axis under direct control.

Commands from the translation control can be initiated simultaneously in the three axes and appear as logic inputs to the automatic reaction control subsystem logic. The logic signals are obtained from switch closures in the control. Translation control is not available after CM/SM separation.

THRUST VECTOR CONTROL

Spacecraft attitude is controlled during a velocity change by positioning the engine gimbals for pitch and yaw control while maintaining roll attitude with the attitude control subsystem. The stabilization and control electronics can be configured to accept attitude sensor signals for automatic control or rotation control signals for manual control. Manual thrust vector control can be selected to use vehicle rate feedback signals summed with the manual signals. A different configuration can be selected for each axis; for example, one axis can be controlled manually while the other is controlled automatically.

In automatic thrust vector control, spacecraft angular rates and attitude errors are sensed by the body-mounted attitude gyros. Attitude error, gimbal position, and gimbal trim signals are summed at the input to an integrator amplifier. The integrator output is then summed with rate, attitude error, gimbal position, and gimbal rate at the servo amplifier input.

Steady-state operation is obtained when the gimbal is positioned so that the thrust vector is

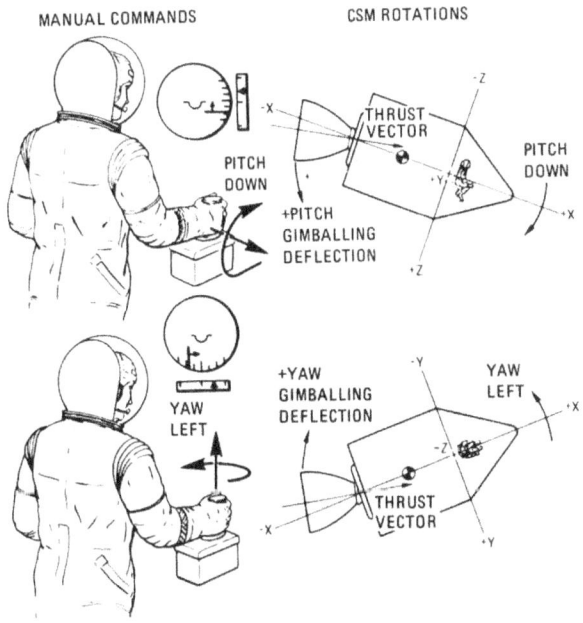

P-248 *Gimballing of service propulsion engine*

aligned through the vehicle center of gravity and the error at both summing points is a constant-zero. The integrator input error is zero when gimbal position minus gimbal trim is equal to the negative of the attitude excursion sensed by gyro assembly No. I. This is the spacecraft/gimbal orientation necessary to obtain and maintain the desired thrust direction. Transients due to center-of-gravity uncertainty errors or shifts during thrusting are forced by the integrator to have the necessary steady-state solution. However, final pointing vector errors will be incurred because of the quadrature accelerations induced during the transient phases. Errors also will result from amplifier gain and component inaccuracies.

The gimbals are trimmed before thrust by turning the trim wheels on the gimbal position indicator. The trim wheel in each axis is mechanically connected to two potentiometers connected with the gimbal servomechanisms. It is desirable to trim before velocity change to minimize the transient duration time and the accompanying quadrature acceleration. The trim wheels also are set before a velocity change controlled by the CM computer so that the stabilization and control subsystem can relocate the desired thrust direction if a transfer is required after engine ignition.

In manual thrust vector control, the signal from the rotation control is sent to a proportional plus integral amplifier. This circuit maintains a gimbal deflection after the rotation control is returned to rest and makes corrections with the control about its rest position, rather than holding a large displacement. Depending on how switches are set, it also can damp out spacecraft rate.

There are two manual thrust vector control configurations: rate command and acceleration command. Rate command is similar to the proportional rate control in the attitude control subsystem except there is no deadband. The thrust

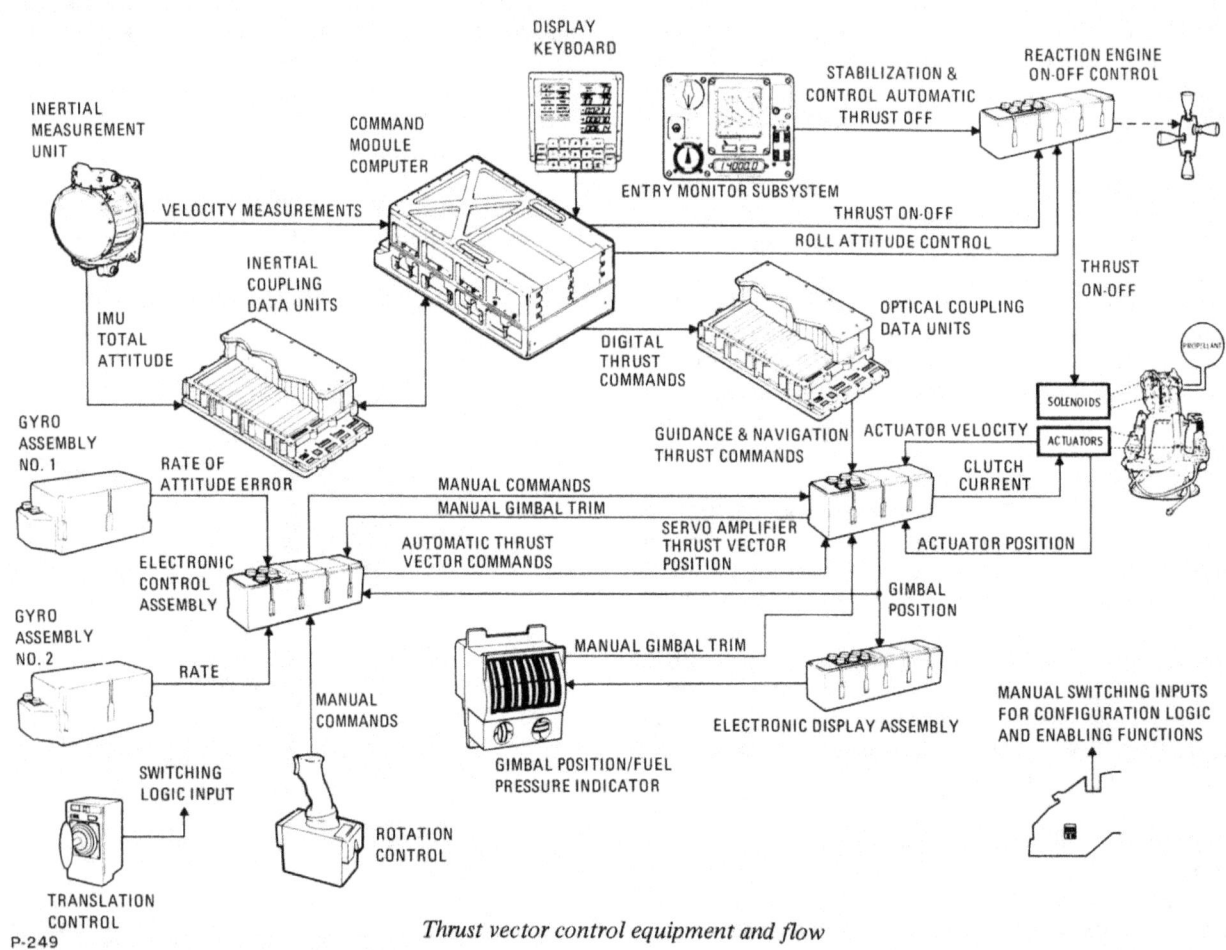

Thrust vector control equipment and flow

vector is under body-mounted attitude gyro control. If there is an initial gimbal center-of-gravity misalignment, an angular acceleration will develop. The gyros, through the proportional gain, will drive the gimbal in the direction necessary to cancel this acceleration. The rate feedback is inhibited in acceleration command; the rotation control input must be properly trimmed to position the thrust vector through the center of gravity. This drives the rotational acceleration to zero but additional adjustments are necessary to cancel residual rates and obtain the desired attitude and positioning vector.

ENTRY MONITOR SYSTEM

The entry monitor system provides a visual display of automatic guidance navigation and control system entry and velocity change maneuvers. It also provides sufficient display data to permit manual entry in case of guidance and control malfunctions and automatic velocity cutoff commands for the stabilization and control subsystem when controlling the service propulsion engine. The velocity display also can be used to cut off thrust manually if the automatic commands malfunction.

The system provides five displays used to monitor an automatic entry or perform a manual entry: threshold indicator, roll attitude indicator, corridor verification indicators, range displays, and the flight monitor.

The threshold indicator, labeled .05G, displays sensed deceleration. The altitude at which this indicator is illuminated depends on entry angle (velocity vector with respect to local horizontal), the magnitude of the velocity vector geographic location and heading, and atmospheric conditions. Bias comparator circuits and timers are used to activate this indicator. The signal used to illuminate the indicator is also used in the system to start the corridor evaluation timer, scroll velocity drive, and range-to-go circuits.

The roll attitude indicator displays the lift vector position throughout entry. During entry, stability axis roll attitude is supplied to the indicator by the gyro display coupler. There are no degree markings on the display, but the equivalent readout will be zero when the indicator points toward the top of the control panel and increases up to 360 in a counterclockwise direction.

The corridor verification indicators are located on the roll attitude indicator. They consist of two lights which indicate the necessity for lift vector up or down for a controlled entry. (The indicators are valid only for spacecraft entering at velocities and angles that will be used on the return from the moon.) The corridor comparison test is performed approximately 10 seconds after the .05G indicator is illuminated. The lift vector up light (top) indicates greater than approximately 0.2G. The lift vector down light (bottom) indicates less than approximately 0.2G. An entry angle is the angle displacement of the CM velocity vector with respect to local horizontal at 0.05G. The magnitude of the entry angles that determines the capture and undershoot boundaries depend on the CM lift-to-drag ratio. Entry angle less than the capture boundary will result in noncapture regardless of lift orientation. Noncapture would result in an elliptical orbit which will re-enter when perigee is again approached. The critical nature of this would depend on CM consumables: power, control propellant, lift support, etc. The CM and crew would undergo excessive G force (greater than 10G's) with an entry angle greater than the undershoot boundary, regardless of lift orientation.

The range display is an electronic readout of inertial flight path distance in nautical miles to predicted splashdown after 0.05G. The predicted range will be obtained from the guidance and control subsystems or ground stations and inserted into the range display before entry. The range display also shows velocity in feet per second during service propulsion engine thrusting.

The flight monitor provides an entry trace of total G level versus inertial velocity. A Mylar scroll has printed guide lines which provide monitor (or control) information during aerodynamic entry. The entry trace is generated by driving a scribe in a vertical direction as a function of G level, while the Mylar scroll is driven from right to left proportional to the CM inertial velocity change. Monitor and control information for safe entry and range potential can be observed by comparing the slope of entry trace to the slope of the nearest guide lines.

In addition to entry functions, the entry monitor system provides outputs related to delta velocity maneuvers during either service propulsion or reaction control engine thrusting. Displays include a lamp which lights any time the service propulsion engine fires and a counter which shows

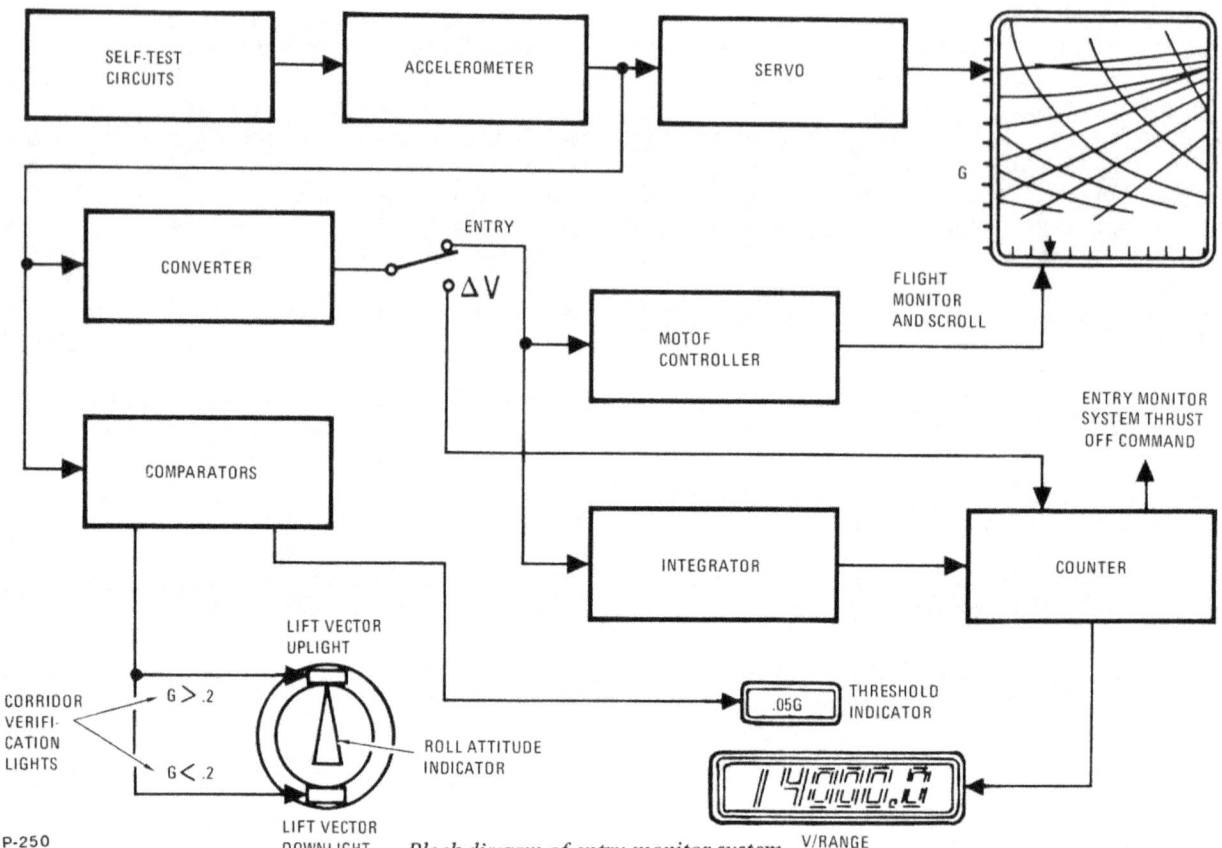

Block diagram of entry monitor system

the velocity remaining to be gained or lost. The latter display can have a range of 14,000 to -1000 feet per second in tenths of a foot per second. The desired velocity change for all service propulsion engine thrusting maneuvers is set in the panel and the display will count up or down. During thrust controlled by the stabilization and control subsystem, the entry monitor system automatically turns off the service propulsion engine when the display reads minus values.

The Mylar scroll in the entry monitor systems flight monitor has ground and flight test patterns together with four entry patterns. Each entry pattern is preceded by two identical flight test patterns and entry instructions that are used to verify operation of the system's entry circuits and set initial conditions. Each entry pattern contains velocity increments from 37,000 to 4,000 feet per second as well as entry guidelines. The entry guidelines are called G on-set or G off-set and range potential lines. During entry the scribe trace should not become parallel to either the nearest G on-set or G off-set lines. If the slope of the entry trace becomes more negative than the nearest G on-set line, the CM should be oriented so that a positive lift vector orientation (lift vector up) exists to prevent excessive G buildup. If the entry trace slope becomes more positive than the nearest G off-set line, the CM should be oriented to produce negative lift (lift vector down). The G on-set and off-set lines are designed to allow a 2-second crew response time and a 180-degree roll maneuver if the entry trace becomes parallel to the target of the nearest guideline.

The range potential lines, shown in hundreds of nautical miles, are used by the crew during entry. They indicate the ranging potential of the CM at the present G level. The crew will compare the range displayed by the range-to-go counter with the range potential indicated at the position of the entry trace. The slope and position of the entry trace relative to a desired ranging line indicates the need for lift vector up or down.

The vertical line on the scroll corresponds to where the CM velocity becomes suborbital; that is, where the velocity has been reduced to less than that required to maintain orbit. The full positive lift profile line represents the steady-state minimum-G entry profile for an entry.

GUIDANCE AND NAVIGATION

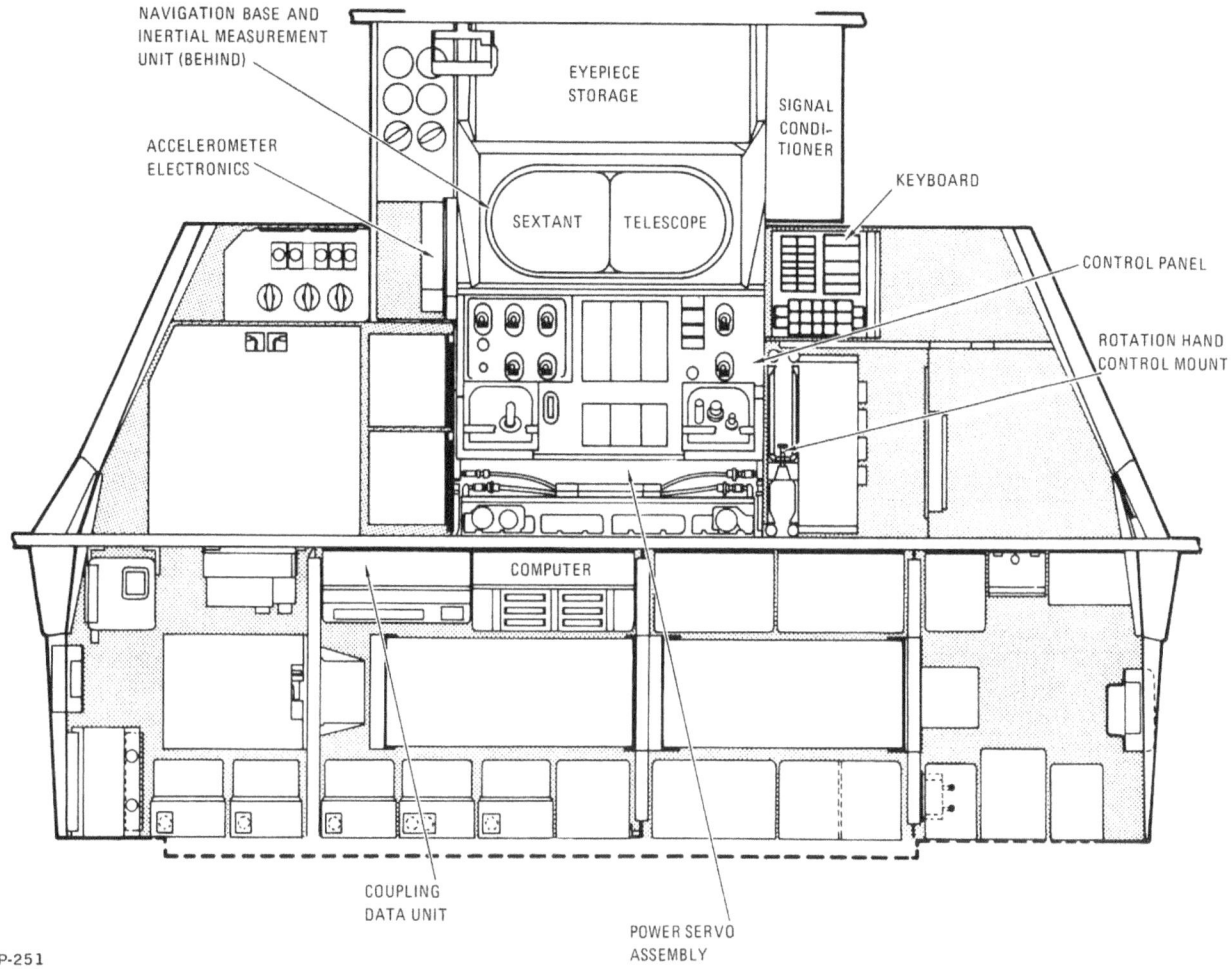

Location of guidance and navigation equipment in lower equipment bay

The guidance and navigation subsystem gives astronauts the ability to navigate the spacecraft on a required course through space. It can be operated either semi-automatically or manually and performs the basic functions of guidance and navigation -- similar to the navigation of an airplane or of a ship at sea. It can also be updated by the ground via telemetry.

While an airplane or ship at sea is concerned with two-dimensional navigation (it is always on or near the surface of the earth), Apollo is faced with exacting three-dimensional navigation as it speeds through deep space. Sightings from the spacecraft of stars and pre-determined landmarks on the earth and moon are used to establish the location and path of the spacecraft in space. The guidance and navigation system is used in conjunction with the stabilization and control, service propulsion, reaction control, electrical power, environmental control, and telecommunications subsystems.

Massachusetts Institute of Technology, an associate contractor to NASA's Manned Spacecraft Center, is responsible for development and design of the subsystem. AC Electronics Division of General Motors is responsible for its production, operation, and integration. The guidance computer is manufactured by the Raytheon Co. and the optics are produced by Kollsman Instrument Co.

There are three main elements in the guidance and navigation subsystem. The inertial guidance subsystem measures changes in the spacecraft position and velocity and assists in generating steering commands. The optical subsystem is used to take precision navigational sightings and provide the computer with measured angles between the stars and landmarks. The computer subsystem consists of a digital computer which takes data from the inertial guidance and optical subsystems and calculates spacecraft position, velocity, and steering commands.

The components of the guidance and navigation subsystem, including the primary controls and displays, are located in the lower equipment bay of the command module. The main display console contains the switches and displays necessary for the astronauts to control the spacecraft while in their couches.

INERTIAL GUIDANCE SUBSYSTEM

This subsystem measures changes in the spacecraft position, assists in generating steering commands, and measures spacecraft velocity changes. Its instruments sense changes in velocity and attitude in a manner similar to the balance system in the human ear.

The main part of this subsystem is an inertial measurement unit mounted on a navigation base. The key part of this mechanism is a device called a stable platform, suspended on gimbals which allow it to incline freely regardless of spacecraft position. It is aligned to star direction references and retains this alignment regardless of the rotational movement of the spacecraft and thus provides a reference against which the movements of the spacecraft can be measured.

Mounted on this stable platform are the actual sensors: three accelerometers and three gyroscopes. The gyros are used to keep the stable platform fixed with respect to some point in space. The accelerometers sense any change in the speed of the vehicle -- forward, backward, up, down, sideways.

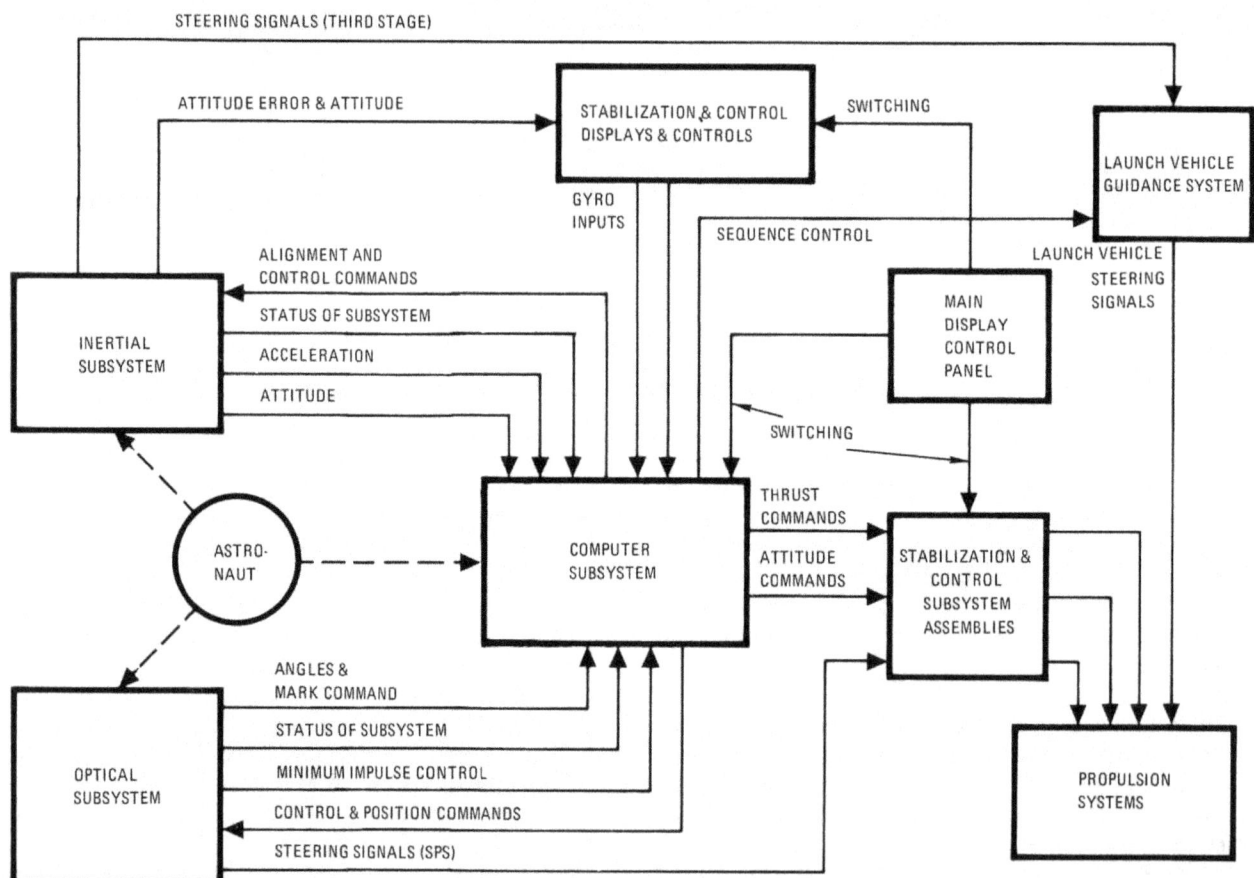

Data flow in guidance and navigation subsystem

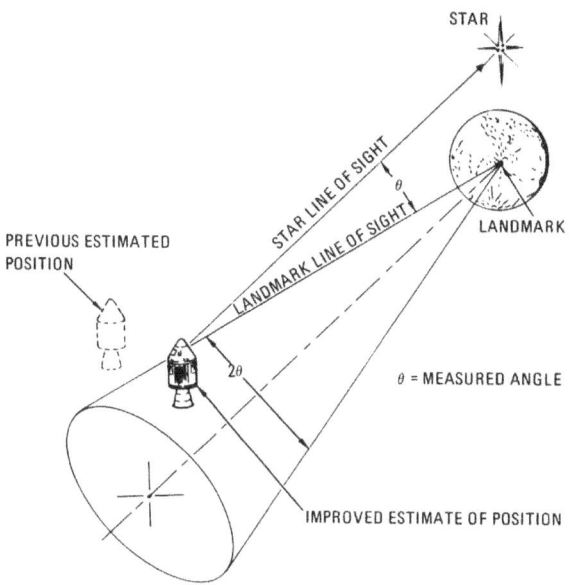

P-253

Determination of midcourse position

60-degree field of view with no magnification and is used to obtain coarse sightings of the stars or landmarks. Because of the telescope's limited field of view, controls are provided so the astronaut can maneuver the entire spacecraft to point the instrument in the general desired direction. The sextant is used to take the precision sightings and has a much smaller field of view (1.8 degrees) but has a magnification of 28.

These two instruments are used to measure the angle between two targets such as stars and earth or moon landmarks. The telescope and then the sextant are manipulated by the astronaut to line up a sighting and enter the reading into the computer.

COMPUTER SUBSYSTEM

This subsystem consists of a digital computer which stores and uses signals from the inertial guidance subsystem and sightings by the optical subsystem with other data. This highly sophisticated computer uses this data to calculate necessary corrections to maintain course. The computer memory contains 38,912 words, or pieces of guidance data, and is divided into non-erasable and erasable sections. The non-erasable memory contains all the basic data necessary to achieve the round trip to the moon. The erasable memory is used by the astronauts when performing the various guidance and navigation computations.

The major functions of the computer are to

Any change from the pre-determined flight path or attitude generates electronic signals which result in firing of the reaction control subsystem engines (for attitude change only) or positioning of the service propulsion engine for flight path changes. The service propulsion engine is fired on signal from the computer or astronauts.

Devices known as resolvers, mounted on the gimbal axes, measure how far the spacecraft has rotated with respect to the stable platform. These measurements are transmitted to the guidance computer. While the gyros are used to maintain the spacecraft in a required attitude, the resolvers are used primarily to orient the spacecraft when firing the service propulsion engine.

The inertial guidance subsystem is controlled automatically by the guidance computer through crew selection of a computer program.

OPTICAL SUBSYSTEMS

This subsystem is used by the astronauts to take navigational sightings of the stars and earth or moon landmarks. It consists of a navigation base, a telescope, a sextant, and equipment to permit operation with the computer and inertial guidance subsystems.

The telescope and sextant can be operated independently but generally are used together to obtain precision navigational sightings. The telescope has a

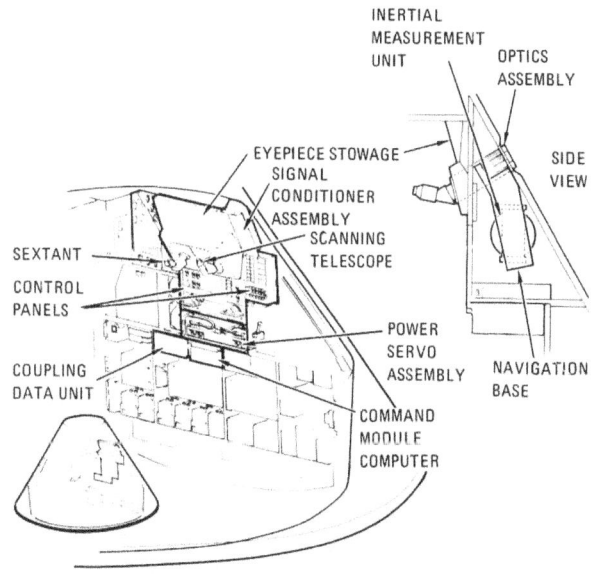

P-254

Optical equipment installation

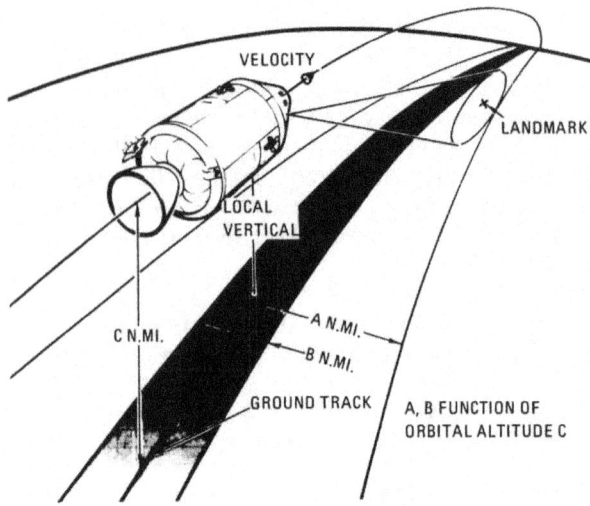

P-255 *Orbital tracking of landmark*

calculate spacecraft position, velocity, and steering data; to calculate the signals for the main engine and attitude control jets necessary to keep the spacecraft on the required flight path and attitude; to position the stable platform in the inertial measurement unit as defined by precise optical measurements; to position the optical unit to celestial objects; to monitor the guidance and navigation subsystems for failure indications; and to supply information to the display and control panel.

The display of information, the results of computations, and the control of the computer by the astronauts is accomplished with two display and keyboard panels. One is located in the lower equipment bay and the other (a duplicate) is on the main display console. Lights on the panels also indicate the detection of malfunctions in guidance and navigation systems.

SEQUENCE OF OPERATIONS

Operation of the guidance and navigation subsystem in specific flight phases, control modes, and critical maneuvers during a typical lunar mission is described here briefly.

To conserve electrical power and fuel, the subsystem is activated only about 20 percent of the time, for specific sightings, alignments, and engine-firing maneuvers. Each time the guidance and navigation subsystem is activated the stable platform must be aligned with respect to a predetermined reference. Before launch the platform is aligned with respect to the earth and during flight it is aligned to the stars.

Launch and Translunar Injection — Guidance of the launch vehicle is monitored during the ascent from earth and firing of the third stage which sends the spacecraft on a trajectory to the moon. At launch, the inertial measurement unit is switched from an earth reference to a space reference frame. The crew then uses the guidance and navigation subsystem to monitor the spacecraft flight profile.

Earth and Lunar Orbit — During these phases, the crew checks the spacecraft position and orbital path with optical sightings. The crew takes these sightings by identifying and tracking landmarks with the optical instruments. The computer records optical sighting data, spacecraft attitude, and the time of the optical sighting.

Midcourse Navigation — This may be performed several times during the translunar and transearth mission phases. Starlandmark sightings are taken and the computer records the angles and time of sighting and determines spacecraft position and velocity.

Course Correction — The spacecraft's course must be corrected during both the translunar and transearth journeys. The computer, after calculating spacecraft position and velocity from navigation sightings, determines the need for a course

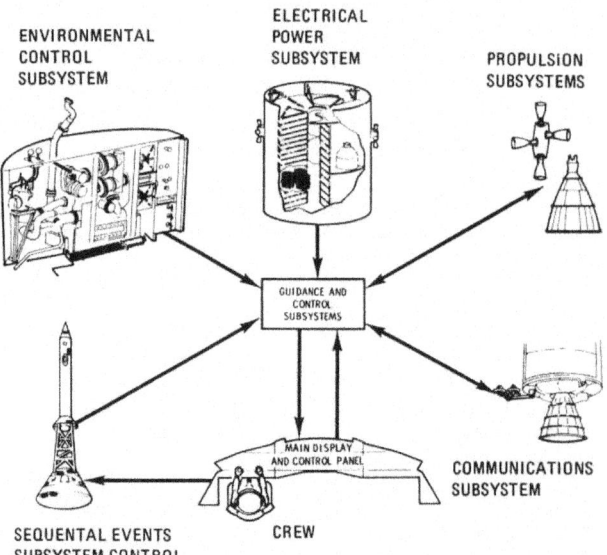

Guidance and control relationship to other subsystems

P-256

EQUIPMENT

The guidance and navigation system occupies a space 4 feet high, 2 feet deep, and about 3 feet across the top and 2-1/2 feet across the bottom. It is in the command module lower equipment bay.

Inertial Measuring Unit (AC Electronics Division of General Motors) — This ball-shaped unit has a diameter of 12.6 inches and weighs 42.5 pounds.

It consists of three gimbals of which the inner gimbal is the stable member, with three gyroscopes and three accelerometers, all can-shaped and mounted onto the stabilized inner member structure, or platform. The gimbals are connected to each other by drive motors and angle resolvers. The unit is pressurized in dry air for good heat transfer. When in operation, the unit requires 217 watts at 28 volts dc. It maintains an inertially referenced coordinate system for spacecraft attitude control and measurement and maintains three accelerometers in this coordinate system for accurate measurement of spacecraft velocity changes.

Navigation Base (AC Electronics) — This 27-by-22-by-4.5-inch unit weighs 17.4 pounds and is made of riveted and bonded anodized preformed sheet aluminum alloys. It is filled with polyurethane foam. It is a rigid supporting structure for the inertial measuring unit and optical equipment.

Power and Servo Assembly (AC Electronics) — It is 2.75 inches high, 23.1 inches wide, and 22.6 inches deep and weighs 49.4 pounds. It contains 37 modules, most of the electronic modules for the inertial and optical subsystem servo loops and power supplies.

Coupling Data Unit (AC Electronics) — It is 20 by 11.3 by 5.5 inches and weighs 36.5 pounds. This sealed unit contains modular packaged solid-state electronics necessary to provide five separate coupling data unit channels for use with inner, middle, and outer inertial measuring unit gimbal resolvers and the shaft and trunnion resolvers of the optical subsystem. It also contains failure detection circuitry for inertial subsystem and optical subsystem. It provides analog-to-digital conversion of the inertial measuring unit gimbal angles and optical subsystem trunnion and shaft angles, and digital-to-analog conversion of

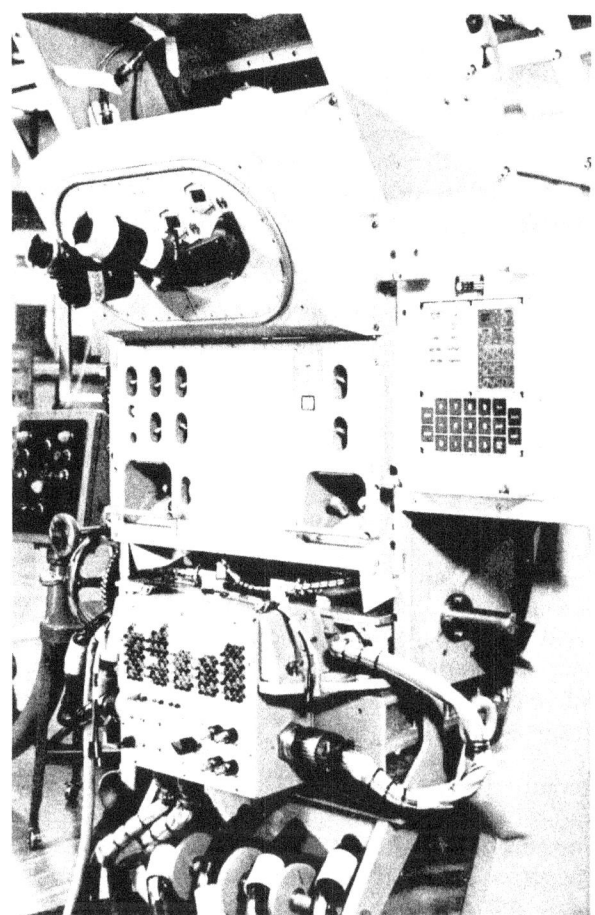

P-257 *Guidance and navigation equipment in test fixture simulating spacecraft installation*

correction by comparing the actual to the required trajectory. If a course correction is necessary, the computer calculates the time of firing and velocity change needed, repositions the spacecraft, and controls the initiation and duration of thrusting of the service propulsion engine.

Lunar Injection and Return to Earth — The guidance and navigation subsystem places the spacecraft into the attitude required for firing the service propulsion engine and controls the time of thrusting required.

Entry — The guidance and navigation subsystem controls the flight path of the command module during entry. The computer determines the proper trajectory and steers the command module by rolling it. This changes the lifting force acting on the command module and thereby varies its trajectory.

computer-derived data to control inertial subsystem and optical subsystem modes of operation using discretes issued by the computer. It also controls service propulsion system engine gimbaling, attitude error display, and, as a backup mode, it controls the attitude of the launch's third stage.

Inertial Reference Integrating Gyros (AC Electronics) — Three gyros are mounted mutually perpendicular to each other on the stable platform of the inertial measuring unit. Each is 2-1/2 inches in diameter. They sense displacement of the inertial measuring unit's stable platform and generate error signals. They are pressurized in an atmosphere of helium to provide good heat transfer.

Accelerometers (AC Electronics) — There are three pulse-integrating pendulous accelerometers mounted perpendicular to each other on the stable member of the inertial measuring unit. Each is can-shaped with a 1.6-inch diameter. They measure velocity changes along all axes of the three-axis inertial measuring unit.

Sextant (Kollsman Instrument Co., Syosset, N.Y.) — This is a highly accurate dual line-of-sight electro-optical instrument with 28X magnification and 1.8 degrees field of view. It can sight two celestial targets simultaneously and measure the angle between them with 10 arc seconds accuracy to determine the position of the spacecraft. It is mounted on the navigation base. One line of sight is fixed along the shaft axis normal to the local conical surface of the spacecraft. It is positioned by changes in the spacecraft attitude. The other line of sight has two degrees of rotational freedom about the shaft axis (plus or minus 270 degrees) and trunnion axis (minus 5 to plus 50 degrees). The variation about the trunnion axis is represented by movement of an indexing mirror.

Scanning Telescope (Kollsman) — It is a single line-of-sight, refracting-type, IX magnification instrument with 60-degree instantaneous field of view. It is similar to theodolite (surveyor's instrument used to measure horizontal and vertical angles). It has a double-dove prism mounted in the head assembly. The operating power for the telescope, sextant, and associated electronic equipment is 94.5 watts at 28 volts dc. The telescope has two axes of rotational freedom, which are normally slaved to the sextant axis. The wide field of view is used for general celestial viewing and recognition of target bodies. It is also used to track landmark points during earth and lunar orbits.

Computer (Raytheon Co.) — This is 24 by 12.5 by 6 inches and weighs 70.1 pounds with six memory modules. It consumes 70 watts of power at 28 volts dc during normal operation. It is a digital computer with fixed and erasable memory. The erasable memory has a capacity of 2048 words; fixed memory has a capacity of 36,864 words. The fixed memory contains programs, routines, constants, star and landmark coordinates, and other pertinent data. The computer solves guidance and navigation problems, provides control information to optical and inertial subsystems as well as other spacecraft systems, provides pertinent information to astronauts and the ground on request, provides means by which astronauts or ground control can directly communicate with the primary guidance navigation control system, provides direct on-off control for reaction control jets and service propulsion engines, and monitors its own operation and other primary guidance navigation control system operations.

Display Keyboard (Raytheon Co.) — This 8-by-8-by-7-inch panel weighs 17.5 pounds. It is made up of a keyboard, power supply, a decoder relay matrix, status and caution circuits, and displays. It is a 21-digit character display and a 16-button keyboard through which crewmen can

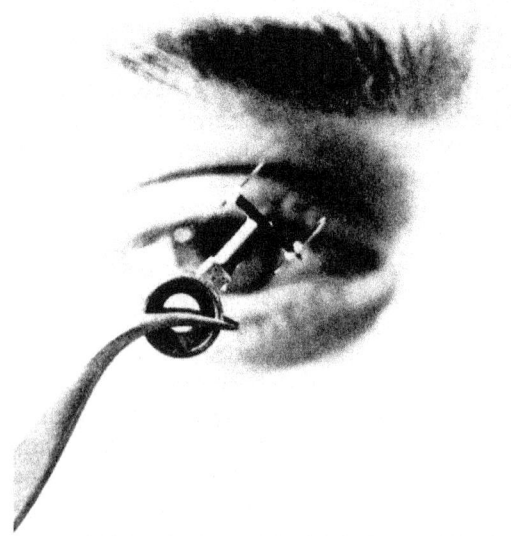

This tiny pendulum is heart of an accelerometer, a key component in guidance and navigation subsystem.
P-258

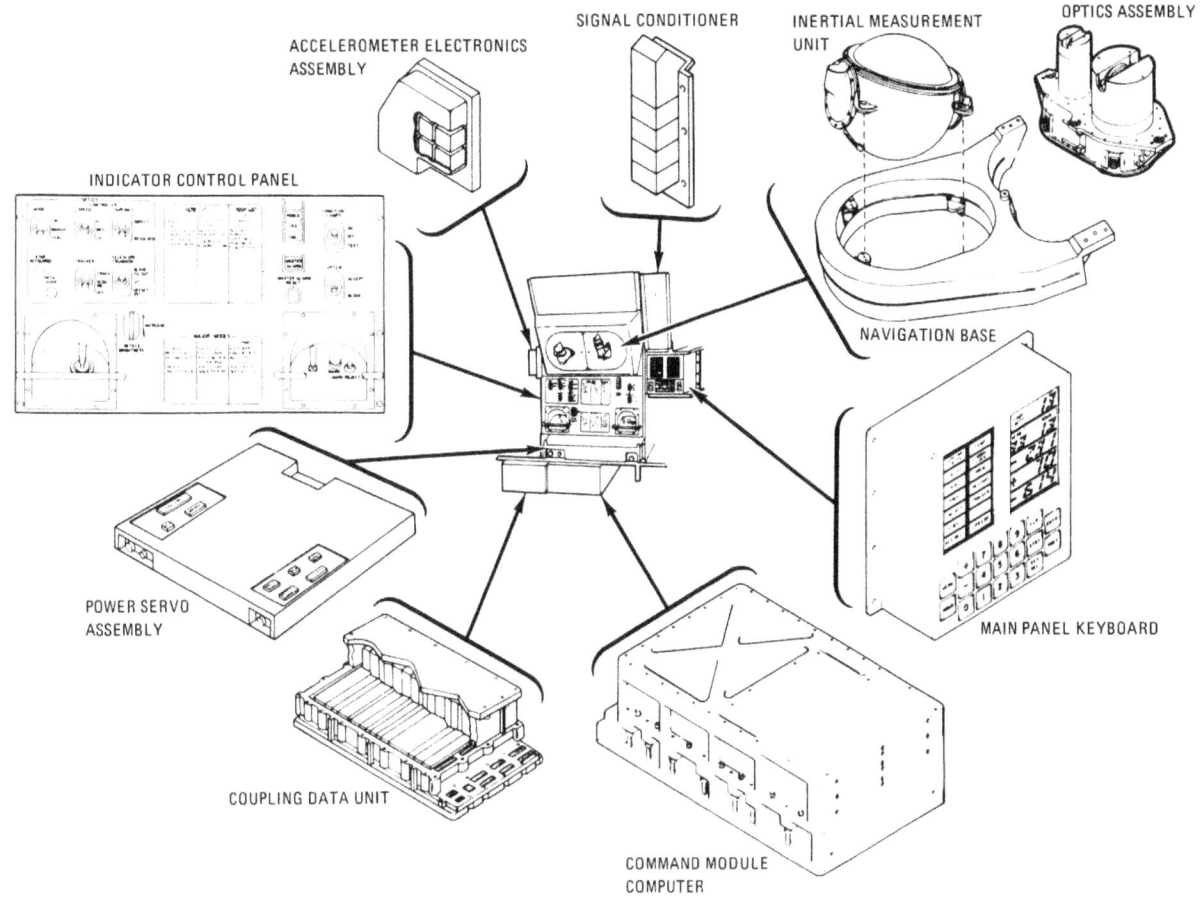

Major guidance and navigation subsystem equipment

communicate in a coded numerical language. Crewman inserts data and commands the computer by punching numbers on the keyboard. They are then displayed to him in electroluminescent counter-type readout windows. The computer communicates with the crewman by displaying numbers in the same windows. When the computer requests the crewman to take some action, numbers flash to attract attention.

Signal Conditioner Assembly — This 3-by 5.7-by-14.3-inch unit weighs 5.8 pounds. It contains encapsulated electronic circuitry to condition primary guidance navigation control system signals so that they are acceptable to the spacecraft telemetry system.

DETAILED DESCRIPTION

INERTIAL SUBSYSTEM

The inertial subsystem provides a space stabilized inertial reference from which velocity and attitude changes can be sensed. It is composed of the inertial measurement unit, the navigation base, parts of the power and servo assembly, parts of the control and display panels, and parts of the coupling data unit.

The navigation base is the rigid supporting structure on which the inertial measurement unit and optical instruments are mounted. It is manufactured and installed to close tolerances to provide accurate alignment of the equipment mounted on it. It also provides shock mounting for the inertial measurement unit and optics.

The inertial measurement unit is the main unit of the inertial subsystem. It is a three-degree-of-freedom stabilized platform assembly, containing three inertial reference integrating gyros, and three pulsed-integrating pendulous accelerometers. The stable member is machined from a solid block of beryllium with holes bored for mounting the accelerometers and gyros.

The stable platform attitude is maintained by the gyros, stabilization loop electronics, and gimbal torque motors. Any angular displacement of the stable platform is sensed by the gyros which generate error signals. These signals are resolved and amplified at the inertial measurement unit and applied to stabilization loop electronics. The resultant signal is conditioned and applied to the gimbal torque motors, which restores the desired attitude.

The stable platform provides a space-referenced mount for the three accelerometers, which sense velocity changes. The accelerometers are mounted orthogonally (each at right angles to the other two) to sense the velocity changes along all three axes. Any translational force experienced by the spacecraft causes an acceleration or deceleration which is sensed by one or more accelerometers. Each generates an output signal proportional to the magnitude and direction of velocity change. This signal, in the form of a pulse train, is sent to the computer which uses it to update the velocity information.

Temperature is controlled by a thermostatic system that maintains the gyro and accelerometer temperatures within their required limits during inertial measurement unit standby and operating modes. Heat is applied by end-mount heaters on the inertial components, stable member heaters, and a temperature control anticipatory heater. Heat is removed by convection, conduction, and radiation. The natural convection used during inertial measurement unit standby modes is changed to blower-controlled, forced convection during operating modes. Inertial measurement unit internal pressure is normally between 3.5 and 15 psia, enabling the required forced convection. To aid in removing heat, a water-glycol solution passes through coolant

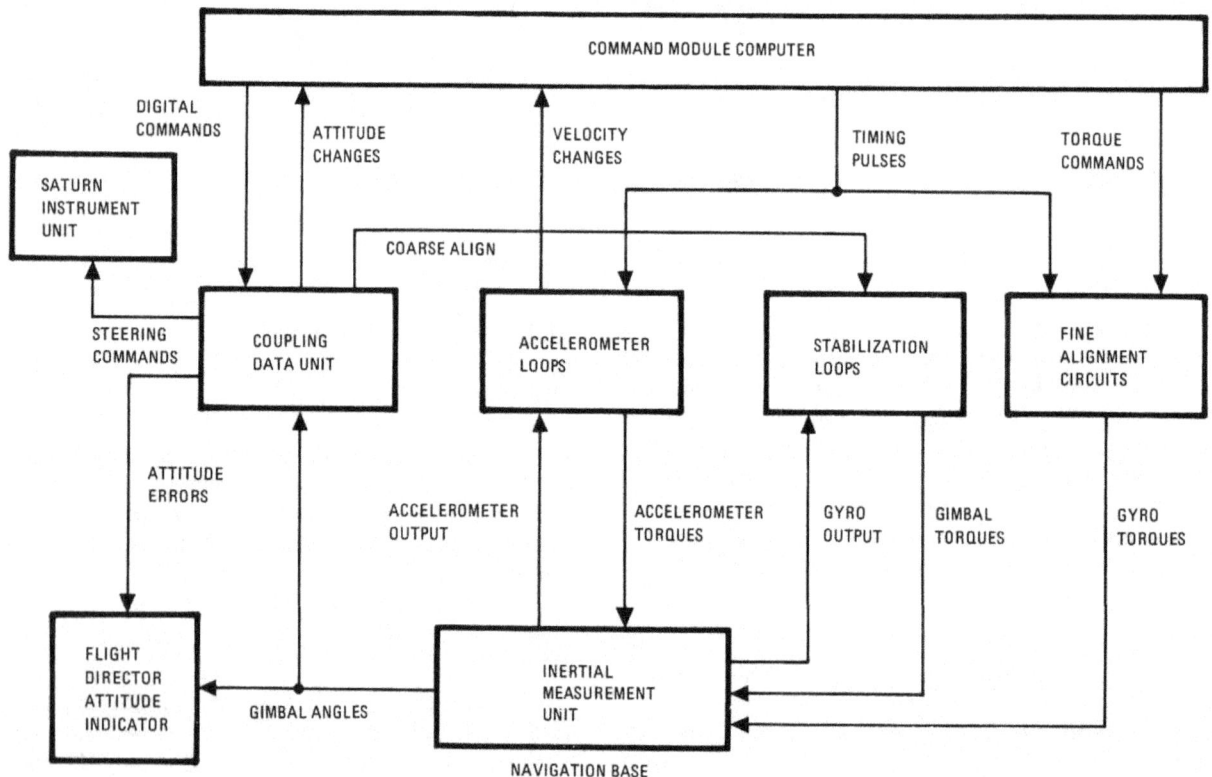

Block diagram of inertial subsystem

Cutaway of inertial measurement unit

passages in the outer case of the inertial measurement unit. Therefore, heat flow is from the stable member to the case and coolant. The temperature control system consists of the temperature control circuit, the blower control circuit, and the temperature alarm circuit.

The coupling data unit is an all-electronic device used to convert and transfer signals among the inertial, computer, and optical subsystems and between the computer and various controls and displays. Each of the five channels has two sections: one acting as an analog-to-digital converter and the other as a digital-to-analog converter. These five channels are: one each for the inner, middle, and outer inertial measurement unit gimbals, and one each for the shaft and trunnion optical axes. Each channel converts inertial or optical gimbal angles from a resolver analog form to a digital form, supplies the computer with this information, and converts digital signals from the computer to either 800-cycle per second or direct current analog signals. The coupling data unit also controls the modes of the inertial and optical subsystems through logical manipulations with the computer.

The power and servo assembly provides a central collection point for most of the guidance and navigation subsystem power supplies, amplifiers, and other electronic components. It is located in the lower equipment bay directly beneath the inertial measurement unit. It consists of 42 modules mounted to a header assembly. Connectors and harnessing are integral to the construction of the header assembly, and guidance and navigation harness branches are brought out from the power servo assembly header. A thin cover plate is mounted on the assembly to provide a hermetic seal for the interior. During flight this permits pressurization of the assembly to remain at 15 psi. Connectors are available for measuring signals at various system test points.

OPTICAL SUBSYSTEMS

The optical subsystem is used to take precise optical sightings of celestial bodies and landmarks. These sightings are used to align the inertial measurement unit and to determine the position of the spacecraft. The system includes the navigational base, two of the five coupling and data units, parts of the power and servo assembly, controls and displays, and the optics, which include the scanning telescope and the sextant.

The optics consist of the telescope and the sextant mounted in two protruding tubular sections of the optical base assembly. The scanning telescope and sextant line of sight may be offset depending on the mode of operation.

The sextant is a highly accurate optical instrument capable of measuring the included angle between two targets and the direction of any single target with respect to the navigation base. Angular sighting between two targets is made through a fixed beam splitter and a movable mirror located in the sextant head. The sextant lens provides 1.8-degree true field-of-view with 28X magnification. The movable mirror is capable of sighting a target to 57 degrees line of sight from the shaft axis. This is reduced to approximately 45 degrees when installed in the spacecraft, however, because of interference from structure. The mechanical accuracy of the trunnion axis is twice that of the line-of-sight requirement due to mirror reflection which doubles any angular displacement in trunnion axis.

The scanning telescope is similar to a theodolite in its ability to measure elevation and azimuth angles of a single target accurately using an established reference. The lenses provide 60-degree true field of view at 1X magnification. This is reduced to about 40 degrees when installed because of obstruction by vehicle structure.

The coupling data unit used in the inertial subsystem is also used as a part of the optical subsystem. Two channels of the unit are used, one for the sextant shaft axis and one for the sextant trunnion axis. These channels repeat the sextant shaft and trunnion angles and transmit angular change information to the computer in digital form.

The angular data transmission in the trunnion channel is mechanized to generate 1 pulse to the computer for 5 arc-seconds of movement of the sextant trunnion which is equivalent to 10 arc-seconds of star line-of-sight movement. The shaft channel issues 1 pulse for each 40 arc-seconds of shaft movement. The location of the sextant shaft and trunnion axes are transmitted to the coupling data units through 16X and 64X resolvers, located on the sextant shaft and trunnion axes. This angular information is transmitted to the units in the form of electrical signals proportional to the sine and cosine of 16X shaft angle and 64X trunnion angle. During the computer mode of operation, the unit provides digital-to-analog conversion of the computer output to generate an ac input to the sextant shaft and trunnion servos. This analog input to the sextant axes will drive the star line of sight to some desired position. In addition, the optical subsystem channels of the coupling data unit perform a second function on a time-sharing basis. During a thrust vector control function these channels provide digital-to-analog conversion of the service propulsion engine gimbal angle command between the computer and the service propulsion subsystem gimbals.

The modes of operation for the optical subsystem are selected by the astronaut using the controls located on the indicator control panel. There are three major modes: zero optics, manual control, and computer control.

During the zero optics mode, the shaft and trunnion axis of the sextant are driven to their zero positions by taking the output of the transmitting resolvers (1X and 64X in trunnion, and 1/2X and 16X in shaft) and feeding them through the two-speed switches to the motor drive amplifier. This in turn drives loops to null positions as indicated

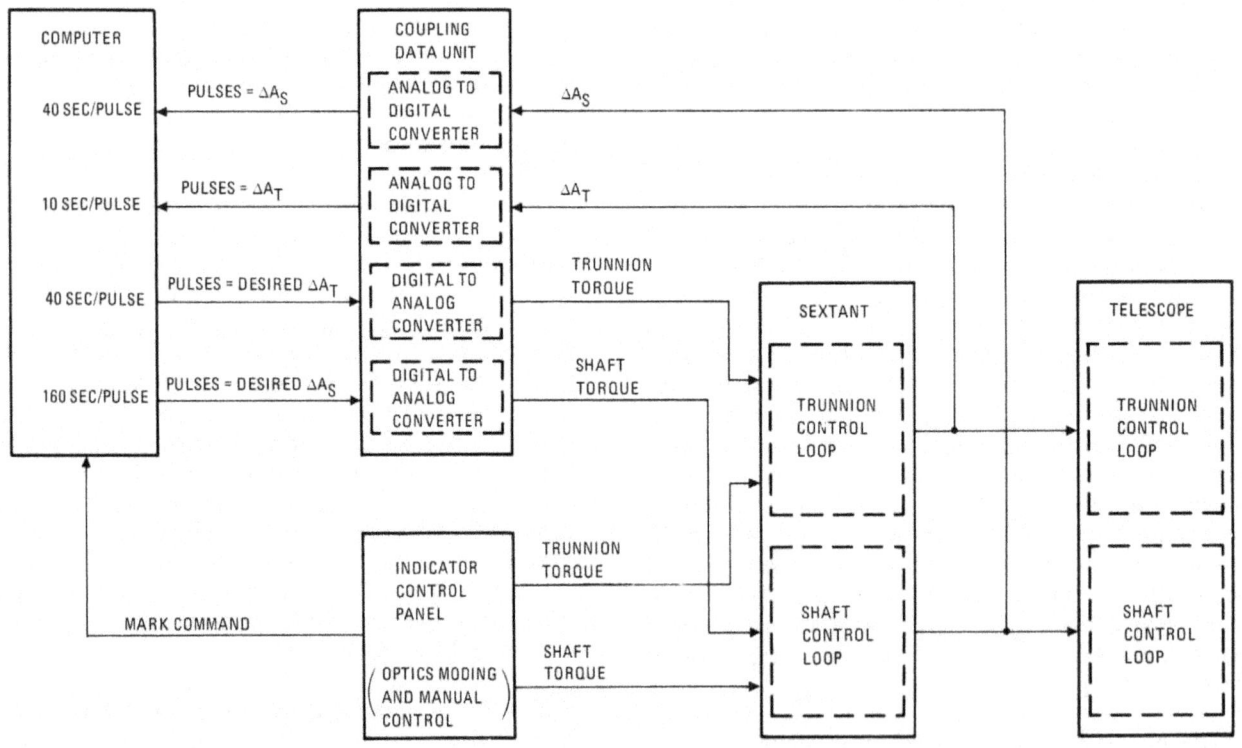

Schematic of optical subsystem

by zero output from the resolvers. The telescope and trunnion axes follow to a zero position. The zero optics mode can be selected either manually by the flight crew or by the computer after the computer control mode of operation has been manually selected.

The manual mode can be selected to operate under either direct hand control or resolved hand control. Independent control of the telescope trunnion is possible in both of these variations.

In direct hand control the hand controller outputs are applied directly to the sextant shaft and trunnion motor drive amplifiers. Forward and back motion of the hand controller commands increasing and decreasing trunnion angles, and right and left motion of the hand controller commands increasing and decreasing shaft angles. The apparent speed of the image motion can be regulated by the flight crew by selecting either low, medium, or high controller speed on the indicator control panel. This regulates the voltage applied to the motor drive amplifier and therefore the shaft and trunnion drive rates.

The slave telescope modes provide for alternate operation of the telescope trunnion while the sextant is being operated manually. Three alternate modes can be selected by a telescope trunnion switch: the telescope trunnion axis slaved to the sextant trunnion (the normal operating position), the telescope trunnion locked in a zero position (by applying fixed voltage to the telescope trunnion 1X receiving resolver, causing this position loop to null in a zero orientation. Therefore, this means the centerline of the telescope 60-degree field-of-view is held parallel to the landmark line of sight of the sextant), and the telescope trunnion axis is offset 25 degrees (the centerline of the 60-degree field of view is offset 25 degrees from the landmark line of sight of the sextant). This last position of the telescope trunnion will allow the landmark to remain in the 60-degree field of view while still providing a total possible field of view of 110 degrees if the telescope shaft is swept through 360 degrees.

In manual resolved operation, the hand controller outputs are put through a matrix transformation before being directed to the shaft and trunnion motor drive amplifiers. The matrix transformation makes the image correspond directly to the hand controller motion. That is, up, down, right, and left motions of the hand controller commands the target image to move up, down, right, and left, respectively, in the field of view.

Buttons on the indicator control panel are used to instruct the computer that a navigation fix has been made and that sextant shaft and trunnion position and time should either be recorded or rejected. The "mark" command is generated manually by the flight crew which energizes the mark relay. The mark relay transmits a mark command to the computer. If an erroneous mark is made, the "mark reject" button is depressed.

The computer-controlled operation is selected by placing the moding switch in computer position. The mechanization of this loop is chosen by the

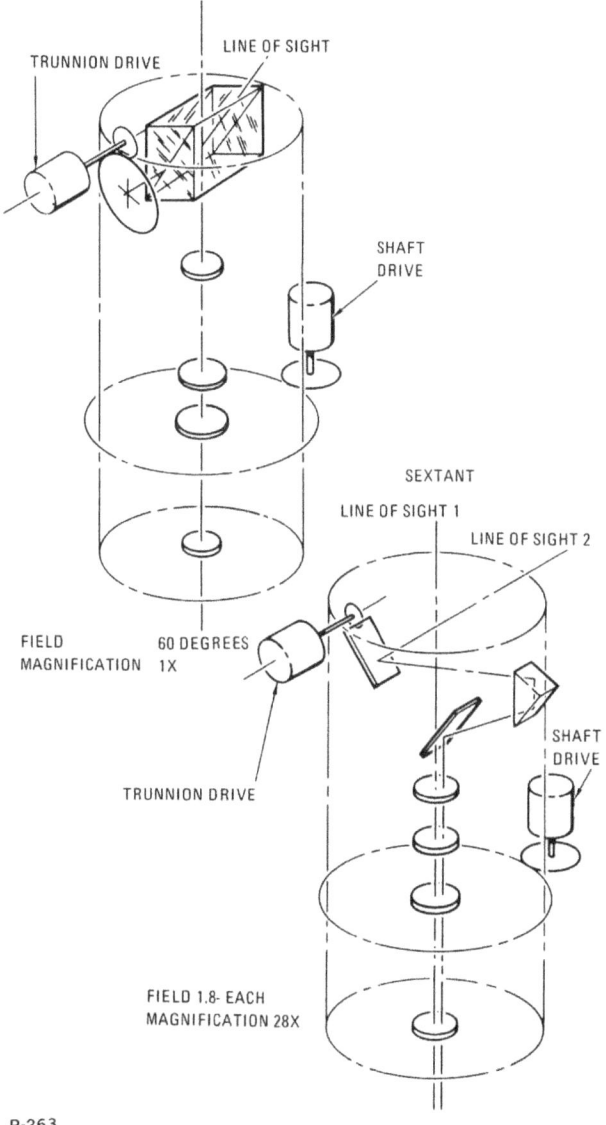

P-263

Schematic of sextant and telescope

computer program that has been selected by the flight crew. The computer controls the sextant by completing the circuit from the coupling data unit digital-to-analog converters to the shaft and trunnion motor drive amplifier. The computer can then provide inputs to these amplifiers via a digital input to the coupling data unit which converts it to an 800-cycle signal that can be used by the motor drive amplifier. This mode is used when it is desired to look at a specific star for which the computer has the corresponding star coordinates. The computer will know the attitude of the spacecraft from the position of the inertial measurement unit gimbals and will, therefore, be able to calculate the position of the sextant axes required to acquire the star.

COMPUTER SUBSYSTEM

The computer subsystem consists of the command module computer and two display and keyboard panels. The computer and one keyboard are located in the lower equipment bay. The other keyboard is located on the main display console. All computer controls and displays are located on the keyboards.

The computer is a core memory digital computer with both fixed and erasable memory. The fixed memory permanently stores navigation tables, trajectory parameters, programs, and constants. The erasable memory stores intermediate information.

The computer processes data and issues discrete control signals, both for guidance and control and for other spacecraft subsystems. It is a control computer with many of the features of a general-purpose computer. As a control computer, it aligns the stable platform of the inertial measurement unit in the inertial subsystem, positions the optical unit in the optical subsystem, and issues control commands to the spacecraft thrusters. As a general-purpose computer, it solves guidance problems required for the spacecraft mission. In addition, it monitors the operation

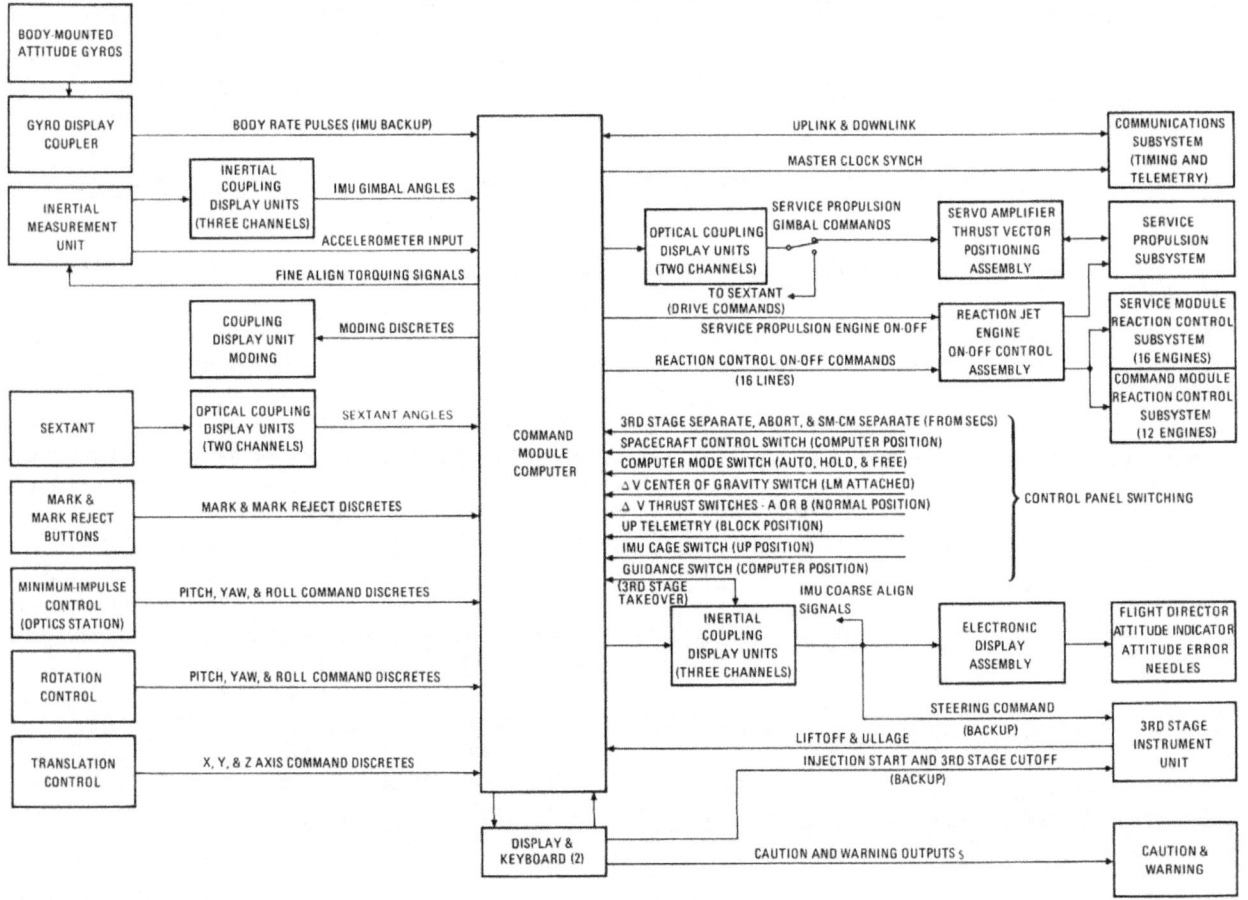

Block diagram of computer relationships

of the guidance and control and other spacecraft subsystems.

The computer computes information about the flight profile that the spacecraft must assume in order to complete its mission. This position, velocity, and trajectory information is used by the computer to solve the various flight equations. The results of various equations can be used to determine the required magnitude and direction of thrust required. Corrections to be made are established by the computer. The spacecraft engines are turned on at the correct time, and steering and engine control signals are generated by the computer to re-orient the spacecraft to a new trajectory if required. The inertial subsystem senses acceleration and supplies velocity changes to the computer for calculating the total velocity. Drive signals are supplied from the computer to coupling data unit and stabilization gyros in the inertial subsystem to align the gimbal angles in the inertial measurement unit. Error signals are also supplied to the coupling data unit to provide attitude error display signals for the flight director attitude indicator. Coupling data unit position signals are fed to the computer to indicate commanded changes in engine gimbal angles. The computer receives mode indications and angular information from the optical subsystem during optical sightings and uses it to calculate present position and orientation and to refine trajectory information. Optical subsystem components can also be positioned by drive signals supplied from the computer.

The computer is functionally divided into seven blocks: timer, sequence generator, central processor, memory, priority control, input-output, and power.

The timer generates the synchronization pulses to assure a logical data flow from one area to another within the computer. It also generates timing waveforms which are used by the computer in alarm circuitry and other areas of the spacecraft for control and synchronization purposes.

The master clock frequency is generated by an oscillator and is applied to the clock divider logic. The divider logic divides the master clock input into gating and timing pulses at the basic clock rate of the computer. Several outputs are available from the pulses at the basic clock rate of the computer. Several outputs are available from the scaler, which further divides the divider logic output into output pulses and signals which are used for gating, to generate rate signal outputs, and for the accumulation of time. Outputs from the divider logic also drive the time pulse generator which produces a recurring set of time pulses. This set of time pulses defines a specific interval (memory cycle time) in which access to memory and word flow takes place within the computer. The start-stop logic senses the status of the power supplies and specific alarm conditions in the computer and generates a stop signal which is applied to the time pulse generator to inhibit word flow. Simultaneously, a fresh-start signal is generated which is applied to all functional areas in the computer.

The sequence generator directs the execution of machine instructions. It does this by generating control pulses which sequence data throughout the computer. The control pulses are formed by combining the order code of an instruction word with synchronization pulses from the timer. The sequence generator contains the order code processor, command generator, and control pulse generator. The sequence generator executes the instructions stored in memory by producing control pulses which regulate the data flow of the computer. The manner in which the data flow is regulated among the various functional areas of the computer and between the elements of the central processor causes the data to be processed according to the specifications of each machine instruction.

The order code processor receives signals from the central processor, priority control, and peripheral equipment (test equipment). The order code signals are stored in the order code processor and converted to coded signals for the command generator. The command generator decodes these signals and produces instruction commands. The instruction commands are sent to the control pulse generator to produce a particular sequence of control pulses, depending on the instruction being executed. At the completion of each instruction, new order code signals are sent to the order code processor to continue the execution of the program.

The central processor performs all arithmetic operations required of the computer, buffers all information coming from and going to memory, checks for correct parity on all words coming from memory, and generates a parity bit for all words written into memory. It consists of flip-flop registers, the write, clear, and read control logic, write amplifiers, memory buffer register, memory address register and decoder, and the parity logic.

Primarily, the central processor performs operations indicated by the basic instructions of the program stored in memory. Communication within the central processor is accomplished through the write amplifiers. Data flows from memory to the flip-flop registers or vice versa, between individual flip-flop registers, or into the central processor from external sources. In all instances, data is placed on the write lines and routed to a specific register, or to another functional area under control of the write, clear, and read logic. The logic section accepts control pulses from the sequence generator and generates signals to read the content of a register onto the write lines, and write this content into another

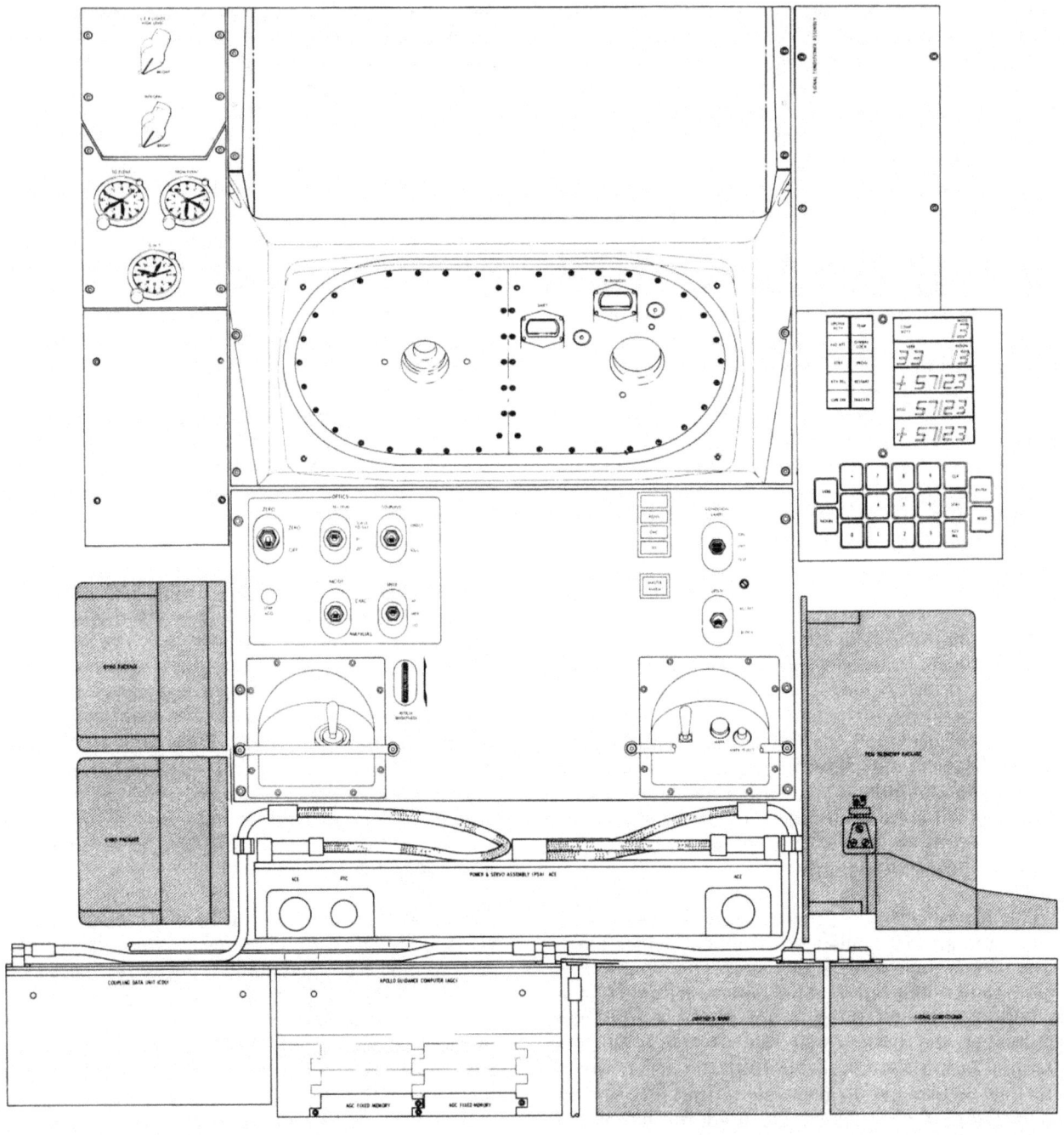

Guidance and navigation station in lower equipment bay

register of the central processor or to another functional area of the computer. The particular memory location is specified by the content of the memory address register. The address is fed from the write lines into this register, the output of which is decoded by the address decoder logic. Data is subsequently transferred from memory to the memory buffer register. The decoded address outputs are also used as gating functions within the computer.

The memory buffer register buffers all information read out or written into memory. During readout, parity is checked by the parity logic and an alarm is generated in case of incorrect parity. During write-in, the parity logic generates a parity bit for information being written into memory. The flip-flop registers are used to accomplish the data manipulations and arithmetic operations. Each register is 16 bits or one computer word in length. Data flows into and out of each register as dictated by control pulses associated with each register. The control pulses are generated by the write, clear, and read control logic.

External inputs through the write amplifiers included the content of both the erasable and fixed memory bank registers, all interrupt addresses from priority control, control pulses which are associated with specific arithmetic operations, and the start address for an initial start condition. Information from the input and output channels is placed on the write lines, and routed to specific destinations either within or external to the central processor.

Memory provides the storage for the computer and is divided into two sections: erasable memory and fixed memory. Erasable memory can be written into or read from; it is destroyed when it is read out (displayed); therefore, information required for later use must be restored. Fixed memory cannot be written into and its readout is nondestructive. Erasable memory stores intermediate results of computations, auxiliary program information, and variable data supplied by the guidance and control and other subsystems of the spacecraft. Fixed memory stores programs, constants, and tables. There is a total of 38,912 sixteen-bit word storage locations in fixed and erasable memories. It should be noted that the majority of the memory capacity is in fixed memory (36,864 word locations). The erasable memory uses planes or ferrite (iron) cores as storage devices. A core is a magnetic storage device having two stable states. It can be magnetized in one or two directions by passing a sufficient current through a wire which pierces the core. The direction of current determines the direction of magnetization. The core will retain its magnetization indefinitely until an opposing current switches the core in the opposite direction. Wires carrying current through the same core are algebraically additive. Sense wires which pierce a switched core will carry an induced pulse. The fixed memory is high-density core rope—tiny nickel-iron cores woven together with thousands of copper wires and encapsulated in plastic. Each core functions as a transformer and storage does not depend on magnetization. The advantages of encapsulated core rope for fixed memory are indestructibility, permanence of data, and storage of a vast amount in a small volume. The technique requires, however, that programs for classes of missions be developed and verified before the rope is woven, because the program determines the wiring sequence.

Priority control establishes a processing priority of operations that must be performed by the computer. These operations are a result of conditions which occur both internally and externally. Priority control consists of counter priority control and interrupt priority control. Counter priority control initiates actions which update counters in erasable memory. Interrupt priority control transfers control of the computer to one of several interrupt subroutines stored in fixed memory.

The start instruction control restarts the computer after a hardware or program failure. The counter instruction control updates the various counters in erasable memory upon reception of certain incremental pulses. The interrupt instruction control forces the execution of the interrupt instruction to interrupt the current operation of the computer in favor of a programmed operation of a higher priority.

The input-output section routes and conditions signals between the computer and other areas of the spacecraft. In addition to the counter interrupt and the program interrupts previously described, the computer has a number of other inputs derived from its interfacing hardware. These inputs are a result of the functioning of the hardware, or an action by the operator of the spacecraft. The counter interrupts, in most cases, enable the computer to process inputs representing such things as changes in velocity. The program interrupt inputs are used to initiate processing that must be done a relatively short time after a particular function is present. The other inputs to the computer, in general,

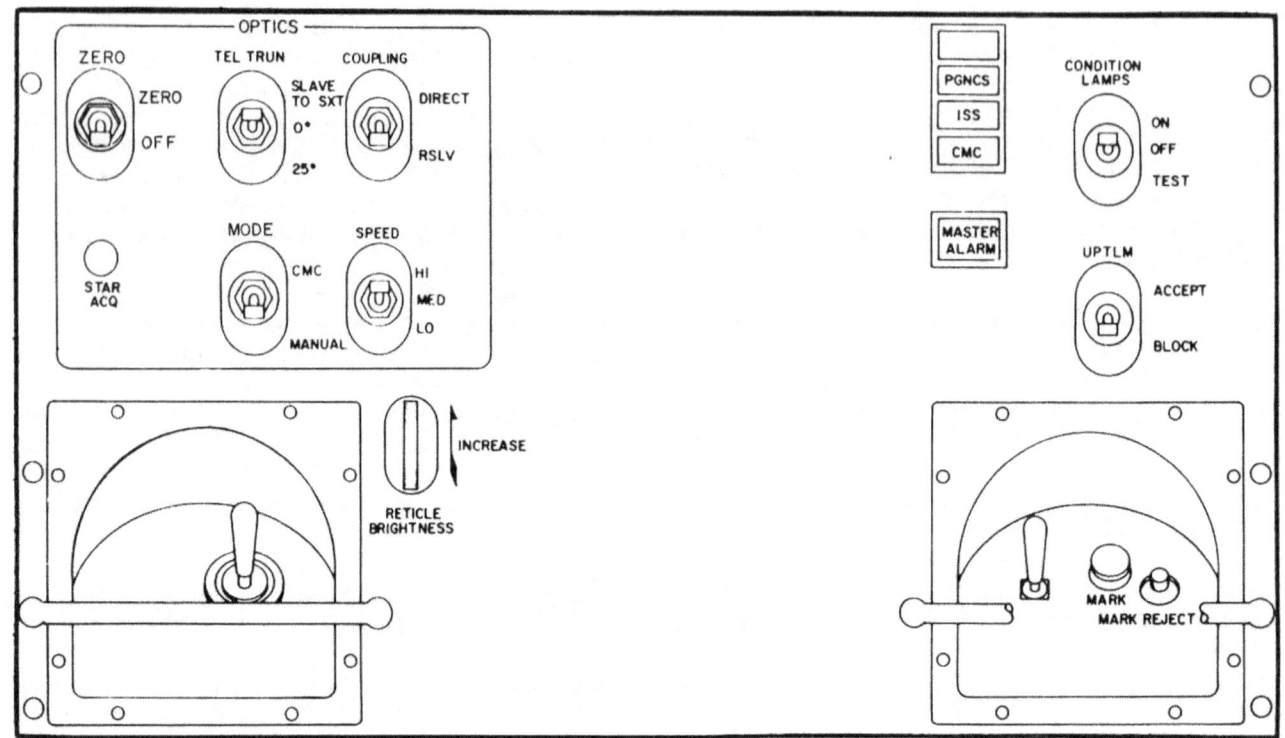

Optics control panel

enable it to be aware of conditions which exist in its environment. These inputs are routed to computer and are available to its programs through the input channels.

The outputs of the computer fall in one of the following categories: data, control, or condition indications. Some of these outputs are controllable through the computer program while others are present as a function of computer circuitry. All of the outputs which are controlled by the computer programs are developed through the computer output channels.

The power section provides voltage levels necessary for the proper operation of the computer. Power is furnished by two switching-regulator power supplies: a +4-volt and a +14-volt power supply which are energized by fuel cell powerplants in the electrical power subsystem.

The power supply outputs are monitored by a failure detector consisting of four differential amplifiers. There are two amplifiers for each power supply, one for overvoltage and one for undervoltage detection. If an overvoltage or undervoltage condition exists, a relay closure signal indicating a power fail is supplied to the spacecraft.

The display and keyboard panels provide communication between the flight crew and the computer. They operate in parallel, with the main display console keyboard providing computer display and control while crewmen are in their couches.

The exchange of data between the flight crew and the computer is usually initiated by crew action; however, it can also be initiated by internal computer programs. The exchanged information is processed by the keyboard program. This program allows the following four different modes of operation:

1. Display of internal data—both a one-shot display and a periodically updating display (called monitor) are provided.

2. Loading external data—as each numerical character is entered, it is displayed in the appropriate display panel location.

3. Program calling and control—the keyboard is used to initiate a class of routines which are concerned with neither loading nor display; certain routines require instructions from the operator to determine whether to stop or continue at a given point.

4. Changing major mode—the initiation of large-scale mission phases can be commanded by the operator.

The data involved in both loading and display can be presented in either octal or decimal form as the operator indicates. If decimal form is chosen, the appropriate scale factors are supplied by the program. Decimal entries are indicated by entering a plus or minus sign.

The basic language of communication between the operator and the computer is a pair of words known as verb and noun. Each of these is represented by a two-character decimal number. The verb code indicates what action is to be taken (operation); the noun code indicates to what the action is applied (operand). Typical verbs are those for displaying and loading. Nouns usually refer to a group of erasable registers within the computer memory. The program, verb, and noun displays provide two-digit numbers which are coded numbers describing the action being performed. The register 1, 2, and 3 displays show the contents of registers or memory locations. These displays are numbers which are read as decimal numbers if a plus or minus sign is present and octal numbers if no sign is used. The register displays operate under program control unless the contents of a specific register or memory location is desired. The crew may request display of the contents of a specific register or memory location by commanding the display from the keyboard. The only other displays are the "activity" lights which indicate whether the computer is computing or accepting telemetry from the ground and status lights.

The keyboard consist of ten numerical keys (pushbuttons) labeled 0 through 9, two sign keys (+ or -) and seven instruction keys: Verb, Noun, Clr (clear), Pro (proceed), Key Rel (key release), Entr (enter) and Rset (reset). Whenever a key is depressed, 14 volts are applied to a diode encoder which generates a unique five-bit code associated with that key. There is, however, no five-bit code associated with the proceed key. The function of the keys is as follows:

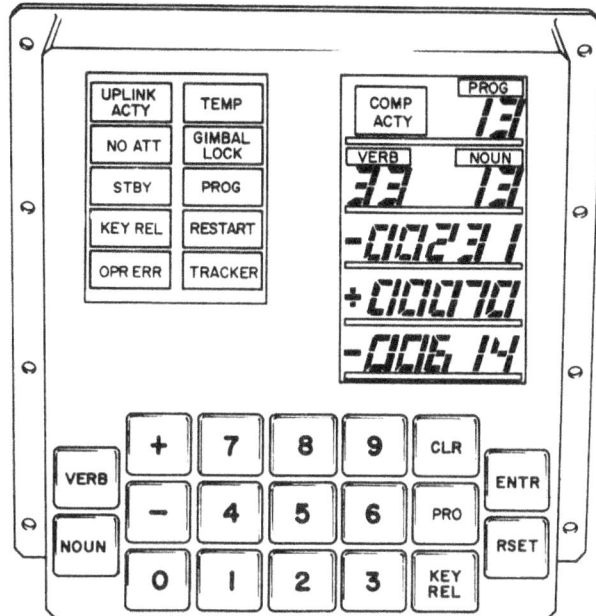

Computer display and keyboard

Pushbutton	Function
0 through 9	Enters numerical data, noun codes, and verb codes into the computer
+ and -	Informs the computer that the following numerical data is decimal and indicates the sign of the data
Noun	Conditions the computer to interpret the next two numerical characters as a noun code and causes the noun display to be blanked
Clear	Clears data contained in the data displays; depressing this key clears the data display currently being used; successive depressions clear the other two data displays
Proceed	Commands the computer to proceed on to the standby mode when depressed, if in standby mode, depression commands the computer to resume regular operation (operate mode)
Key Release	Releases the keyboard displays initiated by keyboard action so that information supplied by the computer program may be displayed
Enter	Informs the computer that the assembled data is complete and that the requested function is to be executed

Reset	Extinguishes the lamps that are controlled by the computer
Verb	Conditions the computer to interpret the next two numerical characters as a verb code and causes the verb display to be blanked

The standard procedure for the execution of keyboard operations consists of a sequence of seven key depressions:

Verb V_2 V_1 Noun N_2 N_1 Enter

Pressing the Verb key blanks the two verb lights on the keyboard and clears the verb code register in the computer. The next two numerical inputs are interpreted as the verb code. Each of these characters is displayed by the verb lights as it is inserted. The Noun key operates similarly with the keyboard noun lights and computer noun code register. Pressing the Enter key initiates the program indicated by the verb-noun combination. Thus, it is not necessary to follow a standard procedure in keying verb-noun codes; it can be done in reverse order or a previously inserted verb or noun can be used without re-keying it. No action is taken by the computer in initiating the verb-noun-defined program until the Enter key is actuated. If an error is noticed in either the verb or noun code before actuation of the Enter key, it can be corrected simply by pressing the corresponding Verb or Noun key and inserting the proper code.

If the selected verb-noun combination requires data from the operator, the Verb and Noun lights flash on and off about once per second after the Enter key is pressed. Data is loaded in five-character words and is displayed character-by-character in one of the five-position data display registers. The Enter key must be pressed after each data word. This tells the program that the numerical word being keyed in is complete.

The keyboard also can be used by internal computer programs for subroutines. However, any operator keyboard action (except error reset) inhibits keyboard use by internal routines. The operator retains control of the keyboard until he wishes to release it. This assures that the data he wishes to observe will not be replaced by internally initiated data displays.

A noun code may refer to a device, a group of computer registers, or a group of counter registers, or it may simply serve to convey information without referring to any particular computer register. The noun is made up of 1, 2, or 3 components, each entered separately as requested by the verb code. As each component is keyed, it is displayed on the display panel: component 1 in Register 1, component 2 in Register 2, and component 3 in Register 3. There are two classes of nouns: normal and mixed. Normal nouns (codes 01 through 39) are those whose component members refer to computer registers which have consecutive addresses and use the same scale factor when converted to decimal. Mixed nouns (codes 40 through 99) are those whose component members refer to non-consecutive addresses or whose component members require different scale factors when converted to decimal, or both.

A verb code indicates what action is to be taken. It also determines which component member of the noun group is to be acted upon. For example, there are five different load verbs. Verb 21 is required for loading the first component of the selected noun; verb 22 loads the second component; verb 23 loads the third component; verb 24 loads the first and second component; and verb 25 loads all three components. A similar component format is used in the display and monitor verbs. There are two general classes of verbs: standard and extended. The standard verbs (codes 01 through 39) deal mainly with loading, displaying, and monitoring data. The extended verbs (codes 40 through 99) are principally concerned with calling up internal programs whose function is system testing and operation.

SPACE SUIT

A space suit is a mobile chamber that houses and protects the astronaut from the hostile environment of space. It provides atmosphere for breathing and pressurization, protects him from heat, cold, and micrometeoroids, and contains a communications link.

The suit is worn by the astronauts during all critical phases of the mission, during periods when the command module is unpressurized, and during all operations outside the command and lunar modules whether in space or on the moon.

The space suit is produced by the International Latex Corp., an associate contractor directly responsible to NASA.

The suit systems must provide an artificial atmosphere (100-percent oxygen for breathing and for pressurization to 3.7 psi), adequate mobility (lunar gravity is one-sixth that of earth), micrometeorite and visual protective systems, and the ability to operate on the lunar surface for periods of 3 hours. Design of the Apollo spacecraft and suits will permit the crew to operate — with certain restraints — in a decompressed cabin for periods as long as 115 hours.

Insulation must protect the astronaut from temperatures varying from 250°F above (lunar day) to 250° below zero (lunar night). Solar heat flux is calculated at 10,000 Btu per hour. Superinsulation must limit heat leak into the suit to approximately 250 Btu per hour during lunar day and heat out to 350 Btu per hour during lunar night.

The astronauts must be protected from meteoroid particles traveling at speeds up to 64,000 miles per hour and from particles ejected by the meteoroid striking the lunar surface. During the lunar day, the crewmen's faces must be protected from solar ultraviolet, infrared, and visible light radiation.

The complete space suit is called the pressure garment assembly. It is composed of a number of items assembled into two configurations: extravehicular (for outside the spacecraft) and intravehicular. The addition of the backpack to the extravehicular space suit makes up the extravehicular mobility unit. The backpack (called the portable life support system) supplies oxygen, electrical power, communications, and liquid cooling.

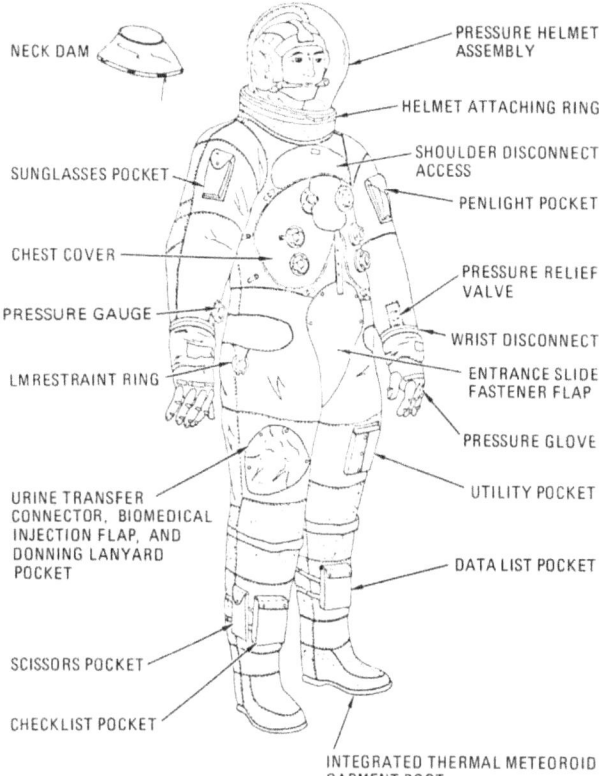

P-268   *Pressure garment assembly (space suit)*

The intravehicular space suit consists of: fecal containment subsystem, constant wear garment, biomedical belt, urine collection transfer assembly, torso limb suit, integrated thermal micrometeoroid garment, pressure helmet, pressure glove, and communications carrier.

In the extravehicular configuration, the constant-wear garment is replaced by the liquid-cooling garment and four items are added to the intravehicular suit: extravehicular visor, extravehicular glove, lunar overshoe, and a cover which fits over umbilical connections on the front of the suit.

The pressure suit is a white, snowsuit-like garment that weighs about 60 pounds with the integrated thermal meteoroid garment. The latter weighs about 19 pounds.

SPACE SUIT

The space suit is in a constant state of change as new and improved designs are developed and as new materials become available. Therefore the description in this section will be generalized and can be considered only as typical.

The basic components of the suit or pressure garment assembly are the torso limb suit, the pressure helmet, the pressure glove, the integrated thermal meteoroid garment, and the extravehicular glove. The constant-wear garment, which is worn under the suit in the intravehicular configuration, is described in the Crew section.

TORSO LIMB SUIT

The torso limb suit is the basic pressure envelope for the astronaut; it encloses all the body except the head and hands. It has three layers: an inner cloth comfort lining, a bladder which serves as the gas-retention layer, and a restraint layer designed to hold elongation to a minimum. The torso section is custom-fitted to each astronaut; the limb sections are graduated in size and adjustable.

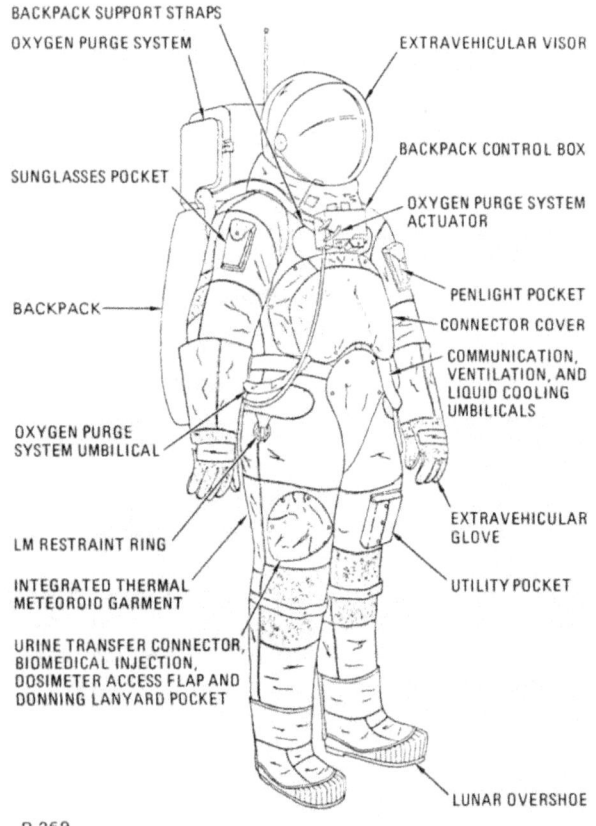

Extravehicular suit

The torso limb suit contains cables to sustain axial limb loads and a block and tackle system to foreshorten the suit for sitting or bending. Ducts on the inner surface of the suit direct oxygen to the helmet for breathing and defogging and also permit flow over the body for cooling. Connectors in the suit include those for oxygen (from the spacecraft's environmental control subsystem or the portable life support system), water (for the liquid cooling garment), and urine (to transfer it to the spacecraft's waste management system). An electrical harness in the suit connects communications and biomedical equipment to either the spacecraft or the portable life support system.

The right wrist of the torso limb suit contains a pressure gauge and the left wrist a pressure relief valve which opens to relieve suit pressure of more than 5.5 psi.

INTEGRATED THERMAL METEOROID GARMENT

The integrated thermal meteoroid garment is a many-layered structure laced to the torso limb suit. It is composed of an inner and outer shell of Beta cloth, seven layers of aluminized Kapton film separated by six layers of Beta Marquisette, and a liner of two layers of Neoprene-coated nylon Ripstop. A layer of Chromel-R (a woven metal) is added to the knee, elbow, and shoulders to protect the suit against abrasion. Chromel-R also is used to protect the garment's boot from abrasion. The boot is attached to the space suit boot by loop tape.

Covers are provided for the shoulder cable disconnect, LM restraints, the entrance slide, and the urine transfer fitting-medical injection area. The cover for the last-named has four snaps at the top and folds down; the inner side has pockets for a radiation dosimeter and for a lanyard.

Pockets include one on the upper left arm (for two pens and a penlight), one on the upper right arm (for sunglasses), and one on the upper right thigh (a utility pocket about 1-3/4 by 6 by 8-1/4 inches). In addition, there are strap-on pockets for both legs. These contain a data list (left leg) and checklist and scissors (right leg).

PRESSURE HELMET

The pressure helmet consists basically of an aluminum neck ring and a transparent shell made of polycarbonate (plastic). The shell is bonded to the neck ring, which fits into and locks with a similar

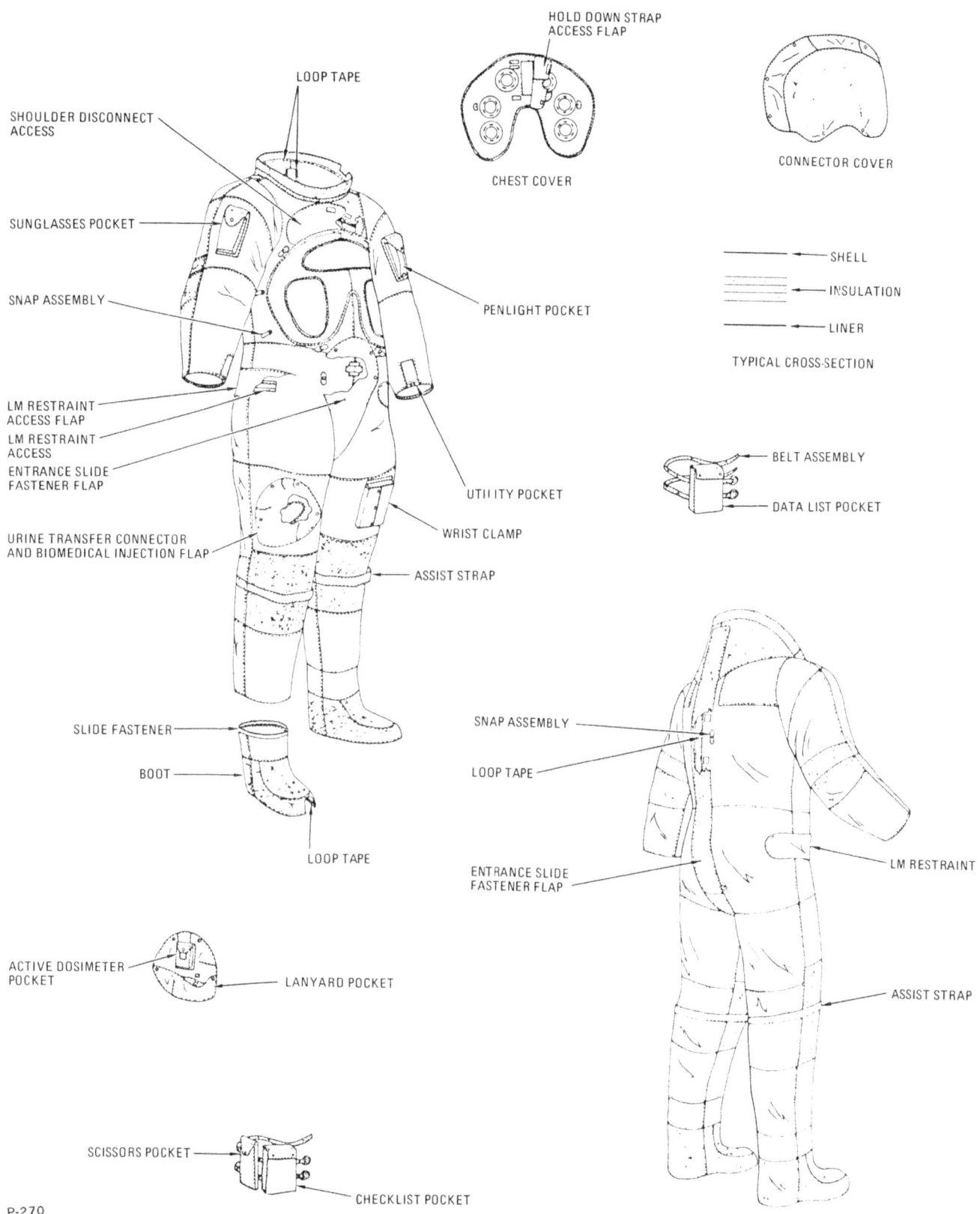

Integrated thermal meteoroid garment

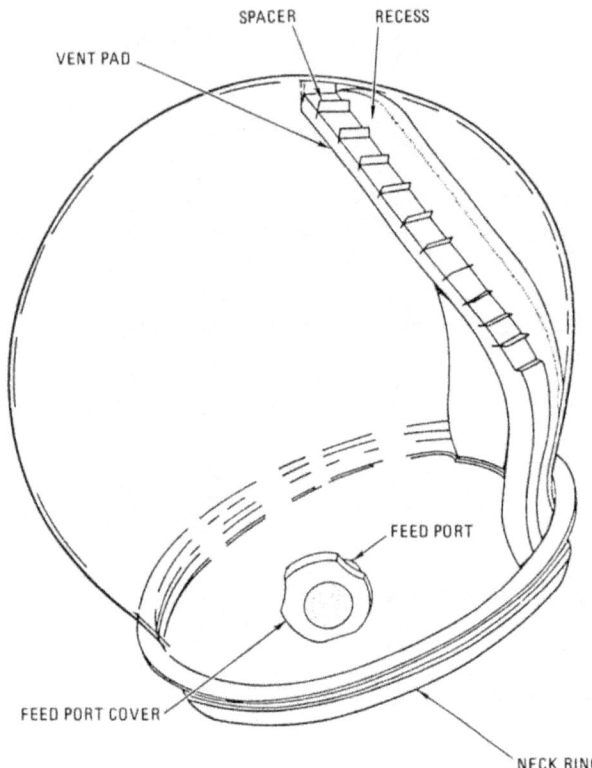

Pressure helmet

EXTRAVEHICULAR GLOVE

This glove is used for extravehicular activities and is for thermal protection. It covers the entire hand and has a cuff that extends well above the joint between the torso limb suit and the pressure glove. The extravehicular glove consists of a modified pressure glove (called the thermal meteoroid pressure glove) to which a thermal insulating shell is secured. The shell is similar in construction to the integrated thermal meteoroid garment, with additional layers of insulating material in the palm and fingers. The Chromel-R is coated with a silicon dispersion compound to improve the grip.

LUNAR OVERSHOE

This fits over both the thermal meteoroid garment boot and the suit boot and is used for extravehicu-

neck ring on the torso limb suit. The helmet also contains a feed port and a vent pad. The former is on the left side of the helmet and provides an air-tight attachment for the water and feed probes and for a purge valve. The vent pad (made of synthetic elastomer foam) is bonded to the back of the helmet and has a recess which acts as a ventilation flow manifold.

PRESSURE GLOVE

The pressure glove is a flexible gas-retaining device which locks to the torso limb suit. It consists of a bladder, a fingerless glove, inner and outer covers, and a restraint system. The bladder is moulded from a cast of each astronaut's hand. The bladder core, made of nylon tricot dipped in a Neoprene compound, is exposed at the inner thumb and fingertips to give the astronauts feeling in those areas. The fingerless glove is a restraint cemented to the bladder. A restraint strap over the palm minimizes ballooning and thus aids in grip control.

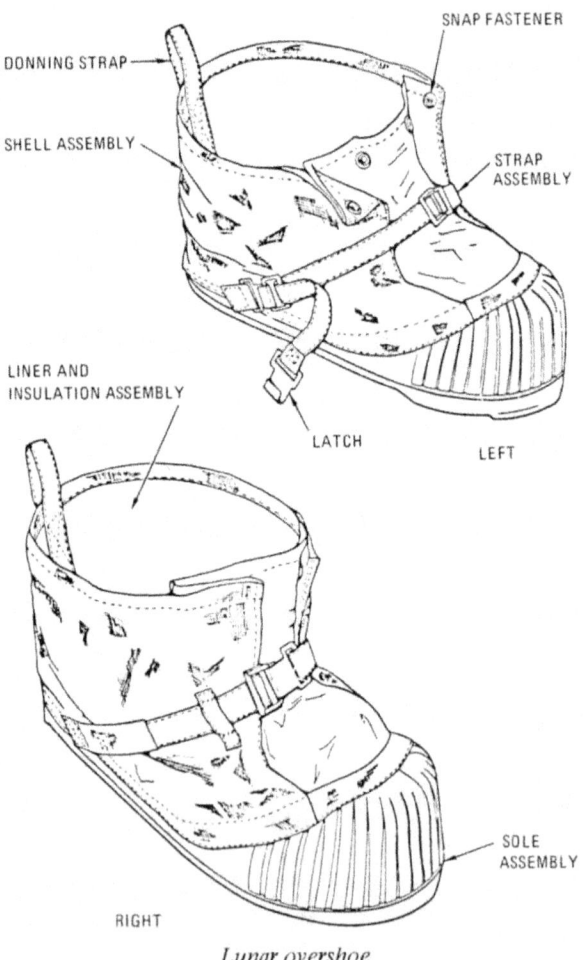

Lunar overshoe

lar activity. It consists of an insulation and liner, and an outer shell. The liner is Teflon-coated Beta cloth and the insulation is 13 layers of aluminized Kapton film separated by 12 layers of Beta Marquisette. The sole portion contains two additional layers of Beta felt interspaced between the uppermost film and spacer layers. The outer shell features a silicone rubber sole sewn to a laminated structure made up of four layers of two-ply Beta Marquisette. Chromel-R is used as the outer layer of the shell, except for the tongue, which is Teflon-coated Beta cloth.

EXTRAVEHICULAR VISOR

The extravehicular visor is used over the pressure helmet to protect the astronaut from light, heat, and micrometeoroids, and to protect the pressure helmet. It consists of a polycarbonate shell to which are attached two pivoting visors, one for micrometeroid protection and one for protection from the sun's rays. Both visors are made of polycarbonate and can be set anywhere from their full-up to full-down positions.

LIQUID-COOLING GARMENT

The liquid-cooling garment is used to cool an astronaut during extravehicular activity. It consists of a nylon Spandex material which supports a network of Tygon tubing through which water from the portable life support system is circulated. The inner surface of the garment is nylon chiffon. The socks attached to the garment do not contain cooling tubes.

PORTABLE LIFE SUPPORT SYSTEM

The portable life support system (backpack) is contained in a fiberglass shell contoured to fit the back. It is 26 inches high, 28 inches wide, and 11 inches thick, and has three control valves, 2 control switches, and a 5-position switch for the radio transceiver. Total weight is about 68 pounds. It is produced by the Hamilton Standard Division of United Aircraft Corp., Windsor Locks, Conn.

The system will assimilate an average crewman output of 1600 Btu per hour with peak rates of 2000 Btu per hour. It will operate for 4 hours in a space environment before replenishment of oxygen and replacement of the battery. There are four subsystems of the system: oxygen, liquid, telecommunications, and electrical.

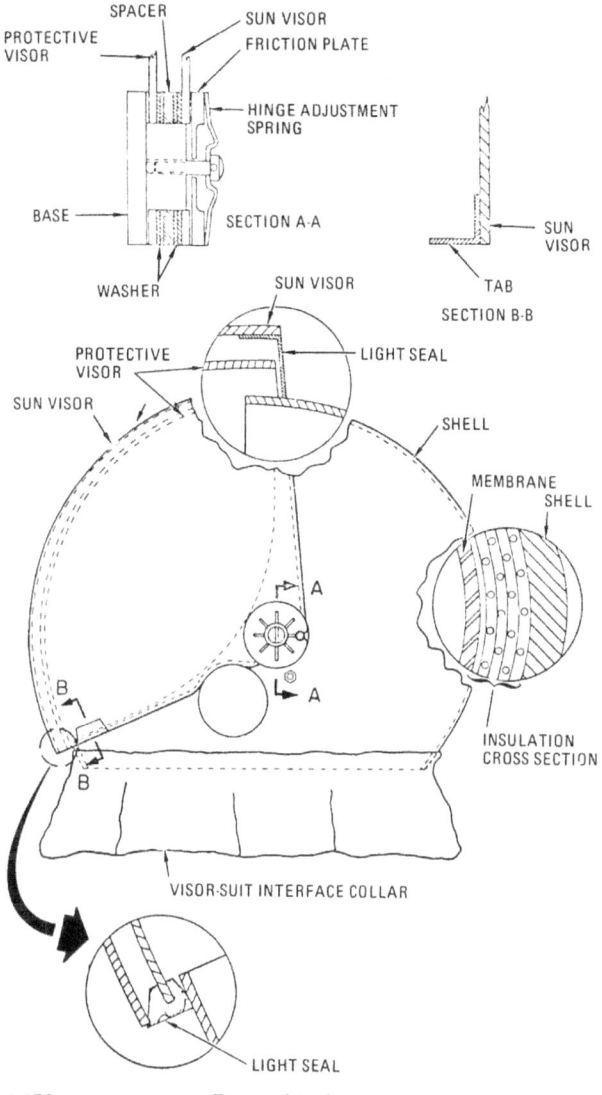

Extravehicular visor

The primary oxygen (1.05 pounds) is supplied from a 46.6-cubic-inch tank pressurized at 900 psi. The system is filled through a quick-disconnect before launch or a CM flex line connection during the mission.

Oxygen flows to the suit via the oxygen supply hose at a temperature of 45 to 50°F and returned from the suit at 80 to 85°F laden with impurities such as carbon dioxide, body odor, and water molecules. It passes through a canister containing deactivated charcoal and lithium hydroxide, which absorbs the carbon dioxide and purifies the gas. The gas is then cooled to 40° to 45°F in the sublimator/heat extractor which, during the process,

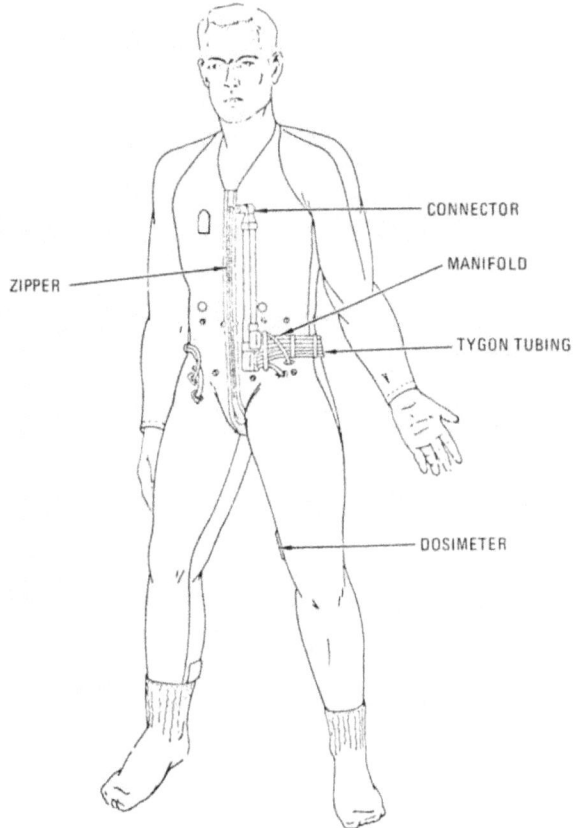

Liquid-cooling garment

The backpack's electrical subsystem distributes power to the other subsystems. It consists of two 16.8-volt batteries, a distribution panel, and power distribution harness. A battery weighs about 4 pounds and has 27 watt-hours of use.

EMERGENCY OXYGEN SYSTEM

The emergency oxygen system provides an immediate supply of oxygen to maintain suit pressure. It is doughnut shaped, weighs 2.9 pounds, and has an actuating mechanism, a pressure gauge, and a regulator. The system stores 7.2 cubic inches of oxygen at 7500 psig. The units are stowed on the back of the left couch leg pan or stowage mounting plates. When preparing for extravehicular activity or crew transfer, a unit is mounted on the left side of the backpack below the contaminant control canister.

will condense the water into droplets. As the flow reaches 90 degrees, the heavier water droplets continue straight and impact the water separator wick. The cool and conditioned gas then begins the cycle again.

The transport water subsystem of the backpack absorbs heat from the body surface of the crewman, transports it to the backpack, and loses heat in the sublimator. It is a closed loop with an operating pressure of 20 psi.

The telecommunications subsystem receives biomedical and communications data from the crewman, transmits it to the LM, and receives communications data from the LM, transmitting it to the crewman. The subsystem consists of primary and secondary transceiver (transmitters and receivers in one unit), interconnecting cable, an antenna, and controls. The cable connects the chest connector on the right side of the suit. The antenna is housed in a small fiberglass dome.

Suited astronaut enters spacecraft at Space Division, Downey, Calif.

CHECKOUT AND FINAL TEST

Before any Apollo spacecraft is launched it has "flown" its complete mission a number of times on the ground. These ground flights are part of an extensive series of tests the completed spacecraft must pass before it is committed to launch.

The final checkout and testing of the Apollo command and service modules takes place in two stages: that conducted at North American Rockwell's Space Division in Downey, Calif., to assure that the vehicles are in condition for delivery to NASA, and that conducted at the Kennedy Space Center to assure that the modules are ready for launch.

NORTH AMERICAN ROCKWELL SPACE DIVISION

Final assembly of the command and service modules takes place in a large "clean room," at Downey. When components reach this room, they have already gone through many hours of severe testing.

Checkout and testing at Downey can be separated into three broad categories: component and subassembly, individual subsystem tests during build-up, and integrated system testing. The first is at the component and subassembly level, before the equipment is installed in the spacecraft. The second follows installation on the spacecraft and is a long and complex series of tests involving such major operations as pressure cell testing, continuity checks of all wiring, and testing of each subsystem after installation in the vehicle.

The third category—integrated systems testing—is the final operation at Downey. It is conducted in two phases: "plugs-in" and "plugs-out" checkout in which the command and service module subsystems are operated together through a complete simulated mission. In addition, a manned suit loop test is performed with the flight crew to check the environmental control subsystem and all other crew equipment.

The individual and integrated system testing is conducted with acceptance checkout equipment (ACE) similar to that at KSC. This equipment interrogates the spacecraft systems to elicit automatic responses as to their status. The responses

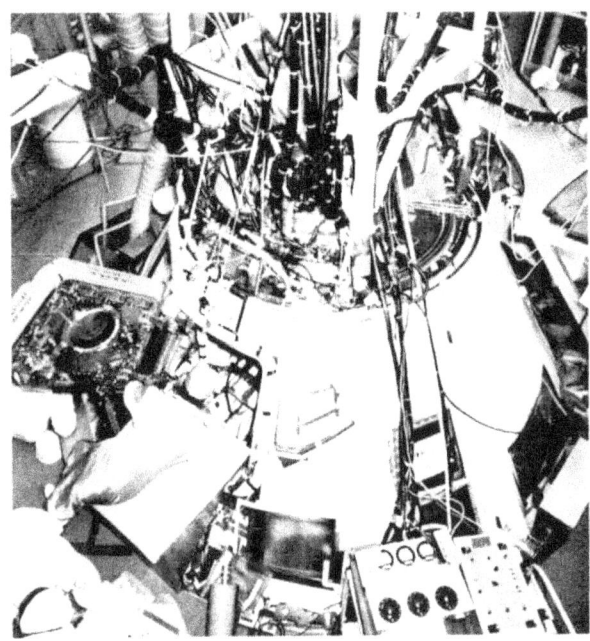

CM undergoes electrical test in Downey clean room
P-275

are automatically gathered, processed, and displayed to test personnel for immediate evaluation and also are recorded and stored for later detailed analysis. The displays are part of a modular system which allows a staff of engineers to monitor more than 25,600 samples per second of spacecraft test data containing about 1500 separate spacecraft and ground support conditions while the test is being performed.

The integrated system tests have a number of objectives: to check the operation of each subsystem under mission conditions and assure that all subsystems work together properly, to verify the electromagnetic compatibility of the subsystems, to assure that all crew equipment functions properly, and to check operation of specific alternate and backup equipment and procedures.

The plugs-in and plugs-out tests are similar. The differences are that the former uses ground power and ACE connected to the spacecraft while the latter uses simulated fuel cell power and the ACE is removed so that responses are sent over spacecraft communications equipment as it would operate during the actual mission.

Checkout at Downey; technicians check cabling to command and service modules (left), while others monitor responses to stimuli on battery of computerized display equipment

P-276

The spacecraft is prepared for these tests exactly as it would be for actual flight except that special test devices are installed in place of some equipment such as ordnance and expendable items.

In each phase the proper operation and interaction of all spacecraft subsystems are checked. The checkout equipment enables both automatic and manual operations to be performed, so that manual backups and overrides to automatic operations can be evaluated. When the series of testing at Downey has been completed, the spacecraft and its subsystems are considered ready. Before it is delivered to NASA, however, additional tests are performed to be sure that all the crew equipment, and particularly the environmental control subsystem, works properly with an actual suited crew aboard.

All loose crew equipment is stowed aboard the spacecraft in its proper place. Three crewmen don and check out the space suits and check ease of entering and leaving the command module. The flight crew then enters the cabin and checks the operation and manipulation of all crew equipment while in a pressurized suit, in a ventilated suit, and in shirtsleeves. Operation of the environmental control subsystem's suit loop is checked as the final step in the test sequence.

After the manned suit loop test the modules are prepared for shipment. This preparation involves demating, final checkout of pressure vessels and the reaction control subsystem, installation of the aft heat shield, application of a thermal coating to the exterior of the command module, installation of earth landing subsystem components and ordnance, a tumble-cleaning, and weight and balance checks.

KENNEDY SPACE CENTER

When the command and service modules arrive at Kennedy Space Center, they begin another long series of tests leading up to launch in which every vital part of the spacecraft is checked once again. Inspection, cleaning, fit checks, functional checks, and leak tests all are part of the pre-launch operations.

Interspersed with these tests are the spacecraft and launch vehicle handling operations. When the command and service modules arrive at Kennedy Space Center, they are taken to the Manned Spacecraft Operations Building where they undergo inspection, and leak and functional tests. The modules are then moved into the altitude chamber and are mated there.

After the chamber and reaction control subsystem tests, the CSM will be moved to a test stand where the service propulsion engine nozzle will be installed and the modules mated with the spacecraft-LM adapter. The latter will already have a lunar module installed.

The spacecraft next is moved to the Vehicle Assembly Building for mating with the launch vehicle and additional tests, including the simulated flight tests. The launch escape subsystem also is mated to the spacecraft in the Vehicle Assembly Building. (On Apollo 7, the first Apollo manned flight, the spacecraft did not go to the Vehicle Assembly Building; instead, it was moved directly to the pad from the Manned Spacecraft Operations Building for testing.)

Two of the most important of the pre-launch operations are the altitude tests and the flight readiness test.

The altitude tests are conducted in a chamber in which conditions at an altitude of more than 200,000 feet can be simulated. The tests consist of four runs: a manned egress test at sea level, an unmanned run at 150,000 feet, and two manned runs at more than 200,000 feet, one with the primary crew and one with the backup crew. All operations in the altitude chamber are televised and recorded.

P-277

Tumbling shakes dust and debris from CM

P-278 *Apollo command and service modules are prepared for thermal-vacuum tests in huge test chamber at NASA's Manned Spacecraft Center*

The altitude tests verify spacecraft operation and integrity at high altitude. Among the specific tests conducted in the chamber runs are suit integrity, manual manipulation of controls and equipment with the cabin depressurized, cabin repressurization (rapid and normal), fuel cell operation, environmental control subsystem cooling and water boiler operation, manual and automatic operation of the guidance and navigation and stabilization and control subsystems, urine dump, emergency breathing system, and entry while both pressurized and unpressurized.

The emergency exit tests are performed in the chamber but before the altitude runs. This test consists of verifying the quick opening of the side hatch after the boost protective cover has been installed.

The countdown demonstration test is one of the final hurdles before launch. In it a complete countdown is performed to verify proper timing and sequence of operations and to check spacecraft-ground communications.

The flight readiness test is the last test of the whole spacecraft before launch. It is designed to check out all subsystems and assure that the craft is ready to proceed to countdown. The electrical power, environmental control, instrumentation, and communications subsystems are first checked to assure proper operation and so that they will support the test. Then the operation of a number of subsystems is checked individually: guidance and navigation, stabilization and control, service propulsion, reaction control, and entry monitor. Next, the integrated operation of these subsystems is checked. Finally, the spacecraft is taken through several simulated aborts and then a complete mission from liftoff to splashdown. The same kind of acceptance checkout equipment used at Downey is used for these tests.

When the flight readiness test is completed satisfactorily, the spacecraft is essentially ready for launch. The only thing left is a final leak and functional test of the service propulsion and reaction control subsystems. After that propellants and cryogenics are loaded and the launch countdown is begun.

Spacecraft is mated with Saturn V

Apollo spacecraft at Kennedy Space Center

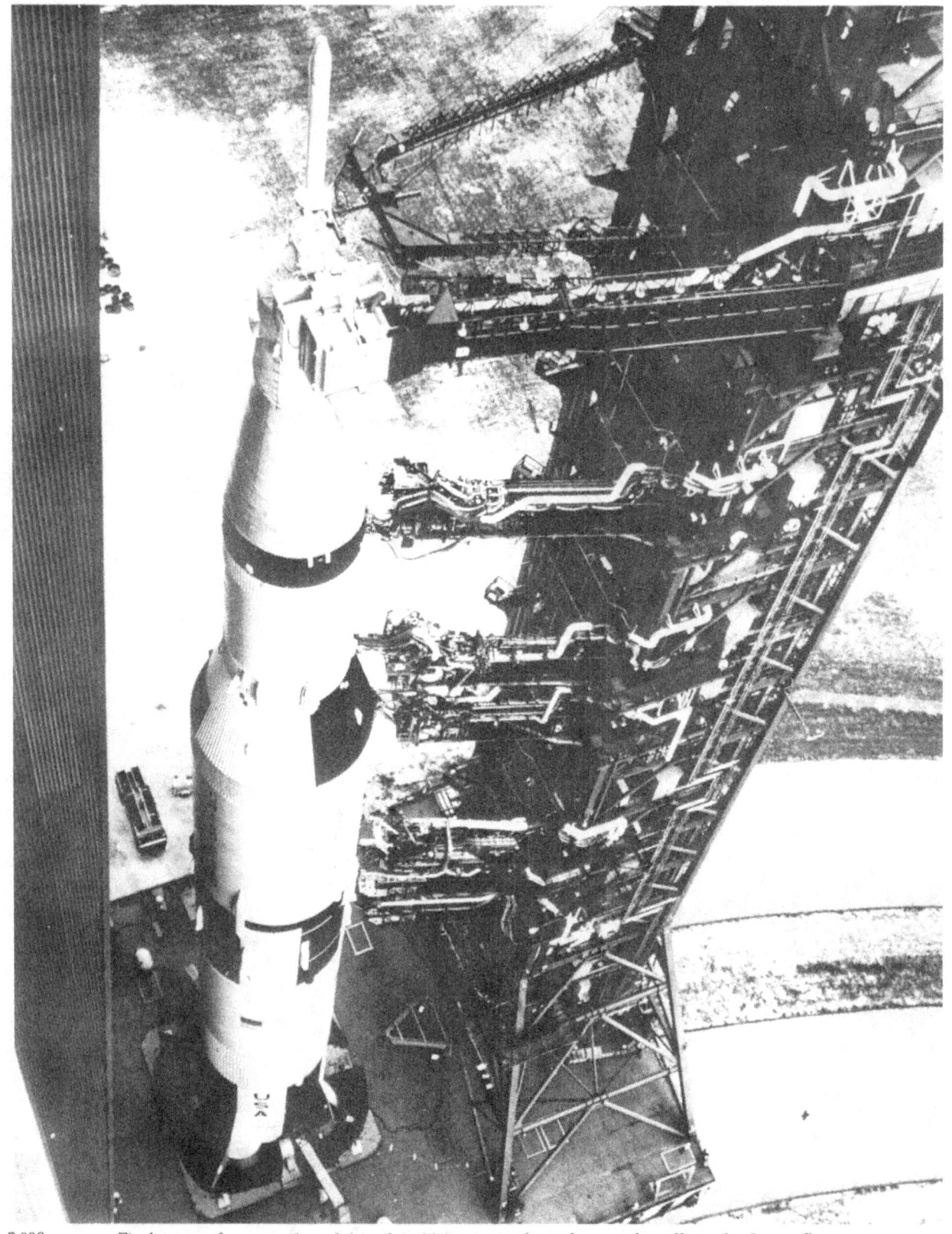

Final tests of spacecraft and launch vehicle are performed on pad at Kennedy Space Center

SAFETY

Apollo safety requirements in space and on the ground required new hardware and procedures in Block II (lunar mission type) spacecraft. Major changes affect the command module's test and pre-launch atmosphere, the hatch, the use of non-metallic materials, cabin emergency oxygen and fire-fighting provisions, wiring protection, and monitoring of crew and command module interior during hazardous ground tests.

GROUND ATMOSPHERE

The atmosphere in the cabin of the command module for tests on the launch pad and at launch will be 60-percent oxygen and 40-percent nitrogen (60/40) rather than pure (100-percent) oxygen. The new mixed-gas atmosphere is supplied by ground equipment. Astronauts breathe pure oxygen in their space suits from Apollo's on-board systems. After launch, the cabin atmosphere is vented at a controlled rate, then replenished with pure oxygen so that in 4 to 6 hours it is approximately 95-percent oxygen. The safety of the modified spacecraft was judged acceptable in the 60/40 mixed-gas atmosphere of 16 psi, and in a pure oxygen atmosphere at the space pressure of about 6 psi after extensive tests at NASA's Manned Spacecraft Center.

SIDE HATCH

A one-piece door replaces the two-cover hatch system on the command module. The side hatch is made of aluminum with fiberglass and ablative material. The door deployment mechanism has a gas-operated counter-balancing device that offsets gravity and permits easy opening on the ground. The hatch can be unlatched and opened by the flight crew in less than seven seconds and by the ground crew in about 10 seconds.

On the ground and during the early part of boost, the command module is shielded from boost heating by the boost protective cover. This cover, which is attached to and jettisoned with the launch escape tower, also has a hatch. As the unified hatch is opened from the inside, it activates a release mechanism between it and the boost protective cover hatch. The mechanism releases the single latch of the cover hatch and the two hatches swing open together.

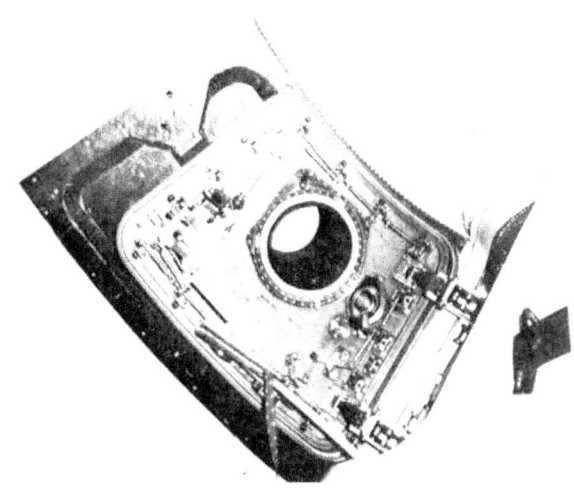

New hatch with boost protective cover hatch opening
P-281

MATERIAL

All materials in the spacecraft command module have been re-evaluated. Non-metallic materials were subjected to a rigid series of flammability tests and were replaced as required.

Among the more important changes are the use of stainless steel tubing instead of aluminum for the astronauts' high-pressure oxygen system. Aluminum solder joints of lines carrying water-glycol liquid for cooling or heating have been reinforced with protective armor where necessary. Protective plates cover coolant lines and also protect wiring against wear or accidental damage. Stowage boxes are made of aluminum.

Flammable materials are stowed in fireproof containers (metal or polyimide fiberglass storage boxes and Beta Cloth stowage bags).

Nylon Velcro material, used to grip or hold objects in the weightlessness of space, has been replaced with a new Teflon and polyester Beta fiberglass product, and wherever practical, mechanical fasteners are used to "button down" or hold equipment. A new flame-resistant material called Ladicote has been introduced and is applied by brush to potted connections.

Significant material changes include:

Old	New
Nylon Velcro	Teflon Beta fiberglass (for the pile); polyester Beta fiberglass (for the hook)
Polyurethane line insulation	Molded glass fibers
Nylon Raschell knit debris trap	Aluminum coverings
Silicone rubber wire bundle antichafe wrap	Teflon sheet
Nomex (nylon) wire bundle spot ties	Teflon-coated Beta fiberglass
Mylar window shades	Aluminum sheeting (not roll-up type)
Silicone heat-shrink wire insulation	Teflon heat-shrink wire insulation
Trilock couch padding	A new fabric couch pad made of Teflon-coated fiberglass
Most plastic knobs and switch levers	Aluminum
Polyolefin coaxial cable	Wrapped with aluminum foil tape; later spacecraft to have Teflon cable
Plastic switches in main display panel	Metal
Silicone oxygen umbilical hose	Covered with Fluorel
Crewman's communications umbilical (silicone rubber)	Molded Fluorel
Epoxy laminate food boxes	Polyimide laminate
Silicone laminate panel scuff covers	Polyimide laminate covers
Electroluminescent panels	Covered with copper overcoat
Silicone rubber spacers	Covered with Beta fabric
Nylon zipper on space-suit bags	Metal
Circuit breakers of diallylphthalate (DAP) and Melamine, both resins	Covered with Ladicote
Epoxy laminated structures	Polyimide structures
Postlanding vent duct (silicone laminate)	Metal and Fluorel impregnated glass fabric

Old	New
Felt filters in lithium-hydroxide canisters	Teflon felt
Uralane foam (cushion material for mirrors, etc.)	Fluorel foam
Fiberglass tape	Aluminized tape
Nylon webbing (such as hook on CO_2 absorber)	Beta webbing
Dacron cloth in the environmental control system	Armalon cloth
Aluminum high-pressure oxygen lines of environmental control system	Stainless steel

CABIN PROVISIONS

An emergency oxygen system with three masks and an independent oxygen supply would protect the crew from toxic fumes. Special fire-fighting provisions include a portable fire extinguisher, protection panels to isolate a fire, and special ports where the extinguisher's nozzle is inserted to douse a flame behind a panel.

WIRING PROVISIONS

A number of changes makes the estimated 15 miles of wiring safer in Block II spacecraft. Some circuit breakers were added and others reduced in capacity to improve wiring protection. Teflon wrapping separates power wires from others in a bundle. Aluminum enclosures protect wire runs in the crew compartment.

Ladicote, a special fire-resistant material which is applied by a brush, coats terminals, metallic electronic components, and circuit breakers. Ladicote was developed by chemists at North American Rockwell's Los Angeles Division.

MONITORING

Hazardous ground tests are more closely controlled by monitoring of biomedical data from the three crew members and observation through closed-circuit television of the command module interior.

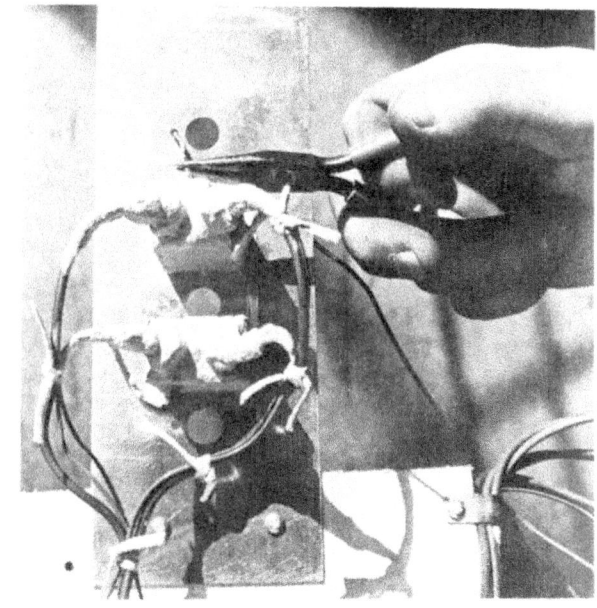

Wire terminals coated with Ladicote fire retardant
P-282

EARTH LANDING SUBSYSTEM

The earth landing parachute system has been modified to handle the increased weight of the command module. Its two drogue parachutes were expanded from 13.7 to 16.5 feet. A dual-reefing feature was added to permit the three main chutes to open more slowly.

Astronaut Wally Schirra leaves CM after Downey test of Block II spacecraft

RELIABILITY AND TRAINING

The Apollo command and service modules have approximately two million functional parts, miles of wiring, and thousands of joints. The operation and integrity of each part and structure must be assured.

To do this, the Apollo spacecraft undergoes exhaustive testing, starting with the smallest component. Systems and subsystems are tested under various simulated mission conditions and in their interaction. All components are tested far beyond the required safety level.

There are 587,500 inspection points on the command and service modules. In addition, the vehicle is checked to make sure it conforms to each of approximately 8,000 drawings and 1,700 manufacturing and engineering specifications.

Integrity of hundreds of feet of weld and the thousands of joints must be verified. The adhesive bonded honeycomb structure of the modules is inspected ultrasonically and the brazed honeycomb heat shield structure is inspected radiographically. Deviation from the stringent requirements results in test to determine the cause, repair or replacement, and a new cycle of final tests.

But reliability is achieved primarily through preventive rather than curative measures. These include such things as conservative design (that is, design with a wide margin of safety) and stringent technical and administrative controls. Reliability assessment of critical components is performed at the end of development, at the end of qualification testing, and before flight.

Tests of Apollo CSM structure and systems have been performed at Downey, White Sands Missile Range, N.M., Manned Spacecraft Center in Houston, Tullahoma, Tenn., and Kennedy Space Center.

In addition, more than 7500 hours of wind-tunnel testing has been conducted in government, university, and industrial facilities to gather data on aerodynamic performance during boost, spacecraft and booster loads, acoustic noise and aerodynamic heating, and lift-to-drag hypersonic velocities. Although the Apollo spacecraft operates in the atmosphere only for a few minutes, it underwent almost twice as much wind-tunnel testing as the X-15

Environmental (vacuum) chamber at Downey

and almost as much as the XB-70. The XB-70 had 11,500 hours of wind tunnel testing.

TESTING FACILITIES

Pressure Test Cell—This cell is used to test pressure and leakage of the service propulsion subsystem at

Impact test facility

1.5 times the maximum working pressure. The environmentally controlled cell is a concrete-lined pit 25 feet deep, 25 feet long, and 32 feet wide. It is separated from other buildings by more than 150 feet to permit test with a hazard rating of up to 50 pounds of TNT. Helium gas is the test medium. CM pressure tests also are conducted in the cell.

Altitude Chamber and Airlock—Called the bell jar, this chamber was used for a 14-day simulated mission with three space-suited engineers in a CM. The chamber contains an environmental control system with an airlock. The chamber can be evacuated to 10^{-4} torr (a hard vacuum), simulating conditions from launch to a 200,000-foot pressure altitude. The airlock contains instruments for the life support system. Ground support equipment was used to supply electrical power, potable water, and oxygen furnished in space flight by the fuel cell powerplants and cryogenic storage system.

Impact Test Facility—This four-legged tower contains a pendulum that was used to swing a full-scale instrumented CM at controlled speeds and angles, dropping it into a water tank or on a special land impact area to simulate parachute landings. Drop tests provided information on how impact affects the spacecraft structure and crew system response. The impact information is relayed and recorded on oscillographs and magnetic tapes and is used to confirm and define spacecraft and equipment design. information is relayed and recorded on oscillographs and magnetic tapes and is used to confirm and define spacecraft and equipment design.

Tower height is 143 feet, height of the catwalk and pendulum pivot is 125 feet, length of pendulum arms is 91 feet, and maximum impact velocity is 40 feet per second vertical and 50 feet per second horizontal.

Space Simulation Facility—This provides a simulated space environment (high vacuum, solar radiation, and temperature extremes) to determine its effect on the spacecraft and its materials. The actual space vacuum (10^{-12} torr) can be achieved in the facility. Supporting test equipment includes temperature measurement, residual gas analysis, leakage measurement, spectrum analysis, and vacuum measurement systems.

Fuel Cell Test Facility—Fuel cells, power sources, power storage, and power distribution designs are tested in this facility. Bus switching techniques for single and parallel powerplant operations can be developed in the facility and transient susceptibility for spacecraft operation in a vacuum can be analyzed.

Structural Test Facility—This facility covers an area of 14,000 square feet and contains hydraulic

Oven-freezer tests CM structural strength by roasting one side at 600° while dousing other side with liquid nitrogen at 320° below zero

equipment, including proportioning units, load cells, and hydraulic struts with loading capacities ranging up to 500,000 pounds, and four 24-foot-high test columns, each with the ability to react to 10,000,000 inch-pounds of moment.

Plasmajet Test Facility—Approximately 1,000 plasmajet tests are conducted on ablative specimens, simulating radiative and convective heat fluxes. Heat fluxes from 5 to 800 British thermal units per square foot per second and gas stream enthalpies from 5,000 to 25,000 British thermal units per pound are produced. Panels of typical CM substructure covered with ablative material are cycled from room temperature through ascent heating temperatures, then down to space flight temperatures, and finally to temperatures simulating entry heating.

Radio Frequency Laboratory—All radio frequency characteristics of spacecraft, radio command, antenna, and telemetry systems are measured in this laboratory.

Climatics Laboratory—Spacecraft components are tested for resistance to elements in the ground and atmosphere environments in this facility. Laboratory equipment exposes equipment to sand, dust, rain, salt spray, and oxygen, individually and in combination.

Acoustics and Data Facilities—All types of dynamic tests (acoustic, vibration, shock, and acceleration) of Apollo components are conducted here. Test findings are recorded on dynamic data equipment (magnetic tape and oscillograph).

Electronic and Electrical Facilities—Electronic and electric circuits, components, and subsystems are

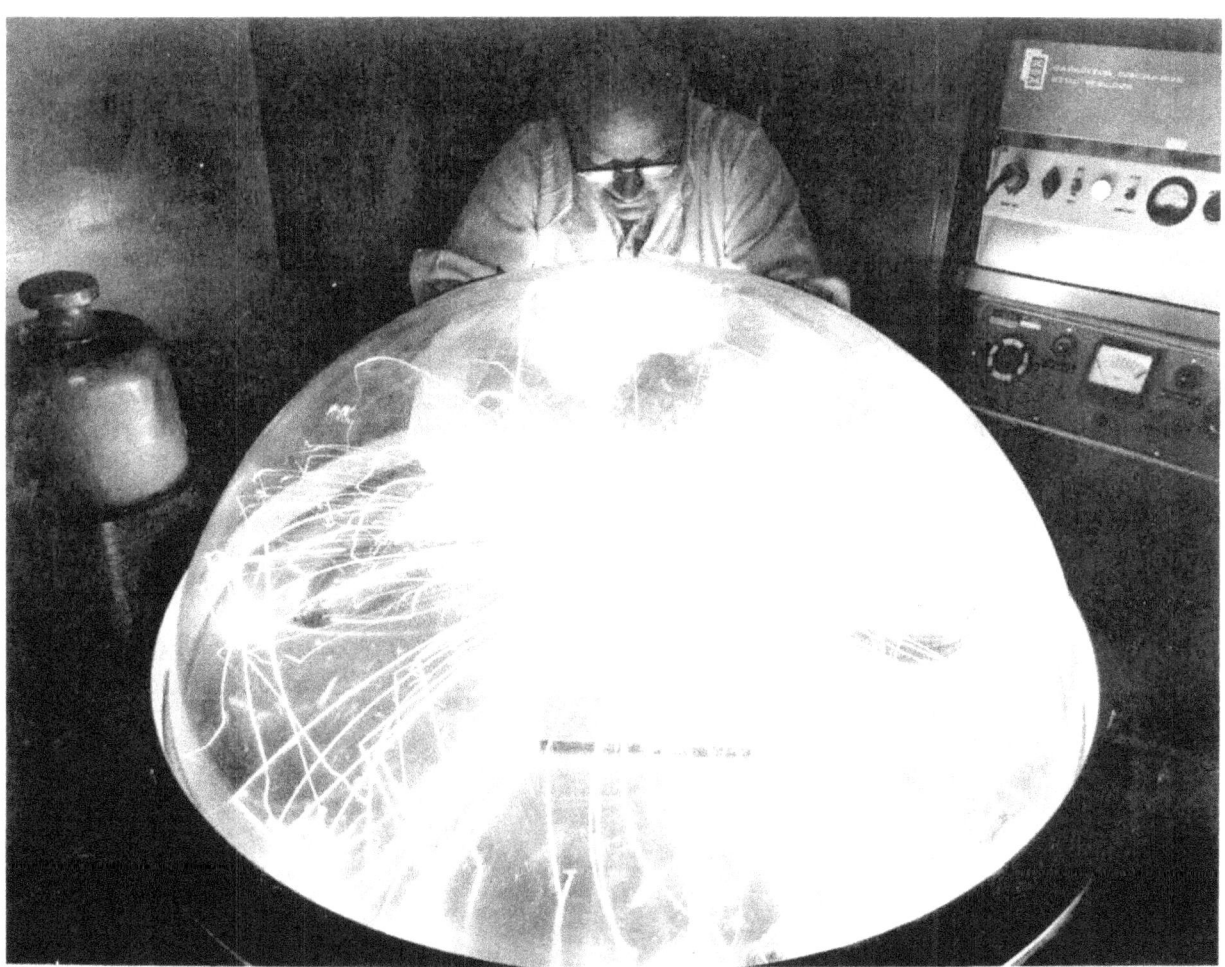

P-287

Technician tests new welding technique for space metals at Downey laboratory

tested and analyzed in these facilities and prototypes are developed and evaluated.

Clean Room—The final assembly and checkout area is in the world's largest known clean room. It contains 45,000 square feet of floor space and 2,500,000 cubic feet of air space. It is 410 feet long, 100 feet wide, and separated into two bays, one 63 feet high and the other 42 feet high. The air is filtered and changed three times an hour; temperature is kept at 73 degrees and humidity at 50 percent. Glassed areas on either side of the clean room are kept at higher levels of cleanliness and used for component assembly. Stringent rules govern the dress and operations of workers in the room. The command and service modules enter the clean room through huge airlocks and are tumbled and vacuum-cleaned to remove dust and debris. Subsystems are installed in the two modules and a number of tests performed, including the final series of checkout tests of the completed modules.

Many ground tests have been conducted during development with full-scale test modules. The major ground tests of combined command and service modules include:

Test Site	Purpose
White Sands Test Range, N.M.	Evaluate service propulsion and reaction control subsystems during malfunction, normal, and mission profile conditions
Downey and Houston	Test CM for earth recovery and land impact
Downey	Verify integrity of CSM structure under critical static and thermal loads
Downey and Gulf of Mexico	Test CM transmissibility (bending loads in free fall), water impact, and flotation
Houston	Test environmental control subsystem in manned and unmanned deep-space environmental chamber
El Centro, Calif.	To test earth recovery system
Houston	To test for launch vibration environment
Tullahoma, Tenn.	To test service propulsion engine altitude starting characteristics

TRAINING EQUIPMENT

The training program for management, staff, flight crew, and test and operations personnel parallels the design and manufacture of Apollo spacecraft.

Special equipment for the training program includes spacecraft evaluators at Downey and mission simulators at the Manned Spacecraft Center in Houston and at Kennedy Space Center.

SPACECRAFT EVALUATORS

Apollo astronauts practice spacecraft procedures and operate the command module's displays and controls at the Space Division in Downey. The evaluators, simulated command modules with crew displays and controls and control system elements similar to the flight version, are connected to a computer complex which controls their operation.

Peripheral equipment includes an earth, stars, and sun as they would appear to the astronauts. The earth, a six-foot globe in which landmarks are scaled to an accuracy of within three miles of their actual position as seen from space, revolves in a manner to simulate its own revolution and the orbit of the command module. The revolution can be controlled to reproduce exactly that which would appear to the astronauts at different velocities and orbit heights.

Astronauts can "fly" the command module through operation of the same controls that are on the flight spacecraft. Data on operation of the evaluator's controls is sent to the computing equipment, which interprets it and relays the proper reaction back to the simulated spacecraft, all in a fraction of a second. Thus the displays in the evaluator respond to the command module controls in the same manner as they would in space.

Two spacecraft evaluators aid astronaut training
P-288

Apollo mission simulator at Houston includes CM, peripheral equipment, control center

CM MISSION SIMULATORS

The command module mission simulators, built by the Link Group of General Precision Systems, Inc., Binghampton, N.Y., under contract to Space Division, are fixed-base trainers capable of simulating characteristics of spacecraft system performance and flight dynamics. In them the astronauts practice operation of spacecraft subsystems, spacecraft control and navigation, and crew procedures for space missions. Malfunctions and degraded performance of spacecraft subsystems also can be simulated.

The interior of the CM mission simulator is a replica of the actual command module, containing all panels, controls, switches, and equipment. The essential life support systems are designed to operate up to 14 days.

An entire lunar mission—except for lunar descent and ascent—can be simulated. Visual and acoustic effects are simulated; everything, in fact, except the sensations of weightlessness and the gravitational forces of launch and earth re-entry. (Training for the lunar descent and ascent is performed in the lunar module simulator.)

The CM mission simulator has four computers integrated into a single complex to provide real-time simulation of all spacecraft subsystems and equations of motion of both the CM and LM. Each of the computers can perform 500,000 mathematical operations per second. The complex has 208,000 memory core locations.

Each simulator is programmed to provide normal, emergency, and abort conditions. More than 1,000 training problems can be inserted into the simulated spacecraft subsystems, enabling the crew to prepare for nearly every situation. The computers also generate telemetry information in actual mission format for transmission to ground station equipment, thus training ground personnel.

The CM mission simulator's visual system, which contains more than five tons of lenses and curved glass, presents realistic external environments that change according to the position of the command module. Objects ranging from six feet to infinity—including earth, moon, sun, stars, and the LM—are duplicated. Separate units simulate the views seen through each of the command module's four windows and through the sextant and telescope.

The simulators are designed to operate independently as full mission trainers for astronauts, as well as to operate in connection with the Mission Control Center and the LM mission simulators.

Astronauts (from left) Tom Stafford, John Young, and Eugene Cernan train in mission simulator

APOLLO MANUFACTURING

The variety and complexity of components in the Apollo command and service modules and the degree of reliability and quality demanded for each imposed many fabrication problems.

Solution of these manufacturing problems required application of skills in such areas as advanced electronics, fire retardant organics, plastics and cryogenic insulation, welding and brazing, adhesive and diffusion bonding, and machining, plus design and development of many tools and fixtures.

In fact, almost all of the tools and fixtures used in fabrication and assembly of the command and service modules were designed especially for the Apollo program.

For the Apollo spacecraft there are five major manufacturing assemblies: the command module, service module, lunar module, launch escape subsystem, and the spacecraft-LM adapter. All but the LM are assembled by North American Rockwell. The CM, SM systems, and launch escape subsystem are at Downey, Calif. The SLA and basic SM structure are produced at North American Rockwell's Tulsa (Okla.) Division. The LM is produced by Grumman Aircraft Engineering Corp., Bethpage, N.Y.

In the original basic mastering programs, conventional airframe mastering techniques were used. Tooling specialists soon realized, however, that while plaster model masters had been satisfactory for constructing aircraft, they could not maintain the tolerances required for critical space hardware. So the technique was conceived of fabricating control masters, masters, and assembly tools from like materials, compatible with the end hardware: for example, aluminum masters and aluminum tools for the aluminum hardware and steel masters and steel tools for the steel hardware. Basic tolerances could be integrated into these tools and were not nullified by differential expansion during operations involving the application of heat. Mainly because of this improved tolerance control, some heat shields have been delivered without any defective weld despite the 718 feet of weld in the crew compartment heat shield and the difficult access to some areas.

Many welding innovations have been developed during the program. One of these was the use of a pressurized portable clean room that enclosed a total weld station to maintain temperature and dust particle control. Another was the development of closed-circuit television for monitoring and controlling manufacturing operations. Miniaturized weld skates were developed for use in inaccessible areas.

One of the most important innovations was an induction brazing method in which a small unit can be moved as far as 600 feet away from its bulky generator. The small unit is used to join stainless steel fluid system components in remote and relatively inaccessible areas of the spacecraft.

In the portable brazing tool, a radio frequency current flows through coils and produces a high-frequency magnetic field around the work piece. This magnetic field produces the induction heating (up to 2,000 degrees) needed for brazing. The brazing substance is a gold alloy inside the sleeve which joins the two ends of a conduit.

Most of the spacecraft plumbing joints are induction-brazed stainless steel. This successful joining process offers a number of advantages. The joints are light (compared with mechanical joints), strong, and low cost. X-ray examinations have determined that more than 97 percent of these braze joints are acceptable. In addition, this system permits joining of tube stubs having widely different wall thickness.

The boost protective cover is an example of problems solved on the program. It is a multi-layer, resin-impregnated fiberglass assembly 11 feet tall and 13 feet in diameter, weighing approximately 700 pounds. It fits over the command module like a glove.

Originally it was concluded that the protective cover would be a standard configuration adaptable to all spacecraft. As the program progressed, however, it was apparent that each cover must be tailored to each heat shield.

In the process, heat shields are mounted on a holding fixture and a mixture of resin and fiberglass blown against the shield to produce a fiberglass female mold identical to the heat shield. Through a series of carefully controlled casting operations, a full-size plaster master is constructed to reproduce the outer moldline of the heat shield.

The plaster simulators match so exactly the actual heat shield that the finished boost protective cover is inspected for a match with the simulator rather than the actual heat shield, eliminating hundreds of hours of inspection and other operations for the spacecraft.

The unified hatch for the command module is probably the most carefully engineered and manufactured door ever built. A system of 12 linked latches seals the door shut.

Many advanced technologies were used to produce this hatch, both in tooling and in the various tool fabricating and assembling areas. One noteworthy innovation was the conversion of an existing fixture to machine three complex components: edge ablators which fit around the periphery of the door and the hatch opening, and the ablator which attached to the inner crew compartment door. In all, about 150 new tools were designed and built for the hatch.

A major element of the environmental control subsystem is the coldplate, a mounting plate through which coolant flows to prevent overheating of electronic components. Originally, coldplates were machined, ladder-type cores that were eutectic-bonded between two face sheets. These were difficult to bond and the rejection rate was prohibitive.

To overcome the problem, a pin-fin configuration was developed which could be machined by electrical discharge and which immeasurably reduced fabrication complexity yet proved more effective in heat dissipation. In addition, heated platens with precise thermal controls were developed to provide the degree of heat, pressure, and flatness necessary to diffusion-bond the coldplates. Although required to function at a pressure of 90 psi, the coldplates now being produced are being tested at 1000 pounds without any evidence of failure.

One of the severest requirements of the Apollo program was for a heat shield that would withstand the intense aerodynamic heating experienced during entry from a lunar mission.

The heat shield is fabricated of a special stainless steel honeycomb sandwich manufactured by the Aeronca Co., Middletown, Ohio, and serves as the outer structure of the vehicle. The shield is assembled from 40 individual panels produced by

P-291 *Unified hatch in final assembly*

means of a special electric-blanket brazing process. The brazing material used to join the steel skins to the honeycomb is a silver-copper-lithium alloy in a nickel matrix. Each panel is subjected to X-ray inspection after brazing to assure quality.

The ablative (heat-dissipating) material is a phenolic-filled epoxy compound developed and applied by the Avco Corp.'s Space Systems Division, Lowell, Mass. The ablative material is dielectrically heated and injected with specially developed guns into each of more than 370,000 cells in the glass-phenolic honeycomb bonded to the outer surface of the three heat shield sections. Each section is X-rayed to assure that all cells are completely filled, then cured in specially designed ovens. For machining the various thicknesses required of the contoured shields, computers operate machining heads of giant lathes. Pore sealer is applied as the final process, and thermal paint is applied to the heat shield.

COMMAND MODULE

The basic command module structure consists of a nonpressurized outer shell (the heat shield) and a pressure-tight inner shell for the crew compartment. The inner compartment is formed of aluminum honeycomb sandwich while the heat shield is formed of stainless steel honeycomb sandwich. The space between the inner and outer structures is filled with a special fibrous insulation (Q felt).

ASSEMBLY

The heat shield structure consists of three basic assemblies: the forward, crew compartment, and aft sections. The complete assembly envelops the the inner crew compartment and provides thermal protection during entry.

The forward assembly of the heat shield consists of four conical-shaped honeycomb panels, one machined aft ring, one forward bulkhead, and four launch escape tower leg fittings. The section is assembled in the following sequence. The tower leg fittings are installed, trimmed, and welded to each of the four honeycomb panels. The panels are installed in a fixture which accommodates all four panels; the panels are trimmed longitudinally, then butt-fusion welded. The welded panels, forward bulkhead, and aft ring are placed in another fixture for circumferential trim and weld. The aft ring and forward bulkhead inside ring are finish-machined after welding. The completed assembly is fit-checked

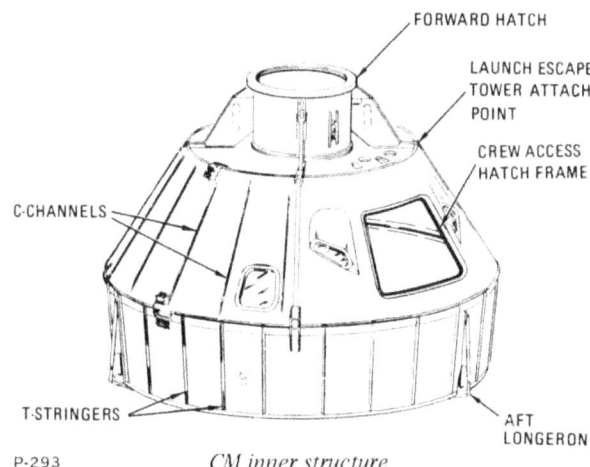

CM inner structure

to the inner crew compartment and crew compartment heat shield, and then removed for application of ablative material.

In addition, the forward heat shield assembly has an outer access door. This door consists of two machined rings that are weld-joined to a brazed honeycomb panel. The inner ring and outer ring are machined after welding. The door closes the forward end of the access tunnel of the crew compartment. It provides thermal and water-tight protection and may be opened from inside or outside.

The crew compartment heat shield is formed from numerous brazed honeycomb panels, numerous machined edge members which provide for door openings, and three circumferential machined rings joined by fusion welding. The panels and rings are installed in a series of jigs for assembly, trimming, and welding. The welded sections are placed in a large fixture for precision machining of the top and bottom rings. The assembly is fit-checked with the inner crew compartment and the forward and aft heat shields, then removed for application of ablative material.

The aft heat shield consists of four brazed honeycomb panels, four spotwelded, corrugated, sheet metal fairing segments, and one circumferential machined ring. The honeycomb panels are joined laterally by fusion welds. The four fairing segments are attached to the honeycomb panels and machined ring using conventional mechanical fasteners. Holes for inner and outer structure attachment points and tension tie locations are cut through the assembly. The complete section is fit-checked with the crew compartment heat shield, then removed for application of ablative material.

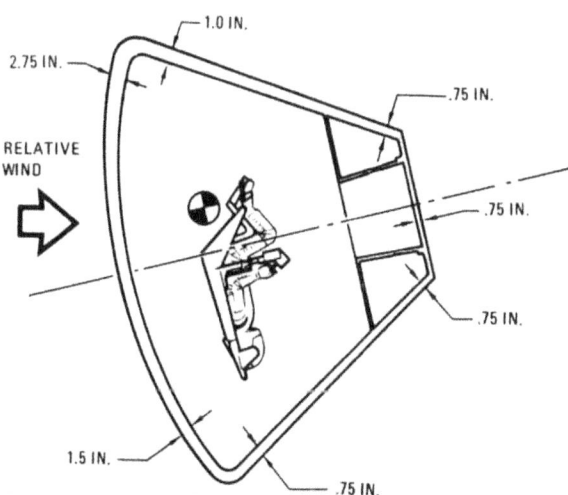

Thickness of CM ablative material

The inner crew compartment is built in two assemblies: the compartment structure and the system support structure. The compartment structure is made of aluminum and is fabricated in two sections. The forward section consists of an access tunnel, a forward bulkhead, and a forward sidewall. The aft section consists of an aft sidewall, an aft bulkhead, and a circumferential machined ring. The two sections, when joined, form the spacecraft's pressure vessel.

The forward section welded inner skin is fabricated from panels, four machined longerons, window frames, a machined circumferential girth ring, and fittings. Aluminum honeycomb core and outer face sheets are thermally bonded to the inner skin and cured in a giant autoclave (similar to a giant pressure cooker). Attachments and fittings are then bonded to the structure for installation of the system support structure, wiring, tubing, and other equipment. The access tunnel, which is bonded to the forward bulkhead, includes a forward ring for mounting the docking ring, the pressure hatch cover, and external frames which absorb loads from parachute deployment and the recovery sling.

The aft section welded inner skin is fabricated from panels, machined ring, and fusion-welded bulkheads. Aluminum honeycomb core and outer face sheets are thermally bonded to the inner skin and cured in a giant autoclave. External frames and internal attachments are bonded to the structure for the system support structure.

The inner crew compartment is completed when the forward and aft assemblies are circumferentially trimmed and fusion welded at the girth ring. The final assembly operation is the bonding of aluminum honeycomb core fillers and facing sheets.

"Egg crate" fixtures developed to locate exactly the CM interior components

The system support structure, which is added after completion of the inner structure, consists of the main display console and the structure for the equipment bays. The bays are fabricated of sheet and machined aluminum panels and vertical frames. Each equipment bay is assembled outside and then transferred into the inner compartment through the crew access hatch. Basically, the final assembly of the command module involves the installation of the heat shield over the inner crew compartment and the mechanical attachment of the two structures. Fibrous insulation (Q felt) is installed between inner and outer structures.

"Egg crate" fixtures were developed for more accurate and efficient installation of CM interior components. These curved tooling structures simulate a bay of the spacecraft and give workers the precise location for brackets, stringers, and other mountings. The attachments are located with the jig, and fixed in place with metallic tape and the egg crate is removed. Then the devices are bonded to their locations. The egg crate tool is used again to determine whether any of the components have moved during bonding. The largest of the egg crate jigs covers about one-quarter of the inside circumference of the CM.

Engineers say the egg crate jig is more flexible in use and more accurate than the "wrap-around" tool that was used for the same purpose but covered the entire circumference of the inside of the module. The old tool was much bulkier and less adaptable for close tolerance work.

SUBSYSTEM INSTALLATION

Subsystems are installed in a giant clean room in Downey. When structural assembly of the command module is complete, it is moved from the main manufacturing area to the clean room. There it goes through an outer airlock and is mounted on a special machine which vacuum-cleans and tumbles it, removing all dust and other particles. After this cleaning operation, it goes through an inner airlock to a station in the clean room for installation of subsystems.

Workers entering the room must pass through an air shower and clean their shoes with an electric buffing machine before entering the anteroom. There they don clean smocks and head coverings and pass through the air shower again before entering the clean room proper.

Even the workers clothing is restricted. Wool is prohibited (too much lint) and leather soles may not be worn. Workers entering the command module must remove everything from their pockets, and even rings and tie tacks, to assure that no foreign material will be left in the module. They also must put on special "booties" to protect the crew compartment. A hatch guard is stationed at the entrance to each command module to check each worker in and out.

Tools used by the clean room workers in installing the spacecraft's wiring and subsystems are issued in specially-designed, fitted boxes. These boxes are checked at the beginning and end of each shift to account for every tool and item of equipment.

When subsystem installation and the many testing operations are completed, the module is moved to another part of the clean room for the acceptance checkout tests described in the section on Checkout and Final Test.

SERVICE MODULE

This is a cylindrical structure consisting of forward and aft honeycomb sandwich bulkheads, six radial beams, four outer honeycomb sandwich panels, four honeycomb sandwich reaction control system

P-295

SM radiator panel after assembly in Tulsa

P-296
Subsystems are installed and checked in command and service modules at Space Division's clean room, Downey, Calif., then shipped by air to Kennedy Space Center, Fla.

panels, aft heatshield assembly, and a payload fairing and radiator assembly.

The outer sector panels are 1 inch thick, and made of aluminum honeycomb bonded between aluminum face sheets. The radial beams, made from milled aluminum alloy plates, separate the module into six unequal sectors around a center section. Maintenance doors are located around the exterior of the module for access to equipment in each sector.

Radial beam trusses on the forward portion of the SM provide the means to connect the CM and SM. Alternate beams (Beams 1, 3, and 5) have compression pads for supporting the CM. The other beams (Beams 2, 4, and 6) have shear-compression pads and tension ties. A flat center section in each tension tie contains explosive charges for SM-CM separation. The six radial beams are machined and Chem-Mill etched (made thinner by chemical action) to reduce weight in areas where there will be no critical stresses.

These beams and separation devices are enclosed within a fairing 26 inches high which seals the joint between the CM and SM. Eight radiators which are part of the spacecraft's electrical power subsystem are alternated with ten honeycomb panels to make up the fairing. Each EPS radiator has three tubes running horizontally to radiate, to space, excess heat produced by the fuel cell powerplants. Two of the four outer honeycomb panels have radiators to dissipate heat produced by the spacecraft's environmental control subsystem. These ECS radiators, each about 30 square feet, are located on opposite sides of the SM.

After its assembly is complete, the service module is mated with the command module for a fit-check and alignment. The modules are then de-mated and the service module follows the same procedures as the command module for installation of subsystems in the clean room.

SPACECRAFT-LM ADAPTER
(SLA)

This structure is a tapered cylinder constructed of eight 2-inch-thick aluminum honeycomb panels (four aft and four forward) joined together with inner and outer doublers. The four forward panels, each about 22 feet tall, are hinged at the bottom. The aft panels are each about 7 feet tall. Other major components of the SLA include devices to separate it from the SM, fold back and jettison the forward panels, and separate the LM from the SLA.

The bonding of the skin to both sides of the honeycomb panels is done in one of the largest autoclaves in the aerospace industry. This autoclave, at North American Rockwell's Tulsa Division, is a huge pressure heater, 20 feet in diameter and 40 feet long, with a heat capacity of 500 degrees and a pressure capability of 110 psi. An epoxy adhesive is used to bond the parts. The autoclave is large enough to accommodate one of four large SLA forward panels at a time. The autoclave also is used to bond all of the service module panels

LAUNCH ESCAPE ASSEMBLY

The basic structure consists of a Q-ball instrumentation assembly (nose cone), a ballast compartment and canard assembly, a pitch control motor, a tower jettison motor, the launch escape motor, a structural skirt, and a latticed tower.

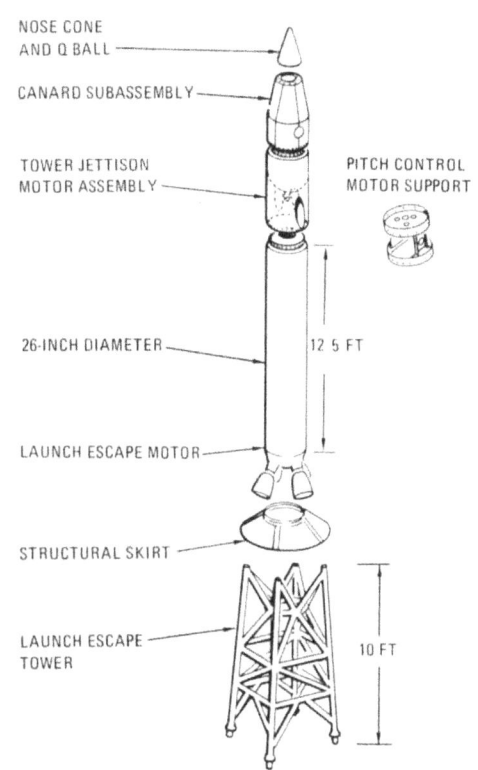

Major launch escape subsystem structure

SLA panel is prepared for bonding in giant autoclave

The nose cone is a little more than 13 inches in diameter at its base and tapers to a rounded apex. Its total height also is a little more than 13 inches. Its skin is made of Inconel (a heat-resistant nickel alloy) and stainless steel riveted together. The cone has four ports to permit the electronic instrumentation inside it to measure pressure changes and the angle of the launch vehicle.

The ballast compartment also is constructed of Inconel and stainless steel and contains lead weights. Two canard subassemblies, each consisting of a thruster, actuating arm, and deployable surface, are faired into the ballast compartment surface.

The pitch control motor assembly is made of nickel alloy steel sheet skins riveted to ring bulkheads and frames. The case for the tower jettison motor is made of high-carbon chrome-molybdenum steel forged.

The launch escape motor is 15 feet long and has a case made of steel. The structural skirt is made of titanium, as is the tubing of the launch escape tower.

P-299 *Command module, nearing completion in Downey clean room, is put on fixture to be moved to new station*

CSM SUBCONTRACTORS

More than $1,078,000,000 has been funded by North American Rockwell's Space Division to Apollo subcontractors and suppliers throughout the United States. Funding to subcontractors with contracts in excess of $500,000 (values are approximate):

Company	Product	Value
Accessory Products Co. Whittier, California	Helium transfer unit, valves, and assemblies	$ 3,469,567
Aerojet General Corp Space Propulsion Division Sacramento, California	Service propulsion engine	99,780,000
Aeronca Manufacturing Co. Middletown, Ohio	Stainless steel honeycomb panels	15,411,779
Air Products & Chemical Co., Inc. Allentown, Pa.	Liquid hydrogen storage units	599,523
Amecom Division Litton Systems, Inc. College Park, Maryland	C-band (Block I) and S-band antennas	2,348,600
Applied Electronics Corp. of New Jersey Metuchen, N.J.	Pulse-code modulation systems	1,185,000
Astrodata, Inc. Anaheim, Calif.	Integrated computer complex	783,000
Avco Corp. Space Systems Division Lowell, Mass.	CM ablative heat shield	47,272,900
Avien, Inc. Woodside, N.Y.	2-gigacycle deep-space antenna	2,628,000
Beckman Instruments, Inc. Scientific & Process Instruments Division Fullerton, Calif.	Data acquisition system and water conditioning	2,838,000
Beech Aircraft Corp. Boulder, Colo.	Cryogenic gas storage system	30,855,876
Bell Aerosystems Company Division of Textron, Inc. Buffalo, N.Y.	Positive expulsion reaction control subsystem propellant tanks	23,084,000
Bendix Corp. Davenport, Iowa	Instrumentation	552,000
Borg-Warner Controls (Formerly Electroplex) Los Angeles, Calif.	Modules and amplifiers	706,781

Company	Product	Value
Calmec Los Angeles, Calif.	Helium pressure relief valves	$ 854,000
Collins Radio Co. Cedar Rapids, Iowa	Communications and data subsystems	130,317,000
Control Data Corp. Government Systems Division Minneapolis, Minn.	Digitial test command system spare parts	9,900,000
Cosmodyne Corp. Torrance, Calif.	Liquid hydrogen, liquid oxygen ground support equipment and unique detail spares of liquid hydrogen and liquid oxygen transfer units	4,348,440
Dalmo Victor Co. (Division of Textron) Belmont, Calif.	S-band high-gain (deep space) antenna	13,072,000
McDonnell Douglas Corp. Douglas Aircraft Div. Long Beach, Calif.	Parachute subsystem testing	904,000
Eagle Picher Joplin, Mo.	Post-entry and storage batteries	720,000
Eckel Valve Co. San Fernando, Calif.	Valves	783,958
Electro-Optical Systems, Inc. Micro Systems, Inc. Pasadena, Calif.	Temperature and pressure transducers, signal conditioners, and electronic control units	8,772,000
Garrett Corp. AiResearch Manufacturing Co. Los Angeles, Calif.	Environmental control subsystem	78,308,748
General Motors Corp. Allison Division Indianapolis, Ind.	Service propulsion fuel and oxidizer tanks	13,389,658
General Precision, Inc. White Plains, N.Y.	Mission simulator trainer, control sequence, heat flux sensor	42,083,340
General Time Corp. Aeronetics Division Rolling Meadows, Ill.	Central timing equipment	6,515,023
Giannini Controls Corp. Pasadena, Calif.	Quantity gauging system	11,073,000
B.H. Hadley Co. Pomona, Calif.	Pressure helium regulator unit and liquid hydrogen tank vent disconnects	1,686,085
Hammond Organ Co. Gibbs Manufacturing and Research Corp. Janesville, Wis.	Mechanical timers and clocks	3,065,151

Company	Product	Value
Honeywell, Inc. Minneapolis, Minn.	Stabilization and control subsystem	$134,145,781
ITT Kellogg Chicago, Ill.	In-flight test systems	2,049,872
Kinetics Corp. Solana Beach, Calif.	Power transfer and motor driver switches	1,804,000
Leach Corp. Azusa, Calif.	Flight qualification recorder	1,286,578
Ling Tempco Vought Dallas, Texas	Selective stagnation indicator system	5,398,542
Lockheed Propulsion Co. Redlands, Calif.	Launch escape motor and pitch control motor	10,011,582
Marquardt Corp. Van Nuys, Calif.	SM reaction control engines	37,894,790
McGraw Edison Daven Division Livingston, N.J.	Rotary switches	1,454,263
Metals & Controls, Inc. A Division of Texas Instruments, Inc. Hollywood, Calif.	Toggle switches	931,585
Microdot, Inc. Instrumentation Division South Pasadena, Calif.	Stress measurement system	1,447,098
Motorola, Inc. Military Electronics Division Scottsdale, Ariz.	Link, digital system, spare parts for up-data link equipment, test equipment, pseudo-random noise ranging test set and digital test command system	13,155,449
National Water Lift Co., Div. Pneumo Dynamics Kalamazoo, Mich.	Solenoid latching valves	1,987,153
Northrop Corp. Ventura Division Newberry Park, Calif.	Earth landing subsystem (parachute canopies) and associated GSE, intercommunication system, pressure measurement system, and ordnance assemblies	51,906,846
Opcalite Santa Ana, Calif.	Panels	1,185,377
Parker Aircraft Los Angeles, Calif.	Cryogenic valve modules	1,742,463
Philco Corp. Western Development Lab. Palo Alto, Calif.	Nuclear particle detection system	1,756,522

Company	Product	Value
Radiation, Inc. Melbourne, Fla.	Single-channel decommutator and time accumulator display (CM) and data processing system	$ 3,529,000
Radio Corpporation of America Astro Division Princeton, N.J.	Television equipment	4,216,038
Remanco, Inc. Santa Monica, Calif.	Rocket engine test set	1,000,000
Rosemount Engineering Co. Minneapolis, Minn.	Transducers and mass flowmeter	1,260,000
Sargent Industries (Formerly Electrada Corp.) El Segundo, Calif.	Pressure vessel	1,335,296
Sciaky Brothers, Inc. Chicago, Ill.	Design, fabrication and installation on weld fixtures for Apollo; heavy duty positioners and carriage fixtures	1,729,000
Scientific Data Systems Santa Monica, Calif.	Real-time simulation system	936,119
Simmonds Precision Products Vergennes, Vt.	Propellant quantity indicating and mixture control system	19,350,154
Stratos Div., Fairchild Stratos Corporation Manhattan Beach, Calif.	Reaction control subsystem helium regulators	1,169,072
Systems Engineering Laboratories, Inc. Ft. Lauderdale, Fla.	Channel data	717,767
Thiokol Chemical Corp. Elkton, Md.	Tower jettison motor	4,550,145
Transco Products, Inc. Venice, Calif.	Telemetry antenna system (R&D)	1,260,000
United Aircraft Pratt & Whitney Aircraft Hartford, Conn.	Fuel cell powerplants	81,123,482
Weber Aircraft Burbank, Calif.	Apollo foldable crew couch system	2,200,000
Westinghouse Electric Corp. Aerospace Electrical Division Lima, Ohio	Static inverters	8,060,695
Weston Instruments, Inc. Newark, N.J.	Electrical indicating meters	2,026,579

CSM CONTRACT

Value of the North American Rockwell Space Division contract with NASA's Manned Spacecraft Center on the Apollo program is approximately $2,996,000,000 ($2 billion, 996 million) as of February 3, 1968. The total is expected to approach $3.3 billion by July 31, 1970.

The Apollo contract is for development and fabrication of 49 manned or test spacecraft command and service modules, 30 boilerplate (engineering test) vehicles, and 23 full-scale mockups, as well as production of accompanying spacecraft-LM adapters, 5 test fixtures, 4 Apollo mission simulators, 3 evaluators, 5 trainers, 2 miscellaneous spacecraft-LM adapters, and tracking and ground support equipment.

The following lists of major end items produced by the Space Division under the Apollo contract is separated into spacecraft, boilerplates, mockups and other items. The numbers (internal designation) are those assigned to each item by North American Rockwell for development and fabrication purpose. The numbers are not consecutive. For example, the list skips from SC 002 to SC 004. In cases such as this, the contract originally called for a SC 003, but the vehicle later was deleted. Any number missing in a sequence was originally in the contract, but later was deleted. A letter designation following a spacecraft number (i.e., SC 002A) indicates a general refurbishment and re-use of that spacecraft).

The use, site, and plan shown for each item is that planned at the time of printing; all of these are subject to change.

SPACECRAFT – BLOCK II

No.	Units	Use	Site	Remarks
2TV-1	CM SM	Thermal-vacuum tests	MSC	Docked mode test; "pogo" test
105AV	CM SM LES SLA #2	Acoustic test	MSC	
101	CM SM LES SLA #5	Manned flight	KSC	First manned flight
102	CM SM LES SLA #6	Pad 34 checkout	Downey	Structure refurbished
103 to 119 (Excluding 105)	CM SM LES SLA	Manned flights		

SPACECRAFT – BLOCK I

No.	Units	Use	Site	Remarks
001	SM	Propulsion tests	WSTF	Tests completed 9-7-68
002	CM SM	Power-on tumbling abort		Modified to SC 002A. Tested 1-20-66; not recovered

SPACECRAFT — BLOCK I

No.	Units	Use	Site	Remarks
	LES			Tested 1-20-66; not recovered
002A	CM	Land drop		Mission cancelled; modified to SC 002B
002B	CM	Land drop	Downey	Assigned to "pogo" test
004	CM SM	Structural test		Modified to SC 004B
004B	CM	Unified hatch model	Downey	
004A	CM	Static & thermal structural test	Downey	In storage
006	CM SM LES	System compatibility	AiResearch, Los Angeles	Block II ECS tests Reassigned as SM 010 Reassigned to BP-14
007	CM SM	Water tests	MSC	Modified to SC 007A Fitcheck of docked model facility
007A	CM	Post-landing tests	MSC	
008	CM SM	Environmental proof test	Downey	Modified to SC 008A In storage
008A	CM	Land tests	Downey	
009	CM SM LES SLA #3	Unmanned reentry flight		Modified to SC 009A Not recovered Not recovered Not recovered
009A	CM	Land impact	Downey	Modified to support structural test
010	CM SM LES	Pad abort		Mission cancelled; modified to SC 004A Not completed Not completed
011	CM SM LES SLA #4	Unmanned reentry flight		Modified to SC 011A Not recovered Not recovered Not recovered
011A	CM	Land impact test	MSC	

SPACECRAFT – BLOCK I

No.	Units	Use	Site	Remarks
012	CM	Was to be first manned flight	Langley Research Center	Damaged
	SM		Downey	In storage
	LES		KSC	
	SLA #5		KSC	Modified for use with SC 101
014	CM	Was to be manned flight		Modified to SC 014A
	SM			Launched on Apollo 6; not recovered
	LES		MSFC	
	SLA #6		Tulsa	Modified for use with SC 103
014A	CM	Land test	MSC	
017	CM	Unmanned reentry (Apollo 4)	Downey	Launched 11/67; slated for Smithsonian Institution
	SM			Damaged by explosion; replaced by SM 020
	LES			Not recovered
	SLA #8			Not recovered
020	CM	Unmanned reentry (Apollo 6)	Downey	Post-recovery test
	SM			Launched on Apollo 4; not recovered
	LES			Launched, not recovered
	SLA #9			Launched, not recovered
2S-1	CM	Impact test		Modified to 2S-1A
2S-1A	CM	Water & land test		Modified to 2S-1C
2S-1C	CM	Water & land test	MSC	
2S-2	CM	Static structural test	Downey	
	SM		Downey	
SLA #7	SLA	LM-1 flight		Launched; not recovered
SLA #7A	SLA	LM-2 flight	KSC	Assigned to "pogo" test

BOILERPLATES

No.	Units	Use	Site	Remarks
BP-1	CM	Land & water impact tests	MSC	
BP-2	CM	Land & water impact tests	MSC	
BP-3	CM	Parachute recovery		Tested 9-6-63; not recovered

BOILERPLATES

No.	Units	Use	Site	Remarks
BP-6	CM LES	Pad abort Pad abort		Modified to B-6A Tested in launch 11-7-63; not recovered
BP-6A	CM	Parachute recovery		Modified to B-6B
BP-6B	CM	Parachute recovery		Modified to B-6C
BP-6C	CM	Parachute recovery	El Centro	
BP-9	CM SM LES	Dynamic test		Modified to B-9A
BP-9A	CM SM LES	Micrometeoroid flight Micrometeoroid flight Micrometeoroid flight		Launched on Pegasus 7-25-65; not recovered
BP-12	CM SM LES	Transonic abort Transonic abort Transonic abort		Modified to B-12A Tested 5-13-64; not recovered Tested 5-13-64; not recovered
BP-12A	CM	Water impact tests		Planned for display
BP-13	CM SM LES	Booster & launch environment compatibility		Launched 5-28-64; not recovered
BP-14	CM SM LES	House spacecraft-Block I	Downey NR Storage	Assigned to BP-23A
BP-15	CM SM LES	Booster & launch environment compatibility		Launched 9-18-64; not recovered
BP-16	CM SM LES	Booster, flight compatibility		Launched 2-16-65; not recovered
BP-18	CM	Structural qualification		CM transferred to BP-30
BP-19	CM	Parachute recovery		Modified to BP-19A
BP-19A	CM	Parachute test Vehicle	Northrup-Ventura	Backup for BP-6C
BP-22	CM SM LES	High-Altitude abort High-Altitude abort High-Altitude abort	MSC	Tested 5-19-65; not recovered Tested 5-19-65; not recovered
BP-23	CM SM LES	High-Q abort High-Q abort High-Q abort		Modified to BP-23A Tested 12-8-64; not recovered Tested 12-8-64; not recovered

BOILERPLATES

No.	Units	Use	Site	Remarks
BP-23A	CM LES	Pad abort Pad abort	MSC	Tested 6-27-65; not recovered
BP-25	CM	Water recovery and handling equipment tests	MSC	
BP-26	CM SM LES	Micrometeoroid flight		Used in Pegasus launch 5-25-65; not recovered
BP-27	CM SM LES SLA #1	Dynamic tests Dynamic tests Dynamic tests Dynamic tests	MSFC MSFC MSFC MSFC	SM 010 (former SM 006)
BP-28	CM	Land impact tests		Modified to BP-28A
BP-28A	CM	Earth landing tests	MSC	
BP-29	CM	Flotation tests		Modified to BP-29A
BP-29A	CM	Flotation tests	MSC	
BP-30	CM SM (015) LES (014) SLA #10	Unmanned LM test Unmanned LM test Unmanned LM test Unmanned LM test	MSC KSC KSC MSC	Backup for SC 020; to return to MSFC Tests with SC 102
BP-1224-1	CM	Flammability tests	MSC	
BP-1250C	CM	Flammability tests	Downey	Cabin pressure vent valve tests

MOCKUPS, TRAINERS, SIMULATORS

No.	Units	Use	Site	Remarks
M-2	CM	Interior arrangement	KSC	Modified to KSC-E
M-3	CM	Interior arrangement	KSC	
M-4	SM (Partial)	Interface studies		Excess - disposed 1-67.
M-5	CM	Exterior arrangement	NR Storage	
M-7	SM	Design studies	MSC	
M-8	CM	Airlock & docking	NR Storage	
M-9	CM	Handling & transportation studies	Tulsa	GSE checkout

MOCKUPS, TRAINERS, SIMULATORS

No.	Units	Use	Site	Remarks
	SM		KSC	Part of M-11 as launch verification vehicle
	LES			Used in first Little Joe launch; not recovered
M-11	CM	Handling & transportation studies	KSC	Launch verification article
	SM		Tulsa	GSE checkout
	LES		KSC	Supported SC 009 launch
M-12	CM (Partial)	Crew support studies		Modified to M-12A
M-12A	CM (Partial)	Lighting studies	Downey	
M-18	CM SM LES	System interface	Downey NR Storage	Modified to MSC-1 Mockup display
	SLA		KSC	Grumman use
M-22	CM	Interior & exterior arrangement		Modified to MSC-2; SC 103 configuration
M-23	CM (Partial) SM (Partial) LES (Partial)	Umbilical tests	MSFC MSFC MSFC	Updated to latest configuration 7-7-67
M-24	CM	Engineering & Manufacturing studies	Downey	Wiring and tubing mockup
M-25	SM	Engineering & manufacturing studies	Downey	Wiring and tubing mockup
M-26	CM	Lower equipment bay	NR Storage	
M-27	CM	Forward compartment	Downey	
M-27A	Docking system	Studies	Downey	Updated to Block II configuration M-27B)
M-27B	Docking System	Studies	Convair, San Diego	Block II; to be shipped to MSC
M-28	CM	Crew compartment reviews	Downey	Converted from CM-B
MSC-1	CM	Configuration review (SC 101)	Downey	Converted from M-18

MOCKUPS, TRAINERS, SIMULATORS

No.	Units	Use	Site	Remarks
MSC-2	CM	Configuration review (SC 103)	MSC	Converted from M-22
KSC-E	CM	Ingress-egress trainer	KSC	Modified from M-2
180/CM/MU	CM (Partial)	Sun interference evaluation	MSC	
CM-A	CM	Engineering simulator	MSC	
CM-B	CM	Engineering simulator	Downey	Converted to a mockup
AMS-1		Mission simulator	MSC	Modified to SC 103 configuration
AMS-2		Mission simulator	KSC	Modified to SC 101 configuration

P-299a

Spacecraft command module for first manned flight readied for shipment

Frogmen secure helicopter lines to command module after successful entry from space flight

BIOGRAPHICAL SUMMARIES

NASA HEADQUARTERS

JAMES E. WEBB — Administrator of National Aeronautics and Space Administration since 1961. Former Director of Bureau of the Budget and former Under Secretary of State. Was Vice President of Sperry Gyroscope Co., Chairman of Board of Republic Supply Co., Director of Kerr-McGee Oil Industries, Inc., and Director of McDonnell Aircraft Co. Born Oct. 7, 1906, in Granville County, N.C. Received education degree from University of North Carolina in 1928 and studied law at George Washington University.

DR. GEORGE E. MUELLER—Associate Administrator in NASA Office of Manned Space Flight. Assumed direction of manned space program in September, 1963. Born July 16, 1918, in St. Louis, Mo. Graduated from Missouri School of Mines in 1939. Holds master's in electrical engineering from Purdue University. Worked at Bell Telephone Laboratories until 1946 when he joined the faculty at Ohio State. Earned Ph.D. in physics in 1951.

SAMUEL C. PHILLIPS—Lt. General on assignment to NASA from the Air Force. Was appointed Director of the Apollo lunar landing program in 1964. Born Feb. 19, 1921, in Springerville, Arizona. Graduated from the University of Wyoming in 1943, holds master's in electrical engineering from University of Michigan. Has an extensive background in ballistic missile systems, including service as Director of Minuteman program.

NASA MANNED SPACECRAFT CENTER

DR. ROBERT R. GILRUTH—Director of the Manned Spacecraft Center since its creation in 1961. Born Oct. 8, 1913, in Nashwauk, Minn. Graduated in 1935 from University of Minnesota. Started at NACA Langley Research Center in 1937 in Flight Research Division. Selected by NACA in 1945 to establish organization and facilities for conducting free-flight experiments with rocket-powered models at supersonic speeds. Named assistant director of Langley Laboratories in 1952. In 1958 was named director of NASA's Space Task Group, which later was reorganized as Manned Spacecraft Center.

GEORGE M. LOW—Manager of Apollo Spacecraft Program since April 1967. Formerly Deputy Director of Manned Spacecraft Center, Deputy Associate Administrator for Manned Space Flight at NASA Headquarters. Chairman of committee which performed original studies leading to manned lunar program. Joined NACA (NASA predecessor) at Lewis Research Center in 1949. Born June 10, 1926, in Vienna, Austria. Graduate of Rensselaer Polytechnic Institute and has master's from RPI. Holds NASA's Outstanding Leadership Medal.

CHRISTOPHER C. KRAFT, JR.—Director of Flight Operations. Formerly Chief of Flight Operations Division. Joined NACA at Langley Research Center in 1945. Born Feb. 28, 1924, in Phoebus, Va. Graduated from Virginia Polytechnic Institute. Holds NASA Distinguished Service Medal, Arthur S. Fleming Award.

KENNETH S. KLEINKNECHT—Manager for Command and Service Modules, Apollo Spacecraft Program. Formerly Deputy Manager of Gemini program and Manager of Mercury Project Office. Joined NACA at Lewis Research Center in 1942, also served at Flight Research Center at Edwards Air Force Base. Born July 24, 1919, in Washington, D.C. Graduated from Purdue University.

NASA MARSHALL SPACE FLIGHT CENTER

DR. WERNHER VON BRAUN—Director of George C. Marshall Space Flight Center, Huntsville, Ala., since its formation in 1960. Director of Army missile development agencies in Redstone Arsenal at Huntsville since 1950. Began his work in rocket field in 1930 (at 18) while at Berlin Institute of Technology. Began full-time research of rocketry in 1934 under sponsorship of German government and from 1937 to 1945 was technical director of Army portion of Peenemuende rocket center where V-2 and antiaircraft guided missiles were developed. Came to United States in 1945 and became U.S. citizen in 1955. Born March 23, 1912, in Wirsitz, Germany. Graduated 1932 from Berlin Institute of Technology and received Ph.D. in 1934.

EDMUND F. O'CONNOR—Director, Industrial Operations, at Marshall Space Flight Center. Brigadier General on special duty to NASA from the Air Force. Technical and administrative manager of Saturn launch vehicle programs and of MSFC's Michoud Assembly Facility at New Orleans and the Mississippi Test Facility. Previously Deputy Director of Air Force's Ballistic Systems Division. Born March 31, 1922, in Fitchburg, Mass. Graduated from the U.S. Military Academy.

NASA KENNEDY SPACE CENTER

DR. KURT H. DEBUS—Director of Kennedy Space Center. Came to United States in 1945 as part of Dr. Von Braun's original group at Fort Bliss, Texas. Supervised development and construction of launch facilities at Cape Kennedy for Redstone, Jupiter, Juno, and Pershing Missiles beginning in 1952. Directed design, development, and construction of NASA's Apollo/Saturn launch facilities. He launched first U.S. ballistic missile (Redstone) and directed launch operations of first U.S. satellite (Explorer I). Born Nov. 29, 1908, in Frankfurt, Germany. Received initial and advanced degrees from Darmstadt University, and honorary Doctor of Laws degree from Rollins College.

MILES ROSS—Deputy Director for Operations at Kennedy Space Center. Responsible for engineering and technical operations at KSC. Joined NASA in September 1967. Previously was project manager for Air Force Thor and Minuteman missile systems for TRW, Inc. Born in New Brunswick, N.J. Holds degrees in mechanical engineering and engineering administration from MIT.

ROCCO A. PETRONE—Director of Launch Operations at Kennedy Space Center. Previously Apollo Program Manager at KSC. Lieutenant Colonel in Army; joined NASA in 1960 from Army General Staff. While with Army participated in development of Redstone missile at Huntsville. Graduated from U.S. Military Academy and has master's degree from MIT. Born March 31, 1926, in Amsterdam, N.Y.

R.O. MIDDLETON—Apollo Program Manager at Kennedy Space Center. Rear Admiral on assignment to NASA from Navy. Formerly Deputy Director for Mission Operations in Office of Manned Space Flight and a mission director in Apollo/Saturn program. Before joining NASA was commanding officer of USS Little Rock and USS Observation Island, chief of staff of Carrier Division 14, and commander of Destroyer Division 142. Born Jan. 23, 1919, in Pomona, Fla. Graduated from U.S. Naval Academy and has master's degree from Harvard. Promoted to Rear Admiral in July, 1967.

NORTH AMERICAN ROCKWELL

J.L. ATWOOD—President and Chief Executive Officer of North American Rockwell Corp. Was Chairman of Board of North American Aviation from 1962 until its merger with Rockwell-Standard Corp. in 1967 and President of NAA since 1948. Joined NAA in 1934 as Chief Engineer and Vice President, became Assistant General Manager in 1938 and First Vice President in 1941. Born Oct. 26, 1904. Graduated from Hardin-Simmons University in 1926 and the University of Texas in 1928. Holds honorary doctor's degree and is recipient of many industry and government awards.

WILLARD F. ROCKWELL, JR.—Chairman of the board of North American Rockwell Corp. Was president of Rockwell-Standard Corporation before its merger with North American Aviation, Inc. Previously President of Rockwell Manufacturing Company. Was on Rockwell-Standard board since 1940. Born March 3, 1914. Holds engineering degree from Pennsylvania State University.

JOHN R. MOORE—President of Aerospace and Systems Group of North American Rockwell Corp. Also Vice President of corporation and member of board. Was Executive Vice President of North American Aviation and, earlier, President of company's Autonetics Division. Born July 5, 1916, in St. Louis. Received engineering degree from Washington University (St. Louis). Holds many awards, including U.S. Navy's Meritorious Public Service Citation and Thurlow Navigation Award as "navigation's man of the year" in 1961.

WILLIAM B. BERGEN—President of North American Rockwell's Space Division since April 1967 and Vice President of corporation. Formerly Vice President of corporation's Space and Propulsion Group. Formerly President of Martin Co., which he joined in 1937. Born March 29, 1915, in Floral Park, N.Y. Graduated from Massachusetts Institute of Technology in 1937. Holds many awards, including Lawrence Sperry Award of Institute of Aeronautical Sciences for aircraft work during World War II and NASA's Public Service Award for contributions to Gemini program.

DALE D. MYERS—Vice President of Space Division since 1960 and Apollo Program Manager. Formerly Program Manager for Hound Dog missile. Background includes aerodynamic and thermodynamic work with missile studies. Born Jan. 8, 1922. Graduated from the University of Washington in 1943.

ROBERT E. GREER—Vice president of Space Division and Saturn S-II (second stage) Program Manager. Joined Space Division in 1965 as Assistant to President and became S-II Program Manager later that year. Formerly Major General in Air Force; among his posts were Director of Special Projects, Deputy Commander for Satellite Programs, and Chief of Staff for Guided Missiles. Born Aug. 7, 1915, in Orange, Calif. Graduated from U.S. Military Academy and later taught electrical engineering at Academy. Also served on faculty at Air War College.

BASTIAN (BUZ) HELLO—Vice President of Space Division for Launch Operations since May 1967. Formerly Director of Maneuverable Spacecraft Programs for Martin Marietta Corp. and Program Director for Air Force Gemini-Titan II launch vehicle. Born Aug. 29, 1923, in Philadelphia. Graduated from University of Maryland in 1947. Received NASA's Public Service Award in 1966 for contributions to Gemini program.

ASTRONAUTS

APOLLO 7 PRIME CREW

WALTER M. SCHIRRA, JR. (Commander)—One of the original seven astronauts. Flew 6-orbit, 9-hour mission in Sigma 7 Mercury spacecraft and was command pilot of Gemini VI mission, which achieved space "first" in rendezvous with orbiting Gemini VII. In both missions was brought aboard recovery ship with his spacecraft. Graduated from U.S. Naval Academy in 1945. Captain in Navy. Flew 90 combat missions in Korea on exchange status with Air Force. Received 2 Distinguished Flying Crosses and 2 Air Medals for Korean service. Born March 12, 1923, in Hackensack, N.J.

DONN F. EISELE (CM Pilot)—Named in third group of astronauts in October 1963. Major in Air Force. Project engineer and experimental test pilot at Kirtland Air Force Base, N.M., flying jets in support of special weapons development. Born June 23, 1930, in Columbus, Ohio. Graduated from the U.S. Naval Academy in 1952 and has master's from U.S. Air Force Institute of Technology.

WALTER CUNNINGHAM (LM Pilot)—Named in third group of astronauts in October 1963. Was research scientist with Rand Corp. working on classified defense studies and problems of earth's magnetosphere. At UCLA, where he received bachelor's and master's degrees, he developed a magnetometer which was flown aboard NASA's first Orbiting Geophysical Observatory satellite. Flew as Marine pilot and was Marine reservist with rank of Major until 1965. Born March 16, 1932, in Creston, Iowa.

APOLLO 7 BACKUP CREW

THOMAS P. STAFFORD (Commander)—Selected in second group of astronauts in 1962, he flew in Gemini 6 flight and participated in first space rendezvous. Served as command pilot of Gemini 9 three-day flight in which spacecraft performed 3 different types of rendezvous with target vehicle. Has logged nearly 100 hours in space in 2 flights. Born Sept. 17, 1930, in Weatherford, Okla. Graduated from U.S. Naval Academy and entered Air Force; is Lieutenant Colonel. Instructor and chief of Performance Branch of Air Force Aerospace Research Pilot School and author of textbooks on performance flight testing.

JOHN W. YOUNG (CM Pilot)—Selected in second group of astronauts in 1962. Pilot of first manned Gemini flight and command pilot of Gemini X, in which Gemini rendezvoused and docked with Agena target, changed orbit, and rendezvoused with another Agena. Born Sept. 24, 1930, in San Francisco. Graduated from Georgia Institute of Technology in 1952. Commander in Navy. Formerly test pilot at Naval Air Test Center and set world time-to-climb records.

EUGENE A. CERNAN (LM Pilot)—Selected in third group of astronauts in 1963 and served as pilot in Gemini flight with Stafford. Born March 14, 1934, in Chicago. Graduated from Purdue University and has master's from U.S. Naval Postgraduate School. Commander in Navy. Served in attack squadrons in Navy.

APOLLO 7 SUPPORT CREW

JOHN L. SWIGERT, JR.—Selected as astronaut in 1966, formerly engineering test pilot for North American Aviation and research engineering test pilot for Pratt and Whitney. Received Octave Chanute Award from AIAA for work on space vehicle landing system. Born Aug. 30, 1931, in Denver. Graduated from University of Colorado in 1953 and has masters from Rensselaer Polytechnic Institute and from University of Hartford. Served as fighter pilot in Air Force.

RONALD E. EVANS—Selected as astronaut in 1966. Lieutenant Commander in Navy, served as pilot on aircraft carrier in Vietnam combat operations. Born Nov. 10, 1933, in St. Francis, Kan. Graduated from University of Kansas in 1956 and has master's from U.S. Naval Postgraduate School. Holds 8 Air Medals, Vietnam Service Medal, and Navy Commendation Medal.

WILLIAM R. POGUE—Selected as astronaut in 1966. Air Force Major. Formerly instructor at Air Force Aerospace Research Pilot School at Edwards Air Force Base and math instructor at Air Force Academy. Test pilot for British Ministry of Aviation in exchange program. Flew 43 combat missions in Korean War. Born Jan. 23, 1930, in Okemah, Okla. Graduated from Oklahoma Baptist University and has master's from Oklahoma State University. Holds several Air Force decorations.

2ND APOLLO PRIME CREW

FRANK BORMAN (Commander)—Selected in second group of astronauts, in September 1962, and was command pilot of Gemini VII, the longest manned space flight (330 hours and 35 minutes) and during which the first space rendezvous occurred, between Gemini VII and Gemini VI. Colonel in the Air Force. Was instructor at USAF Aerospace Research Pilots School. Born Gary, Ind., March 14, 1928. Graduate of U.S. Military Academy and has Master of Science degree in aeronautical engineering from Caltech.

JAMES A. LOVELL, JR. (CM Pilot)—Selected in second group of astronauts, in September 1962. Flew Gemini VII mission with Frank Borman, which was longest space mission (330 hours and 35 minutes) and during which there was the first space rendezvous, between Gemini VII and Gemini VI. Was command pilot on Gemini 12 mission. Captain in Navy. Was test pilot at Naval Air Test Center, Patuxent River, Md. Born in Cleveland, Ohio, March 25, 1928. Graduate of United States Naval Academy.

WILLIAM A. ANDERS (LM Pilot)—Chosen in third group of astronauts, in October 1963. Born in Hong Kong, October 17, 1933. Was nuclear engineer and instructor pilot at Air Force Weapons Laboratory at Kirtland Air Force Base, N.M. Graduate of U.S. Naval Academy and has Masters Degree in nuclear engineering from Air Force Institute of Technology. Major in Air Force.

EDWIN E. ALDRIN, JR. (CM Pilot)—Selected in third group of astronauts, in October, 1963. Has Doctor of Science degree in astronautics from Massachusetts Institute of Technology. Was pilot of Gemini XII mission and conducted EVA. Lieutenant Colonel in Air Force. Worked in Gemini Target Office of the Air Force Space Systems Division, Los Angeles, Calif., and at the USAF Field Office at the Manned Spacecraft Center. Flew 66 combat missions in Korea and received Distinguished Flying Cross. Born in Montclair, N.J., Jan. 20, 1930. Graduate of the U.S. Military Academy (third in his class).

FRED W. HAISE, JR. (LM Pilot)—Selected for astronaut training in April 1966. Research pilot at the NASA Flight Research Center at Edwards, California. Now a civilian, he was military pilot in Marine Corps, Air Force, and Oklahoma Air National Guard. Born in Biloxi, Miss., Nov. 14, 1933. Was graduated from University of Oklahoma.

2ND APOLLO BACKUP CREW

NEIL A. ARMSTRONG (Commander)—Named in second group of astronauts, in September, 1962, and was first civilian to fly in space. Was command pilot of Gemini VIII, which saw the first docking of spacecraft. Was naval aviator and flew 78 combat missions in Korea. Was aeronautical research pilot at NASA High Speed Flight Station, Edwards Air Force Base, Calif., and flew the X-15 at altitude greater than 200,000 feet and speed of about 4000 mph. Born Wapakoneta, Ohio, Aug. 5, 1930. Graduate of Purdue University. Received 1962 Institute of Aerospace Sciences Octave Chanute Award.

2ND APOLLO SUPPORT CREW

THOMAS K. MATTINGLY II—Selected for astronaut training in April 1966. Born in Chicago, March 17, 1936. Graduate of Auburn University. Lieutenant Commander in Navy. Is unmarried. Was student at Air Force Aerospace Research Pilots School. Flew carrier aircraft before that.

GERALD P. CARR—Selected for astronaut training in April 1966. Born in Denver, Aug. 22, 1932. Graduate of the University of Southern California, and has Master of Science degree in aeronautical engineering from Princeton University. Major in Marine Corps. Was with Marine air control squadron.

VANCE D. BRAND—Selected for astronaut training in April 1966. Born in Longmont, Colo., May 9, 1931. Holds degrees in business and aeronautical engineering from University of Colorado and a Master of Science degree in business administration from UCLA. Served as a jet pilot in Marine Corps and later as flight test engineer and experimental test pilot for Lockheed Aircraft Corp. Graduated from U.S. Naval Test Pilot School at Patuxent River, Md.

DAVID R. SCOTT (CM Pilot)—One of third group of astronauts named in October, 1963. Was pilot in Gemini VIII, which achieved first space docking. Lieutenant Colonel in Air Force. Served as research and test pilot in Air Force. Born June 6, 1932, in San Antonio, Texas. Graduated from U.S. Military Academy and has master's and engineering degrees from Massachusetts Institute of Technology.

RUSSELL L. SCHWEICKART (LM Pilot)—One of the third group of astronauts named in October, 1963. Served in Air Force and Air National Guard from 1956 to 1963. Research scientist at Experimental Astronomy Laboratory at MIT, working on upper atmosphere physics, applied astronomy, star tracking, and stabilization of star images. Born Oct. 25, 1935, in Neptune, New Jersey. Holds Bachelor and Master of Science degrees from MIT.

3RD APOLLO PRIME CREW

JAMES A. McDIVITT (Commander)—Named in second group of astronauts in September 1962. Was command pilot for Gemini IV, the 66-orbit, four-day mission from June 3-7, 1965, on which Edward H. White II made first "walk in space." Lieutenant Colonel in Air Force. Flew 145 combat missions during Korean War and received 4 Distinguished Flying Crosses, 5 Air Medals, and a South Korean medal. Later served as experimental test pilot at Edwards Air Force Base in California. Born June 10, 1929, in Chicago. Graduated from University of Michigan.

3RD APOLLO BACKUP CREW

CHARLES CONRAD, JR. (Commander)—Selected in second group of astronauts, in September 1962. Flew, with Gordon Cooper, in Gemini V the first extended manned space flight (127 orbits, 190.9 hours, August, 1965). Was command pilot for the 3-day Gemini XI mission and executed rendezvous and docking in less than one orbit. Commander in the Navy. Was project test pilot at Navy Test Pilot School, Patuxent River, Maryland, and also flight instructor and performance engineer at the school. Born in Philadelphia, Pa., June 2, 1930. Was graduated from Princeton University.

RICHARD F. GORDON, JR. (CM Pilot)—Named in third group of astronauts, in October, 1963. Was pilot for Gemini XI mission and conducted extravehicular activity. Commander in Navy. Was project test pilot for F4H Phantom II and test pilot for other naval aircraft. Born in Seattle, Wash., Oct. 5, 1929. Graduate of University of Washington.

ALAN L. BEAN (LM Pilot)—Named in third group of astronauts, October 1963. Lieutenant Commander in Navy. Attended Navy Test Pilot School, Patuxent River, Maryland. Born in Wheeler, Tex., March 15, 1932. Was graduated from University of Texas.

3RD APOLLO SUPPORT CREW

ALFRED M. WORDEN—One of 19 astronauts selected in April 1966. Major in Air Force. Was instructor at Aerospace Research Pilots School, and is a graduate of the Empire Test Pilots School, in Farnborough, England. Born in Jackson, Mich., Feb. 7, 1932. Graduate of U.S. Military Academy and has Master of Science degree in astronautical and aeronautical engineering and instrumentation engineering from the University of Michigan.

EDGAR D. MITCHELL—Selected for astronaut training in April, 1966. Has Doctor of Science degree in aeronautics/astronautics from Massachusetts Institute of Technology. Commander in Navy. Was graduated first in his class from Air Force Aerospace Research Pilot School. Was Chief, Project Management Division, of the Navy Field Office for Manned Orbiting Laboratory. Born in Hereford, Tex., Sept. 17, 1930. Was graduated from Carnegie Institute of Technology and U.S. Naval Postgraduate school.

JACK R. LOUSMA—Selected with 18 others for astronaut training in April 1966. Born in Grand Rapids, Mich., Feb. 29, 1936. Graduate of University of Michigan and of U.S. Naval Postgraduate School, with degrees in aeronautical engineering. Served as Marine Corps pilot from 1959 until his assignment to astronaut program.

APOLLO CHRONOLOGY

1960

July 29 - Project Apollo, an advanced spacecraft program to land men on the moon, was announced by NASA.

Oct. 25 - NASA selected General Dynamics, General Electric, and Martin to conduct individual feasibility studies of an advanced manned spacecraft as part of the Apollo project.

1961

Jan. - NASA studies, by a committee headed by George Low (present Apollo spacecraft program manager), of a manned lunar-landing program were completed. Both a direct-ascent trajectory using large Nova-type launch vehicles and an earth-orbit rendezvous technique using Saturn-type launch vehicles were considered.

May 15 - Final reports on Project Apollo study contracts were submitted by General Dynamics, GE, and Martin.

May 25 - President Kennedy presented a plan to Congress for accelerating the space program based on a national goal of landing a man on the moon before the end of the decade.

July 28 - NASA issued a request for proposal to 12 companies for development of the Apollo spacecraft.

Aug. 9 - NASA selected MIT's Instrumentation Laboratory to develop the guidance and navigation system for the Apollo spacecraft.

Sept. 19 - NASA announced that the recently established Manned Spacecraft Center would be located at Houston, Tex.

Nov. 28 - NASA announced that a contract had been awarded to North American's Space Division for the Apollo spacecraft program.

Dec. 21 - The first four major Apollo subcontractors were announced: Collins Radio, telecommunications systems; Garrett Corporation's AiResearch Division, environmental control equipment; Honeywell Inc., the stabilization and control system; and Northrop Corporation's Ventura Division, parachute earth landing system.

1962

Jan. 22 - The first Apollo engineering order was issued, for fabrication of the first mockups of the Apollo command and service modules.

Feb. 9 - NASA announced that GE had been awarded a contract to provide integration analysis of the total Apollo space vehicle, including launch vehicle and spacecraft, to assure reliability of the entire system. GE was also named to develop and operate equipment to check out the Apollo systems.

Feb. 13 - Lockheed Propulsion Company was selected to design and build the solid-propellant launch-escape motor for Apollo.

Mar. 2 - Marquardt Corp. was selected to design and build the reaction-control rocket engines for the Apollo spacecraft.

Mar. 3 - Aerojet-General Corp. was named as subcontractor for the Apollo service propulsion system.

Mar. 9 - Pratt and Whitney was selected to build the Apollo fuel cell.

Mar. 23 - Avco Corp. was selected to design and install the ablative material on the spacecraft outer surface.

April 6 - Thiokol Chemical Corp. was selected to build the solid-propellant rocket motor to be used to jettison the Apollo launch escape tower.

July 11 - NASA announced that the lunar rendezvous mode would be used for the moon mission. This new plan called for development of a two-man lunar module to be used to reach the surface of the moon and return the astronauts to the lunar-orbiting command module. NASA administrator James Webb said this method was the most desirable from the standpoint of "time, cost, and mission accomplishment."

July 16 - Beech Aircraft Corp., was selected to build the spacecraft storage tanks for supercritical gases.

Aug. 22 - The length of the Apollo service module was increased from 11 feet 8 inches to 12 feet 11 inches to provide space for additional fuel.

Sept. 7 - Apollo command module Boilerplate I was accepted by NASA and delivered to a Space Division laboratory for land and water impact tests.

Nov. 7 - Grumman Aircraft was named by NASA to design and build the LM.

1963

Mar. 12 - Apollo Boilerplate 13, the first flight-rated boilerplate to be completed, was accepted by NASA and shipped to MSFC.

July 23 - Dr. George E. Mueller was named director, NASA's Office of Manned Space Flight.

Oct. 8 - Dr. Joseph Shea, previously with NASA Headquarters, was named Apollo program manager at MSC.

Nov. 7 - The first launch test - a pad-abort test of Boilerplate 6 - was conducted at White Sands.

1964

February - A boost protective cover was added to the launch escape system in order to protect the windows of the CM and the heat shield surfaces from soot from the LES motor.

May 13 - The second test flight of the Apollo program occurred at White Sands when Boilerplate 12 was launched by a Little Joe II vehicle during a high-stress, high-speed abort test. The launch escape system worked as planned, except that one of the three parachutes cut loose. The CM was landed without damage.

May 28 - Apollo command module Boilerplate 13 was placed in orbit from Cape Kennedy following launch by a Saturn I booster. This was the first Apollo vehicle to be placed in orbit, and the third Apollo test flight.

Sept. 18 - Apollo Boilerplate 15 was successfully orbited at Cape Kennedy by a Saturn I two-stage launch vehicle. This was the fourth Apollo test flight.

Dec. 8 - The fifth Apollo test flight occurred at White Sands when Boilerplate 23 was lifted off the pad by a Little Joe II in a high Q abort test.

1965

Feb. 16 - Apollo Boilerplate 16 was launched from Cape Kennedy in a micrometeoroid test. A Pegasus satellite was carried aloft in a modified Apollo SM. All equipment functioned as planned. This was the sixth Apollo test flight.

May 19 - Apollo Boilerplate 22 was launched at White Sands in a planned high-altitude test of the launch escape system. The Little Joe II disintegrated at low altitude, resulting in an unscheduled but successful low-altitude abort test. This was the seventh test flight.

May 25 - The second Pegasus satellite was put into orbit at Cape Kennedy during the Saturn I launch of Apollo Boilerplate 26. This was the eighth Apollo test flight.

June 29 - Apollo Boilerplate 23A was successfully launched at White Sands during a pad abort test. All systems functioned as planned. This was the ninth Apollo test flight, and the fifth abort test. This boilerplate module, previously designated Boilerplate 23, had been launched at White Sands during a high Q test.

July 30 - Apollo Boilerplate 9A was launched at Cape Kennedy and was used to place the third Pegasus satellite into orbit.

Oct. 20 - The first actual Apollo spacecraft, SC 009, was accepted by NASA and subsequently shipped to Cape Kennedy. All previously completed Apollo vehicles had been boilerplate and mockup articles.

Dec. 26 - Apollo SC 009 was mated with the Saturn IB at the Kennedy Space Center.

Dec. 31 - Command modules accepted by NASA by the end of 1965 included 18 mockups, 18 boilerplates, and 2 spacecraft.

1966

Jan. 20 - A power-on tumbling abort test of the launch escape system was conducted at White Sands with the launch of SC 002. This was the sixth and final launch escape test; the LES was then declared qualified.

Feb. 26 - First unmanned flight of Apollo spacecraft (SC 009) was conducted to test command module's ability to withstand entry temperatures, determine adequacy of command module for manned entry from low orbit, test command and service module reaction control engines and test service module engine firing and restart capability. Recovery was in the South Atlantic, 5300 miles downrange, near Ascension Island.

Aug. 25 - Second unmanned test of Apollo spacecraft (SC 011) was conducted to test command module's ability to withstand entry temperatures under high heat load. After three-quarters of an orbit the spacecraft, which reached an altitude of 700 miles, was recovered 260 statute miles from Wake Island.

Oct. - The first Apollo Block II parachute qualification test was conducted at El Centro, Calif.

1967

Jan. 27 - During a manned ground test of an Apollo spacecraft (SC 012) while the vehicle was atop the Saturn IB booster, a flash fire in the command module resulted in the deaths of Astronauts Gus Grissom, Ed White, and Roger Chaffee. NASA immediately established a review board to determine the cause of the fire and the changes which would be necessary to prevent such fires in the future.

Apr. 9 - The review board presented its findings to the NASA administrator. While the exact cause of the fire was not determined conclusively, the board recommended a number of changes, including the elimination of most of the combustible materials in the spacecraft, the protection of wires in the spacecraft, and the installation of a quick-opening hatch. These and other changes were incorporated in later spacecraft.

Nov. 9 - The Apollo 4 mission, the first using the Saturn V launch vehicle, was considered. The spacecraft reached an altitude of 11,234 miles, entered the atmosphere at a speed of 24,917 mph, and splashed down in the Pacific six miles from the recovery ship after a flight of eight hours 37 minutes. This flight qualified the heat shield for lunar flight.

1968

Jan 22 - The lunar module was tested during the flight designated Apollo 5. A wrong number in the guidance logic caused immediate shutdown of the descent engine, and led to a series of abnormal events. The LM performed very well, however, and accomplished most of its objectives, including its ability to abort a landing on the moon and to return to the command module during its orbiting lunar flight.

Apr. 4 - Apollo 6, the second test of the Saturn V launch vehicle, although problems developed with the launch vehicle, the spacecraft's accomplishments were impressive. These included the longest single burn in space of the service propulsion engine, proper control of the engine by the guidance and navigation subsystem, and another successful test of the heat shield.

Apr. 29 - NASA announced that next Saturn V/Apollo flight would be manned, and would take place during the latter part of 1968. The next scheduled Apollo flight, designated Apollo 7, will be manned and will use a Saturn IB launch vehicle.

Apollo spacecraft are shipped to Kennedy Space Center, Fla., by specially converted aircraft from Long Beach Airport

APOLLO BRIEFS

The possibility of a micrometeoroid as big as a cigarette ash striking the command module during an 8-day lunar mission has been computed as 1 in 1230. If a meteoroid did strike the module, it would be at a velocity of 98,500 feet per second. The probability of the command module getting hit is 0.000815. The probability of the command module not getting hit is 0.999185.

* * *

The heat leak from the Apollo cryogenic tanks, which contain hydrogen and oxygen, is so small that if one hydrogen tank containing ice were placed in a room heated to 70 degrees F, a total of 8-1/2 years would be required to melt the ice to water at just above freezing temperature. It would take approximately 4 years more for the water to reach room temperature. The gases in the cryogenic tanks are utilized in the production of electrical power by the Apollo fuel cell system and provide oxygen for the use of the crew.

* * *

When the Apollo spacecraft passes through the Van Allen belts on its way to the moon, the astronauts will be exposed to radiation roughly equivalent to that of a dental X-ray.

* * *

With gravity on the moon only one-sixth as strong as on earth, it is necessary that this difference be related to the Apollo vehicle. A structure 250 feet high and 400 feet long in which cables lift five-sixth of the spacecraft vehicle weight is being used in tests to simulate lunar conditions and their effect on the vehicle.

* * *

The command module panel display includes 24 instruments, 566 switches, 40 event indicators (mechanical), and 71 lights.

* * *

The command module offers 73 cubic feet per man as against the 68 cubic feet per man in a compact car. By comparison, the Mercury spacecraft offered 55 cubic feet for its one traveler and Gemini provided 40 cubic feet per man.

* * *

The angular accuracy requirement of midcourse correction of the spacecraft for all thrusting maneuvers is one degree.

* * *

If your car gets 15 miles to a gallon, you could drive 18 million miles or around the world about 400 times on the propellants required for the Apollo/Saturn lunar landing mission. The Saturn V launch vehicle contains 5.6 million pounds of propellant (or 960,000 gallons).

* * *

When the Apollo re-enters the atmosphere it will generate energy equivalent to approximately 86,000 kilowatt hours of electricity - enough to light the city of Los Angeles for about 104 seconds; or the energy generated would lift all the people in the USA 10-3/4 inches off the ground.

* * *

The fully loaded Saturn V launch vehicle with the Apollo spacecraft stands 60 feet higher than the Statue of Liberty on its pedestal and weighs 13 times as much as the statue.

* * *

During its 3.5 second firing, the Apollo spacecraft's solid-fuel launch escape rocket generates the horsepower equivalent of 4,300 automobiles.

* * *

The engines of the Saturn V launch vehicle that will propel the Apollo spacecraft to the moon have combined horsepower equivalent to 543 jet fighters.

* * *

The Apollo environmental control system has 180 parts in contrast to the 8 for the average home window air conditioner. The Apollo environmental control system performs 23 functions compared to 5 for the average home conditioner. There are 23 functions of the environmental control system, which include: air cooling, air heating, humidity control, ventilation to suits, ventilation to cabin, air filtration, CO_2 removal, odor removal, waste management functions, etc.

The 12-foot-high Apollo spacecraft command module contains about fifteen miles of wire, enough to wire 50 two-bedroom homes.

* * *

The astronaut controls and monitors the stabilization and control system by means of two handgrip controllers, 34 switches, and 6 knobs.

* * *

The command system of the acceptance checkout equipment can generate up to 2048 separate stimuli or 128 analog signals, or combinations of both, and route them to spacecraft and other checkout systems at a million bits per second. In contrast, hand-operated commercial teletype generates 45 bits per second and automatically, over voice channel, it generates 2400 bits per second.

* * *

The Apollo command module can sustain a hole as large as 1/4 inch in diameter and still maintain the pressure inside for 15 minutes, which is considered long enough for an astronaut to put on a spacesuit.

* * *

The boost protective cover will protect the command module from temperatures expected to reach 1200 degrees during the launch phase.

* * *

The power of one Saturn V is enough to place in earth orbit all U.S. manned spacecraft previously launched.

* * *

Here is an analogy pertaining to the benefits of the multistage concept as opposed to the single-stage, brute-force method. If a steam locomotive pulling three coal cars carries all three cars along until all fuel is exhausted, the locomotive could travel 500 miles. By dropping off each car as its coal is expended the locomotive could travel 900 miles.

* * *

The F-1's fuel pumps push fuel with the force of 30 diesel locomotives.

* * *

Enough liquid oxygen is contained in the first stage tank to fill 54 railroad tank cars.

* * *

The five F-1 engines equal 160,000,000 horsepower, about double the amount of potential hydroelectric power that would be available at any given moment if all the moving waters of North America were channeled through turbines.

* * *

The interior of each of the first stage propellant tanks is large enough to accommodate three large moving vans side by side.

* * *

The Saturn V's second stage construction is comparable to that of an eggshell in efficiency, the amount of weight and pressure constrained by a thin wall.

* * *

Total amount of propellant (fuel and oxidizer) in the Saturn V launch vehicle, service module, and lunar module is 5,625,000 pounds.

* * *

The Apollo spacecraft, including the command and service modules and the adapter which housed the lunar module, is 82 feet tall, only 13 feet shorter than the entire Mercury-Atlas space vehicle used in John Glenn's orbital mission.

* * *

The ratio of propellant to payload in Saturn V is 50 to 1.

* * *

The main computer in the command module occupies only one cubic foot.

* * *

While an automobile has less than 3,000 functional parts, the command module has more than 2,000,000 not counting wires and skeletal components.

The command module uses only about 2000 watts of electricity, similar to the amount required by an oven in an electric range.

* * *

The heat shield and its ablator must resist heat twice as great as that encountered by Gemini and Mercury.

* * *

The configuration of Apollo is designed to give it aerodynamic lift so that it is possible to "fly" it during re-entry. The lift-over-drag ratio is about 0.35.

* * *

The honeycomb aluminum used in Apollo's inner crew compartment is 40-percent stronger and 40-percent lighter than ordinary aluminum.

* * *

There are 50 engines aboard the Apollo spacecraft: 16 reaction control engines on the service module, 16 reaction control engines on the lunar module, 12 reaction control engines on the command module, the service propulsion engine, the lunar module ascent and descent engines, the launch escape motor, the tower jettison motor, and the pitch control motor. The last three are solid-propellant engines and the other 47 all burn a hypergolic liquid propellant composed of nitrogen tetroxide and hydrazine. A hypergolic propellant is one composed of an oxidizer and a fuel which ignite and burn on contact.

* * *

The tanks which hold the cryogenic (ultra-cold) liquid oxygen and liquid hydrogen on the Apollo spacecraft come close to being the only leak-free vessels ever built. If an automobile tire leaked at the same rate that these tanks do, it would take the tire 32,400,000 years to go flat.

* * *

There are approximately 2-1/2 million solder joints in the Saturn V launch vehicle. If just 1/32 of an inch too much wire were left on each of these joints and an extra drop of solder was used on each of these joints, the excess weight would be equivalent to the payload of the vehicle.

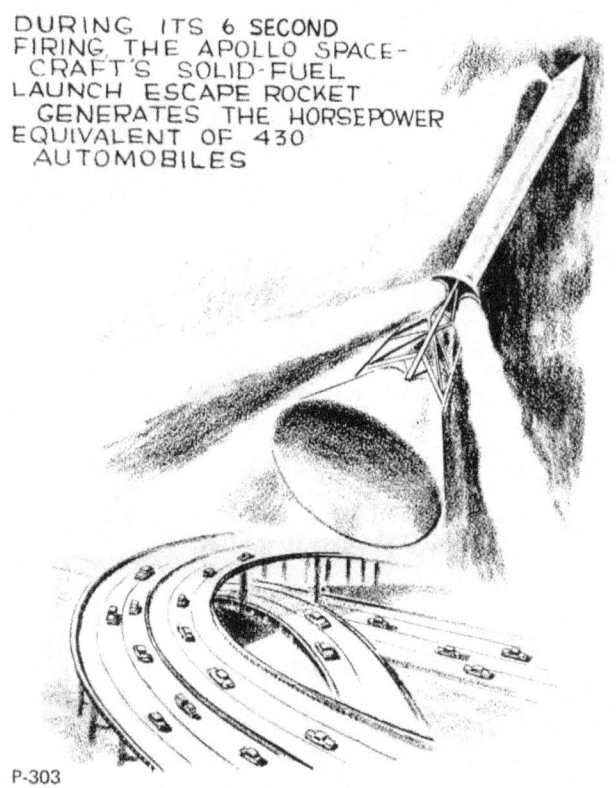

DURING ITS 6 SECOND FIRING THE APOLLO SPACECRAFT'S SOLID-FUEL LAUNCH ESCAPE ROCKET GENERATES THE HORSEPOWER EQUIVALENT OF 430 AUTOMOBILES

P-303

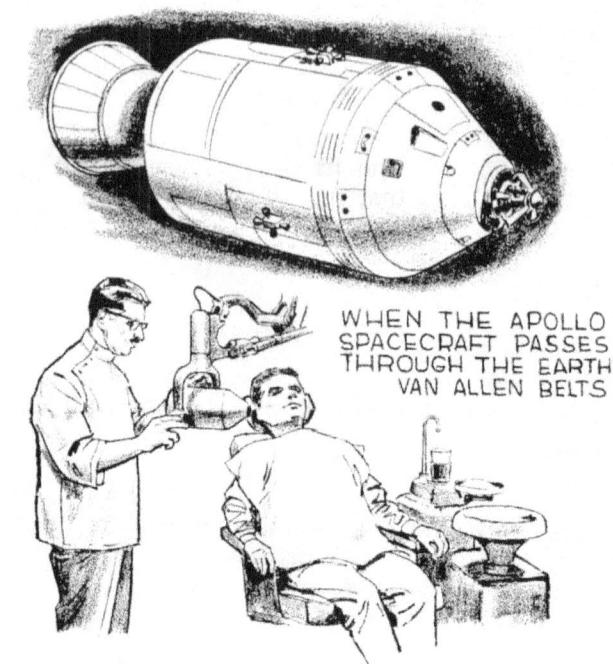

WHEN THE APOLLO SPACECRAFT PASSES THROUGH THE EARTH'S VAN ALLEN BELTS EN ROUTE TO THE MOON, ITS TRIO OF ASTRONAUT CREWMEN WILL BE EXPOSED TO RADIATION EQUIVALENT TO THAT OF A DENTAL X-RAY.

P-305

THE FULLY LOADED APOLLO SATURN V LUNAR VEHICLE STANDS 363 FEET TALL -- 60 FEET HIGHER THAN THE STATUE OF LIBERTY ON ITS PEDESTAL...

AND WEIGHS MORE THAN SIX MILLION POUNDS -- 13 TIMES MORE THAN THE FAMED FIGURE.

P-304

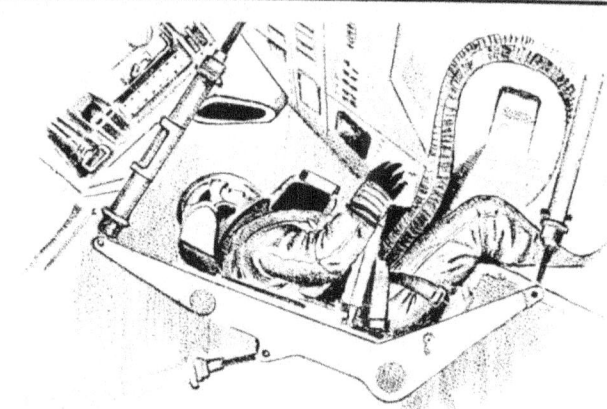

THE APOLLO SPACECRAFT COMMAND MODULE WHICH WILL CARRY U.S. ASTRONAUTS TO AND FROM THE MOON USES ONLY **2000 WATTS** OF ELECTRICITY, ABOUT THE SAME AS THAT REQUIRED BY THE OVEN IN AN ELECTRIC RANGE.

P-306

THE ENGINES OF THE SATURN V LAUNCH VEHICLE THAT WILL PROPEL THE APOLLO SPACECRAFT TO THE MOON HAVE THE COMBINED HORSEPOWER EQUIVALENT TO APPROXIMATELY 500 JET FIGHTERS

P-307

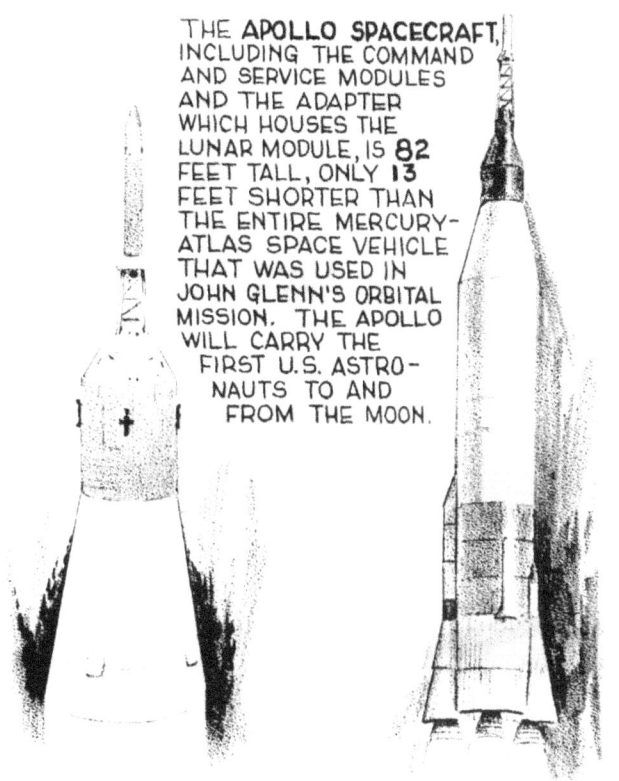

THE **APOLLO SPACECRAFT**, INCLUDING THE COMMAND AND SERVICE MODULES AND THE ADAPTER WHICH HOUSES THE LUNAR MODULE, IS **82** FEET TALL, ONLY **13** FEET SHORTER THAN THE ENTIRE MERCURY-ATLAS SPACE VEHICLE THAT WAS USED IN JOHN GLENN'S ORBITAL MISSION. THE APOLLO WILL CARRY THE FIRST U.S. ASTRONAUTS TO AND FROM THE MOON.

P-309

THE F-1 ENGINES THAT BOOST THE FIRST STAGE OF THE SATURN V LUNAR LAUNCH VEHICLE INTO SPACE GENERATE 160 MILLION HORSEPOWER, ABOUT DOUBLE THE AMOUNT OF POTENTIAL HYDROELECTRIC POWER THAT WOULD BE AVAILABLE AT ANY GIVEN MOMENT IF ALL THE MOVING WATERS OF NORTH AMERICA WERE CHANNELED THROUGH TURBINES.

P-308

THE FUEL PUMPS OF THE F-1 ENGINES THAT POWER THE FIRST STAGE OF THE SATURN V LUNAR LAUNCH VEHICLE PUSH FUEL WITH THE FORCE OF 30 DIESEL LOCOMOTIVES.

P-310

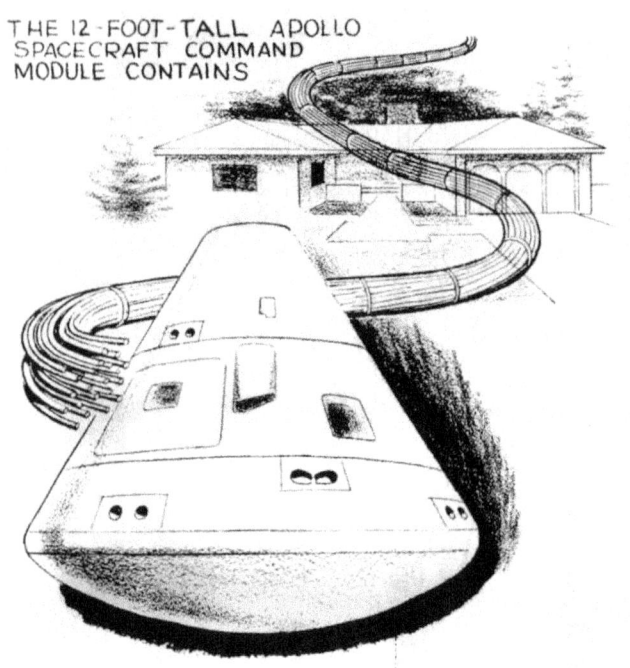

ALMOST 15 MILES OF WIRE, ENOUGH TO WIRE 50 TWO-BEDROOM HOMES

THE 12-FOOT-TALL APOLLO SPACECRAFT COMMAND MODULE CONTAINS

P-311

AT ITS PEAK, MORE THAN 20,000 INDUSTRIAL FIRMS, EMPLOYING MORE THAN 350,000 PERSONS, WERE PRODUCING EQUIPMENT FOR THE U.S. APOLLO/SATURN SPACE PROGRAM UNDER CONTRACTS WITH THE NATIONAL AERONAUTICS AND SPACE ADMINISTRATION.

P-313

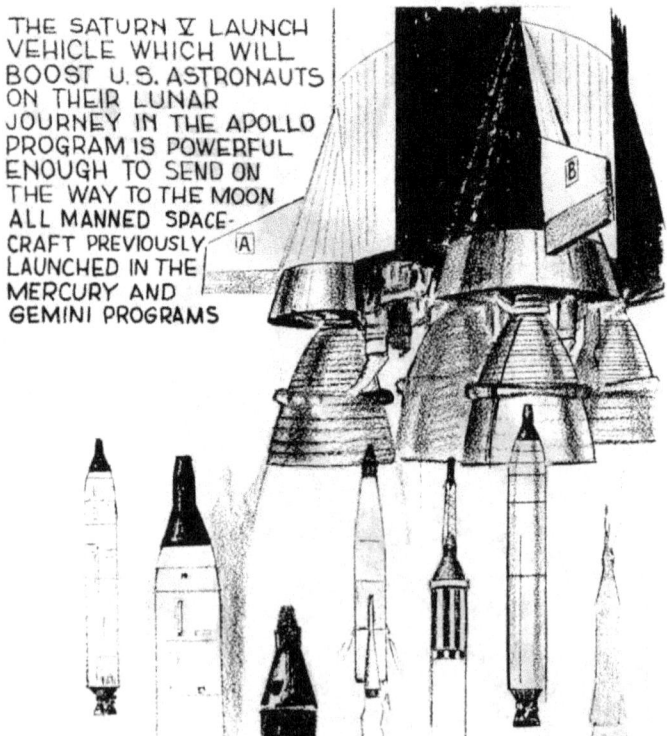

THE SATURN V LAUNCH VEHICLE WHICH WILL BOOST U.S. ASTRONAUTS ON THEIR LUNAR JOURNEY IN THE APOLLO PROGRAM IS POWERFUL ENOUGH TO SEND ON THE WAY TO THE MOON ALL MANNED SPACECRAFT PREVIOUSLY LAUNCHED IN THE MERCURY AND GEMINI PROGRAMS

P-312

IT WOULD TAKE 96 TANK CARS TO HOLD ENOUGH ROCKET PROPELLANT TO FILL THE MIGHTY SATURN V —

THE 363 FT LAUNCH VEHICLE DESTINED TO TRANSPORT U.S. ASTRONAUTS TO THE SURFACE OF THE MOON

P-314

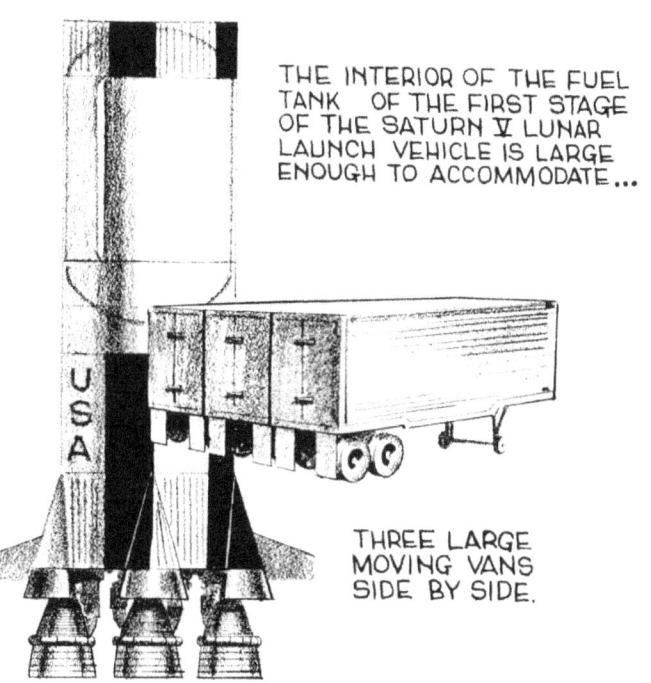

THE INTERIOR OF THE FUEL TANK OF THE FIRST STAGE OF THE SATURN V LUNAR LAUNCH VEHICLE IS LARGE ENOUGH TO ACCOMMODATE...

THREE LARGE MOVING VANS SIDE BY SIDE.

P-315

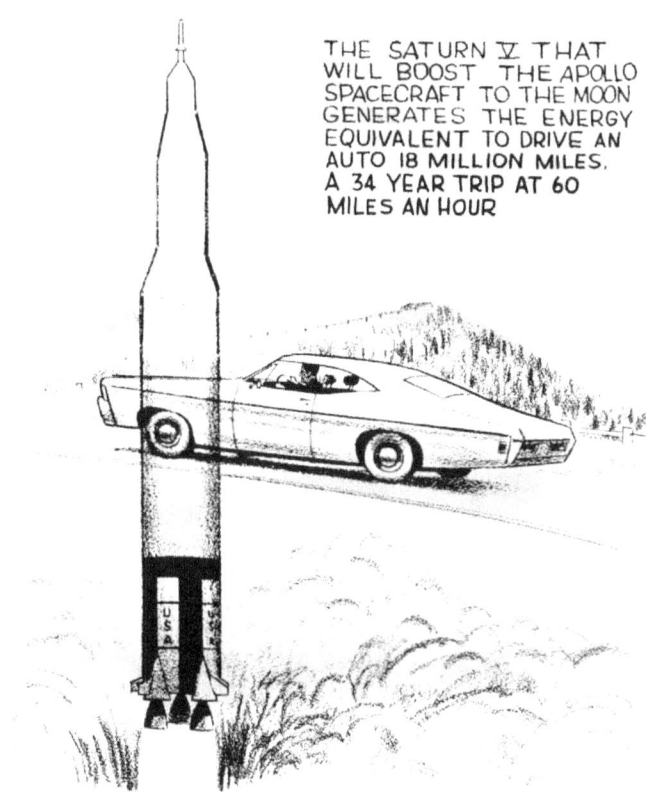

THE SATURN V THAT WILL BOOST THE APOLLO SPACECRAFT TO THE MOON GENERATES THE ENERGY EQUIVALENT TO DRIVE AN AUTO 18 MILLION MILES. A 34 YEAR TRIP AT 60 MILES AN HOUR

P-317

THE FIVE 225,000-POUND THRUST J-2 ENGINES THAT POWER THE SECOND STAGE OF THE SATURN V LUNAR LAUNCH VEHICLE GENERATE THRUST EQUAL TO ABOUT 95.4 BILLION WATTS, OR THE POWER OF 72 HOOVER DAMS.

P-316

IF YOUR CAR GETS 15 MILES TO A GALLON, YOU COULD DRIVE 10 MILLION MILES OR AROUND THE WORLD ABOUT 400 TIMES ON THE PROPELLANTS REQUIRED FOR THE APOLLO/SATURN LUNAR LANDING MISSION

P-318

WHILE AN AUTOMOBILE HAS LESS THAN 2000 FUNCTIONAL PARTS, THE APOLLO SPACECRAFT COMMAND MODULE WHICH WILL CARRY U.S. ASTRONAUTS TO AND FROM THE MOON, HAS NEARLY **TWO MILLION** PARTS, NOT COUNTING WIRE AND SKELETAL COMPONENTS.

P-319

WHEN THE APOLLO REENTERS THE ATMOSPHERE, IT WILL DISSIPATE ENERGY EQUIVALENT TO APPROXIMATELY 86,000 KILOWATT HOURS OF ELECTRICITY, ENOUGH TO LIGHT THE CITY OF LOS ANGELES FOR ABOUT 104 SECONDS

P-321

THE TANKS THAT HOLD THE SUPER-COLD FUEL IN THE APOLLO SPACECRAFT SERVICE MODULE ARE SO WELL INSULATED THAT ICE CUBES PLACED INSIDE THE TANKS WOULD TAKE EIGHT AND ONE-HALF YEARS TO MELT.

P-320

THE APOLLO COMMAND MODULE IN WHICH THREE U.S. ASTRONAUTS WILL RIDE TO AND FROM THE MOON, OFFERS 73 CUBIC FEET OF SPACE PER MAN AGAINST THE 68 CUBIC FEET PER PERSON AVAILABLE IN A COMPACT CAR.

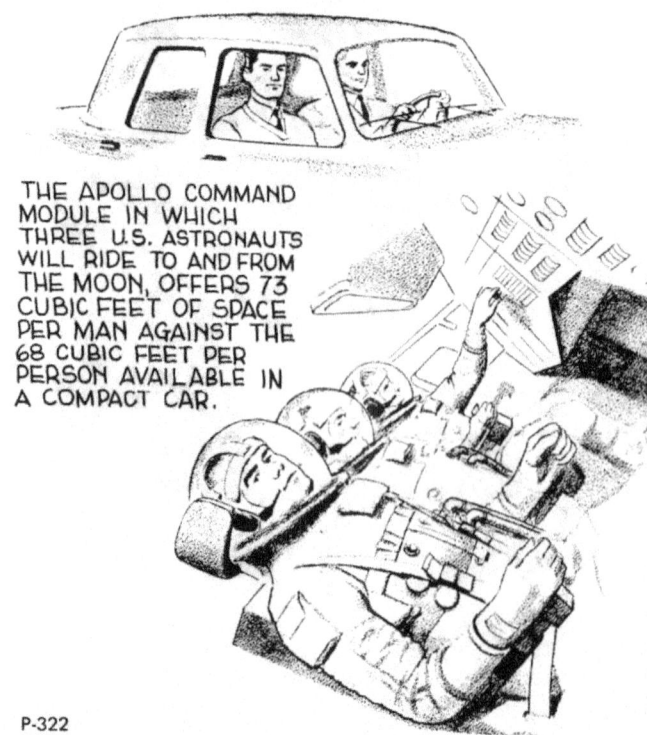

P-322

GLOSSARY

Ablating Materials—Special heat-dissipating materials on the surface of a spacecraft that can be sacrificed (carried away, vaporized) during re-entry.

Ablation—Melting of ablative heat shield materials during re-entry of spacecraft into earth's atmosphere at hypersonic speeds.

Abort—The cutting short of an aerospace mission before it has accomplished its objective.

Accelerometer—An instrument to sense accelerative forces and convert them into corresponding electrical quantities usually for controlling, measuring, indicating, or recording purposes.

Acceptance Test—A test or series of tests to demonstrate that performance is within specified limits.

Acquisition and Tracking Radar—A radar set which searches for, acquires, and tracks an object by means of reflected radio frequency energy from the object, or tracks by means of a radio-frequency signal emitted by the object.

Actuators—Devices which transform an electrical signal into a mechanical motion using hydraulic or pneumatic power.

Adapter Skirt—A flange or extension of a stage or section that provides a ready means of fitting another stage or section to it.

Aerothermodynamic Border—An altitude of about 100 miles in which the atmosphere becomes so rarefied that there is no longer any significant heatgenerating air friction or thermal influence on the skin of fast-moving vehicles.

Airborne Data—Data obtained from space systems during flight.

Ambient—Environmental conditions such as pressure or temperature.

Amorphous—Without definite form; in reference to supercooled liquids and colloidal substances, without real or apparent crystalline form.

Anacoustic Zone—The zone of silence in space; the region above 100 miles altitude where the distance between the rarefied air molecules is greater than the wavelength of sound, and sound waves can no longer be propagated.

Analog Computer—A computing machine that works on the principle of measuring, as distinguished from counting, in which the measurements obtained, as voltages, resistances, etc., are translated into desired data.

Aphelion—Point on an elliptical orbit around sun which is greatest distance from sun. (Earth's aphelion is about 94,500,000 miles from sun.)

Apocynthion—The point at which a satellite (e.g., a spacecraft) in its orbit is farthest from the moon; differs from apolune in that it is an earth-originated orbit.

Apogee—The point at which a moon or artificial satellite in its orbit is farthest from earth.

Apolune—The point at which a satellite (e.g., a spacecraft) in its orbit is farthest from the moon; differs from apocynthion in that the orbit is originated from the moon.

Asteroid—One of the many thousands of minor planets which revolve around the sun, mostly between the orbits of Mars and Jupiter.

Astrogation—Navigating in space.

Astronaut—One who flies or navigates through space.

Astronautics—The art or science of designing, building, or operating space vehicles.

Astronics—The science of adapting electronics to aerospace flight.

Astrobiology—A branch of biology concerned with the discovery or study of life on planets.

Astronomical Unit—Mean distance of earth from the sun, equal to 92,907,000 miles.

Astrophysics—Application of laws and principles of physics to all aspects of steller astronomy.

Atmosphere—The envelope of gases which surrounds the earth and certain other planets.

Atmosphere Refraction—Refraction of light from a distant point by the atmosphere, caused by its passing obliquely through varying air densities.

Attenuator—An adjustable resistive network for reducing the amplitude of an electrical signal without introducing appreciable phase or frequency distortion.

Attitude—The position of an aerospace vehicle as determined by the inclination of its axes to some frame of reference; for Apollo, an inertial, space-fixed reference is used.

Axis—Any of three straight lines, the first running through the center of the fuselage lengthwise, the second at right angles to this and parallel to the horizontal airfoils, and the third perpendicular to the first two at their point of intersection (aircraft).

Azimuth—An arc of the horizon measured between a fixed point (e.g., true north) and the vertical circle through the center of an object.

Backout—Reversing the countdown sequence because of the failure of a component in the vehicle or a hold of unacceptable duration.

Ballistic Trajectory—The curved portion of a vehicle trajectory traced after the propulsion force is cut off.

Biatomic Oxygen—The normal oxygen molecule, consisting of two oxygen atoms, which exists in the lower layers of the atmosphere. It constitutes nearly 21 percent of the atmospheric air and is the essential agent in respiration.

Binary Star—Two stars revolving around a common center of gravity.

Bioastronautics—Astronautics considered for its effect on animal or plant life.

Biosphere—That part of the earth and its atmosphere in which animals and plants live.

Bit—A unit of information carried by an identifiable character, which can exist in either of two states - a "one" or a "zero." An abbreviation of binary digit.

Blanketing—When a desired signal is blanketed, or eliminated, from reception by the presence of an overriding, stronger undesired signal.

Bleed-Cycle Operation—Refers specifically to liquid-propellant rocket engines in which a turbopump is driven by hot gases bled from the combustion chamber of the main thrust chamber.

Blip—A spot of light or other indicator on a radar scope (cathode-ray tube).

Blowoff—Separation of an instrument section or package from the remainder of the rocket vehicle by application of an explosive force.

Blow-Out Disc—A mechanism, consisting of a thin metal diaphragm, used as a safety device to relieve excessive gas pressure.

Boilerplate—A full-size mockup that has all of the mechanical characteristics of the true item but none of the functional features.

Booster—An engine that assists the normal propulsive system of a vehicle or other system of a vehicle.

Bootstrap—A self-generating or self-sustaining process.

Boresight Tower—A tower on which there are mounted a visual target and an electrical target (antenna fed from a signal generator); these targets are used for the parallel alignment of the electrical axis of a receiving antenna and the optical axis of a telescope mounted on that antenna.

Braking Ellipses—A series of orbital approaches to a planet's atmosphere to slow a rocket before landing.

Burnout—The point when combustion ceases in a rocket engine.

Burst Diaphragm—Same as a blow-out disc.

Canard—A short, stubby wing-like element affixed to an aircraft or spacecraft to provide better stability.

Capsule—A small pressurized cabin with an acceptable environment, usually for containing a man or animal for extremely high-altitude flights, orbital space flight, or emergency escape.

Captive Firing—Test firing of a complete vehicle where all or any part of the propulsion system is operated at full or partial thrust while the missile is restrained in the test stand.

Captive Test—A test conducted while the vehicle is secured to a test stand; primarily intended to verify proper operation of the propulsion and flight control subsystems under full-thrust conditions.

Capture—(1) The act of a central force field capturing a passing or colliding body or particle. (2) Of a central force field, as of a planet: to overcome the velocity or centrifugal force of a passing or colliding body or particle and bring its behavior under control of the force field or integrate the body's mass into the force field.

Cavitation—The rapid formation and collapse of vapor pockets in a flowing liquid under very low pressures; a frequent cause of serious structural damage to rocket components.

C Band—A radio frequency band of 3.9 to 6.2 gigacycles per second.

Celestial Guidance—The guidance of a vehicle by reference to celestial bodies.

Celestial Mechanics—The science that deals primarily with the effect of force as an agent in determining the orbital paths of celestial bodies.

Celestial Sphere—Imaginary sphere of infinite radius, assumed for navigational purposes and center of which coincides with the center of earth.

Center of Mass—Commonly called the center of gravity, it is the point at which all the given mass of a body or bodies may be regarded as being concentrated as far as motion is concerned.

Centrifugal Force—A force which is directed away from the center of rotation.

Centrifuge—A large motor-driven apparatus with a long rotating arm used to produce centrifugal force.

Centripetal Force—A force which is directed toward the center of rotation.

Characteristic Length—In propulsion, the ratio of the chamber volume to its nozzle throat area. A measure of the length of travel available for the combustion of propellants.

Characteristic Velocity—Sum of all velocities that have to be obtained or overcome for purposes of braking by a rocket intended for a particular journey.

Checkout—A sequence of operational and calibrational tests to determine the condition and status of a system.

Chemical Fuel—(1) A fuel that depends on an oxidizer for combustion or for development of thrust, such as liquid or solid rocket fuel, jet fuel, or internal-combustion-engine fuel. Distinguished from nuclear fuel. (2) An exotic fuel that uses special chemicals.

Chemosphere—A stratum of the atmosphere marked for its photochemical activity. (By some meteorologists, the chemosphere is considered to be an extension of the stratosphere.)

Chuffing—The characteristic of some rockets to burn intermittently and with an irregular puffing noise.

Circular Velocity—Critical velocity at which a satellite will move in a circular orbit, it is extremely difficult to attain because of the accuracy of control needed.

Circumlunar—Trips or missions in which a vehicle will circle the moon.

Cislunar Space—Space between the earth and the orbit of the moon.

Closed Ecological System—A system that provides for the metabolism of the body in a spacecraft cabin by means of a cycle in which exhaled carbon dioxide, urine, and other waste matter are converted chemically or by photosynthesis into oxygen and food.

Closed Loop—Automatic control units linked together with a process to form an endless chain.

Closed Respiratory Gas System—A completely self-contained system within a sealed cabin, capsule, or spacecraft that will provide adequate oxygen for breathing, maintain adequate cabin pressure, and absorb the exhaled carbon dioxide and water vapor.

Cloud Chamber—The path of subatomic particles are made visible in this kind of chamber by depositing a "cloud" of water particles on them.

Cluster—Two or more engines bound together so as to function as one propulsive unit.

Comet—A loose body of gases and solid matter revolving around the sun.

Command—A pulse or signal initiating a step or sequence.

Companion Body—A nose cone, last-stage rocket, or other body that orbits along with an earth satellite.

Comparator—An electronic processing instrument that compares one set of data with another.

Condensation Trail (Contrails or Vapor Trails)—A visible cloud streak, usually brilliantly white in color, which trails behind a vehicle in flight under certain conditions; caused by the formation of water droplets or sometimes ice crystals due to sudden compression, then expansive cooling, of the air through which the vehicle passes, and of introduction of water vapors through condensation of certain fuels.

Console—Term applied to a grouping of controls, indicators, and similar electrical or mechanical equipment.

Constellation—Any one of the arbitrary groups of fixed stars, some 90 of which are now recognized. A division of the heavens in terms of any one of these groups.

Control Rocket—A rocket used to guide, accelerate, or decelerate a launch vehicle or spacecraft.

Control System—A system that serves to maintain attitude stability during forward flight and to correct deflections.

Controlled Leakage System—A system that provides for the body's metabolism in an aircraft or spacecraft cabin by a controlled escape of carbon dioxide and other waste.

Converter—A unit that changes the language of information from one form to another.

Coriolis Effect—The deflection of a body in motion due to the earth's rotation, diverting horizontal motions to the right in the northern hemisphere and to the left in the southern hemisphere.

Corpuscular Cosmic Rays—Primary cosmic rays from outer space which consist of particles, mainly atomic nuclei (protons) of hydrogen and helium, positively charged and possessing extremely high kinetic energy.

Corpuscular Radiations—Consisting of a flux of small particles.

Cosmic Rays—Extremely fast particles continually entering the upper atmosphere from interstellar space; atomic nuclei which have very great energies because of their enormous velocities; potentially dangerous to life during extended exposure.

Creep—The property of a metal which allows it to be permanently deformed when subjected to a stress.

Cryogenics—The subject of a physical phenomena in the temperature range below about -50 degrees C. More generally, cryogenics or its synonym cryogery refers to methods of producing very low temperatures.

Cyclic Testing—Repeated testing of an object at regular intervals to be assured of its reliability.

Damping—Restraining.

Data Link Equipment—Electronic equipment that coordinates data collection, reduction, and analysis.

Deadband—In a control system, the range of values through which the measure can be varied without initiating an effective response.

Declination—In astronomy and celestial navigation, the angular distance of a celestial body from the celestial equator measured through 90 degrees and named "north" or "south" as the body is north or south of its celestial equator measured on an hour circle.

Deep Space—Used to refer to any space other than that in the vicinity of earth.

Delta V (ΔV)—Velocity change

Destruct—The deliberate action of detonating or otherwise destroying a missile or other vehicle after launch.

Dielectrically Heated—Heating while producing power (i.e., the fuel cell).

Diffusion Process—The exchange of molecules in gas mixtures or solutions across a border line between two or more different concentrations.

Digital Computer—A computer in which quantities are represented numerically and which can be used to solve complex problems.

Doppler Drift—The drift of a vehicle as determined through use of Doppler's (German mathematician Christian Doppler) principle by means of radar.

Doppler Effect—The apparent change in frequency of vibrations, as of sound, light, or radar, when the observed and observer are in motion relative to one another.

Doppler Principle—A principle of physics that, as the distance between a source of constant vibrations and an observer diminishes or increases, the frequencies appear to be greater or less.

Doppler Shift—A shift of a luminous body's line in a spectrum toward the red, indicating an increase in distance.

Dosimeter—An instrument that measures the amount of exposure to nuclear or X-ray radiation; also called an intensitometer or dosage meter.

Down-Link—The part of a communication system that receives, processes, and displays data from a spacecraft.

Drag—The aerodynamic force in a direction opposite to that of flight and due to the resistance of the body to motion in air.

Drift Error—A change in the output of an instrument over a period of time, usually caused by random wander or by a condition of the environment.

Drogue—The hollow (female) part of a connector into which a probe (male) part fits.

Dry-Fuel Rocket—A rocket that uses a mixture of fast-burning power. Used especially as a booster rocket.

Dual Thrust—A rocket thrust derived from two propellant grains using the same propulsion section of a missile.

Dual Thrust Motor—A solid rocket motor built to obtain dual thrust.

Earth-Fixed Reference—An oriented system using some earth phenomena for positioning.

Eccentric—Of an orbit, deviating from the line of a circle so as to form an ellipse.

Ecliptic—Plane of the earth's orbit around the sun, used as a reference for other interplanetary orbits; also the name for the apparent path of the sun through the constellations as projected on the celestial sphere.

Ecosphere—The great circle on the celestial sphere which describes the apparent path of the sun in the course of the year.

Effective Atmosphere—That part of the atmosphere which effectively influences a particular process of motion.

Effective Exhaust Velocity—The velocity of an exhaust stream after the effects of friction, heat transfer, non-axially directed flow, and other conditions have reduced it.

Effector—The mechanical means of maneuvering a vehicle during flight: an aerodynamic surface, a gimbaled motor, or an auxiliary jet.

Electrojet—Current sheet or stream moving in an ionized layer in the upper atmosphere of a planet.

Electrolyte—A substance in which the conduction of electricity is accompanied by chemical action; the paste which forms the conducting medium between the electrodes of a dry cell, storage cell, or electrolytic capacitor.

Emissivity—The relative power of a surface or a material composing a surface to emit heat by radiation.

Entry Corridor—The final flight path of the spacecraft before and during earth re-entry.

Ephemeris—A publication giving the computed places of the celestial bodies for each day of the year, or for other regular intervals.

Escape Orbit—One of various paths that a body or particles escaping from a central force field must follow in order to escape.

Escape Velocity—The speed a body must attain to overcome a gravitational field, such as that of earth; the velocity of escape at the earth's surface is 36,700 feet per second.

Event Timer—An instrument that times an event and records time taken to perform the cycle or event; can record several events simultaneously.

Exerciser—A machine that simulates the strains and vibrations to which a missile is subjected, and used to test for structural integrity.

Exhaust Stream—The stream of gaseous, atomic, or radiant particles that emit from the nozzle of a rocket or other reaction engine.

Exosphere—The outermost fringe or layer of the atmosphere, where collisions between molecular particles are so rare that only the force of gravity will return escaping molecules to the upper atmosphere.

Exotic Fuel—Unusual fuel combinations for aircraft and rocket use.

Explosive Bolts—Bolts surrounded with an explosive charge which can be activated by an electrical impulse.

Explosive Bridge Wire—Wire which heats to a high temperature and burns, thus igniting a charge.

Extension Skirt—Adapter used to connect elements of the spacecraft.

Extravehicular—Indicates that an element, such as an antenna, is located outside the vehicle.

Fairing—A piece, part, or structure having a smooth, streamlined outline, used to cover a nonstreamlined object or to smooth a junction.

Fallaway Section—Any section of a rocket vehicle that is cast off and falls away from the vehicle in flight.

Final Trim—Action that adjusts a vehicle to the exact direction programmed for its flight.

Flash Point—The temperature at which the vapor of a fuel or oil will flash or ignite momentarily.

Float Bag—A collar located around the spacecraft used to keep the spacecraft upright in the water and prevent sinking.

Free-Flight Rocket—A rocket without electronic control or guidance.

Free-Flight Trajectory—The part of a ballistic missile's trajectory that begins with thrust cutoff and ends at re-entry.

Free Gyro—Sometimes referred to as space reference gyro in that the free gyro will maintain its orientation with respect to the stars rather than with respect to the earth.

Frequency Spectrum—The area encompassed by frequencies, from very low to very high, in terms of cycles (vibration) in a unit of time.

Free-Return Trajectory—A return to earth without power; this trajectory would be used in the event of a failure of the spacecraft propulsion system.

Fuel Cell—An electrochemical generator in which the chemical energy from the reaction of air (oxygen) and a fuel is converted directly into electricity.

G or G Force—Force exerted upon an object by gravity or by reaction to acceleration or deceleration, as in a change of direction: one G is the measure of the gravitational pull required to move a body at the rate of about 32.16 feet per second.

Galaxy—(1) The group of several billion suns, stars, star clusters, nebulae, etc., to which the earth's sun belongs; (2) any of several similar groups of stars forming isolated units in the universe.

Gamma Radiation—Electromagnetic radiation, similar to X rays, originating from the nucleus and having a high degree of penetration.

Gas Chromatograph—An oscillating filter-photometer that separates and analyzes gasses.

Geocentric—Relating to or measured from the center of the earth: having, or relating to, the earth as a center.

Geodetic—Pertaining to or determined by that branch of mathematics which determines the exact positions of points and the figures and areas of large portions of the earth's surface, or the shape and size of the earth and the variations of terrestrial gravity.

Geophysical Constant—A quantity that expresses a fixed value for a law or magnitude that applies to the physics of the earth.

Geophysics—The physics of the earth, or science treating of the agencies which modify the earth.

Gimbal—Mechanical frame containing two mutually perpendicular intersecting axes of rotation (bearing and/or shafts).

Gimballed Motor—A rocket motor mounted on gimbal; i.e., on a contrivance having two mutually perpendicular axes of rotation, so as to obtain pitching and yawing correction moments.

Glycol—Ethylene glycol, a coolant mixed with water in varying proportions, depending on rate of cooling desired.

Grain—The body of a solid propellant used in a rocket, fashioned to a particular size and shape so as to burn smoothly without severe surges or detonations.

Gravitation—Force of attraction that exists between all particles of matter everywhere in the universe.

Gravity—That force which tends to pull bodies toward the center of mass; that is, to give bodies weight.

Gravity Anomalies—Deviations between theoretical gravity and actual gravity due to local topographic and geologic conditions.

Gravity Simulation—Use of centripetal force to simulate weight reaction in a condition of free fall.

Ground Trace—The theoretical mark traced on the surface of the earth by a flying object or satellite as it passes over the surface.

Guidance System—A system which measures and evaluates flight information, correlates this with target data, converts the result into the conditions necessary to achieve the desired flight path, and communicates this data in the form of commands to the flight control system.

Guidance Tapes—Magnetic or paper tapes that are placed in the computer and on which there previously has been entered information needed in guidance.

Gyro-Compassing—Use of gyro with axle pointed due north in directional guidance.

Gyroscope—A device consisting of a wheel so mounted that its spinning axis is free to rotate about either of two other axes perpendicular to itself and to each other; once set in rotation, its axle will maintain a constant direction, even when the earth is turning under, when its axle is pointed due north, it may be used as a gyro compass.

Heat Exchanger—A device for transferring heat from one substance to another, as by regenerative cooling.

Heat Sink—A contrivance for the absorption or transfer of heat away from a critical part or parts, as in a nose cone where friction-induced heat may be conducted to a special metal for absorption.

Heaviside—Kennelly Layer—Region of the ionosphere that reflects certain radio waves back to earth.

Heliocentric—Measured from the center of the sun.

Honeycomb Sandwich—A type of construction in which the space between the upper and lower surfaces is occupied by a strengthening material of a structure resembling a honeycomb mesh.

Horizon Photometer—An instrument to determine the distinction between the sky and the horizon; thus, measures light by means of monitoring the infrared emanations.

Hydrosphere—The aqueous (watery) envelope of a planet.

Hydrostatic Effects—The pressures exerted by a column of liquid (water, blood, etc.) under normal gravitational conditions on the surface of the earth or in a gravitational field during an acceleration.

Hyperacoustic Zone—The region in the upper atmosphere between 60 and 100 miles where the distance between the rarefied air molecules roughly equals the wave length of sound, so that sound is transmitted with less volume than at lower levels. Above this zone, sound waves cannot be propagated.

Hyperbola—A conic section made by a place intersecting a cone of revolution at an angle smaller than that of a parabola.

Hypergolic—Refers to bipropellant combinations which ignite spontaneous upon contact or mixing.

Hypersonic—Speeds faster than Mach 5 or five times the speed of sound.

Hypoxia—Oxygen deficiency in the blood cells or tissues of the body in such a degree as to cause psychological and physiological disturbances.

Ice Frost—A thickness of ice that gathers on the outside of a rocket vehicle over surfaces supercooled by liquid oxygen or hydrogen inside the vehicle.

Incidence Angle—The angle between earth and the path of a vehicle.

Inertia—The tendency of an object to remain put or if moving to continue on in the same direction.

Inertial Guidance—A sophisticated automatic navigation system using gyroscopic devices, etc., for high-speed vehicles. It absorbs and interprets such data as speed, position, etc., and automatically adjusts the vehicle to a predetermined flight path. Essentially, it knows where it's going and where it is by knowing where it came from and how it got there. It does not give out any signal so it cannot be detected by radar or jammed.

Inertial Orbit—The type of orbit described by all celestial bodies, according to Kepler's laws of celestial motion. This applies to all satellites and spacecraft provided they are not under any type of propulsive power, their driving force being imparted by the momentum at the instant propulsive power ceases.

Inertial Space—An assumed stationary frame of reference. A non-rotating set of coordinates in space relative to which the trajectory of a space vehicle is calculated.

Injection—The process of injecting a spacecraft into a calculated orbit.

Integrating Accelerometer—A mechanical and electrical device which measures the forces of acceleration along the longitudinal axis, records the velocity, and measures the distance traveled.

Intergalactic Space—That part of space conceived as having its lower limit at the upper limit of interstellar space, and extending to the limits of space.

Interior Ballistics—That branch of ballistics concerned with behavior, motion, appearance, or modification of a rocket when acted upon by ignition and burning of a propellant. Sometimes called "internal ballistics." In rocketry, interior ballistics deals with the missile's behavior in reaction to gas pressures inside the rocket, escapements, shift in the center of gravity as propellants are consumed, etc.

Interleaver—The act of combining computer data to produce, from several sources, a single result.

Interplanetary Space—That part of space conceived, from the standpoint of the earth, to have its lower limit at the upper limit of translunar space, and extending to beyond the limits of the solar system several billion miles.

Interstellar Flight—Flight between stars; strictly, flight between orbits around the stars.

Interstellar Space—That part of space conceived, from the standpoint of the earth, to have its lower limit at the upper limit of interplanetary space, and extending to the lower limits of intergalactic space.

Inverter—A device that changes dc current to ac, or vice versa.

Ion Engine—A type of engine in which the thrust to propel the missile or spacecraft is obtained from a stream of ionized atomic particles, generated by atomic fusion, fission or solar energy.

Ionic Conduction Path—That part of the vehicle where radio communication is not possible due to the ionization of the air - the transmitting medium. The ions interfere with the radio frequency signal.

Ionization—Formation of electrically charged particles; can be produced by high-energy radiation, such as light or ultra-violet rays, or by collision or particles in thermal agitation.

Ionized Layers—Layers of increased ionization within the ionosphere. Believed to be caused by solar radiation. Responsible for absorption and reflection of radio waves and important in connection with communication and tracking of satellites and other space vehicles.

Ionosphere—An outer belt of the earth's atmosphere in which radiations from the sun ionize, or excite electrically, the atoms and molecules of the atmospheric gases. The height of the ionosphere varies with the time of day and the season, but its lower limit is generally considered to lie between 25 and 50 miles. It is divided into several layers with respect to radiation and reflective properties. A characteristic phenomenon is its reflection of certain radio waves.

Iostatic—Under equal pressure from every side.

Isothermal Region—The stratosphere considered as a region of uniform temperature.

Jet Steering—The use of fixed or movable jets on a space vehicle, ballistic missile, or sounding rocket to steer it along a desired trajectory, during both propelled flight (main engines) and after thrust cutoff.

Kelvin Scale—(After Baron Kelvin, English physicist and inventor.) A temperature scale that uses centigrade degrees but makes the zero degree signify absolute zero.

Keplerian Trajectory—Elliptical orbits described by celestial bodies (and satellites) according to Kepler's first law of celestial motion.

Kepler's Law—The three laws of planetary motion discovered by Kepler:
(1) The orbit of every planet about the sun is an ellipse, the sun occupying one focus. (2) A line from each planet to the sun sweeps over equal area in equal times. (3) The squares of the times required for the different planets to complete their orbits are proportional to the cubes of their mean distances from the sun.

Leveled Thrust—A rocket power plant equipped with a programmer or engine control unit that maintains the output at a relatively constant thrust.

Lift-Drag Ratio—The ratio of lift to drag, obtained by dividing the lift by the drag or the lift coefficient by the drag coefficient.

Light Year—Distance traveled in one year by light, which covers 186,284 miles in one second; equal to 5,880,000,000,000 miles.

Linear Explosive Charge—The shaping of a charge; shaping the explosive pattern of charge to achieve an explosive profile.

Liquified Gases—These are gases which have been converted to liquids under certain pressure and temperature conditions.

Liquid Hydrogen (LH_2)—A liquid rocket fuel that develops a specific impulse, when oxidized by liquid oxygen, ranging between 317 and 364 seconds depending upon the mixture ratio. Hydrogen gas becomes liquid at 423 degrees below zero.

Liquid Oxygen (LOX)—Oxygen supercooled and kept under pressure so that its physical state is liquid. Oxygen gas becomes liquid at 279 degrees below zero.

Loxing—Vernacular term for the task of loading liquid oxygen into fuel tanks of a missile from a ground supply.

Lunar Base—A projected installation on the surface of the moon for use as a base in scientific or military operations.

Lunar Gravity—The attraction of particles and masses towards the gravitational center of the moon.

Mach—(After Ernst Mach, 1858 - 1916, Austrian physicist.) A unit of speed measurement for a moving object equal to the speed of sound in the medium in which the object moves.

Mass—A measure of the quantity of matter in a body.

Mass Ratio—Initial mass of a vehicle at the instant of liftoff divided by the final mass at some point of the powered ascent or at burnout and thrust cutoff.

Mechanical Border—That layer in the atmosphere where air resistance and friction become negligible (from 120 to 140 miles altitude).

Mesosphere—Applied to two different layers on the upper atmosphere: (1) a layer that extends approximately from 19 to 50 miles above the earth's surface; (2) a layer that extends approximately from 250 to 600 miles, lying between the ionosphere and the exosphere.

Metabolism—Chemical and physical processes continuously going on in living organism; assimilated food built up into protoplasm, used, and broken down into waste matter.

Micrometeoroid—Meteoroids less than 1/250th of an inch in diameter.

Miniaturized Data Interleaving System—Where several results are combined to indicate one single result - as in computers; a transistorized version.

Mission Time—Period of time for completing a mission.

Monopropellant—A rocket propellant in which the fuel and oxidizer are premixed ready for immediate use.

Moon—The natural celestial body that orbits as a satellite above the earth, revolving around it about once every 29-1/2 days, reflecting the sun. The moon's mean distance from the earth is about 238,857 miles. The moon's diameter is about 2160 miles and its mass about 1/81 that of earth and the volume about 1/49. Its mean velocity is about 2285 statute miles per hour, its apogee 252,710 miles, perigee 221,463 miles.

Multiplexing—The simultaneous transmission of two or more signals within a single channel. The three basic methods of multiplexing involve the separation of signals by time division, frequency division, and phase division.

Noise (Radio Transmission)—The noise behind the signal, caused by the signal, but not including the signal; can be man-made or atmospheric.

Nose Cone—The shield that fits over, or is, the nose of an aerospace vehicle.

Nova—A star which undergoes a sudden and enormous increase in brightness; about twenty-five appear every year in our galaxy. Supernova is a star which explodes with a liberation of most of its energy into space.

Null-Circle—Theoretical point in space where gravitational attraction of one planet balances that of another planet. There can be no real null point, circle, or region because the solar system is dynamic; parts of it are always moving in relation to other parts.

Omnidirectional—All-directional, not favoring any one direction (also called nondirectional).

Optical Navigation—Navigation by optical means, as opposed to mathematical methods.

Orbital Curve—One of the tracks on a primary body's surface traced by a satellite that orbits about it several times a day in a direction other than true east or west, each successive track being displaced to the west by an amount equal to the degrees of rotation of the primary body between each orbit.

Oxidizer—In a rocket propellant, a substance such as liquid oxygen or nitric acid that yields oxygen for burning the fuel.

Ozone Layer—Layer in the atmosphere about 20 miles above sea level which strongly absorbs solar ultraviolet radiation.

Ozonosphere—A stratum in the upper atmosphere at an altitude of approximately 40 miles having a relatively high concentration of ozone.

Parabola—A conic section made by a plane intersecting a cone parallel to an element of the cone.

Parabola of Escape—Critical orbit in a central force field; the parabolic orbit is such that a body has escape velocity at every point along it.

Parallax—The apparent displacement of an object, or the apparent difference in its direction of motion, if viewed from two different points.

Pendulous Accelerometer—A device employed to determine linear acceleration.

Pericynthion—The point at which a satellite (e.g., a spacecraft) in its orbit is closest to the moon; differs from perilune in that the orbit is earth-originated.

Perigee—The point at which a moon or an artificial satellite in its orbit is closest to the earth

Perilune—The point at which a satellite (e.g., a spacecraft) in its orbit is closest to the moon: differs from pericynthion in that the orbit is moon-originated.

Perihelion—That point on an elliptical orbit around the sun which is nearest to the sun.

Photon—Minute particles which form streams to become light rays. These streams theoretically may be harnessed to power a spacecraft.

Photon Engine—A projected species of reaction engine in which thrust is to be obtained from a stream of light rays.

Photosphere—The outermost luminous layer of the sun's gaseous body.

Pitch—The movement of a space vehicle about an axis (Y) that is perpendicular to its longitudinal axis.

Pitchup—A correction movement of a missile in which it assumes a vertical ascent.

Planetoid—A starlike body, one of the numerous small planets nearly all of whose orbits lie between Mars and Jupiter (also called asteroid and minor planet).

Plasmajet—High-temperature jet of electrons and positive ions that has been heated and ionized by the magneto-hydro-dynamic effect of a strong electrical discharge.

Plasma Physics—The science dealing with the study of fully ionized gases.

Premodulation Processor—Part of the communications system; processed data for further use.

Primary—The body around which a satellite orbits.

Propagation—In missile terminology, to describe the manner in which an electromagnetic wave such as a radar signal, timing signal, or ray of light travels from one point to another.

Propellant Utilization System—The automatic electromechanical system that is installed to control precisely the mixture ratio of the liquid propellants, as they are consumed during a firing.

Pseudo-Random Noise—Noise produced by a definitely calculated process, while satisfying one or more of the standard tests for statistical randomness.

Pulse-Code Modulation Telemetry (PCM)—Pulse modulation in which the signal is sampled periodically, and each sample is quanitized and transmitted as a digital code.

Pyro Batteries—Batteries used to fire pyrotechnic elements.

Pyro Cartridges—Pyrotechnic cartridges.

Q Band—See radio frequencies.

Q-Ball—A device for measuring the angle of attack of a vehicle.

Radial Beam Extensions—Connecting links between command and service modules.

Radial Velocity—The velocity of approach or recession between two bodies, especially between an observer and a source of radiation in a line connecting the two.

Radio Command—A radio signal to which a guided missile, drone, or the like, responds.

Radio Frequencies—Normally expressed in kilocycles per second at and below 30,000 kc/s, and megacycles per second above this frequency. Frequency subdivisions are: very low frequency (VLF), below 30 kc/s; low-frequency (LF), 30 to 300 kc/s; medium frequency (MF), 300 to 3000 kc/s; high frequency (HF), 3000 to 30,000 kc/s; very high frequency (VHF), 30 to 300 mc/s; ultra high frequency (UHF), 300 to 3000 mc/s; super high frequency (SHF), 3000 to 30,000 mc/s; extremely high frequency (EHF), above 30,000 mc/s. During World War II, radio frequency bands were designated by letters (e.g., K band, L band, P band, Q band, S band, V band, and X band). These designations were used originally to maintain military secrecy but currently have no official standing.

Radio Telescope—A radio receiving station for detecting radio waves emitted by celestial bodies or by space probes in space.

Rate Gyro Signals—Signals that indicate the rate of angular motion.

Reaction Engine—An engine or motor that derives thrust by expelling a stream of moving particles to the rear.

Receiver-Decoder—A combination receiver that accepts the signal and then decodes to a given command.

Re-entry—The return of a spacecraft that re-enters the atmosphere after flight above it.

Regenerative Cooling—The cooling of a rocket engine by circulating the fuel or oxidizer fluid in coils about the engine prior to use in the combustion chamber.

Remaining Body—That part of a missile or other vehicle that remains after the separation of a fallaway section or companion body.

Redundant—A second means for accomplishing a given task.

Resolver—(1) A means for resolving a vector into two mutually perpendicular components; (2) A transformer, the coupling between primary and secondary of which can be varied; (3) A small section with a faster access than the remainder of the magnetic-drum memory in a computer.

Reticle Pattern—Pattern established by the crew alignment sight. Used in docking procedure.

Retrofit—To add on or modify.

Retrograde Impulse—The impulse employed to slow a spacecraft or vehicle by applying a thrust in an opposite direction from the direction of motion of the spacecraft.

Retrograde Motion—Orbital motion opposite in direction to that normal to spatial bodies within a given system.

Retrorocket—A rocket that gives thrust in a direction opposite to the direction of the object's motion.

Reverse Thrust—Thrust applied to a moving object in a direction opposite to the direction of the object's motion.

Roentgen—A unit used in measuring radiation, as of X rays.

Roll—The movements of a space vehicle about its longitudinal (X) axis.

Rope—Reflectors of electromagnetic radiation consisting of long strips of metal foil.

RP-1 Fuel—Kerosene-like fuel.

S Band—A radio-frequency band of 1550 to 5200 megacycles per second.

Scintillating Counter—An instrument that measures radiation indirectly by counting the light flashes emitted when radiation particles are absorbed into any of several phosphors.

Scrub—To cancel out a scheduled launch either before or during countdown.

Second of Arc—A measure of an angle 1/60th of a minute.

Seeker—A guidance system which moves on energy emanating or reflected from a target or station.

Selenoid—A lunar satellite.

Sensible Atmosphere—That part of the atmosphere that may be felt, i.e., that offers resistance.

Sensor—A sensing element. In a navigational system, that portion which perceives deviations from a reference and converts them into signals.

Sequencer—A mechanical or electronic device that may be set to initiate a series of events and to make the events follow in a given sequence.

Servos—A short term for servomechanism or servomotors.

Serial Command Words—Specific instructional data to the up-link system.

Shear-Compression Pads—Pads which are sheared during separation of the service and command modules.

Sideband—Two frequencies, located on both sides of the carrier frequency; upper sideband, lower sideband.

Sidereal—A measurement of time. A sidereal day, for example, is the time it takes the earth to make a complete revolution.

Solar Corona—Outer atmospheric shell of the sun.

Solar Noise—Electromagnetic radiation which radiates from the atmosphere of the sun at radio frequencies.

Solenoid—A coil of wire, which, when current flows through it will act as a magnet and tend to pull an iron core that is movable to a central position; used for switching.

Sounding Rocket—A research rocket used to obtain data on the upper atmosphere.

Space Biology—A branch of biology concerned with life as it may come to exist in space.

Space-Fixed Reference—An oriented reference system in space independent of earth phenomena for positioning.

Space Platform—Large satellite with both scientific and military applications, conceived as a habitable base in space.

Space-Time Dilemma—According to Einstein's theory of relativity, time slows down increasingly in systems (e.g., extremely high-performance spacecraft) moving at velocities approaching the speed of light, relative to other systems in space (e.g., the earth). This slowdown is not apparent to the inhabitants of the moving system (the spacecraft) until they return to the redundant system in space from which they started (the earth).

Spatiography—The "geography" of space.

Specific Impulse—A means of determining rocket performance. It is equivalent to the effective exhaust velocity divided by gravity expressed in pounds per second.

Sphygmomanometer—An instrument for measuring arterial blood pressure.

Squib—A small explosive device whose primary function is to produce heat; usually used to achieve ignition in a larger combustible process.

Stabilized Gyro—Normally refers to stabilization to effect coincidence between the vertical axis of the gyro and the vertical established by an earth-seeking pendulum. In another axis, the gyro may be stabilized with respect to the electromagnetic field surrounding the earth, or with the true north direction through appropriate computers.

Stabilized Platform—Major part of an all-inertial guidance system, composed of an assembly of gimbal frames that hold three accelerometers in a fixed position in relation to inertial space. The accelerometers are mounted perpendicular to each other to measure accelerations along the three reference axes. These accelerations can be fed to a computer to determine instantaneous velocity and position in space.

Star Tracker—A telescopic instrument on a missile or other flightborne object that locks onto a celestial body and gives guidance to the missile or other object during flight. A star tracker may be optical or radiometric.

Stationary Orbit—In reference to earth known as a 24-hour orbit; a circular orbit around a planet in the equatorial plane and having a rotational period equal to that of the planet. For earth, the stationary orbit is about 26,000 miles in radius. A body moving in a stable stationary orbit appears fixed in the sky to an observer on the surface of the planet.

Step Rocket—A rocket with two or more stages.

Stratosphere—A calm region of the upper atmosphere characterized by little or no temperature change in altitude.

Sunseeker—Two-axis device actuated by servos and controlled by photocells to keep instruments pointed toward the sun despite rolling or tumbling of an aerospace vehicle in which instruments are carried.

Subsonic—Speed less than that of sound.

Sustainer Rocket—A rocket engine used as a sustainer, especially on an orbital glider or orbiting spacecraft that dips into the atmosphere at its perigee.

Synergic Curve—A curve plotted for the ascent of an aerospace vehicle determined to give the missile or other vehicle maximum economy in fuel with maximum velocity.

Telemetering—A system for taking measurements within an aerospace vehicle in flight and transmitting them by radio to a ground station.

Thrust Vector—The directional line of thrust of the spacecraft.

Torquing Commands—A command given to the gyros to maintain attitude.

Transceiver—A unit combining the radio or radar transmitter and receiver, such as used in a transponder.

Transducer—A device by means of which energy can be made to flow from one or more transmission systems to other transmission systems.

Transearth Coast—The flight, under no power, between moon and earth.

Transistor—An electronic device that controls an electron current by the conducting properties of germanium or like material.

Translunar Space—That part of space conceived as a spherical layer centered on the earth, with its lower limits at the distance of the orbit of the moon, but extending to several hundred thousands of miles beyond.

Translation—For Apollo, movement of the spacecraft along the X axis acceleration.

Translational Control—A joystick located in the crew compartment to enable the pilot to control flight.

Transponder—A radio transmitter-receiver which transmits identifiable signals automatically when the proper interrogation is received.

Tumbling—An unsatisfactory attitude situation in which a vehicle continues on its flight, but turns end over end about its center of gravity with its longitudinal axis remaining in the plane of flight.

Ullage—The volume in a closed tank or container above the surface of a stored liquid. Also the ratio of this volume to the total volume of the tank.

Ultrasonic—Very high sound waves; not audible to humans.

Umbilical Cord—A cable fitted to a vehicle with a quick-disconnect plug, through which electrical power, oxygen, etc., is transmitted.

Up-Link Data—Telemetry information from the ground.

Van Allen Radiation Belts—Two doughnut-shaped belts of high-energy particles trapped in the earth's magnetic field which surround the earth; first reported by Dr. James A. Van Allen of the University of Iowa.

Vectory Steering—Vernacular for a steering method where one or more thrust chambers are gimbal-mounted so that the thrust force may be tilted in relation to the center of gravity of the spacecraft to produce a turning moment.

Wicking (or Wicking Axis)—Capillary action where fluid travels along a path.

X Axis—A designation for the longitudinal axis in a coordinate system of axes.

Yaw—Displacement of a space vehicle from its vertical (Z) axis.

Y Axis—A designation for the lateral axis in a coordinate system of axes.

Z Axis—A designation for the vertical axis in a coordinate system of axes.

ABBREVIATIONS

ac	Alternating current	CO_2	Carbon dioxide
ACCEL	Accelerometer or acceleration	COAS	Crew optical alignment sight
ACE	Acceptance checkout equipment	COAX	Coaxial
ACK	Acknowledge	COI	Contingency orbit insertion
ACP	Audio control panel	COMM	Communications
ACS	Attitude control subsystem	COMPR	Compressor
A/D	Analog to digital	COMPEN	Compensator
AGC	Automatic gain control	COND	Condenser or conditioner
AGE	Aerospace ground equipment	CONT	Control
AH	Ampere hour	CPLR	Coupler
ALT	Altitude	CPS	Cycles per second
AM	Amplitude modulation or ammeter	CRYO	Cryogenic
AMPL	Amplifier	CSC	Cosecant computing amplifier
ANL	Analog	CSM	Command and service modules
AOA	Angle of attack	CSS	Computer subsystem
ARS	Attitude reference subsystem	C&W	Caution and warning subsystem
ASCP	Attitude set control panel	CW	Clockwise or continuous wave
ASI	Apollo standard initiator	CTE	Central timing equipment
ATT	Attitude or attenuator	CWG	Constant wear garment
AUTO	Automatic		
AUX	Auxiliary	D/A	Digital to analog
AVC	Automatic volume control	DAC	Digital-to-analog converter
		DAP	Digital autopilot
BAT	Battery	DB	Deadband
BCD	Binary coded decimal	db	Decibel
BCN	Beacon	dc	Direct current
BECO	Booster engine cutoff	DECR	Decrease
BMAG	Body-mounted attitude gyro	DEG	Degree
BPC	Boost protective cover	DEMOD	Demodulate
bps	Bits per second	DET	Detector or digital event timer
BTU	British thermal unit	DISCR	Discriminator
BU	Backup	DRI	Data rate indicator
BUR	Backup rate	DSE	Data storage equipment
		DSIF	Deep Space Instrumentation Facility
CB	Circuit breaker	DSKY	Display and keyboard
CCFT	Controlled current feedback transformer		
		E	Elevation angle
CCTV	Closed-circuit television	ECA	Electronic control assembly
CCW	Counterclockwise	ECO	Engine combustion or engine cutoff
C&D	Controls and displays	ECS	Environmental control subsystem
CDF	Confined detonating fuse	ECU	Environmental control unit
CDU	Coupling data unit	EDA	Electronic display assembly
cfm	Cubic feet per minute	EDS	Emergency detection subsystem
CG	Center of gravity	ELECT	Electronic
CHAN	Channel	ELS	Earth landing subsystem
CKT	Circuit	ELSC	Earth landing sequence controller
CL	Centerline	EMER	Emergency
CLM	Core logic module	EMI	Electromagnetic interference
CM	Command module	EMS	Entry monitor subsystem
CMC	Command module computer	EMU	Extravehicular mobility unit
CMD	Command	ENC	Encode
C/O	Checkout	ENG	Engine

EOS	Emergency oxygen system	IF	Intermediate frequency
E&PL	Entry & post-landing	IFN	In-flight maintenance
EPS	Electrical power subsystem	IGA	Inner gimbal angle
ERR	Error	IGN	Ignition
ETR	Eastern test range	IMP	Impulse
EU	Electronic unit	IMU	Inertial measurement unit
EVA	Extravehicular activity	INCR	Increase
		IND	Indicator
FC	Fuel cell	INV	Inverter
f_c	Center frequency	IPB	Illuminated push button
FCSM	Flight combustion stability monitor	IPS	Instrumentation power subsystem or inches per second
FDAI	Flight director attitude indicator		
FDT	Full duplex teletype circuit	IRIG	Inertial rate integrating gyro
F/F	Flip-flop	ISOL	Isolation
FHS	Forward heat shield	ISS	Inertial subsystem
FLT	Flight	IU	Instrument unit
FLSC	Flexible linear shaped charge		
FM	Frequency modulation	JETT	Jettison
FOV	Field of vision		
FQR	Flight qualification recorder	kbs	Kilobits per second
fs	Full scale	kc	Kilocycles
FSK	Frequency shift-keyed	kHz	Kilohertz
FWD	Forward	KOH	Potassium hydroxide
		KSC	Kennedy Space Center
G	Gravity	kw	Kilowatt
GA	Gyro assembly	LAT	Latitude
gc	Gigacycles	LCC	Launch Control Center
G&C	Guidance and control	LDEC	Lunar docking events controller
GDC	Gyro display coupler	LEA	Launch escape assembly
GET	Ground elapsed time	LEB	Lower equipment bay
GFE	Government-furnished equipment	LEM	Launch escape motor (also lunar excursion module, old name for lunar module)
GMBL	Gimbal		
GN_2	Gaseous nitrogen		
GND	Ground	LES	Launch escape subsystem
GNCS	Guidance, navigation, and control subsystem	LET	Launch escape tower
		LEV	Launch escape vehicle
GPI	Gimbal position indicator	LF	Low frequency
GSE	Ground support equipment	LH_2	Liquid hydrogen
GSFC	Goddard Space Flight Center	LHEB	Left-hand equipment bay
		LHFEB	Left-hand forward equipment bay
ha	Apogee altitude	LM	Lunar module
H_2	Hydrogen	LMK	Landmark
He	Helium	LO	Low
HF	High frequency	LOR	Lunar orbit rendezvous
Hg	Mercury	LOS	Line of sight, loss of signal
HGA	High-gain antenna	LOX	Liquid oxygen
HI	High	LSB	Lower sideband
hp	Perigee altitude	LSC	Linear-shaped charge
HR	Hydrogen relief or hour	LSSC	LM separation sequence controller
HTR	Heater	LV	Launch vehicle or lift vector
Hz	Hertz (cycle per second)		
		MAN	Manual or manifold
IC	Intercom	MAX	Maximum
ICDU	Inertial coupling data unit	MAXQ	Maximum dynamic pressure
IECO	Inboard engine cutoff	MCC	Mission Control Center

MDC	Main display console	PCM	Pulse code modulation or pitch control motor
MDF	Mild detonating fuse		
MED	Medium	PCVB	Pyro continuity verification box
MESC	Master events sequence controller	PDM	Pulse duration modulation
MGA	Middle gimbal angle	PF	Pulse frequency or powered flight
mHz	MegaHertz	PGA	Pressure garment assembly
MIKE	Microphone	PH	Phase
mil	1/1000	pH	Hydrogen ion concentration
MIN	Minimum	PIPA	Pulsed integrating pendulous accelerometer
ML	Moldline		
MMH	Monomethylhydrazine	PLBK	Playback
MNA	Main bus A	PLSS	Portable life support system
MNB	Main bus B	PM	Phase modulation
MOD	Modulator	PMP	Premodulation processor
MOT	Motor	POS	Positive
MS	Motor switch	POT	Potentiometer
MSC	Manned Spacecraft Center	PPM	Parts per million or pulse position modulation
MSFC	Marshall Space Flight Center		
MSFN	Manned Space Flight Network	PPS	Pulses per second
MTVC	Manned thrust vector control	PRF	Pulse repetition frequency
mv	Millivolt	PRI	Primary
mw	Milliwatt	PRN	Pseudo-random noise
		PROP	Propellant
N_2	Nitrogen	PS	Pressure switch
NAV	Navigation	PSA	Power servo assembly
NB	Navigation base or narrow band	PSI	Pounds per square inch
NEG	Negative	PSIA	Pounds per square inch absolute
NEUT	Neutral	PSIG	Pounds per square inch gauge
n.mi.	Nautical mile	PSK	Phase shift-keyed
NO.	Number	PSO	Pad safety officer
N.O.	Normally open	PTT	Push to talk
NON-ESS	Non-essential	PU	Propellant utilization
NORM	Normal	PUG	Propellant utilization gauging
NRZ	Non-return to zero	PWR	Power
NSIF	Near-Space Instrumentation Facility	PYRO	Pyrotechnic
O_2	Oxygen	R	Range
OCDU	Optics coupling data unit	RAD	Radiation dosage or radiator
OECO	Outboard engine cutoff	RC	Rotation control or range command
OGA	Outer gimbal angle	RCDR	Recorder
OH	Hydroxyl ion	RCS	Reaction control subsystem
O/L-RC	Overload - reverse current	RCSC	Reaction control subsystem controller
OMNI	Omni-directional	RCV	Receive
OPT	Optics	RCVR	Receiver
OR	Oxygen relief	RECO	Rough engine cutoff
ORDEAL	Orbit rate drive electronics Apollo LM	RECT	Rectifier
OSC	Oscillator	R&D	Research and development
OSS	Optics subsystem	REG	Regulator
O/V	Overvoltage	REGEN	Regenerator
OXID	Oxidizer	REL	Release
		REV	Reverse
PA	Power amplifier	RF	Radio frequency
PAM	Pulse amplitude modulation	RFI	Radio frequency interference
PB	Push button	RGA	Rate gyro assembly
		RHC	Rotation hand control

RHEB	Right-hand equipment bay		TLC	Translunar coast
RHFEB	Right-hand forward equipment bay		TLI	Translunar injection
RJD	Reaction jet driver		TLM	Telemetry
RJ/EC	Reaction jet and engine control		TMG	Thermal meteoroid garment
RMS	Root mean square		TPAC	Telescope precision angle counter
RNG	Range		T/R	Transmit-receive
RNDZ	Rendezvous		TRNFR	Transfer
ROT	Rotation		TTE	Time to event
RRT	Rendezvous radar transponder		TV	Thrust vector or television
RSI	Roll stability indicator		TVC	Thrust vector control
RSO	Range safety officer		TVSA	Thrust vector position servo amplifier
RTC	Real-time command		TWR	Tower
RTTV	Real-time television		TWT	Traveling wave tube
RUPT	Interrupt			
RZ	Return to zero		UCD	Urine collection device
			UDL	Up-data link
S/C	Spacecraft		UDMH	Unsymmetrical dimethyl hydrazine
SCE	Signal conditioning equipment		UHF	Ultra high frequency
SCI	Scientific		UPTL	Up-link telemetry
SCO	Subcarrier oscillator		USBE	Unified S-band equipment
SCS	Stabilization and control subsystem		U/V	Undervoltage
SCT	Scanning telescope			
SEC	Second or secondary		V	Voice, volt, or velocity
SECO	S-IVB (third stage) engine cutoff		VAC	Volts alternating current
SECS	Sequential events control subsystem		V_c	Circular velocity
SENS	Sensitivity		VCO	Voltage-controlled oscillator
SEP	Separation or spacecraft electronic package		VDC	Volts direct current
			VGP	Vehicle ground point
SEQ	Sequencer		VHF	Very high frequency
SIG	Signal		VHF/AM	Very high frequency/audio modulated
SLA	Spacecraft-LM adapter		VM	Voltmeter or measured velocity
SLOS	Star line of sight		VO	Initial velocity
SM	Service module		VOL	Volume
SMJC	Service module jettison controller		VOX	Voice-operated relay
SNSR	Sensor		VSWR	Voltage standing wave ratio
SOV	Shutoff valve			
SPEC	Specification		W/G	Water-glycol
SPS	Service propulsion subsystem or samples per second		WMS	Waste management system
			WPM	Words per minute
SSB	Single sideband		WTR	Western test range
STBY	Standby			
SW	Switch		X_C	Command module station
SXT	Sextant		XCVR	Transceiver
SYNC	Synchronization		XDUCER	Transducer
TB	Talkback indicator		XFMR	Transformer
TC	Translation control		XMIT	Transmit
T/C	Telecommunications subsystem		XMTR	Transmitter
T/D	Time delay		XPONDER	Transponder
TEC	Transearth coast		X_S	Service module station
TEI	Transearth injection			
TEMP	Temperature		ZN	Zinc
TFL	Time from launch			
THC	Translation hand control		ΔP	Differential pressure
TIGN	Time of ignition		ΔV	Differential velocity
TJM	Tower jettison motor		ϕ	Phase

INDEX

Ablative heat shield	42, 245	Docking:		
Abort	144	System	87	
Accelerometer	203, 210	Ring	87, 91	
Accumulator	120	Latches	91	
Aldrin, E.E.	29, 270	Tunnel	44, 63, 91	
Amplifier, S band	177	Seals	91	
Anders, W.A.	270	Dosimeters	76, 80, 82	
Antennas	59, 62, 97, 175, 176, 178, 187	Drogue	63, 91	
Armstrong, N.A.	29, 270	Eisele, D.F.	268	
Ascent stage	61, 62	Emergency detection system	146	
Attitude gyro accelerometer assembly	192	Emergency oxygen system	228	
Atwood, J.L.	267	Engine:		
Audio center	176	SPS	162	
		RCS	58, 150	
Batteries	40, 63, 101, 105	LM	63	
Battery charger	101, 113	LV	9	
Bays, equipment	48	LES	139	
Beacon, recovery flashing	96, 178	Entry	23, 42, 60, 85, 93	
Bean, A.L.	272	Entry monitor system	85, 203	
Belyayev, P.	30	Environmental control unit	117, 119	
Bergen, W.B.	267	Evaluators, spacecraft	242, 261	
Bioinstrumentation harness	69, 76, 78	Evans, R.E.	269	
Biomedical equipment	76	Evaporator	120	
Block I	3, 4	Extension latch assembly	90	
Body-mounted attitude gyros	192			
Boost protective cover	140	F-1 engine	12	
Borman, F.	29, 269	Fairing	50, 55	
Bykovsky, V.	30	Fecal subsystem	74	
		Feokistov, K.	30	
Cabin	44, 62	Fire	27, 235	
Cameras	75, 78	Fire extinguisher	75, 80	
Canards	141	Flights, Apollo	33, 35	
Capture latches	90	Flight director attitude indicator	192	
Carpenter, S.	28	Flight combustion stability monitor	172	
Carr, G.P.	270	Flight coveralls	69, 80	
Caution and warning system	85	Food	70	
Central timing equipment	176	Fuel:		
Cernan, E.A.	29, 269	Service propulsion	159	
Chaffee, R.	27	Launch vehicle	10	
Clean room	242, 249	LM	63	
Clothing	69	Reaction control	147	
Coldplates	121	Fuel cell powerplants	57, 101, 106	
Collins, M.	29	Fuel tanks	57, 149	
Command module	2, 4, 15, 39			
Communications soft hat	70, 78	Gagarin, Y.	27, 30	
Compartments	44	Gemini	28, 29, 31	
Compressors	120	Gilruth, Dr. R.R.	265	
Computer	19, 189, 207, 210	Gimbal actuator	164, 202	
Conrad, C.	29, 271	Gimballing	164, 201, 213	
Console, main display	83	Gimbal position indicator	193	
Constant-wear garment	69, 78	Glenn, J.	27, 28	
Controls	72, 83, 185, 193, 211	Glycol	117, 122	
Controllers	48, 60, 83, 95, 139	Gordon, R.F.	29, 271	
Cooper, G.	28, 29	Greer, R.E.	268	
Couches	41, 71, 78	Grissom, V.	27, 28, 29	
Coupling display units	209	Guidance and navigation	205	
Crewman optical alignment sight	92	Gyros	192, 210	
Crew titles	41, 83	H-1 engine	10	
Cunningham, W.	268	Haise, F.W.	270	
		Hatches	45, 63, 91	
Data files	80	Heaters	103, 129, 167	
Data system	174	Heat exchangers	119, 120	
Debus, K.H.	266	Heat shields	42, 55, 93, 245	
Descent stage	61, 63	Helium tanks	44, 58, 63, 149, 162	
Digital ranging generator	176			

Hello, B.	268
Helmet, pressure	224
Hydrogen, liquid tanks	57, 100, 102
Impact attenuation	43
Inertial measurement unit	18, 208, 209
Inertial subsystem	189, 206
Instrument unit	10, 66
Insulation	42, 55
Integrating gyros	210
Inverters	100, 102, 110
J-2 engine	10
Jettison:	
Stages	16, 17
Launch escape tower	144
SLA panels	66
SM	60
Kennedy Space Center	231
Kits	81
Kleinknecht, K.S.	266
Komarov, V.	27, 30
Kraft, C.C.	265
Landing gear	63, 93
Launch escape:	
Motor	139
Subsystem	137
Tower	140
Launcher, mobile	14
Launching pad	9, 15, 231
Leonov, A.	30
Lights	60
Liquid-cooling garment	227
Lithium-hydroxide canister	120
Little Joe	34
Lovell, J.A.	29, 270
Low, G.M.	265
Lunar module	61
Lunar orbit coast	19
Lunar orbit insertion	19
Manned Spacecraft Center	2, 239
Manned space flight network	17
Manufacturing	245
Marshall Space Flight Center	2, 9
Mattingly, T.K.	270
McDivitt, J.A.	29, 271
Medical equipment	76, 80
Mercury	27, 28, 31
Middleton, R.O.	267
Mirrors	49, 81
Missions	15
Mitchell, E.D.	272
Moore, J.R.	267
Mortars:	
Drogue	96
Pilot	96
Forward heat shield	96
Muller, Dr. G.E.	265
Myers, D.D.	267
Navigation base	209
Nikolayev, A.	30
Nose cone	137
Nuclear particle detection system	77, 80

O'Connor, E.F.	266
Orbit:	
Earth	15
Lunar	16
Ordnance	44, 60, 64, 66, 90, 96, 142, 148
Oxidizer	149
Oxidizer tanks	44, 56, 63, 135, 149, 150, 162
Oxygen:	
Emergency system	228
Masks	75, 80
Tanks	57, 100, 102, 122
Surge tank	120
Panels:	
Protection	48, 237
SLA	17, 66
Parachutes:	
Drogue	24, 94
Main	24, 95
Pilot	24, 95
Personal Hygiene	73, 80
Petrone, R.A.	266
Phillips, S.C.	265
Pitch arms	89
Pitch control motor	140
Pitch engine	147
Pogue, W.R.	269
Popovich, P.	30
Portable life support system	227
Power and servo assembly	209
Premodulation processor	176
Probe	41, 89
Pulse-code modulation telemetry	176
Pyrotechnic batteries	101, 105
Q-ball	141
Quads	59, 147
Quantity measurement system	104
Quantity sensing system	162
Radiation survey meter	77, 81
Radiators	58, 107, 121, 132
Ratchet assembly	90
Rate gyros	192
Reaction control:	
Engines	17, 44, 58, 62, 147, 193, 217
System	147
Reefing lines	95
Reefing line cutters	95
Reliability	239
Rendezvous radar transponder	187
Resolvers	206
Restraints	71, 81
Retraction system	90
Retrobraking	19
Rockwell, W.F.	267
Ross, M.	266
Rotation control	83, 192, 215
Russian manned space program	28, 30
Safety	235
Saturn IB	2, 10, 33
Saturn V	12
Schirra, W.M.	28, 29, 268
Schweickart, R.L.	271
Scott, D.R.	29, 271
Sea marker	96
Sectors, S.M.	56
Separation:	
Stages	16, 17
CM-SM	49, 60
CSM-LM	88
Launch escape tower	139
SLA	66
Service module	53
Service propulsion:	
Engine	18, 58, 160, 162, 201, 217
Subsystem	159
Sextant	210
Shepard, A.	27, 28
Shock attenuation	89
Signal conditioner	177, 211, 261
Simulators, mission	246
Sleeping bag	73, 81
Spacecraft-LM adapter	65
Space suit	223
Spring thruster	17, 66
Squib valve	148, 155
Stafford, T.	29, 268
Structure:	
CM	41, 239, 245
LM	62, 63
SLA	66, 251
SM	55, 249
Surge tank	120
Survival equipment	77, 81
Swigert, J.L.	269
Swimmers' umbilical	82, 96
Tanks:	
Cryogenic	102, 122
Helium	150, 162
Propellant	145, 150, 160, 162, 170
Water	44, 121
Targets, docking	92
Telemetry	18, 31, 69, 173
Telescope	210
Television camera	21, 177
Tereshkova, V.	29, 30
Tests	15, 33, 35, 71, 229, 239
Thermal meteoroid garment	224
Thrust chamber	162, 218
Thruster, spring	17, 66
Titov, G.	30
Tools	75, 82
Tower, launch escape	140
Tower jettison motor	140
Transearth coast	22
Transearth injection	22
Translation control	83, 192
Translunar coast	17
Translunar injection	17
Trays, wire	49, 177
Triplexer	178
Tunnel, docking	44, 63, 91
Umbilical:	
Cable	180
CM-LM	91
Hoses	79
LM-SLA	67
SM-CM	49, 60
Swimmers	82, 96
Unified hatch	235
Unified S band	177
Up-data link	178
Uprighting system	95
Urine subsystem	74
Vacuum chamber	240
VHF/AM transmitter-receiver	178
VHF recovery beacon	81, 97, 178
Visor, extravehicular	227
Volume:	
CM	3
LM	61
SLA	65
Von Braun, W.	266
Waste bags	74
Waste management	74
Water:	
Chiller	71, 120
-glycol system	117, 122, 131
Reservoir	117, 119
System	107, 117, 128
Tanks	121
Webb, J.E.	265
White, E.	27, 29
Windows	47, 63
Worden, A.M.	272
Workshelf	74
Weight:	
CM	3, 39
LM	3, 61
Saturn IB	10
Saturn V	12
SM	3, 53
Yegorov, B.	30
Young, J.W.	29, 269

NASA PROJECT GEMINI

FAMILIARIZATION MANUAL
Manned Satellite Capsule

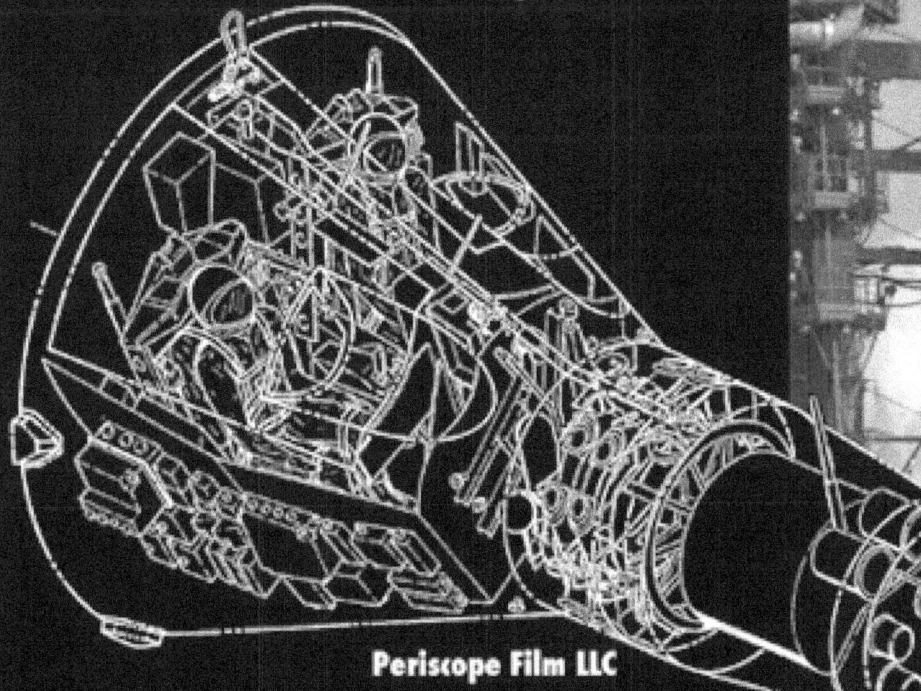

Periscope Film LLC

LMA 790-1

PROJECT APOLLO

lem
LUNAR EXCURSION MODULE

NOW AVAILABLE!

FIRST MANNED LUNAR LANDING
FAMILIARIZATION MANUAL

GRUMMAN AIRCRAFT ENGINEERING CORPORATION • BETHPAGE, L. I., N. Y.

PROJECT MERCURY

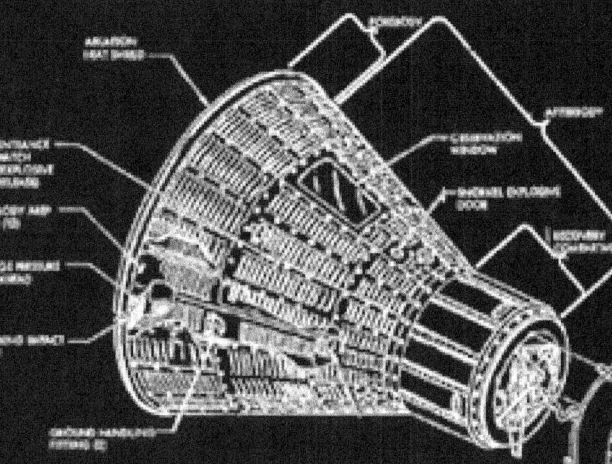

FAMILIARIZATION MANUAL

Manned Satellite Capsule

Periscope Film LLC

©2011 Periscope Film LLC
All Rights Reserved
ISBN #978-1-937684-99-0
www.PeriscopeFilm.com

www.ingramcontent.com/pod-product-compliance
Lightning Source LLC
Chambersburg PA
CBHW082026300426
44117CB00015B/2362